Forensic Anthropology

Forensic Anthropology
A Comprehensive Introduction
Second Edition

Edited by
Natalie R. Langley
MariaTeresa A. Tersigni-Tarrant

CRC Press
Taylor & Francis Group
Boca Raton London New York

CRC Press is an imprint of the
Taylor & Francis Group, an **informa** business

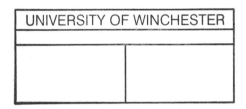

CRC Press
Taylor & Francis Group
6000 Broken Sound Parkway NW, Suite 300
Boca Raton, FL 33487-2742

© 2017 by Taylor & Francis Group, LLC
CRC Press is an imprint of Taylor & Francis Group, an Informa business

No claim to original U.S. Government works

Printed and bound in India by Replika Press Pvt. Ltd.

Printed on acid-free paper
Version Date: 20170119

International Standard Book Number-13: 978-1-4987-3612-1 (Hardback)

This book contains information obtained from authentic and highly regarded sources. Reasonable efforts have been made to publish reliable data and information, but the author and publisher cannot assume responsibility for the validity of all materials or the consequences of their use. The authors and publishers have attempted to trace the copyright holders of all material reproduced in this publication and apologize to copyright holders if permission to publish in this form has not been obtained. If any copyright material has not been acknowledged please write and let us know so we may rectify in any future reprint.

Except as permitted under U.S. Copyright Law, no part of this book may be reprinted, reproduced, transmitted, or utilized in any form by any electronic, mechanical, or other means, now known or hereafter invented, including photocopying, microfilming, and recording, or in any information storage or retrieval system, without written permission from the publishers.

For permission to photocopy or use material electronically from this work, please access www.copyright.com (http://www.copyright.com/) or contact the Copyright Clearance Center, Inc. (CCC), 222 Rosewood Drive, Danvers, MA 01923, 978-750-8400. CCC is a not-for-profit organization that provides licenses and registration for a variety of users. For organizations that have been granted a photocopy license by the CCC, a separate system of payment has been arranged.

Trademark Notice: Product or corporate names may be trademarks or registered trademarks, and are used only for identification and explanation without intent to infringe.

Visit the Taylor & Francis Web site at
http://www.taylorandfrancis.com

and the CRC Press Web site at
http://www.crcpress.com

Contents

Preface	vii
Acknowledgments	ix
Editors	xi
Contributors	xiii

SECTION I Forensic Anthropology and the Crime Scene — 1

1. Forensic Anthropology in the United States: Past and Present — 3
   MariaTeresa A. Tersigni-Tarrant and Natalie R. Langley

2. Skeletal Remains as Evidence — 23
   Marin A. Pilloud and MariaTeresa A. Tersigni-Tarrant

3. Forensic Archaeology: Survey Methods, Scene Documentation, Excavation, and Recovery Methods — 35
   Denise To

4. Forensic Taphonomy — 57
   James T. Pokines and MariaTeresa A. Tersigni-Tarrant

SECTION II The Skeleton and Skeletal Documentation — 79

5. Human Osteology — 81
   MariaTeresa A. Tersigni-Tarrant and Natalie R. Langley

6. Human Odontology and Dentition in Forensic Anthropology — 111
   Debra Prince Zinni and Kate M. Crowley

7. Skeletal Examination and Documentation — 125
   Lee Meadows Jantz

SECTION III Skeletal Individuation and Analyses — 141

8. Sex Estimation of Unknown Human Skeletal Remains — 143
   Gregory E. Berg

9. Ancestry Estimation: The Importance, The History, and The Practice — 163
   M. Katherine Spradley and Katherine Weisensee

10. Age Estimation Methods — 175
    Natalie R. Langley, Alice F. Gooding, and MariaTeresa A. Tersigni-Tarrant

11. Stature Estimation — 195
    Natalie R. Langley

12. Pathological Conditions as Individuating Traits in a Forensic Context — 205
    David R. Hunt and Kerriann (Kay) Marden

13. Analysis of Skeletal Trauma — 231
    Natalie R. Langley

14. Introduction to Fordisc 3 and Human Variation Statistics — 255
    Richard L. Jantz and Stephen D. Ousley

SECTION IV Human Identification and Advanced Forensic Anthropology Applications — 271

15 Time Since Death Estimation and Bone Weathering: The Postmortem Interval — 273
Rebecca J. Wilson-Taylor and Angela M. Dautartas

16 Methods of Personal Identification — 313
Angi M. Christensen and Bruce E. Anderson

17 Mass Fatalities, Mass Graves, and the Forensic Investigation of International Crimes — 335
Pierre Guyomarc'h and Derek Congram

18 Advanced Scene Topics—Fire and Commingling — 347
Joanne Bennett Devlin and Nicholas P. Herrmann

Appendix A: Application of Dentition in Forensic Anthropology — 365
Debra Prince Zinni and Kate M. Crowley

Appendix B: Age Estimation in Modern Forensic Anthropology — 381
Bridget F. B. Algee-Hewitt

Glossary — 421
Index — 433

Preface

The discipline of forensic anthropology has grown considerably since the founding of the Physical Anthropology Section of the American Academy of Forensic Sciences in 1972. Forensic anthropologists have gone from being expert consultants working primarily in academia to full-time board-certified employees in the medico-legal system (although a number of forensic anthropologists still hold dual appointments as professors and consultants). The formal organization of the field spurred the formation of academic programs for educating and training researchers and practitioners. Proper training in the theory, methods, and best practices of forensic anthropology begins at the undergraduate level and continues into graduate education, preferably at the PhD level. An undergraduate anthropology education is typically broad-based and establishes familiarity with three or four of the subdisciplines in the field (physical anthropology, cultural anthropology, archaeology, and linguistic anthropology). Specialization in applied areas such as forensic anthropology usually does not begin until the graduate level; however, more and more programs are offering undergraduate courses in forensic anthropology.

As researchers, practitioners, and educators, the editors of this volume found it challenging to develop introductory courses in forensic anthropology because educational texts in forensic anthropology tend to be targeted at more advanced audiences. Hence, the motivating force behind the development of the first edition of this text was to provide a book designed primarily for an introductory course that would prepare the next generation of forensic anthropologists to face the important issues of best practice being addressed currently by the Scientific Working Group for Forensic Anthropology (SWGANTH) and the issues brought forward in the National Academy of Sciences report *Strengthening Forensic Science in the United States: A Path Forward* (2009). To this end, these chapters are written by content experts in the field who have done novel research in forensic anthropology and helped to provide the foundations for best practice.

The second edition, *Forensic Anthropology: A Comprehensive Introduction*, has been reorganized considerably to provide a cohesive and user-friendly text for a semester course in forensic anthropology. The authors incorporated book reviews, feedback from users, and anonymous peer reviews of the first edition into the design of the second edition and anticipate that the new product will serve the educational community more effectively. Each chapter has learning objectives, review questions aligned with these objectives, a chapter summary, and glossary definitions for advanced terms or jargon that are not defined in the chapter text. The second edition has been condensed from 22 chapters to 18 chapters with more specialized and advanced topics added as appendices. The text is organized into four main sections plus two appendices: I. Forensic Anthropology and the Crime Scene, II. The Skeleton and Skeletal Documentation, III. Skeletal Individuation and Forensic Anthropological Analyses, and IV. Human Identification and Advanced Forensic Anthropology Applications.

The first section, Forensic Anthropology and the Crime Scene, contains Chapters 1 through 4. Chapter 1 provides an overview of forensic anthropology in the United States, including a history of the discipline from the earliest application of physical anthropology in a medico-legal setting to the issues faced by forensic anthropologists today. Chapter 2 is new to this edition and addresses how anthropologists handle skeletal remains as evidence. Chapter 3 discusses how archaeological methods are used to recover and document crime scenes. Chapter 4 discusses how taphonomic processes affect the formation, recovery, and interpretation of crime scenes and skeletal remains.

The second section, The Skeleton and Skeletal Documentation, contains Chapters 5 through 7 and covers human osteology, odontology, and the first and last steps of doing a forensic anthropology case (processing and the case report). Chapter 5 presents basic human osteology that will be helpful for understanding the content in the remaining chapters in this book. This chapter is not meant to replace a human osteology class, but simply to provide a quick reference for less experienced readers; advanced reading and reference suggestions are provided for those requiring more in-depth osteology. Chapter 6 provides an overview of the human dentition and dental numbering systems used in forensic case reports. Chapter 7 details how to process remains and prepare a forensic case report and includes sample case reports for reference.

The third section, Skeletal Individuation and Analyses, consists of Chapters 8 through 14 and covers how forensic anthropologists develop a biological profile from skeletal remains (sex, age, ancestry, and stature) and analyze skeletal pathology and trauma. The importance of choosing proper methods and understanding the advantages and limitations of available methods is highlighted throughout these chapters. In addition, the authors call attention to areas of burgeoning research and areas where additional research is needed. Chapters 8 through 11 deal with the primary elements of the biological profile. Chapter 12 is new to this edition and provides an overview of individualistic skeletal features and skeletal responses to pathology and trauma. Chapter 13 discusses the biomechanics and interpretation of low- and high-velocity skeletal trauma and nascent research in the field of skeletal trauma interpretation. Chapter 14 provides a detailed explanation of the Fordisc program used by forensic anthropologists in the United States to estimate sex, ancestry, and stature from metric data, including a brief history of the software development, its limitations, and common criticisms.

Chapters 15 through 18 comprise the fourth section, Human Identification and Advanced Forensic Anthropology Applications. Chapter 15 provides a comprehensive discussion of time since death (TSD) and the many processes that are factored into determining the postmortem interval (environmental/meteorological, individual, and cultural/case-specific factors), the variety of methods available for TSD estimation (entomology, botany, soil chemistry, and morphoscopic techniques), and decomposition research (past, present, and future directions). Chapter 16 details acceptable techniques for tentative, circumstantial, and positive personal identification and their implications for forensic practice, research, and the investigative process. Chapter 17 deals with forensic anthropology in large-scale situations, namely human rights violations and mass disasters, and provides considerations for handling these unique and sensitive situations. Chapter 18 explicates the effects of fire on the human body and details how to approach fire scenes, thermal trauma, and situations in which remains are commingled.

This book concludes with two appendices. Appendix A, Application of Dentition in Forensic Anthropology, discusses the use of the dentition in forensic anthropology, including age, ancestry, and sex determination and assessing dental anomalies and pathologies. Appendix B, Age Estimation in Modern Forensic Anthropology, provides an advanced look at the topic of age estimation, including an examination of statistical methods, bias, and the problems with age-at-death estimation.

Although the chapters are written so that less experienced readers can understand the content, the editors feel that readers from a wide variety of experience levels will find the material both interesting and useful. As the chapters are thoroughly referenced and give special attention to current research and best practice, the text can serve as a useful addition to any forensic anthropology laboratory or library. We encourage its use as part of a forensic anthropology curriculum, as a reference for the laboratory, and as a review of the many complex subject areas that have come to encompass forensic anthropology. Thank you for choosing to purchase this text, and we look forward to receiving input from our reading audience on this revised edition.

Natalie R. Langley and MariaTeresa A. Tersigni-Tarrant

Acknowledgments

The editors thank the contributing authors for their time, patience, and effort to make the needed improvements to this second edition. We also thank Mark Listewnik at Taylor & Francis Group for his invaluable suggestions, efficient responses to all matters, and undivided attention during the preparation of this text. We acknowledge the support of our respective employers (Saint Louis University, Missouri, LMU-DeBusk College of Osteopathic Medicine, Tennessee, and Mayo Clinic School of Medicine, Arizona Campus) during this time-consuming endeavor and the patience of our friends, families, and colleagues as we worked many nights and weekends. Finally, we are thankful for the friendship we have in one another that makes a task such as this possible and all the more pleasant.

Editors

Natalie R. Langley, PhD, D-ABFA, is a faculty member in the Department of Anatomy at the Mayo Clinic College of Medicine and Science. She is an adjunct faculty member in the University of Tennessee Anthropology Department and the DeBusk College of Osteopathic Medicine Anatomy Department (previous name: Natalie R. Shirley). She received her BA (1998) and MA (2001) in anthropology from Louisiana State University and her PhD (2009) from the University of Tennessee, Knoxville. Dr. Langley is a diplomate of the American Board of Forensic Anthropology (#110) and consults for the Georgia Bureau of Investigation. She is a fellow of the American Academy of Forensic Sciences, and a member of the American Association of Anatomists, American Association of Clinical Anatomists, and American Association of Physical Anthropologists. In 2007, the AAFS Forensic Science Foundation awarded her the Emerging Forensic Scientist Award for her research in skeletal maturation. Dr. Langley has published articles in the *Journal of Forensic Sciences, American Journal of Physical Anthropology, Forensic Science International, International Journal of Legal Medicine, Anatomical Sciences Education,* and *Clinical Anatomy,* and has written numerous book chapters. Her research interests include skeletal maturation in modern populations, age and sex estimation from the human skeleton, secular changes in skeletal biology, currency of forensic standards, skeletal trauma, and anatomy education. During the spring of 2012, she was the forensic anthropologist on an eight-episode television series airing on the National Geographic Channel titled *The Great American Manhunt.*

MariaTeresa A. Tersigni-Tarrant, PhD, D-ABFA, is an associate professor in the Department of Surgery's Center for Anatomical Science and Education (CASE) at the Saint Louis University School of Medicine (SLU SOM), Missouri, where she serves as the director of Gross Anatomy Laboratory and Morgue Operations. Dr. Tersigni-Tarrant teaches gross anatomy, embryology, and histology to medical students, graduate students, and allied health students at SLU SOM. Dr. Tersigni-Tarrant is also an adjunct associate professor in the SLU SOM Department of Pathology where she serves as the director of forensic education, organizing, developing, and implementing curriculum for basic and advanced medico-legal death investigator training courses that are offered multiple times per year. She received both her BS degrees (microbiology and anthropology) from Michigan State University (2000). She received her MA (2002) and PhD (2005) in physical anthropology from the University of Tennessee, Knoxville. She completed a postdoctoral fellowship at the laboratory formally known as JPAC-CIL (2005–2006), where she helped to create the standard operating procedures for histological analysis of osseous remains. Dr. Tersigni-Tarrant is a diplomate of the American Board of Forensic Anthropology (#98), and she provides anthropological consulting services for the city of St. Louis (Missouri), the Missouri Disaster Response System, and the Georgia Bureau of Investigation. Dr. Tersigni-Tarrant is a fellow of the American Academy of Forensic Sciences and is a member of the American Association of Physical Anthropologists and American Association of Clinical Anatomists. She serves on the Missouri State Anatomical Board, Missouri Disaster Response Team, and the St. Louis Area Regional Response System. Her research interests include human and nonhuman histology, child abuse: patterned fractures and timing of healing, long bone fracture patterning, bone pathology, and developmental anatomy.

Contributors

Bridget F. B. Algee-Hewitt, PhD, is a biological anthropologist in the Department of Biology at Stanford University, California. Dr. Algee-Hewitt was awarded her PhD in anthropology at the University of Tennessee, Knoxville and her master's degree from Bryn Mawr College, Pennsylvania, in classical and near eastern archaeology. She has completed postdoctoral work in both skeletal biology and human genetics, as the Haslam Post-Doctoral Fellowship for Research in Forensics at the University of Tennessee, and is a fellow in the Stanford Center for Computational, Evolutionary and Human Genomics. She has also served as an assistant professor in biomedical sciences, teaching clinically oriented anatomy, physiology, and advanced human dissection. Dr. Algee-Hewitt's current research program includes developing novel computational approaches to biological data analysis, addressing, in particular, population inference from a dual skeletal and genetic perspective. The breadth of her interests in skeletal biology, population genetics, and biostatistics has allowed her to contribute to research that has theoretical implications for the larger study of human variation, evolution, and population history in anthropology while also making practical contributions to the applied fields of human identification in the forensic sciences. Dr. Algee-Hewitt has recently received a National Institute of Justice grant in support of newly developed computational methods for age-at-death estimation from skeletal indicators using laser scan data.

Bruce E. Anderson, PhD, D-ABFA, is a forensic anthropologist with the Pima County Office of the Medical Examiner (PCOME), in Tucson, Arizona. Dr. Anderson received his PhD degree in 1998 from the University of Arizona, where he is currently an adjunct assistant professor of anthropology. Prior to his hiring by the PCOME in 2000, he served as a senior anthropologist for the U.S. Army's Central Identification Laboratory in Hawaii (CILHI) where his principal duties were the field recovery and laboratory analyses leading toward identification of human remains associated with past U.S. military conflicts. Dr. Anderson currently mentors anthropology students in the Forensic Anthropology Internship Program at the PCOME and works with postdoctoral fellows as part of the PCOME's Forensic Anthropology Fellowship Program. He is a fellow in the American Academy of Forensic Sciences (AAFS), is certified as a diplomate by the American Board of Forensic Anthropology (ABFA), was a founding member of the Scientific Working Group in Forensic Anthropology (SWGANTH), and served as a forensic anthropologist during the development and initial launch of the National Missing and Unidentified Persons System (NamUs).

Gregory E. Berg, PhD, D-ABFA, is a laboratory manager and forensic anthropologist at the Defense POW/MIA Accounting Agency Central Identification Laboratory in Hawaii, where he works on the recovery and identification of missing U.S. service personnel. He earned his BA in anthropology at the University of Arizona in 1993, his MA in the bioarchaeology program at Arizona State University in 1999, and his PhD at the University of Tennessee, Knoxville, in 2008. He has more than 25 years of field experience in archaeology and physical anthropology and has presented or published numerous articles and papers in the *Journal of Forensic Sciences, Journal of Archaeological Science, Optometry*, and at various annual meetings. His recent research has concentrated on ancestry and sex determination, trauma analysis, aging techniques, human identification and eyewear, and intra- and interobserver error studies, which have been particularly focused on aging and population determination methods used in human identification. Dr. Berg is a member of the American Academy of Forensic Sciences, and is a diplomate of the American Board of Forensic Anthropology (#93).

Angi M. Christensen, PhD, D-ABFA, attended the University of Washington, earning a BA in anthropology, and received her MA and PhD in anthropology from the University of Tennessee, Knoxville. She currently works as a forensic anthropologist for the Federal Bureau of Investigation (FBI) Laboratory in Quantico, Virginia. She assisted with the development of the FBI's Forensic Anthropology Program, whose mission is to provide casework services, operational assistance, consultation, research, training, and outreach to law enforcement at federal, state, and local levels. Her research interests have included methods and issues related to personal identification, error and error rates in forensic contexts, scientific admissibility issues, human remains in aquatic forensic contexts, and the interpretation of skeletal blast trauma. She also serves as an active member of the Anthropology Subcommittee of NIST's Organization of Scientific Area Committees, assisting to develop standards and best practices for the discipline of forensic anthropology.

Derek Congram, PhD (Simon Fraser University, Canada), teaches and conducts research at the Munk School of Global Affairs, University of Toronto, Canada, and also consults as an archaeologist/biological anthropologist in medico-legal and humanitarian contexts. He has worked for governments, universities, nongovernmental, and international organizations in 20 countries. His principal research interests are spatial analysis and GIS-modeling of unmarked burials in conflict contexts, bioarchaeology of the Spanish Civil War, professional ethics, and public (bio) archaeology. He is the editor of and a contributor to the publication *Missing Persons; Multidisciplinary Perspectives on the Disappeared* (Canadian Scholars' Press), 2016.

Kate M. Crowley, DMD, D-ABFO, is a board certified forensic odontologist on staff with the Office of the Chief Medical Examiner in Boston, Massachusetts. Dr. Crowley is also a periodontist working in private practice with offices located in Weymouth and Fall River, Massachusetts.

Joanne Bennett Devlin, PhD, D-ABFA, is a faculty member in the Anthropology Department at the University of Tennessee, Knoxville. She previously worked as an archaeologist in the Great Basin, the mid-Atlantic, and the Southeast. Currently, as an assistant director of the Forensic Anthropology Center (FAC), she oversees graduate student training, and conducts casework while also engaging in teaching in FAC sponsored short training courses. Dr. Devlin is a certified firefighter in Knox County, where she has served for more than 10 years. She is also a diplomate of the American Board of Forensic Anthropology and a consulting forensic anthropologist with the Hamilton County Cold Case Task Force. Dr. Devlin also works with the National Center for Missing and Exploited Children (NCMEC) on cases of unidentified subadults. Her research interests include taphonomic signatures of fire and geophysical methods of clandestine grave discovery.

Angela M. Dautartas, MA, received her BS in anthropology from Radford University, Virginia, in 2005 and her MA in anthropology from the University of Tennessee, Knoxville, in 2009. She is currently a doctoral candidate in the Anthropology Department and research assistant with the Forensic Anthropology Center at the University of Tennessee, Knoxville. Her current research explores the use of animal models in taphonomic studies. Her research interests include human decomposition, time since death, and paleopathology. She is also an adjunct instructor in the Natural and Behavioral Sciences Department at Pellissippi State Technical and Community College and in the Division of Social Sciences at Maryville College, Tennessee.

Alice F. Gooding, MS, is an assistant professor of anthropology at Kennesaw State University (KSU). She received her MS from the University of Georgia (UGA) in 2011. Her research at UGA tested age-at-death estimation methods on a modern sample and explored the application of these methods on cold cases from the Georgia Bureau of Investigation. She is finishing her PhD at the University of Tennessee, Knoxville, in biological anthropology. Her dissertation examines the relationships between skeletal senescence and biomechanical loading in the adult limbs. Gooding is also a clinical instructor with the Philaphelphia College of Osteopathic Medicine. Her current research tests the biomechanical response of bone to blast force trauma. She is a member of the FBI Atlanta Regional Forensic Specialist Working Group and works closely with local high schools to develop curriculum for forensic courses and teacher training.

Pierre Guyomarc'h, PhD, received his BA in archaeology from Paris Sorbonne and Besancon Universities in France, and his MSc and PhD in biological anthropology from the University of Bordeaux, France, in 2011. After working for two years as a forensic anthropologist at the Joint POW/MIA Accounting Command Central Identification Laboratory (JPAC-CIL, HI), he started to work as a forensic adviser with the International Committee of the Red Cross (ICRC) in Burundi, Ukraine, and Lebanon. His research interests include the development of identification methodologies through morphometrics and medical imaging. Within the ICRC, he focuses on forensic humanitarian action, promoting a proper management of the dead and best practices for the identification of missing persons after armed conflicts or disasters.

Nicholas P. Herrmann, PhD, is an associate professor of anthropology at Texas State University in the Department of Anthropology and the Forensic Anthropology Center at Texas State (FACTS). Dr. Herrmann works extensively with human skeletal remains from archaeological, forensic, and anatomical contexts. His research focuses on the biological variation of human populations both past and present. His research is based on prehistoric, historic, and modern populations from Greece, Cyprus, Honduras, and across the United States. He has directed or codirected projects examining stature estimation, geophysical prospection, commingled human remains, geospatial mapping applications, archaeological surveys, historic cemeteries, and stable isotope and trace element variation in modern U.S. populations.

David R. Hunt, PhD, D-ABFA, is the collections manager in the Physical Anthropology Division of the Smithsonian's National Museum of Natural History, Washington, DC. He received two BA degrees in 1980 from University of Illinois, Urbana, in physical anthropology and classical archaeology, and earned his MA in 1983 and PhD in 1989 at University of Tennessee, Knoxville, in skeletal biology and forensic anthropology. He is a diplomate of the American Board of Forensic Anthropologists. His areas of research include human skeletal biology/variation, paleopathology, forensic anthropology, and nondestructive investigation using computer-assisted tomography and mummies of the world. He has a professorial position at George Washington University and is a physical/forensic anthropology consultant for the Office of Chief Medical Examiner, Northern Virginia Region, and is a consultant for Project Alert and the Forensic Imaging Unit at the National Center for Missing and Exploited Children. Hunt is involved in U.S. archeological excavations as well as international archeological excavations and skeletal biological research in Mongolia, Egypt, and Rwanda. He consults on human remains curation in museum and institutional settings.

Richard L. Jantz, PhD, was educated at the University of Kansas, Lawrence, and has been a faculty member at the University of Tennessee, Knoxville, since 1971. His research interests mainly include quantitative human variation with an emphasis on American populations, both early and recent. In the mid-1980s, he established the forensic anthropology data bank, which serves as a resource to study skeletal variation among modern Americans. He, along with Steve Ousley, developed Fordisc, software that automates estimation of sex, ancestry, and height from skeletal measurements. His primary research activity in forensic anthropology deals with improving estimates of sex, ancestry, and height, and documenting the changes occurring in the American population during the twentieth century. He retired from his active faculty position in 2009 and as the director of the Forensic Anthropology Center in 2011, a position he had held since 2000. He is now an emeritus professor and is enjoying not going to meetings and working on research that has been put off for decades.

Kerriann (Kay) Marden, PhD, is currently an assistant professor of anthropology and the director of the Forensic Science Program at Eastern New Mexico University, and regularly consults on forensic anthropology cases for the New Mexico Office of the Medical Investigator. Her research has spanned topics ranging from bioarchaeology to forensic anthropology. She was granted three Smithsonian Institution Fellowships, where she conducted paleopathological research of Ancestral Puebloan remains under Don Ortner and David R. Hunt. Kay served as an assistant to Marcella Sorg, a forensic anthropology consultant for the Offices of the Chief Medical Examiner of Maine, New Hampshire, and Rhode Island. Kay continues to collaborate with Dr. Sorg on taphonomic research projects and on a forthcoming edited volume. Kay holds doctoral and master's degrees in anthropology from Tulane University, Louisiana, and a master's degree in applied sociology from the University of Maryland at Baltimore County. She was previously certified by the American Board of Medico-legal Death Investigators, and she is an active member of D-MORT.

Lee Meadows Jantz, PhD, received her BA, MA, and PhD in anthropology from the University of Tennessee, Knoxville. As associate director of the Forensic Anthropology Center, Dr. Meadows Jantz is responsible for the body donation program and curation of the William M. Bass Donated and Forensic Skeletal Collections. She is a senior lecturer and adjunct associate professor in the Department of Anthropology at University of Tennessee, Knoxville, and serves as instructor for the outdoor recovery, taphonomy, and human identification courses offered through the Forensic Anthropology Center. Dr. Meadows Jantz is a founding member of the SWGANTH board and serves as a forensic anthropologist on the federal DMORT-WMD team. She consults with federal, state, and local law enforcement agencies on human identification cases. Her research interests include skeletal biology (past and recent populations), forensic anthropology, and human growth and development.

Stephen D. Ousley, PhD, attended the University of Tennessee, Knoxville. As a graduate research assistant, he worked as database administrator and data collector for the Forensic Data Bank, a repository of skeletal data from modern Americans. His master's thesis evaluated theories of the peopling of the New World using Native American anthropometrics, and his dissertation was a quantitative genetics analysis of fingerprints. For nine years, he was the director of the Repatriation Osteology Laboratory in the Repatriation Office of the National Museum of Natural History at the Smithsonian Institution. In 2007, he started teaching at Mercyhurst College, Pennsylvania. He is best known for coauthoring Fordisc, a computer program that aids in the identification of unknown human remains using various statistical methods. His research interests focus on forensic anthropology, human growth and development, human variation, geometric morphometrics, and dermatoglyphics.

Marin A. Pilloud, PhD, D-ABFA, RPA, is currently an assistant professor of anthropology at the University of Nevada, Reno. Dr. Pilloud's research is broadly focused on the application of dental morphology and metrics to answering research questions in both bioarchaeology and forensic anthropology. She has been active in bioarchaeological research programs in Neolithic Anatolia and prehistoric California, and, within forensic anthropology, she is primarily interested in the use of teeth in the estimation of ancestry.

James T. Pokines, PhD, D-ABFA, is an assistant professor in the Forensic Anthropology Program, Department of Anatomy and Neurobiology, Boston University, Massachusetts, and is the forensic anthropologist for the Office of the Chief Medical Examiner, Commonwealth of Massachusetts, in Boston. He served for more than a decade as a forensic anthropologist and laboratory manager at the Joint POW/MIA Accounting Command, Central Identification Laboratory, in Hawaii, after completing a postdoctoral fellowship there in forensic anthropology administered by the Oak Ridge Institute for Science and Education. Dr. Pokines received his BA in anthropology and archaeology from Cornell University, New York, his MA and PhD degrees in anthropology from the University of Chicago, and ABFA Board certification in forensic anthropology. His research also includes comparative vertebrate osteology, zooarchaeology, taphonomy, and paleoecology, and he has ongoing archaeological projects in the Bolivian Andes (Tiwanaku and related sites), South Africa (taphonomy), and northern Jordan (including a natural faunal trap and multiple Paleolithic sites). He has also conducted research in the Upper Paleolithic of Cantabrian Spain, modern predator taphonomy in Kenya, and zooarchaeology in the Nile Delta (Tell Timai).

M. Katherine Spradley, PhD, is an associate professor in the Department of Anthropology at Texas State University and faculty in the Forensic Anthropology Center at Texas State. Dr. Spradley is the Director of Operation Identification (OpID). OpID aims to facilitate the identification and repatriation of human remains found along or within close proximity of the South Texas border through community outreach, scientific analysis, and collaboration with governmental and nongovernmental organizations. Her current research focuses on metric data from human skeletons to address identification methods in forensic anthropology, to track population migrations when there is little or no historical documentation, and to explore the skeletal morphological changes associated with human migrations and changing environments (e.g. climate, nutrition, and health).

Denise To, PhD, D-ABFA, RPA, is a forensic anthropologist and forensic archaeologist with the Defense POW/MIA Accounting Agency where she currently serves as a laboratory manager. Dr. To received her master's degree in anthropology with an emphasis on bioarchaeology from Arizona State University and her PhD in physical anthropology from the same school in 2008. She has more than 22 years of experience as an archaeologist, which includes excavation of the Pyramid of the Moon at Teotihuacan. In addition to numerous sites in the United States, she has worked in Belgium, Bulgaria, France, Germany, Laos, Mexico, Palau, Papua New Guinea, Philippines, Saipan, Solomon Islands, South Africa, South Korea, Vanuatu, Vietnam, and Wake Island. Dr. To has taught human remains recovery courses in Hawaii for Chaminade University, Hawaii, as well as in Bogota, Colombia, for the U.S. Department of Justice's International Criminal Investigative Training Assistance Program. Dr. To received the U.S. Joint Civilian Service Achievement Award in 2011 for her efforts in Kathmandu, Nepal, in response to the crash of Agni Air 101, and was also deployed to Louisiana in response to Hurricane Katrina while as a member of the Disaster Mortuary Operational Response Team. She is certified as a diplomate by the American Board of Forensic Anthropology, is a member of the Register of Professional Archaeologists, and is a fellow of the American Academy of Forensic Sciences.

Katherine Weisensee, PhD, is currently an associate professor in the Department of Sociology and Anthropology at Clemson University, South Carolina. She received her BA in anthropology from Brandeis University, Massachusetts, and her MA and PhD in anthropology from the University of Tennessee, Knoxville. Dr. Weisensee teaches a variety of undergraduate courses, including forensic anthropology, human variation, biological anthropology, and introduction to anthropology. She acts as a forensic anthropology consultant for several counties in the upstate of South Carolina. Her research interests include geometric morphometrics, paleodemography, population genetics, and time since death.

Rebecca J. Wilson-Taylor, PhD, received her PhD in anthropology from the University of Tennessee, Knoxville, in 2013. Currently, she is a forensic anthropologist in the Central Identification Laboratory at the Defense POW/MIA Accounting Agency (DPAA), which seeks to fulfill the nation's promise to recover and identify unaccounted-for U.S. service members. Prior to joining DPAA, Dr. Wilson-Taylor worked as the assistant coordinator for the University of Tennessee's Forensic Anthropology Center where she conducted research at the Anthropology Research Facility, developed continuing education opportunities, and helped to manage the body donation program. She is also a member of the federal DMORT-WMD team. Her research interests include forensic taphonomy, human decomposition, sexual dimorphism, and paleopathology.

Debra Prince Zinni, PhD, D-ABFA, is the forensic anthropology laboratory manager at the Defense POW/MIA Accounting Agency, Joint Base Pearl Harbor-Hickam, Hawaii. Dr. Zinni received her BS in anthropology and psychology at James Madison University, Virginia, her MA in anthropology from The George Washington University, and her PhD from the University of Tennessee, Knoxville. Prior to her current position, she was an assistant professor at Boston University School of Medicine, Massachusetts, and the state forensic anthropologist for the Commonwealth of Massachusetts. Dr. Zinni is a fellow in the Anthropology Section of the American Academy of Forensic Sciences and is certified as a diplomate by the American Board of Forensic Anthropology (ABFA). Dr. Zinni serves as a treasurer on the Board of Directors of the ABFA.

Section I

Forensic Anthropology and the Crime Scene

1. Forensic Anthropology in the United States: Past and Present
2. Skeletal Remains as Evidence
3. Forensic Archaeology: Survey Methods, Scene Documentation, Excavation, and Recovery Methods
4. Forensic Taphonomy

Chapter 1

Forensic Anthropology in the United States
Past and Present

MariaTeresa A. Tersigni-Tarrant and Natalie R. Langley

CONTENTS

Forensic anthropology in the public eye	7
What we are "not": Debunking Hollywood myths	7
Educational and employment opportunities	8
Brief history of forensic anthropology	8
Formative period (early 1800s–1938)	8
Consolidation period (1939–1971)	10
Modern period (1972–present)	11
Anthropology research facilities	12
Documented donated skeletal collections	14
Ethics in practice and research	15
Best practice in forensic anthropology: From SWGANTH to OSAC	16
Forensic anthropology in the international arena	17
Summary	18
Review questions	19
Glossary	19
References	20
Additional information	21
Useful websites	21
Nonfiction forensic anthropology resources	22
Nonfiction books	22
Textbooks	22

> **LEARNING OBJECTIVES**
>
> 1. Explain how forensic anthropology fits into the larger discipline of anthropology and define the major subdisciplines of anthropology.
> 2. Describe how training in each of the anthropology subfields contributes to a forensic anthropologist's understanding of human skeletal biology.
> 3. Recount the events that began and ended each of the three historical periods in forensic anthropology.
> 4. Explain the role of outdoor decomposition research facilities in forensic anthropology practice, research, and education.
> 5. Discuss the importance of documented modern skeletal collections for forensic anthropology research, practice, and education.
> 6. Explain the significance of the National Academy of Sciences (NAS) report to the forensic sciences and forensic anthropology. Name the organizations and committees that resulted from this report, and list their primary objectives.

The discipline of **anthropology** seeks to understand the many intricate aspects of what it means to be human. Derived from the Greek word *anthropos*, meaning "human," and *logia*, referring to the "study of," anthropology seeks to shed light on human behavior, biology, language, and culture in past and present contexts. Anthropology is a holistic discipline that encompasses multiple subdisciplines. The four most common subdisciplines are archaeology, sociocultural anthropology, linguistic anthropology, and physical/biological anthropology. These subdisciplines are not mutually exclusive, and each seeks to define and interpret various aspects of the human condition. Archaeology reconstructs the history of past populations through contextual analysis of the artifacts and structures (i.e., material culture) that these populations have left behind. Sociocultural anthropology uses observation and interviews of participants to understand cultural groups or subcultures. Linguistic anthropology investigates the origins and use of language, as well as language changes over time. Physical/biological anthropology studies human biological origins, adaptation, and variation in an evolutionary context, as well as the life histories of our nonhuman primate relatives. Each subdiscipline of anthropology is broken down further into smaller, more specialized subfields or applied areas of study that focus on specific aspects of what it means to be human. Figure 1.1 illustrates several common subfields of archaeology, cultural anthropology, linguistic anthropology, and physical/biological anthropology.

Forensic anthropology is an applied subdiscipline of physical/biological anthropology. Forensic anthropologists use their knowledge of modern human skeletal variation to help law enforcement identify unknown decedents and, if possible, provide information about the circumstances surrounding a death. The American Board of Forensic Anthropology defines forensic anthropology as "the application of the science of physical or biological anthropology to the legal process," adding that "[p]hysical or biological anthropologists who specialize in forensics primarily focus their studies on the human skeleton" (www.theabfa.org).

Forensic anthropologists employ the principles of skeletal growth, development, degeneration, and variation to ascertain biological information about an individual, such as age, sex, ancestry, and stature. These four components are collectively referred to as the **biological profile**. If the remains are human, modern, and of forensic significance, a forensic anthropologist constructs a biological profile to assist law enforcement in identifying the unknown decedent. Forensic anthropologists may also use their understanding of bone biomechanics (i.e., the way in which bone behaves under certain loads or forces) and/or bone healing to evaluate skeletal trauma. (The interpretation of skeletal trauma is discussed in detail in Chapter 13.) Furthermore, forensic anthropologists apply the principles of forensic taphonomy and bone weathering to determine what happened to the remains in a given depositional environment. Forensic taphonomy encompasses animal activity and bone weathering due to environmental factors such as sun, soil, plants, and humidity and is discussed further in Chapter 15.

Nationally, anthropology is recognized as one of the forensic sciences by the **American Academy of Forensic Sciences** (AAFS). The AAFS has 11 sections, each representing a different subdiscipline of forensic science. The AAFS has specific requirements for membership, as do each of the sections within the academy. Requirements have been developed for various levels of membership, from student and trainee affiliate members to associate and full members and fellows. Promotion from one level of membership to the next is contingent on completing the necessary requirements

Figure 1.1 Subfields within the four major subdisciplines of anthropology.

for promotion, which vary by section. Figure 1.2 presents the membership of each AAFS section as a percentage of the total AAFS membership. Figure 1.3 imparts the notion that, although anthropology is not the largest section, it is the section with the highest percentage of student or trainee affiliates. This may represent a significant trend for the future membership of the Anthropology Section of the AAFS.

Of the AAFS-recognized forensic science disciplines, anthropologists work most closely with forensic pathologists and forensic odontologists. Forensic pathologists are medical doctors with specialized training in pathology. Pathologists conduct forensic autopsies and determine cause of death. Forensic pathologists consult forensic anthropologists to examine skeletal, badly decomposed, severely burned, fragmentary, or commingled remains and submit a case report. Forensic odontologists are also referred to as forensic dentists. Forensic anthropologists rely on forensic odontologists to certify a positive identification of human remains by using the dentition. In these instances, the odontologist compares antemortem records (typically radiographs) with postmortem radiographs to make the identification.

Most forensic anthropologists are affiliated with the AAFS, as well as with the **International Association of Forensic Sciences (IAFS)**. Forensic anthropologists also disseminate research at the **American Association of Physical Anthropologists (AAPA's)** annual meetings. In addition, several regional forensic anthropology organizations exist

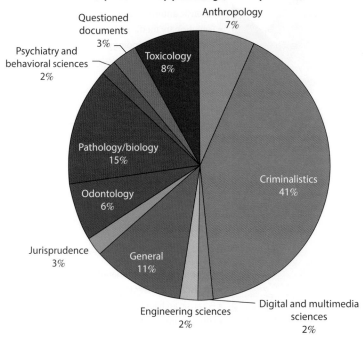

Figure 1.2 Membership in the sections of the American Academy of Forensic Sciences.

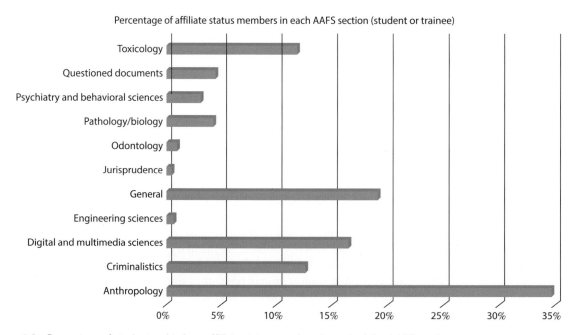

Figure 1.3 Percentage of student and trainee affiliate status members in each of the AAFS sections.

throughout the United States: Mountain, Swamp, and Beach Forensic Anthropologists (MS&B—Southeast United States); Mountain, Desert, and Coastal Forensic Anthropologists (MD&C—Southwest United States); Northeast Forensic Anthropology Association (NEFAA—Northeast United States); and Midwest Bioarchaeology and Forensic Anthropology Association (BARFAA—Midwest United States). The annual meetings of these organizations, though often a slightly less formal affair than the AAFS annual meetings, allow for the presentation of anthropological research and symposium discussions that facilitate the dissemination of research and scholarship in burgeoning areas of physical and forensic anthropology.

FORENSIC ANTHROPOLOGY IN THE PUBLIC EYE

The rising popularity of the forensic sciences as a result of popular crime novels and television (TV) dramas such as *CSI* and *Bones* has led to a misconception about the true nature of forensic anthropology. Forensic cases are not solved in a day, and, unfortunately, many cases go unsolved for years or even decades. Furthermore, although DNA is an excellent way to identify unknown persons, it is expensive and time-consuming, often taking six months or more to get results. Whenever possible, a forensic anthropologist seeks dental records or other radiographs to make a positive identification, because these methods are quick and inexpensive.* This also helps alleviate DNA laboratories of excessive backlogs, so that they can process other pertinent crime scene material. Nonetheless, there are some informative nonfiction texts and TV shows that give an accurate depiction of forensic anthropology. For a list of some of these resources, please see the Additional Information section of this chapter.

WHAT WE ARE "NOT": DEBUNKING HOLLYWOOD MYTHS

Some popular TV dramas give inaccurate depictions of the forensic sciences. In most crime scenes, multidisciplinary collaborations between various branches of the forensic sciences are necessary to solve a crime. For example, criminalists process scene evidence such as fingerprints and blood spatter, toxicologists process evidence from bodily fluids, pathologists conduct a forensic autopsy on the body, and anthropologists assist with skeletal trauma analysis. Each of these tasks is accomplished by trained professionals within each discipline and not by a single person with training in all areas of the forensic sciences. In other words, the forensic sciences are splintered into a number of highly specialized fields. This ensures that all aspects of a crime scene are analyzed by qualified experts and then assembled to give the most complete picture possible. Certainly, fictional TV dramas use their artistic license in order to present a case in a one-hour time slot, using a limited number of actors. Nonetheless, a number of nonfiction programs give more accurate depictions of how forensic scientists work together to solve crimes (e.g., *Forensic Files*, *The New Detectives*, and *FBI Files*).

Another misconception about forensic anthropology involves the educational requirements to practice forensic anthropology. There is no formal degree in "forensic anthropology" in the United States; instead, forensic anthropologists pursue a degree in anthropology, with a concentration in physical or biological anthropology and further specialization in forensic anthropology. Forensic anthropologists are proficient in skeletal biology, anatomy, and modern human variation. In addition, many forensic anthropologists are knowledgeable in bone injury biomechanics, pathology, and taphonomy, as well as statistics and archaeological recovery methods. Anthropologists emphasize a **four-field approach** and receive training in each subfield of anthropology. As a result of this multifaceted educational training, forensic anthropologists bring a unique perspective to understanding human variation in that they consider the biological and sociocultural determinants of human skeletal biology. For example, characteristic skeletal features indicate whether a skeleton is male or female, but other skeletal markers can reveal aspects of an individual's identity, such as socioeconomic status, health history, and occupation. These clues give law enforcement a more complete picture of an unidentified person and assist greatly in the search for identity.

Unfortunately, with the increasing number of TV dramas and movie plots based on the forensic sciences, the line between fiction and reality has become effectively blurred. This has resulted in general public being increasingly

---

* Methods of positive identification are discussed in Chapter 16.

aware of the forensic sciences but not necessarily well versed on what can actually be achieved through each type of analysis. Consequently, potential jurors and members of the public may have expectations of forensic analyses that are unattainable and unwarranted. Moreover, the fictionalization of forensic practitioners' ability to be able to analyze and report on multiple aspects of a crime (such as the fingerprint analysis, ballistics, and autopsy report) can cause confusion when the data are presented to families or in the courtroom. In sum, while these shows are intended to intrigue the audience and present a plausible (however unlikely) crime scene, it would behoove the curious viewer to perform a critical analysis of what is fact, what is fiction, and what is a stretch with regard to the forensic sciences, or for that matter, any discipline that is represented in a TV drama or movie.

EDUCATIONAL AND EMPLOYMENT OPPORTUNITIES

In order to practice forensic anthropology, a master's or doctorate in physical or biological anthropology with additional training and experience in forensic anthropology methods is recommended. Students interested in this career path should consult the website of **American Board of Forensic Anthropology** (http://www.theabfa.org/) for additional information. The website of **Scientific Working Group for Forensic Anthropology** also offers a list of journal articles and publications concerning forensic anthropology practice and theory that was created by the Society of Forensic Anthropologists (SOFA); this resource can be found at http://www.swganth.org/news-resources.html. Although universities do not offer a formal "forensic anthropology" degree, a number of anthropology programs offer a focus in forensic anthropology. In addition to an anthropology curriculum, students may benefit from coursework in genetics and biomechanics. According to a 2009 survey of the AAFS's Anthropology[*] section members, nearly half of practicing forensic anthropologists (44.5%) are employed at an academic institution and do forensic casework on a consulting basis (Agostini and Gomez 2009). Within academia, forensic anthropologists are typically employed as physical anthropologists, anatomists, osteologists, and/or skeletal biologists. Other common employment agencies are medical examiner and coroner's offices (19.1%), federal agencies (12.7%), private consulting firms (i.e., self-employed consultants) (5.5%), museums (2.7%), and nonprofit organizations (2.7%). The remaining 12.7% of the members surveyed were retired or students. The survey by (Agostini and Gomez 2009) also revealed that the majority of forensic anthropologists report an annual income of $50,000–$100,000.

It is highly recommended that students interested in pursuing a career in forensic anthropology or any of the forensic sciences, whether applied or research-oriented, read the recent National Academy of Sciences (NAS) report entitled "Strengthening Forensic Science in the United States: A Path Forward" (2009), as this report will likely affect training, policy, and practice in the forensic sciences for years to come (see section on the NAS report later in this chapter).

BRIEF HISTORY OF FORENSIC ANTHROPOLOGY

Forensic anthropology is a relatively young subfield within biological anthropology. The development of forensic anthropology is divided into three periods that are divided by events that arguably changed the path of the field: the formative period (early 1800s–1938), the consolidation period (1939–1971), and the modern period (1972–present).

FORMATIVE PERIOD (EARLY 1800s–1938)

The origin of forensic anthropology is said to lie within the twisted tale of the Parkman murder in 1849. Dr. George Parkman was a physician who donated the land to Harvard, on which the medical school was built. Dr. Parkman was murdered by Harvard's Chemistry Professor John Webster in the medical building. Dr. Webster had purportedly borrowed money from Parkman and killed him to avoid paying back the debt. The local newspapers reported the salacious details of the case, suggesting that Webster mutilated Dr. Parkman's body, put parts of it in the anatomy laboratory and in a septic tank, and burned the head in the furnace. Harvard's anatomy professors Oliver Wendell Holmes and Jeffries Wyman were asked to aid in the investigation of Dr. Parkman's death. Wyman and Holmes were able to put the pieces back together and to suggest that the skeleton belonged to a person whose description was consistent with George Parkman. Webster was eventually convicted of the murder when dentures found in the furnace were shown to match a mold of Parkman's teeth that his dentist had used to make the dentures.

---

[*] When the study was published, the Anthropology Section of the AAFS was known as the Physical Anthropology section.

The first avid practitioner of the applied study of forensic anthropology was Thomas Dwight (1843–1911). Dwight was considered the father of forensic anthropology in the United States (Stewart 1979), because he was one of the first Americans to discuss how to identify remains using information obtained from the human skeleton. In 1878, Dwight submitted an essay to the Massachusetts Medical Society entitled, "The Identification of the Human Skeleton. A Medico-Legal Study" (Dwight 1878). Dwight succeeded Oliver Wendell Holmes in the Parkman Professorship of Anatomy position at Harvard. As a trained anatomist, Dwight recognized the need for research on methods to determine age, sex, and stature from the skeleton.

George Dorsey (1839–1931) learned from Dwight's research at Harvard. Dorsey received his doctorate from Harvard in anthropology in 1894 and became curator of the Field Museum of Natural History in Chicago. It was at this museum that Dorsey tested his theory that the articular surfaces of long bones could be used as an indicator of sex; he concluded that humeral head diameter was a better diagnostic tool for sex estimation than the femoral head diameter, a point later confirmed by (Dwight 1905). Dorsey was asked to consult on the case of the Leutgert murder, in which the Chicago sausage maker Adolph Leutgert was accused of killing his wife by placing her in a vat of potash, which effectively dissolved most of the body, except four small pieces of bone and the ring that she normally wore. Dorsey was able to say that the fragments came from a human rib, hand, and foot. Leutgert was later convicted, but Dorsey faced harsh criticism from other anatomists of the time for his testimony and conclusions in this case. This criticism is said to have caused Dorsey to abandon further pursuit of forensics, although (Stewart 1979) believes that Dorsey's assertions about the skeletal remains were correct. At the beginning of World War I, Dorsey removed himself from academe and joined the U.S. Navy.

Harris H. Wilder (1864–1928) was a contemporary of George Dorsey, whose most notable contribution to forensic anthropology dealt with personal identification. Wilder was a European-trained zoologist, who became interested in physical anthropology while teaching at Smith College late in his career. Wilder's physical anthropological focus was on dermatoglyphics (fingerprint analysis) and facial reconstruction by using skulls. Wilder and Bert Wentworth published a book in 1918 entitled *Personal Identification: Methods for the Identification of Individuals, Living or Dead* (Wilder and Wentworth 1918). However, (Stewart 1979) points out that this text has no mention of Dwight's previous research on identification, suggesting that at that point, Dwight's research may not have found its way to other scholars within and outside of the field.

Another American anatomist whose work had implications for forensic anthropology was Paul Stevenson (1890–1971). Stevenson spent a good deal of his career studying in China. He contributed two important publications dealing with age determination based on epiphyseal union (Stevenson 1924) and stature estimation by using long bones in a Chinese population (Stevenson 1929). However, it is unclear if Stevenson was aware of the impact of these contributions on forensic anthropology.

According to (W.M. Krogman 1976), Aleš Hrdlička (1869–1943) was the "founding father of American Physical Anthropology." Hrdlička was born in Bohemia in 1869 and came to New York in 1882, at the age of 13 years, where he worked as a cigar maker and went to school at night. After a bout with typhoid at the age of 19 years, Hrdlička began to study medicine at the Eclectic Medical College of New York City, where he graduated in 1892. He went on to study at the New York Homeopathic College for two years, and he took an internship at the New York Homeopathic Hospital for the Insane, where he published on the somatometry of adult patients with various types of insanity. In 1896, Hrdlička received his training in physical anthropology in Paris under Manouvrier, where he learned how to measure the skeleton quantitatively. In 1897, Hrdlička began to study human skeletons at the College of Physicians and Surgeons in New York City. In 1899, Hrdlička embarked on a series of trips to study the American Indians of the southwestern United States and northern Mexico for the American Museum of Natural History. In 1903, he became a part of the Division of Physical Anthropology at the U.S. National Museum in Washington, DC (now known as the Smithsonian), and became curator in 1910. Hrdlička was a giant in the field of physical anthropology. He founded the *American Journal of Physical Anthropology* (AJPA) in 1918 and the American Association of Physical Anthropologists (AAPA) in 1928, two contributions for which he is often remembered most. Hrdlička was the editor of AJPA from 1918 until 1942.

Another physical anthropologist whose work on human variation shaped the theoretical foundations of biological and forensic anthropology was Earnest A. Hooton (1887–1954). Hooton received his doctorate in liberal arts at the University of Wisconsin in 1911. In 1912, he received his diploma in anthropology from Oxford University, and the following year, he joined the anthropology department at Harvard. Hooton's research focused on human variation with respect to human origins and adaptation. Under his direction, Harvard became an important center for training physical anthropologists, and his students became prominent practitioners in the field.

T. Wingate Todd (1885–1938) was also a prolific researcher in anatomy and physical anthropology during the latter part of the formative period. Todd's work influences forensic anthropology even today. Todd was interested in skeletal aging methods and growth and development. He was trained in England as an anatomist and moved to the United States in 1912 to take Dr. Carl Hamann's vacated teaching position at Western Reserve University (Dr. Hamann had become Dean of the medical school). A recent change in Ohio state law permitted professors to retain cadavers that the medical students dissected, and Drs. Todd and Hamann had the foresight to begin an anatomical collection that would soon surpass any other in existence in terms of number of specimens and level of documentation (i.e., age, sex, ancestry, stature, weight, cause of death, and case history). In addition, Todd took anthropometric measurements and photographs of most of the cadavers. By the time of his death in 1938, Todd had managed to build a skeletal research collection containing over 3000 documented individuals. He used the specimens in the collection for numerous anatomical and anthropological studies, and the Hamann–Todd Collection continues to be an important resource for skeletal biology research today.

Todd's contributions to anthropology are numerous and include documentation of differences in limb proportions between American Blacks and Whites; establishment of the usefulness of endo- and ectocranial suture closure for age estimation; development of an age estimation method based on the pubic symphysis; establishment of principles of epiphyseal union; and extensive documentation of human postcranial and craniofacial growth, development, and maturation. During his career, Todd authored nearly 200 publications in anatomy and physical anthropology, many of which have significant implications for forensic anthropology. In addition, two of his students, Wilton Krogman and Montague Cobb, went on to make important contributions to physical anthropology. We discuss Krogman in the following sections, but his legacy includes the landmark bulletin *Guide to the Identification of Human Skeletal Material* (Krogman 1939), as well as the close mentoring of William M. Bass. Cobb was the first African-American to earn a Ph.D. in physical anthropology and left behind a legacy reminiscent of his mentor. The W. Montague Cobb Human Skeletal Collection at Howard University contains approximately 700 skeletons and serves many of the same research purposes as the Hamann–Todd Collection in Cleveland.

Robert J. Terry (1871–1966) was also an anatomist who had the foresight to curate a research collection of skeletal remains. Dr. Terry was an anatomy professor and department head at Washington University Medical School in St. Louis, Missouri. In the same manner as Todd, Terry began collecting skeletal remains from medical school cadavers during the 1920s. Terry Collection cadavers have associated morgue records with the individual's name, sex, age, ancestry, date, and cause of death. Terry also took photographs and anthropometric measurements of most of the cadavers, as well as skin and hair samples; however, only the hair samples remain today. In addition, plaster death masks were made of 836 of the cadavers. Terry retired in 1941, and Dr. Mildred Trotter (1899–1991) assumed his anatomy teaching position and continued to build the collection, until she retired 26 years later. Today, the nearly 2000 skeletons of the Terry Anatomical Collection are housed in the Smithsonian Institution in Washington, DC.

CONSOLIDATION PERIOD (1939–1971)

It has long been posited that the end of the initial period of forensic anthropology (here termed the formative period) and the beginning of the consolidation period were marked by the publishing of Wilton Marion Krogman's *Guide to the Identification of Human Skeletal Material* in the Federal Bureau of Investigation's (FBI's) Law Enforcement Bulletin in 1939. This publication summarized all that had been discovered about the identification of skeletal remains until that time. The significance of this publication is that, for the first time, an article pertaining to forensic identification had been written by an anthropologist and was included in a journal focused on forensics, as opposed to anatomy or the broader discipline of physical anthropology.

Wilton Marion Krogman's prestigious career began in the spring of 1925, where he first lectured in physical anthropology at the University of Chicago. Krogman insisted that it was this experience, coupled with a term paper assignment on the anthropology of teeth given by his professor Dr. Fay-Cooper Cole, that focused his entire professional career:

> That did it! Teeth lead to jaws, jaws to face, face to head, head to body: in other words a coordinated whole. But more than that, it led from statics to the dynamics of age progress. Thus launched my life-work in growth and development, comparative and human anatomy (Krogman 1976).

The term paper introduced Krogman to paleontology, orthodontia, and the work of T. Wingate Todd. Krogman submitted a reworked version of this term paper to the First District Dental Society of New York City's Morris L. Chaim Prize, where T. Wingate Todd was one of the judges. Krogman won the prize, and Todd saw promise in the paper and in Krogman. On a stopover in Chicago, Todd sought out Krogman, who indicated that he would like to do graduate work with Todd. In 1928, Krogman was awarded the Cleveland Foundation Fellowship in anatomy, thanks to Todd's arrangements. Krogman wrote his dissertation under Todd's direction. In 1929, Krogman instructed at the University of Chicago, and in 1930, he took a fellowship at the Hunterian Museum at the Royal College of Surgeons in London. In 1931, Krogman became an associate professor of anatomy and physical anthropology at Western Reserve University. Krogman's appointment at his mentor's department put him in contact with the foremost physical anthropologists of the time. As Krogman (1976) explained, "Todd's department was a magnet for the physical anthropologists of the 1930s."

Krogman expanded upon his article in the FBI bulletin and produced the first textbook in Forensic Anthropology, entitled *The Human Skeleton in Forensic Medicine* (Krogman 1962). The textbook focused on the practical application of human osteology to forensics. Krogman's text became the primary reference for physical anthropologists practicing forensic anthropology, much like his 1939 article had been at the time of its publication. The theme of the book was human variation. Krogman emphasized that the methods identified within the text did not present hard and fast rules; instead, they were meant to be guidelines for assessing remains, with the understanding that humans represented a wide range of morphological variability. Krogman's dedication to research helped push forensic anthropology forward. He imparted a great deal of his wisdom on his graduate students. One of these students, William M. Bass, undoubtedly had the greatest impact on the modern era of forensic anthropology. We discuss the influence and legacy of Bass in greater detail in the following sections.

MODERN PERIOD (1972–PRESENT)

The founding of the Physical Anthropology Section of the American Academy of Forensic Sciences (AAFS) in 1972 is often referred to as the beginning of the modern period in forensic anthropology. At this time, the term "forensic anthropologist" began to be used on a regular basis to refer to practitioners in the field. The section founding was thanks in a large part to Ellis R. Kerley (1924–1998). Kerley had joined AAFS in 1968 as part of the pathology/biology section. Through his encouragement, other physical anthropologists also joined the AAFS. By 1972, with 14 anthropologists as members of the academy, Kerley and colleagues had exceeded the minimum number of members required to establish a new section. Thus, the Physical Anthropology Section of AAFS was born, and this formal organization of the field provided an appropriate stage for the presentation of new ideas. In addition, the academy's flagship journal, *Journal of Forensic Sciences*, was well suited for research concerning new methods in the identification of skeletal remains.

Ellis Kerley was also instrumental in establishing the American Board of Forensic Anthropologists (ABFA) in 1977. This board certifies forensic anthropologists in a similar manner to the certification of physicians by their various boards, using a rigorous application and examination process to ensure that each diplomate of the board is qualified and competent to undertake forensic anthropology casework. In 1987, Kerley became the forensic anthropology consultant and scientific director of the United States Army Central Identification Laboratory in Hawaii, where he oversaw the identification of repatriated war remains. He also served as president of the AAFS from 1990 to 1991. In 2000, the Ellis R. Kerley Forensic Sciences Foundation was established in his memory. The foundation is dedicated to furthering the development of forensic anthropology by assisting students in the field of anthropology and continuing the research in forensic identification of the skeleton (http://www.kerleyfoundation.org/). Each year, the foundation issues at least one scholarship to a graduate student who is enrolled in a physical or forensic anthropology program and who is involved with the AAFS or the ABFA. In addition, the foundation hosts a reception at the AAFS annual meetings, where it presents an award to recognize innovative efforts to continue research in human identification.

As mentioned earlier, William M. Bass was Kerley's student who went on to have the most significant impact on forensic anthropology. In the 1960s, Bass established a graduate program in physical anthropology at the University of Kansas and recruited Ellis Kerley and Thomas McKern to be a part of the department. This graduate program produced some of the foremost physical/forensic anthropologists, including Douglas Ubelaker, Walter Birkby, Judy Suchey, Linda Klepinger, and Richard Jantz.

In 1971, Dr. Bass moved from Kansas to Tennessee, and in doing so, he began an anthropology program in the eastern United States that has produced more forensic anthropologists than any other program to date. By the time Dr. Bass retired in 1994, he had trained over 20 practitioners, including Bill Rodriguez, Anthony Falsetti, Hugh Berryman, Steve Symes, Murray Marks, Doug Owsley, Stephen Ousley, Emily Craig, and Walter Birkby (Rhine 1998). In fact, nearly 40% of practicing forensic anthropologists can trace some element of their academic lineage through Bass (Marks 1995). Many of this new generation of graduates went on to change the face of forensic anthropology by taking it from a strictly academic discipline, in which practitioners acted as consultants, to an applied field that became incorporated into the medical examiner setting. Hugh Berryman was the first forensic anthropologist to be employed full-time outside of academia. In 1980, he was hired as morgue director of the Shelby County Medical Examiner and University of Tennessee Hospital Morgue in Memphis. Over three decades, that single position grew to over 20 full-time positions nationwide and continues to grow today (Berryman 2009) (Figure 1.4). While the majority of these positions are filled by PhDs, a little more than one-third of these employees have master's degrees only. Several factors that contributed to the increase in full-time medical examiner/coroner positions over the last 30 years include a favorable economy, the realization that forensic anthropologists' skills include more than just skeletal analysis, and a number of mass disasters that required anthropological expertise (i.e., the Oklahoma City bombing, the 9/11 attacks on the Pentagon and World Trade Center, and the Tri-State Crematorium situation in Noble, Georgia) (Berryman 2009).

Although a number of forensic anthropologists attained full-time positions in coroner's and medical examiner's offices, many maintained academic appointments and continued to train new practitioners. For example, one of Bass' students, Richard Jantz, has mentored a number of forensic anthropologists, including Stephen Ousley, with whom he created the Fordisc program. Fordisc stands for "forensic discrimination" and is currently in its third version. The Fordisc program has made it possible for forensic anthropologists to determine the sex, ancestry, and stature of unknown remains from cranial and/or postcranial measurements with the click of a button. We discuss this program in detail in Chapter 14.

ANTHROPOLOGY RESEARCH FACILITIES

Perhaps, Bass' most notorious accomplishment is the establishment of the first outdoor research facility devoted to the study of human decomposition. The Anthropology Research Facility (ARF) was founded in 1980 in response to Dr. Bass' realization that the forensic community needed information about the postmortem interval that was based on controlled scientific research on human cadavers. The need for this type of research became evident to Dr. Bass in 1977, when he was asked to inspect decomposing remains that were discovered in a disturbed burial (Bass 1984; Bass and Jefferson 2003). Based on his field observations of the state of decomposition, Dr. Bass estimated that the remains had been in the grave for approximately 1 year. However, after exhuming the remains and bringing them back to his laboratory for closer inspection, he noticed aspects of the clothing that indicated that they belonged to a Civil War colonel. To his dismay, Bass had underestimated the time since death (TSD) by 113 years! To Bass' credit, Colonel Shy's remains had been uncharacteristically well preserved for that era owing to embalming and a high-quality iron casket. Nonetheless, a survey of the literature exposed a dearth of research on decomposition changes, so Dr. Bass resolved to remedy the situation. Fortunately, he had the full support of an open-minded UT administration and an available plot of land on which to establish the unconventional research facility.

The first ARF consisted of a 16-square-foot concrete slab enclosed on all sides and above by a chain-link fence. The first donation arrived in 1981, and three more followed that year. Many of the early donations were unclaimed bodies from the State of Tennessee Medical Examiner, but as the donation program grew in popularity with increasing publicity, the donor population also grew in diversity. In 2009, the ARF received its 1000th donation and has received nearly 1800 donations since inception. Over 3500 individuals have filed paperwork to donate their bodies to the ARF for scientific research. The pre-registered donor paperwork requests information about donors, such as birth date, sex, ancestry, height, weight, number of children, medical and dental history, occupation, habitual activities, handedness, shoe size, education level, childhood socioeconomic status, and photographs. In addition, hair and blood samples are taken as donations arrive, and fingerprints and laser scans are soon to be added to that protocol. As a result of the increase in body donations, the ARF has expanded its initial 16-foot plot to a 1.3-acre tract of land. Today, the ARF is part of a larger establishment within the UT Anthropology Department—the Forensic Anthropology Center (FAC). The FAC oversees all activities at the ARF, including donor relations, law

Forensic Anthropology in the United States 13

1980s

- 1980 Hugh Berryman, PhD, D-ABFA — Shelby County ME & UT Chattanooga, Morgue Director
- 1981 Donna Fontana, MS — New Jersey State Police, Forensic Anthropologist
- 1982 David Wolf, PhD — Kentucky ME (first State Forensic Anthropologist)
- 1983 Craig Lahren, MA — Shelby County ME, Assistant Morgue Director
- 1984 William Rodriguez, PhD, D-ABFA — Caddo & Bossier Parish, LA, Deputy Chief Coroner
- 1986 William Rodriguez, PhD, D-ABFA — Onondaga, NY ME , Forensic Anthropologist & Chief of Operations
- 1986 Craig Lahren, MA — Hamilton County, TN ME, Coordinator of Forensic Services
- 1986 Robert Mann, PhD, D-AFBA — Shelby County ME, Assistant Morgue Director
- 1987 Steve Symes, PhD, D-AFBA — Shelby County ME, Assistant Morgue Director
- 1988 William Haglund, PhD — King County, WA ME , Chief Medical Investigator
- 1989 William Rodriguez, PhD, D-ABFA — Office of the Armed Forces Medical Examiner, Chief Forensic Anthropologist and Deputy Chief ME

1990s

- 1994 Emily Craig, PhD, D-AFBA — Kentucky State Forensic Anthropologist
- 1994 Gwen Haugen, MA — St. Louis County MEO, Forensic Investigator/Forensic Anthropologist (1994-97; 2007-present)
- 1990s Deborah Gray — Riverside County, CA Sheriff's Department, Forensic Anthropologist/Archaeologist/CoronerSergeant
- 1995 Charles Cecil, PhD — San Francisco OCME, Medical Inverstigator II/Forensic Anthropologist
- 1995 Nici Vance, PhD — Oregon State Police Forensic Laboratory, Forensic Anthropologist; OSME, State Forensic Anthropologist (since 2006)
- 1996 Dana Austin, PhD, D-AFBA — Tarrant County, TX ME, Forensic Anthropologist
- 1996 Ann Marie Mires , PhD — Massachusetts State OCME, Forensic Anthropologist
- 1997 Craig Lahren, MA — North Dakota State Forensic Examiner's Office, Administrator and Forensic Anthropologist
- 1997 Tom Bodkin, MA — Hamilton County, TN ME, Coordinator of Forensic Services
- 1997 Laura Fulginiti, PhD, D-ABFA — Maricopa County, AZ ME, Forensic Anthropologist
- 1997 Fran Wheatley, MA — Metropolitan Nashville/Davidson County ME, Death Investigator/Forensic Anthropologist
- 1998 Clyde Gibbs, BS, AS — North Carolina OCME, Chapel Hill, Medical Examiner Specialist
- 1999 Amy Mundorff, PhD — New York City OCME, Forensic Anthropologist

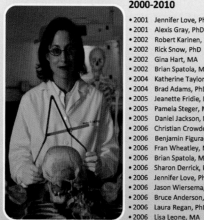

2000-2010

- 2001 Jennifer Love, PhD, D-ABFA — Regional Forensic Center, Memphis
- 2001 Alexis Gray, PhD — San Bernadino County Sheriff's Department, Coroner's Division, Forensic Anthropologist
- 2002 Robert Karinen, MA — Ada County Coroner's Office, Boise, ID, Forensic Supervisor
- 2002 Rick Snow, PhD — Georgia Bureau of Investigation, Forensic Anthropologist
- 2002 Gina Hart, MA — Newark Regional ME, Forensic Anthropologist/Death Investigator
- 2002 Brian Spatola, MA — Washington D.C. OCME, Mortuary Supervisor and Anthropologist
- 2004 Katherine Taylor , PhD — King County, WA ME, Forensic Anthropologist
- 2004 Brad Adams, PhD, D-ABFA — New York City OCME, Director of Forensic Anthropology
- 2005 Jeanette Fridie, MA — New York City OCME, Forensic Anthropologist
- 2005 Pamela Steger, MS — Travis County Medical Examiner's Office, Medicolegal Death Investigator
- 2005 Daniel Jackson, MA — Travis County Medical Examiner's Office, Forensic Anthropologist/Death Investigator
- 2006 Christian Crowder, PhD, D-ABFA — New York City OCME, Deputy Director of Forensic Anthropology
- 2006 Benjamin Figura, MA — New York City OCME, Forensic Anthropologist/Director of Identification
- 2006 Fran Wheatley, MA — Shelby County ME, Administrator and Anthropologist
- 2006 Brian Spatola, MA — Central Virginia OCME, Morgue Supervisor and Anthropologist
- 2006 Sharon Derrick, PhD, D-ABFA — Harris County, TX ME, Agency Coordinator and Anthropologist
- 2006 Jennifer Love, PhD, D-ABFA — Harris County, TX ME, Forensic Anthropology Director
- 2006 Jason Wiersema, PhD, D-ABFA — Harris County, TX ME, Mass Disaster Coordinator
- 2006 Bruce Anderson, PhD, D-ABFA — Tuscon, AZ Forensic Science Center, Forensic Anthropologist
- 2006 Laura Regan, PhD, Lt. Col. — Armed Forces Medical Examiner System, Director of Operations and Deputy Chief Forensic Anthropologist
- 2006 Lisa Leone, MA — Greenville North Carolina Regional ME, Pathologist Assistant/Forensic Anthropologist/Death Investigator
- 2007 Chris Rainwater, MS — New York City OCME, Forensic Anthropologist/Director of Photography
- 2007 Kristen Hartnett, PhD — New York City OCME, Forensic Anthropologist/Assistant Director of Forensic Anthropology
- 2007 Dominique Semeraro, MS — Rhode Island OSME, Senior Medical Examiner Agent, RI State Forensic Anthropologist
- 2008 Murray Marks, PhD, D-ABFA — UT Regional Forensic Center, Forensic Anthropologist
- 2008 Lauren Zephro, PhD — Santa Cruz County Sheriff's Department, Latent Print Examiner, Forensic Anthropologist
- 2008 Hilary Sheaves, MA — North Carolina OCME, Chapel Hill, Autopsy Technician
- 2008 Olivia Alley, MA — Travis County Medical Examiner's Office, Forensic Anthropologist/Death Investigator
- 2009 Debra Prince Zinni, PhD, D-ABFA — Commonwealth of Massachusetts OCME, State Forensic Anthropologis

Figure 1.4 Forensic anthropologists working in the medical examiner's setting between 1980 and 2010.

enforcement training and education, public outreach, and human identification services. The FAC offers hands-on training in forensic taphonomy, human identification, and human remains recovery to a variety of professionals, including the FBI's Emergency Response Teams, the National Forensic Academy, and participants in the FAC's regular short courses.

Since its inception 30 years ago, the ARF has fulfilled Bass' initial goal. Anthropologists now understand the various stages of human decomposition and the variables affecting the rate at which the decomposition process occurs.[*] This understanding has enabled forensic anthropologists to give more accurate TSD estimates. While much of the early research focused on general decomposition changes at the macroscopic level, some of the more recent projects have begun to explore decomposition at the microscopic level. For instance, Dr. Arpad Vass isolated chemical signatures produced by volatile fatty acids during the decomposition process; he used this information to predict TSD with a high degree of accuracy (Vass et al. 1992, 2002). Vass also developed a Decompositional Odor Analysis (DOA) Database for surface and buried remains; this database can be used for training cadaver dogs as well as in the development of portable analytical instruments to locate human remains in shallow burial sites (Vass et al. 2004, 2008). Other researchers have studied DNA recovery from various parts of decomposing remains (i.e., bone, hair, and fingernails) in different environmental conditions (i.e., open ground, shaded ground, buried, and aqueous) (Opel et al. 2006). In addition, projects have explored the effects of factors such as clothing and covers, indoor versus outdoor environments, insect and animal activity, and aqueous environments. One could say that for every question answered by research at the ARF, several more have surfaced. Many of these questions are related to the effects of different geographic locations and climates on the decomposition process (i.e., desert versus woodland). As a result, outdoor research facilities have been established at other geographic locations, and their research is helping forensic anthropologists obtain a more complete picture of the human decomposition process in various environments.

The first of the new generation of decomposition research facilities was opened in 2006 at Western Carolina University (WCU) in Cullowhee, North Carolina, as part of the Western Carolina Human Identification Laboratory. The WCU facility is smaller than the ARF, but it allows for the study of decomposition in a mountain environment. In 2008, Dr. Jerry Melbye at Texas State University in San Marcos obtained funding and land for the Forensic Anthropology Research Facility (FARF). The FARF is under the direction of Dr. Daniel Wescott, a former student of Dr. Richard Jantz, and employs other faculty, including Dr. Michelle Hamilton, a former student of Dr. Bass; Drs. Kate Spradley and Ashley McKeown, former students of Dr. Richard Jantz; and Dr. Nicholas Herrmann, a former student of Dr. Lyle Konigsberg. The facility sits on a 26-acre section of the 4200-acre Freeman Ranch in Texas Hill Country and has an adjoining laboratory to help with facility operations and processing. The ranch has already proven a valuable location to study vulture scavenging—a previously unaddressed topic in the literature (Reeves 2009; Spradley et al. 2012). Much like the ARF, the FARF serves as a resource for forensic anthropology students and law enforcement agencies. Furthermore, the FARF is contributing much-needed data about decomposition in warm, arid climates.

A nearby facility has also opened recently at Sam Houston State University. The Southeast Texas Applied Forensic Science Facility (STAFS) is an outdoor research and training facility located on a 247-acre parcel of land near the Sam Houston National Forest. The research facility proper is a 1-acre plot surrounded by security fencing and an adjoining 8 acres of minimum security reserved for training activities. The STAFS has a variety of simulated environmental conditions, including a fluvial environment, and webcams are located to monitor the facility from computers. In addition, the STAFS has a morgue building with coolers, freezers, digital radiograph, and microscope capabilities (http://www.cjcenter.org/stafs/). Human decomposition outdoor research facilities have been proposed in Illinois, Nevada, and even India, and decomposition research on pigs has begun in Montana, as well.

DOCUMENTED DONATED SKELETAL COLLECTIONS

Research facilities have made it possible to assemble modern skeletal collections, as well. While the Hamann-Todd, Terry, and Montague Cobb Collections are invaluable contributions to physical anthropology in the United States, the demographic profile of these collections is primarily of European American and African American individuals

---

[*] Chapter 15 provides a detailed discussion of the decomposition process and the variables affecting decomposition rates.

from the early to mid-twentieth century. The human skeletal form is in a constant state of flux owing to the changing demographics of the American population and changes in diet and healthcare (to name a few factors). In order for forensic anthropologists to develop methods and standards to use in modern forensic casework, we need modern skeletal samples, and the selfless individuals who donate their remains to forensic anthropology research facilities have made this possible.

For example, after bodies are used for research at the University of Tennessee's ARF, the skeletal remains are collected carefully, cleaned, and curated in the William M. Bass Donated Skeletal Collection, where they are used for teaching and research purposes. Since record keeping began in 1994, researchers ranging from universities to biomedical research companies to law enforcement agencies such as the FBI have visited the Bass Collection (Shirley et al. 2011). The Bass Collection is the largest collection of modern human skeletal remains in the country, and it has given researchers an unparalleled opportunity to gain an understanding of modern human skeletal variation. Researchers have documented differences between twentieth and twenty-first century Americans and their nineteenth century predecessors, including increases in stature and changes in cranial form (Jantz 2001, 1999, 2000; Meadows and Jantz 1995; Ousley and Jantz 1997). These studies have demonstrated the importance of using modern skeletal samples to establish criteria for estimating age, sex, ancestry, and stature from skeletal remains. The Bass Collection is predominantly of American White males, but the demographic profile is increasing in diversity each year. Presently, the collection contains Americans of European, African, Hispanic, Native American, and Japanese ancestries. It comprises ~70% males and 30% females, with ages ranging from fetal to 101 years. To date, 800 of the skeletons have been scanned using computed tomography (CT) scanning, and these scans have been used in a variety of research projects in anthropology as well as biomedical engineering, including the development of a female knee implant design by Dr. Mohamed Mahfouz and colleagues in the UT Department of Mechanical, Aerospace, and Biomedical Engineering (Shirley et al. 2011). The Forensic Anthropology Research Facility (FARF) at Texas State University follows the ARF model and accessions donor skeletal remains into the Texas State Donated Skeletal Collection for scientific research and education purposes.

Another important collection of modern human skeletal remains is the Maxwell Museum Documented Skeletal Collection at the University of New Mexico. This collection was established in 1984, and it contains over 200 individuals with documented age, sex, ancestry, and cause of death. In addition, health and occupational data are available for donations dating 1995 and later. The Maxwell Donated Collection is housed in the Osteology Repository directly above the Maxwell Museum Laboratory of Human Osteology and serves as a resource for training, education, and research. During his tenure at University of New Mexico (UNM), Dr. Stan Rhine often consulted the collection to answer questions about forensic cases that he consulted on for the New Mexico Office of the Medical Investigator (Rhine 1998). In addition, the collection has been used for instructional seminars in forensic anthropology and for research concerning bone response to repetitive motions, skeletal manifestations of disease, skeletal trauma, identification of handedness, and modern human variation.

Finally, a number of forensic anthropology laboratories curate remains from unsolved cases, as well as remains donated by the families of victims. The Louisiana State University FACES Laboratory is one such laboratory. Although these collections do not have the extensive documentation that is associated with the donated collections at the University of Tennessee and the University of New Mexico, they serve as valuable research resources.

ETHICS IN PRACTICE AND RESEARCH

Unfortunately, when judged by today's ethical standards, physical anthropology has a tainted past when it comes to handling skeletal remains. Some of the existing skeletal collections (particularly collections of Native American remains) were obtained in an opportunistic manner that would be considered unethical today. However, at the time that the collections were amassed, there were no laws or regulations governing the collection and handling of human skeletal remains. Laws such as the **Native American Graves Protection and Repatriation Act** (NAGPRA, 1990) set a precedent for addressing ethical concerns that have spread beyond the handling of Native American skeletal material. Physical anthropologists have made great strides in correcting the misdoings of early practitioners, and in doing so, they have created an environment in which research and education are conducted with the utmost respect and reverence. As researchers and curators of human skeletal material, physical anthropologists take their moral responsibility seriously. Indeed, a number of physical anthropologists have become involved in human rights organizations around the globe. Forensic anthropologist bring valuable skills to the table in dealing with human rights violations, including the ability to sort commingled remains in mass graves, establish a biological profile from unidentified skeletal remains, evaluate skeletal trauma, ascertain manner and/or cause of death, and handle the sociocultural factors affiliated with this type of work. Forensic anthropologists have done work in Bosnia, Kosovo, Rwanda, Argentina, Peru, Colombia, and Iraq, to name a few countries. This type of work and the organizations under which it is performed are discussed in Chapter 17.

BEST PRACTICE IN FORENSIC ANTHROPOLOGY: FROM SWGANTH TO OSAC

Germane to the topics covered in this chapter is a brief review of several important developments within the larger forensic science field. This includes the National Academy of Sciences (NAS) report entitled "Strengthening Forensic Science in the United States: A Path Forward" (2009) as well as the work of the Scientific Working Group for Forensic Anthropology (SWGANTH) and the Organization of Scientific Area Committees (OSAC).

The NAS report is a final grant report published by the U.S. Department of Justice and contains recommendations for improving research, practice, and training in the forensic sciences. (For more information on the NAS report, please go to https://www.ncjrs.gov/pdffiles1/nij/grants/228091.pdf.) The report proposed the creation of an independent federal entity, to be named the National Institute of Forensic Science, to serve as an administrator and advisory board to enforce best practice in the forensic sciences. Many predicted that the NAS report suggestions would restructure the way in which many forensic anthropologists operate, particularly with respect to mandated laboratory accreditation and individual certification. An initial response to this report was the independent creation of Scientific Working Groups (SWGs) by many of the forensic science disciplines. The working group for forensic anthropologists is the Scientific Working Group for Forensic Anthropology (SWGANTH).

Mirrored after similar scientific working groups in other disciplines, the SWGANTH was charged with identifying "best practices" in forensic anthropology. The SWGANTH was established by the Department of Defense Central Identification Laboratory (DoD CIL) and the Federal Bureau of Investigation (FBI) and consists of a 20-member board that comprises forensic anthropologists from various backgrounds and agencies. In an effort to remove the veil that often cloaks forensic anthropology methods and practice, the SWGANTH board established committees to review multiple aspects of forensic anthropology and to establish a "best practice model" for each of these areas. These committees drafted best practice suggestions in areas pertinent to the practice of forensic anthropology, including proficiency testing, ethics, qualifications, laboratory management, age estimation, sex assessment, stature estimation, commingled remains, and many other aspects of case analyses. (For a full listing of the guidelines established by the SWGANTH, go to www.swganth.org.) Despite the thorough nature of the guidelines, the SWGANTH does not have the ability to enforce these guidelines within the field or regulate forensic anthropology practitioners. Instead, the onus is on the practitioners to be certain that they are aware of these "best practices" and are striving to achieve these goals.

The NAS report's recommendation to create an independent federal entity that enforces best practice in the forensic sciences became a reality in 2013. The Department of Justice (DOJ) partnered with the National Institute of Standards and Technology (NIST) to create the **National Commission on Forensic Science**. The commission seeks to promote scientific validity, reduce fragmentation, and improve federal coordination of forensic science. The commission consists of federal, state, and local forensic science service providers, research scientists and academics, law enforcement officials, prosecutors, defense attorneys, and judges. The NIST, in coordination with the forensic science community and its practitioners, has created the **Organization of Scientific Area Committees** (OSAC), which will help develop or coordinate the development of guidelines and standards for the forensic sciences. The Scientific Area Committees will replace the independent SWGs and develop national standards for the forensic science disciplines.

These guidelines and standards, similar to those created by the SWGs, are intended to ensure that the highest-quality science is used in each of the forensic sciences and that conclusions are derived through research and testing and are based on rigorous scientific facts. Guidelines and standards are provided for research, testing, reporting and testimony for each of the recognized forensic disciplines. The OSAC has a governing board called the Forensic Science Standards Board, which comprises research representatives, professional association representatives,

Table 1.1 Current listing of the Scientific Area Committees and the Forensic Science Disciplines covered by each committee.

Scientific Area Committee	Forensic Science Disciplines covered by committee
Biology/DNA	Biological data interpretation and reporting Biological methods Wildlife forensics
Chemistry/Instrumental Analysis	Fire debris and explosives Geological materials Gunshot residue Materials (trace) Seized drugs Toxicology
Crime Scene/Death Investigation	Anthropology Disaster victim identification Dogs and sensors Fire and explosion investigation Medico-legal death investigation Odontology
Digital/Multimedia	Digital evidence Facial identification Speaker recognition Video/imaging technology and analysis
Physics/Pattern Interpretation	Blood stain pattern analysis Firearms and toolmarks Footwear and tire Forensic document examination Frictionridge

Scientific Area Committee (SAC) chairs, and an NIST ex officio member. This group of individuals is charged with overseeing all the committees and subcommittees of the OSAC, approving scientific standards, and facilitating communication among the OSAC committees and subcommittees and between OSAC and the forensic science community. Within the OSAC, the forensic science disciplines are separated into five SACs: Biology/DNA; Chemistry/Instrumental Analysis; Crime Scene/Death Investigation; Digital/Multimedia; and Physics/Pattern Interpretation. (See Table 1.1 for a complete listing of the forensic science disciplines encompassed by each SAC.) Anthropology falls under the Crime Scene/Death Investigation SAC. In practice, each SAC develops scientific standards for the forensic disciplines covered by that committee. Then, the OSAC Forensic Science Standards Board approves these standards. Once accepted, the guidelines are disseminated to practitioners in the forensic science disciplines affected by these guidelines. More information on the OSAC can be found at: http://www.nist.gov/forensics/osac/index.cfm.

FORENSIC ANTHROPOLOGY IN THE INTERNATIONAL ARENA

Although the primary focus of this chapter and much of this textbook is practice and research in forensic anthropology in the United States, the discipline has an international presence as well. Chapter 17 summarizes several important ways in which forensic anthropologists are employed in the global arena. The above-mentioned International Association of Forensic Sciences (IAFS) is a worldwide association of academics and practicing professionals from various disciplines in forensic science; it holds meetings every three years. *Forensic Science International* is an English journal that publishes original contributions in the forensic sciences and has an international audience. In addition, a number of non-English journals publish forensic research and case reports.

The Forensic Anthropology Society of Europe (FASE) also offers a certification process for forensic anthropologists. It is structured differently than that provided by the American Board of Forensic Anthropologists. The FASE certification has three levels: one for practitioners with a master's degree (level II); a more advanced certification for practitioners with more experience and an MD or PhD (level II), and the Honoris Causa certification for more experienced and established practitioners, who have been practicing for at least 15 years.

Skeletal collections (both modern and archaeological) are also available for research in various countries throughout the world, including South Africa, Colombia, Portugal, England, France, Germany, China, and Japan. However, to the knowledge of these editors, outdoor human decomposition research facilities are not known outside of the continental United States.

Readers interested in the practice of forensic anthropology outside of the United States are encouraged to consult Dr. Douglas Ubelaker's (2013) edited volume *Forensic Science: Current Issues, Future Directions*.

SUMMARY

In summary, forensic anthropology can be described as an applied subdiscipline of physical/biological anthropology. In practice, forensic anthropologists assist the medico-legal community in the identification of unknown remains by developing a biological profile by using information gleaned from a skeletal examination. They are often involved in trauma analysis of the skeleton, postmortem interval estimation, single and mass burial excavations, mass disasters, and human rights work. The unique multi-disciplinary training of forensic anthropologists enables them to bring a unique perspective to understanding human variation in that they consider the biological and sociocultural determinants of human skeletal biology.

Forensic anthropologists in the United States are affiliated with the American Academy of Forensic Sciences (AAFS), International Association of Forensic Sciences (IAFS), and American Association of Physical Anthropologists (AAPA). Many forensic anthropologists also attend regional conferences. The Anthropology Section of the AAFS was founded in 1972 and comprises 7% of the AAFS membership; it has the largest percentage of student members of any of the sections.

The history of forensic anthropology can be subdivided into three distinct periods, each initiated by an important publication or event. The Formative Period (early 1800s–1938) began with the earliest practitioners and researchers, most of whom were anatomists. The Consolidation Period (1939–1971) began with Krogman's 1939 publication in the Federal Bureau of Investigation's (FBI's) Law Enforcement Bulletin (*Guide to the Identification of Human Skeletal Material*). The Modern Period (1972–present) began with the founding of the Physical Anthropology Section of the American Academy of Forensic Sciences (AAFS) in 1972. Key events and influential figures in each of these periods helped shape forensic anthropology into the discipline that it is today.

Documented skeletal collections and outdoor research facilities serve as the laboratories for much of the research in forensic anthropology and as a data source for developing and validating methods. The University of Tennessee's Anthropology Research Facility (ARF) was established in 1981. It is the oldest of the human decomposition outdoor research facilities in the United States. Texas State University's Forensic Anthropology Research Facility (FARF) was founded in 2008. Other facilities exist (e.g., Sam Houston State University's Southeast Texas Applied Forensic Science Facility [STAFS] and a facility at Western Carolina University), and still others are proposed each year. These facilities contribute much-needed data about decomposition in various climates and geographic regions. They also offer training courses to law enforcement, educators, practitioners, and researchers, and they have made it possible to assemble modern skeletal collections to study modern skeletal variation.

The National Academy of Sciences (NAS) report "Strengthening Forensic Science in the United States: A Path Forward" (2009) forever changed the face of the forensic sciences. The initial response was the creation of Scientific Working Groups (SWGs) to make recommendations for best practice. The Scientific Working Group for Forensic Anthropology (SWGANTH) was established by the Department of Defense Central Identification Laboratory (DoD CIL) and the Federal Bureau of Investigation (FBI) and consists of a 20-member board that comprises forensic anthropologists from various backgrounds and agencies. SWGANTH committees have drafted best practice suggestions in areas pertinent to the practice of forensic anthropology, including proficiency testing, ethics, qualifications, laboratory management, age estimation, sex assessment, stature estimation, commingled remains, and other aspects of case analyses.

The NAS report's recommendation to create an independent federal entity that enforces best practice in the forensic sciences was instituted in 2013, when the Department of Justice (DOJ) partnered with the National Institute of Standards and Technology (NIST) to create the National Commission on Forensic Science. The commission seeks to promote scientific validity, reduce fragmentation, and improve federal coordination of forensic science. The NIST has created the Organization of Scientific Area Committees (OSAC), which will coordinate the development of guidelines and standards for the forensic sciences. The Scientific Area Committees (SACs) will replace the independent SWGs and develop national standards for the forensic science disciplines. Within the OSAC, the forensic science disciplines are separated into five SACs: (1) Biology/DNA, (2) Chemistry/Instrumental Analysis, (3) Crime Scene/Death Investigation, (4) Digital/Multimedia, and (5) Physics/Pattern Interpretation. Anthropology falls under the Crime Scene/Death Investigation SAC. The OSAC has a governing board (Forensic Science Standards Board) that comprises research representatives, professional association representatives, Scientific Area Committee (SAC) chairs, and an NIST ex officio member. In practice, each SAC develops scientific standards for the forensic disciplines covered by that committee, and then, the OSAC Forensic Science Standards Board approves these standards. Once accepted, the guidelines are disseminated to practitioners in the forensic science disciplines affected by these guidelines.

Although this text takes a U.S.-centric approach to forensic anthropology practice and research, readers should be aware that the discipline has a global presence. Professional organizations, scientific journals, certification processes, and skeletal collections are present in many countries outside of the United States. Readers interested in the practice of forensic anthropology outside of the United States should consult Dr. Douglas Ubelaker's (2013) edited volume *Forensic Science: Current Issues, Future Directions*.

Review questions

1. List and define the subdisciplines of anthropology.
2. Describe the general principles used by forensic anthropologists to develop a biological profile.
3. Identify the AAFS and describe its structure.
4. Identify the three main historic periods of the history of forensic anthropology and describe the salient event that initiated each period.
5. Describe the contributions of T. Dwight, A. Hrdlička, and W. M. Krogman to physical/forensic anthropology.
6. What do anthropologists hope to learn from outdoor research facilities?
7. Name three documented modern skeletal collections. Why are these important to the discipline of forensic anthropology?
8. What is NAGPRA and why is it important to forensic anthropology?
9. What did the NAS report conclude regarding the state of forensic science in the United States? What has been done to address this?
10. Describe the SWGANTH and its purpose.
11. Define OSAC. What is the principle charge of this group? Where does forensic anthropology fit into the SACs?

Glossary

American Academy of Forensic Sciences (AAFS): the professional organization with which most forensic anthropologists are affiliated in the United States. The AAFS is comprised of 11 sections and publishes the *Journal of Forensic Sciences*.

American Association of Physical Anthropologists (AAPA): the leading professional organization for physical anthropologists consisting of paleoanthropologists, primatologists, and forensic anthropologists. The AAPA publishes the *American Journal of Physical Anthropology* and the *Yearbook of Physical Anthropology*.

American Board of Forensic Anthropology (ABFA): a nonprofit organization that provides a program of certification in forensic anthropology. Diplomates must demonstrate an ongoing record of practice and research in the field of forensic anthropology and engage in continuing education.

Anthropology: the discipline that studies all aspects of what it means to be human (culture, language, history and origins, and biology).

Biological profile: the four primary components of a person's physical identity (phenotype) that forensic anthropologists ascertain from the skeleton: age, sex, ancestry, and stature. The biological profile helps law enforcement in the search for missing persons.

Four-field approach: the study of the four subfields of anthropology in order to gain a more holistic understanding of humans and our ancestors (cultural anthropology, biological anthropology, linguistic anthropology, and archaeology).

International Association of Forensic Sciences (IAFS): the only worldwide association of academics and practicing professionals from various forensic science disciplines. The IAFS holds meetings every three years.

National Commission on Forensic Science: a commission created by the Department of Justice (DOJ) and the National Institute of Standards and Technology (NIST) upon the recommendation of the NAS report (2009). The commission seeks to promote scientific validity, reduce fragmentation, and improve federal coordination of forensic science.

Native American Graves Protection and Repatriation Act (NAGPRA): An Act enacted in 1990 that requires federal agencies and institutions that receive federal funding to return Native American remains and cultural items to lineal descendants and culturally affiliated Indian tribes. NAGPRA also establishes processes for the excavation or discovery of Native American cultural items and makes it a crime to traffic in Native American human remains without the right of possession.

Organization of Scientific Area Committees (OSAC): the overarching committee that consists of five scientific area committees (SACs). The OSAC coordinates development of standards and guidelines to improve quality and consistency of work in the forensic science community.

Scientific Working Group for Forensic Anthropology (SWGANTH): a scientific working group consisting of a number of committees that recommend and disseminate guidelines for best practice, quality assurance, and quality control in forensic anthropology.

References

Agostini GM and Gomez EJ. 2009. Forensic anthropology academic and employment trends. *Proceedings of the American Academy of Forensic Sciences Annual Meetings*, Denver, CO, February 16–21. Vol. 15, p. 346.

Bass WM. 1984. Time interval since death. In: TA Rathbun and JE Buikstra (Eds.), *Case Studies in Forensic Anthropology*. Springfield, IL: Charles C. Thomas, pp. 136–147.

Bass WM and Jefferson J. 2003. *Death's Acre: Inside the Legendary Forensic Lab "The Body Farm" Where the Dead Do Tell Tales*. New York: G.P. Putnam's Sons.

Berryman HE. 2009. Full time employment of forensic anthropologists in medical examiner's/coroner's offices in the United States—A history. *Proceedings of the American Academy of Forensic Sciences Annual Meeting*, Denver, CO, Vol. 15, p. 328.

Committee on Identifying the Needs of the Forensic Sciences Community, National Research Council. 2009. Strengthening forensic science in the United States: A path forward, A Report. Document No. 228091. NIJ Award No. 2006-DN-BX-0001.

Dwight T. 1878. The identification of the human skeleton: A medico-legal study. *Mass Med Soc* 12:165–218.

Dwight T. 1905. The size of the articular surfaces of long bones as characteristics of sex; an anthropological study. *Am J Anat* 4:19–32.

Jantz R. 2001. Cranial change in Americans. *J Forensic Sci* 46(4):784–787.

Jantz L and Jantz R. 1999. Secular change in long bone length and proportion in the United States, 1800–1970. *Am J Phys Anthropol* 110:57–67.

Jantz R and Jantz ML. 2000. Secular change in craniofacial morphology. *Am J Hum Biol* 12:327–338.

Krogman WM. 1939. A guide to the identification of human skeletal material. *FBI Law Enforcement Bulletin* 8(8):3–31.

Krogman WM. 1962. *The Human Skeleton in Forensic Medicine*. Springfield, IL: Charles C. Thomas.

Krogman WM. 1976. Fifty years of physical anthropology: The men, the material, the concepts, the method. *Annu Rev Anthropol* 5:1–14.

Marks MK. 1995. William M. Bass and the development of forensic anthropology in Tennessee. *J Forensic Sci* 40(5):741–750.

Meadows L and Jantz R. 1995. Allometric secular change in the long bones from the 1800s to the present. *J Forensic Sci* 40(5):762–767.

Native American Graves Protection and Repatriation Act (NAGPRA). 1990. PL 101–601, 104 Stat 3048, United States.

Opel KL, Chung DT, Drabek J, Tatarek NE, Jantz LM, and McCord BR. 2006. The application of miniplex primer sets in the analysis of degraded DNA from human skeletal remains. *J Forensic Sci* 51(2):351–356.

Ousley SD and Jantz RL. 1997. The forensic data bank: Documenting skeletal trends in the United States. InKJ Reichs(Ed.), *Forensic Osteology: Advances in the Identification of Human Remains*. Springfield, IL: Charles C. Thomas, pp. 442–458.

Reeves NM. 2009. Taphonomic effects of vulture scavenging. *J Forensic Sci* 54(3):523–528.

Rhine S. 1998. *Bone Voyage: A Journey in Forensic Anthropology*. Albuquerque, NM: University of New Mexico Press.

Shirley NR, Wilson RJ, and Jantz ML. 2011. Cadaver use at the University of Tennessee's Anthropological Research Facility. *Clin Anat Rev* 24(3):372–380.

Spradley MK, Hamilton MD, Giordano A. 2012. Spatial patterning of vulture scavenged human remains. *Forensic Sci Int* 219:57–63.

Stevenson PH. 1924. Age order of epiphyseal union in man. *Am J Phys Anthropol* 7:53–93.

Stevenson PH. 1929. On racial differences in stature long bone regression formulae for the Chinese (with editorial note by Karl Pearson). *Biometrika* 21:303–318.

Stewart TD. 1979. *Essentials of Forensic Anthropology, Especially as Developed in the United States*. Springfield, IL: Thomas, Vol. 17, p. 300.

Ubelaker DH(Ed.). 2013. *Forensic Science: Current Issues, Future Directions*. West Sussex, UK: Wiley Blackwell.

Vass AA, Barshick SA, Sega G, Caton J, Skeen JT, Love JC, and Synstelien JA. 2002. Decomposition chemistry of human remains: A new methodology for determining the postmortem interval. *J Forensic Sci* 47(3):542–553.

Vass AA, Bass WM, Wolt JD, Foss JE, and Ammons JT. 1992. Time since death determinations of human cadavers using soil solution. *J Forensic Sci* 37(5):1236–1253.

Vass AA, Smith RR, Thompson CV, Burnett MN, Dulgerian N, and Eckenrode BA. 2008. Odor analysis of decomposing buried human remains. *J Forensic Sci* 53(2):384–391.

Vass AA, Smith RR, Thompson CV, Burnett MN, Wolf DA, Synstelien JA, Dulgerian N, and Eckenrode BA. 2004. Decompositional odor analysis database. *J Forensic Sci* 49(4):760–769.

Wilder HH and Wentworth B. 1918. *Personal Identification: Methods for the Identification of Indivduals, Living or Dead*. Boston: Gorham.

Additional information

Useful websites

American Academy of Forensic Sciences (www.aafs.org)

American Association of Physical Anthropologists (www.physanth.org)

American Board of Forensic Anthropology (www.theabfa.org)

Maxwell Museum of Anthropology Documented Skeletal Collection (http://www.unm.edu/~osteolab/coll_doc.html)

Midwest Bioarchaeology and Forensic Anthropology Association (http://archlab.uindy.edu/barfaa/index.php)

Mountain, Desert, and Coastal Forensic Anthropologists (http://foil.ucsc.edu/mdc/)

National Commission on Forensic Science (https://www.justice.gov/ncfs)

National Institute of Standards and Technology (http://www.nist.gov)

Organization of Scientific Area Committees (http://www.nist.gov/forensics/osac.cfm)

Scientific Working Group for Forensic Anthropology (www.swganth.org)

Skeletal Collections Database website—includes a list of collections from around the world (http://skeletal.highfantastical.com/)

Southeast Texas Applied Forensic Science Facility (http://www.cjcenter.org/stafs/)

Texas State University San Marcos Forensic Anthropology Center (http://www.txstate.edu/anthropology/facts/)

The Ellis Kerley Foundation (www.kerleyfoundation.org)

The Hamann-Todd Osteological Collection (http://www.cmnh.org/site/ResearchandCollections/PhysicalAnthropology/Collections/Hamann-ToddCollection.aspx)

The NAS Report (http://www.ncjrs.gov/pdffiles1/nij/grants/228091.pdf)

The Robert J. Terry Anatomical Collection (http://anthropology.si.edu/cm/terry.htm)

The University of Tennessee Forensic Anthropology Center (http://web.utk.edu/~fac/)

Nonfiction forensic anthropology resources
Nonfiction books

Bass, William M. 2004. *Death's Acre: Inside the Legendary Forensic Lab the Body Farm Where the Dead Do Tell Tales*. New York: Berkley Books.

Maples WR and Browning M. 1995. *Dead Men Do Tell Tales: The Strange and Fascinating Cases of a Forensic Anthropologist*. New York: Broadway Books.

Rhine S. 1998. *Bone Voyage: A Journey in Forensic Anthropology*. Albuquerque: University of New Mexico Press.

Textbooks

Bass WM. 2004. *Human Osteology: Laboratory and Field Manual, 5th edn.*. Columbia:Missouri Achaeological Society.

Burns KR. 2012. *Forensic Anthropology Training Manual*. New York: Routledge. 384 pp.

France DL. 2011. *Human and Nonhuman Bone Identification: A Concise Field Guide, 6th edn.*. Boca Raton: CRC Press.

Schaefer M, Black S and Scheuer L. 2009. *Juvenile Osteology: A Laboratory and Field Manual*. London: Academic Press.

Steadman DW (Ed). 2008. *Hard Evidence: Case Studies in Forensic Anthropology*. New York: Routledge. p. 360.

Tersigni-Tarrant MA and Shirley NR (Eds.). 2012. *Forensic Anthropology: An Introduction*. Boca Raton: CRC Press. p. 486.

Ubelaker DH (Ed). 2013. *Forensic Science: Current Issues, Future Directions*. West Sussex, UK: Wiley Blackwell.

White TD. 2005. *The Human Bone Manual*. London: Academic Press. p. 488.

CHAPTER 2

Skeletal Remains as Evidence

Marin A. Pilloud and MariaTeresa A. Tersigni-Tarrant

CONTENTS

Introduction	23
Is it bone?	24
Is it human?	25
How many individuals are present?	27
Is the material modern?	28
Is the material of forensic significance?	28
Remains of fetuses	29
Chain of custody	31
Conclusions	31
Summary	31
Review questions	31
Glossary	32
References	32

LEARNING OBJECTIVES

1. Describe the process by which forensic anthropologists first assess whether the material in question is bone.
2. Explain how forensic anthropologists ascertain whether bone is human.
3. Describe how forensic anthropologists determine if human skeletal remains are modern.
4. Explain how forensic anthropologists determine whether modern human skeletal remains are of forensic significance.

INTRODUCTION

When a forensic anthropologist is presented with remains suspected to be human skeletal material, it is necessary to assess the materials for evidentiary value. Forensic anthropologists utilize a **biocultural** approach to determine **medico-legal** significance. Most forensic anthropologists are trained to evaluate the cultural, archaeological, and physical evidence (both bone and other material) to determine if materials have forensic value. Forensic anthropologists ask a series of questions to conclude if a forensic investigation is warranted. These questions include: (1) Is the material bone? (2) Is the material human bone? (3) Is the material modern? and (4) Is the material of interest to the medico-legal community; that is, could it be related to a civil or criminal matter?

These are not mutually exclusive categories; for example, nonhuman or nonmodern bone may still be of medico-legal interest. For example, in 2009, the Michigan State University Forensic Anthropology Laboratory assisted with a legal dispute between two parties regarding the disappearance of 160 heads of cattle. The forensic anthropologists were asked to determine the number of cattle and to estimate a postmortem interval (Megyesi et al. 2011). Additionally, archaeological remains may be of interest in antiquities dealings or other illegal activities. Native American human remains are protected under the Native American Graves Protection and Repatriation Act (NAGRPA) enacted in 1990 ("Native American Graves Protection and Repatriation Act" 1990). One of the provisions of NAGRPA is that criminal penalties may result from the illegal trafficking of remains protected under this provision (Section 4).

Regardless of the outcome of these questions, each one must be asked by the forensic anthropologist. The response to each will guide future analyses and aid law enforcement or civil parties in deciding how to proceed with the case. The processes involved in addressing each of these questions are outlined below.

IS IT BONE?

Many materials can look like bone to the untrained eye. Qualified biological anthropologists use their skills as osteologists to determine if the material is **osseous** in nature. If the fragment is large enough, it should be relatively easy to determine the nature of the material. Smaller pieces may be more difficult to assess, particularly if the material is natural, as small pieces of wood, plants, charcoal, and shell can be mistaken for bone. Other materials such as rocks, plastic, and ceramics are also commonly mistaken for bone. Any circumstance where these materials may become intermixed with bone or bone fragments requires the assistance of an anthropologist to differentiate osseous from nonosseous materials (Figure 2.1).

There are several means to distinguish human from nonhuman material if gross examination cannot answer the question. Microscopic examination is the first step (Ubelaker 1998). Bone has a distinct surface that can be detected at even low magnification. Skeletal material has a compact cortical outer surface that appears dense and smooth. This layer covers a trabecular structure that looks like disorganized strands of bone under low magnification. The forensic anthropologist should have known exemplars of bone in the laboratory with which to compare the unknown specimen. A comparison of visual properties aids in determining if a material is osseous.

Radiographs are another means to determine if a material is osseous. Skeletal material is radiopaque, in that it absorbs and scatters radiation, giving it a white appearance on film (Fleckenstein and Tranum-Jensen 2001). A radiograph may quickly differentiate radiopaque material from radiolucent material, as a crude means to identify dense materials similar to bone in a large sample of material. For example, a good practice for a forensic anthropologist is

Figure 2.1 An example of separating osseous material from nonosseous material.

to radiograph a body bag after skeletal material has been sorted and removed for analysis. An examination of the radiographs helps the analyst identify any skeletal material (such as a phalanx or tooth) that may have been overlooked. This practice may be particularly important if the bag contained other materials or debris, such as wood, that obscured the skeletal material.

Finally, elemental analysis may aid in determining if a material is bone. A scanning electron microscope with energy dispersive spectroscopy (SEM-EDS) or an X-ray fluorescence (XRF) analyzer can identify the elemental composition of an object and help determine if the material is osseous or nonosseous (Christensen et al. 2012; Ubelaker et al. 2002). Both tools utilize characteristic X-ray emissions to determine the elements that are present (Houck 2015). Bone is primarily made of hydroxyapatite, which is a mineral composed of calcium, phosphorus, hydrogen, and oxygen (Kuhn 2007). SEM-EDS and XRF are best at detecting heavier elements (Houck 2015); therefore, when identifying bone, the analyst is most interested in identifying the presence of calcium and phosphorus, and hydrogen and oxygen are of less interest.

IS IT HUMAN?

If the material is found to be bone or teeth, the next step is to assess if the remains are human. Determining human versus nonhuman bone and teeth requires a trained osteologist. All aspects of the material must be scoured for evidence of human morphology, especially if the material is fragmentary or taphonomically compromised. Given the proximity of humans to other animals, particularly large mammals, it is common for the forensic anthropologist to receive a mixed assemblage of **faunal** and human remains for examination. In those cases, it is important to separate all nonhuman and human remains.

If the specimen in question is relatively large or contains diagnostic morphology, a determination of whether the bone is human or not can usually be made through gross visual examination. The unique size and mode of locomotion of humans result in distinct **articular** surfaces and bone shapes (Figure 2.2). However, several bones from large mammals can closely resemble human skeletal material. For example, a highly decomposed bear paw may surprisingly approximate the appearance of a human hand. A trained forensic anthropologist, however, can differentiate the two based on distinct morphological differences. Given these similarities in mammalian species, forensic anthropologists frequently make assessments of human versus nonhuman bone via photographs (with scales) provided by law enforcement (Figure 2.3).

If the osseous remains are highly fragmentary, it may be difficult to use either gross morphology or size to determine if the remains are human. In these cases, it is useful to examine a histological sample of the remains by using a light microscope to assess whether the remains have characteristic human micromorphology. Typical adult

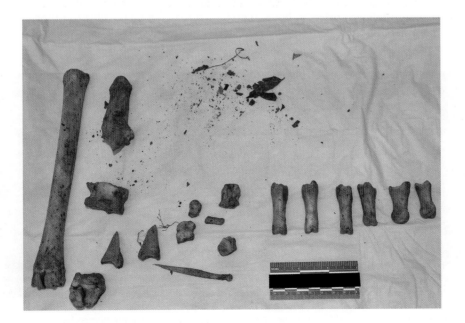

Figure 2.2 Faunal remains received for examination.

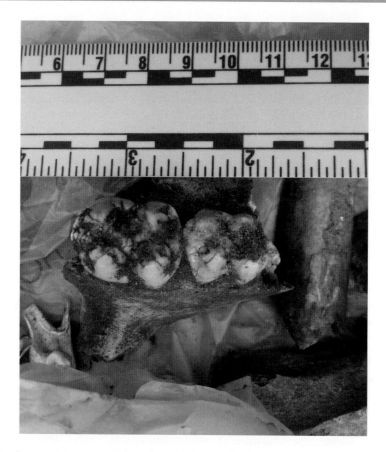

Figure 2.3 An example of a photograph sent from law enforcement to determine if the remains are human. In this case, the teeth were not human.

human micromorphology contains osteonal bone throughout the entirety of the bone section. (See Chapter 5 for a more detailed discussion on human bone micromorphology.) Other nonhuman mammals have varied micromorphology, including plexiform bone, which is easily distinguished from human remains (Figure 2.4). However, some mammals have full-thickness osteonal bone; therefore, it may not be possible to differentiate human from nonhuman based on micromorphology. In these cases, additional analysis may be required. (See Chapter 4, Box 4.3, for further details.)

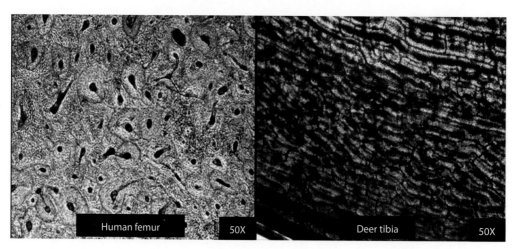

Figure 2.4 Differentiating human from nonhuman bone based on histology. Note that the human bone has round osteons, while the deer bone has brick-shaped plexiform bone.

Finally, it is possible to determine if a piece of bone is human or not through an analysis of DNA. Studies have achieved success at species identification through tests of the highly conserved regions of mitochondrial 12S and 16S ribosomal RNA (Yang et al. 2014).

How many individuals are present?

Once bone and teeth are determined to be human, the next step is to identify how many individuals are present. This assessment is referred to as the determination of the **minimum number of individuals** (MNI). A detailed discussion of MNI assessment is provided in Chapter 18 of this volume. Briefly, an MNI estimate is made by assessing present skeletal elements or fragments of skeletal elements to look for duplicates or duplicating segments of bone. If there are two complete *right femora*, then you may assume that the MNI would be two, as each human only has one right femur.

Elements can also be segregated by age to aid in the estimation of the MNI. For example, if a right femur with an *unfused* femoral head is present (representing a skeletally immature individual) and a left femur with a *fused* femoral head is also present (representing a skeletally mature individual), the assumption could be made that these femora represent two different people. Large differences in size also aid in determining the MNI (Figure 2.5). Small differences should be approached cautiously, as slight **bilateral asymmetry** is not uncommon in a single individual. Taphonomic differences may also provide clues as to number of individuals present. Elements that have been in dissimilar postdepositional environments may look very different from each other.

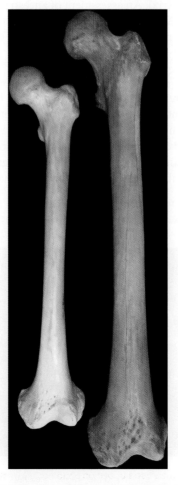

Figure 2.5 An example of human remains (left femora) that can be sorted into distinct individuals based on size.

IS THE MATERIAL MODERN?

Human remains are often recovered in remote locations, during construction activities, or even among personal belongings. When human remains are discovered, they are delivered to law enforcement or to the coroner. Depending on the state or the county, the coroner is usually not qualified to make a determination about forensic significance, so a nearby medical examiner or forensic anthropologist, or both, must evaluate the remains. This assessment involves looking at the following characteristics of the material: state of preservation, **antemortem** body modifications, associated personal belongings, and the conditions of **interment**.

Initially, a consideration of postmortem interval may be appropriate. If soft tissue is in a relatively fresh state and adheres to the remains, the remains are probably relatively recent. However, if remains are mummified and tissues are **desiccated**, then environmental conditions must be considered carefully, since, in the right conditions, soft tissue can preserve for an extended period of time. See Chapter 15 for a discussion on the postmortem interval.

If remains are completely skeletonized, it may be difficult to determine the modernity of the elements. Various clues indicate the age and potential origin of remains. For example, working in the United States may require determining if remains are of archaeological significance and/or prehistoric (i.e., Native American). Several skeletal and dental characteristics and other physical evidence may indicate the antiquity of the remains or if they are Native American. Such an analysis includes a consideration of the color of the remains, as archaeological skeletal material will often be of a brownish color, as bones take on coloration from the surrounding matrix in which they are buried. The condition of the teeth may also be an indicator of antiquity. Prehistoric skeletons typically have severely worn teeth (Figure 2.6). Cultural indicators, such as grave goods and artifacts associated with the remains, may also be indicators of antiquity. Moreover, if known, the circumstances of the interment can indicate the age (e.g., if the burial was in a flexed position or found in an archaeological matrix). An assessment of ancestry also assists in determining if remains are of archaeological significance and Native American. If the antiquity cannot be definitively ascertained through gross examination, radiocarbon dating can be employed to estimate the date of the skeletal material.

IS THE MATERIAL OF FORENSIC SIGNIFICANCE?

Once it is determined that the remains are human and relatively modern, it is necessary to determine if the remains are forensically important. The forensic anthropologist can determine whether the remains are of medico-legal interest, that is, relevant to a civil or criminal matter. There are several other reasons human remains may be encountered that are of no interest to the medico-legal community. These alternative scenarios are outlined below and should be eliminated as possible sources of the remains, before a full forensic investigation can begin. Nonetheless, the remains may still be of forensic significance, even if found to be prehistoric or faunal.

On occasion, historic skeletal remains are recovered that are not of forensic significance. Indicators of historic remains might include evidence of embalming, such as embalming artifacts and heightened levels of embalming fluid in the

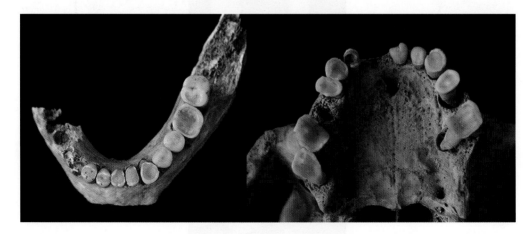

Figure 2.6 An example of archaeological teeth (Left: mandible, Right: maxilla). The teeth are severely worn, with little enamel remaining; teeth in this condition may indicate that the remains are not modern.

tissues (Berryman et al. 1991). Other evidence can be in the form of a coffin, such as nails or pieces of wood. In addition, remains can have coffin wear on the posterior surface, where they have been in long-term contact with the coffin in a supine position (Rogers 2005). The skeletal tissue will be worn flat in areas of protuberance, such as along the scapular spine and the posterior femoral condyles. This taphonomic signature can be related to slight movements and settling within the coffin over time, once the remains skeletonized.

It is also not uncommon to encounter anatomical specimens. These specimens can be identified based on the presence of drill holes or springs, cut cranial vaults with attached hinges (to permit viewing of internal anatomical structures), or bleaching preparation (Figure 2.7). These remains may show evidence of curation such as shelf wear (skeletal elements that are damaged or eroded from sitting on a shelf for an extended period of time; see Figure 2.8), dust, evidence of handling (polished areas), and the presence of glue or tape to hold elements together. The fact that the remains in question were anatomical specimens at one time may help direct the remainder of the investigation.

There may be other reasons that human skeletal remains come into an individual's personal possession or are found by law enforcement. Some remains may have been collected as trophy skulls from previous military involvement (Sledzik and Ousley 1991). Others may have been part of religious practices, most commonly the belief systems of Santeria and Palo Mayombe (Gill et al. 2009; Wetli and Martinez 1981). When remains that are associated with these practices are discovered, forensic anthropologists are usually consulted. These remains are typically faunal, anatomical, or historic in nature and can easily be identified by the forensic anthropologist.

Remains of fetuses

Two primary questions are asked when fetal remains are discovered: 1) Was the fetus **viable?** (i.e., was the fetus able to survive outside the womb?) 2) Is there evidence of any trauma that cannot be explained by childbirth?

It can be difficult to determine from the skeletal remains if an individual survived the birthing process, particularly if that individual died shortly thereafter. Forensic anthropologists will attempt to approximate viability by assessing the developmental age of the fetus based on the metric evaluation of the skeletal remains. Forensic anthropologists take measurements of the observable skeletal remains and compare these measurements to known dry bone size standards (Fazekas and Kosa 1978; Schaefer et al. 2009) to come up with a developmental age of the individual (Figure 2.9). This developmental age can assist the medical examiner in determining whether the fetus could have survived outside of the womb. In addition, it is possible to examine the dental tissue to estimate time since birth. The birthing process leaves a distinct mark in the histological structure of the tooth, called the *neonatal line*.

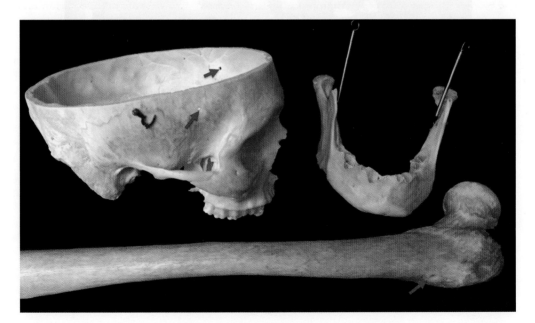

Figure 2.7 Examples of skeletal remains prepared as anatomical specimens. Note that the drill holes (arrows), indicative of a prepared specimen, would be evident in the remains, even if the metal hardware was missing.

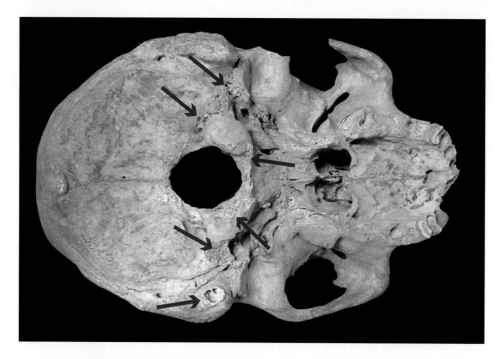

Figure 2.8 Skeletal remains exhibiting shelf wear (indicated with red arrows). Note the cortical erosion of the skeletal material on areas where the cranium was resting on a shelf for display.

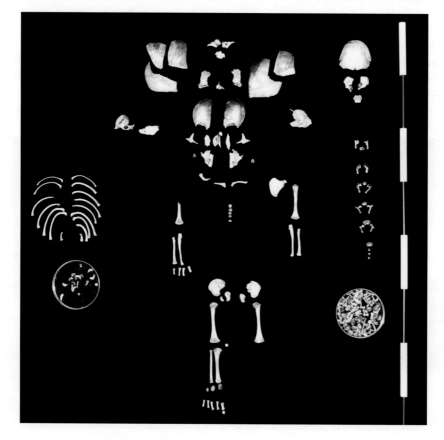

Figure 2.9 Example of the skeletal remains from a 40-week gestation fetus. Note the large number of osseous remains that can be used for metric assessment of gestational age.

The presence of such a line and associated dental development *after* the creation of this line may aid in determining the age at death, in addition to proving that an individual survived childbirth (Aka et al. 2015; Janardhanan et al. 2011).

The determination of viability can be important in the charging phase of the justice process. While 38 of the 50 states in the United States have fetal homicide laws, only 23 of these states have laws that apply to any age of gestation or development from conception onward (National Conference of State Legislatures website). In the remaining 15 states with fetal homicide laws that do not have this distinction, and in the 12 states without fetal homicide laws, determining the charges to bring against the accused may depend on the potential viability of the fetus.

In terms of fetal skeletal trauma analysis, forensic anthropologists use their understanding of normal bone growth and development, bone fracture mechanics, and the normal fracture-healing process to ascertain if the trauma could be related to the birth process or may have been inflicted intentionally. This information is relayed to the medical examiner, who makes the ultimate decision as to whether the trauma resulted in the demise of the fetus. For more information on fetal and child trauma assessment, especially in case of abuse, see Love et al. (2011).

CHAIN OF CUSTODY

It is imperative to maintain a chain of custody, regardless of the origin of the material (see Chapter 3). In many cases, a chain of custody will already be established before the forensic anthropologist's involvement, as these materials may have been **accessioned** into custody once the coroner or law enforcement became involved. However, if the material is sent to a forensic anthropology laboratory, a chain of custody should be started as soon as the materials are received and should be maintained until all materials are returned to the requesting agent, regardless of forensic significance.

CONCLUSIONS

Once skeletal and dental remains are determined to be human, modern, and/or of forensic importance, the medical examiner or law enforcement requests a biological profile and/or trauma assessment from the forensic anthropologist. The following chapters describe how forensic anthropologists use skeletal remains to estimate sex (Chapters 8 and 14), ancestry (Chapters 9 and 14), age (Chapter 10), and stature (Chapters 11 and 14) from skeletal remains, as well as assess antemortem skeletal changes (Chapter 12), perimortem trauma (Chapter 13), and postmortem damage (Chapters 4, 15, 17, and 18), and prepare a forensic case report (Chapter 7).

SUMMARY

This chapter discussed how forensic anthropologists become involved in skeletal analyses and the initial steps required to ascertain if the materials presented to them warrant a forensic investigation. If materials are osseous, human, modern, and of interest to the medico-legal community, then a forensic anthropologist generates a case report that contains information about the biological profile, skeletal trauma, pathology, taphonomy, and/or postmortem interval. A number of specialized tools may be required to ascertain the forensic significance of materials if gross morphological examination is not adequate, including radiographs, histological examination, energy dispersive spectroscopy (SEM-EDS), and X-ray fluoroscopy (XRF). The forensic anthropologist's expertise in ascertaining human versus nonhuman bone, gestational age of fetal remains, antiquity of remains, and unique taphonomic signatures caused by handling, coffins, or interment makes them invaluable resources in cases that may otherwise go unanswered.

Review questions

1. How do forensic anthropologists determine if a material is osseous?
2. How can the difference between human and nonhuman bone be determined?
3. How are remains determined to be modern, historic, or prehistoric?

4. How is the minimum number of individuals determined?
5. Under which circumstances would you use light microscopy to aid in the determination of human versus nonhuman remains? What information may be gleaned from this analytical technique with regard to determination of human versus nonhuman remains?
6. Identify some reasons that a set of remains might *not* be of forensic significance.
7. Describe the type of information that forensic anthropologists provide about fetal skeletal remains.
8. Give an example of when nonhuman or archaeological remains might still have forensic significance?

Glossary

Accessioned: to record the addition of; in forensic matters or cases, the addition of new evidentiary material.
Antemortem: occurring before death.
Articular: of or relating to joints.
Bilateral asymmetry: different or unequal size or shapes on the right versus left side of the body.
Biocultural: incorporation of a biology and culture to understand how the two influence each other and interpretations within forensic anthropology.
Desiccated: dried out, dehydrated, or mummified.
Faunal: pertaining to animals.
Interment: the method of burial of a corpse.
Medico-legal: something involving both medical and legal aspects.
Minimum number of individuals: a calculation of the minimum number of individuals represented in a skeletal assemblage.
Osseous: consisting of bone.
Viable: capable of working successfully, in this case, capable of sustaining life.

References

Aka PS, Yagan M, Canturk N, Dagalp R. 2015. *Primary Tooth Development in Infancy: A Text and Atlas*. Boca Raton, FL: CRC Press.

Berryman HE, Bass WM, Symes SA, Smith OC. 1991. Recognition of cemetery remains in the forensic setting. *J Forensic Sci* 36(1):230–237.

Christensen AM, Smith MA, Thomas RM. 2012. Validation of X-Ray fluorescence spectrometry for determining osseous or dental origin of unknown material. *J Forensic Sci* 57(1):47–51.

Fazekas G, Kosa K. 1978. *Forensic Fetal Osteology*. Budapest: AkademiaiKiado.

Fleckenstein P, Tranum-Jensen J. 2001. *Anatomy in Diagnostic Imaging*. Denmark: Balckwell Publishing.

Gill JR, Rainwater CW, Adams BJ. 2009. Santeria and Palo Mayombe: Skulls, mercury, and artifacts. *J Forensic Sci* 54(6):1458–1462.

Houck MM. 2015. *Forensic Chemistry*. San Diego, CA: Academic Press.

Janardhanan M, Umadethan B, Biniraj KR, Kumar RBV, Rakesh S. 2011. Neonatal line as a linear evidence of live birth: Estimation of postnatal survival of a new born from primary tooth germs. *J Forensic Dental Sci* 3(1):8–13.

Kuhn LT. 2007. Bone mineralization. In JW Martin (Ed.), *Concise Encyclopedia of the Structure of Materials*. Amsterdam, The Netherlands: Elsevier, pp. 47–54.

Love JC, Derrick SM, Wiersema JM. 2011. *Skeletal Atlas of Child Abuse*. New York: Springer Science and Business Media.

Megyesi MS, Jenny LL, Bird C, Michael A, Soler A, Wankmiller J. 2011. *Taphonomy of a mass grave in Mid-Michigan: The case of the missing cattle*. Chicago, IL: American Academy of Forensic Sciences, p. 328.

National Conference of State Legislatures Website. http://www.ncsl.org/research/health/fetal-homicide-state-laws.aspx.

Native American Graves Protection and Repatriation Act. 1990.25 U.S.C §§ 3001–3013.

Rogers TL. 2005. Recognition of cemetery remains in a forensic context. *J Forensic Sci* 50(1):5–11.

Schaefer M, Black SM, Scheuer L. 2009. *Juvenile Osteology: A Laboratory and Field Manual*. Amsterdam: Academic.

Sledzik PS, Ousley, S. 1991. Analysis of six Vietnamese trophy skulls. *J Forensic Sci* 36(2):520–530.

Ubelaker D. 1998. The evolving role of the microscope in forensic anthropology. In KJ Reichs (Ed.), *Forensic Osteology: Advances in the Identification of Human Remains*. 2nd edn. Springfield: Charles C. Thomas, pp. 514–532.

Ubelaker DH, Ward DC, Braz VS, Stewart J. 2002. The use of SEM/EDS analysis to distinguish dental and osseus tissue from other materials. *J Forensic Sci* 47(5):940–943.

Wetli CV, Martinez R. 1981. Forensic sciences aspects of Santeria, a religious cult of African origin. *J Forensic Sci* 26(3):506–514.

Yang L, Tan Z, Wang D, Xue L, GuanMx, Huang T, Li R. 2014. Species identification through mitochondrial rRNA genetic analysis. *Scientific Reports* 4:4089.

Chapter 3
Forensic Archaeology
Survey Methods, Scene Documentation, Excavation, and Recovery Methods

Denise To

CONTENTS

Introduction: General principles	36
In the past	37
Controlled deconstruction	38
Scene documentation: protecting the scene integrity	39
Processing the crime scene	41
Understanding site formation	43
Mapping	44
Excavation and recovery	45
Evidence management	50
Summary	52
Review questions	52
Glossary	53
References	56

LEARNING OBJECTIVES

1. Define "forensic archaeology," and explain how it differs from "archaeology" and how is used to process crime scenes.
2. Describe the difference between a systematic and an unsystematic approach to processing a crime scene.
3. Explain the concept of "controlled deconstruction," and provide examples as to how it applies to crime scene processing.
4. List the various aspects and methods of crime scene documentation, and discuss why documentation is imperative to scene processing and reconstruction.
5. Discuss the various methods of searching for human remains.
6. Explain why forensic anthropologists must understand site formation processes.
7. Describe the three stages of excavating remains.

INTRODUCTION: GENERAL PRINCIPLES

The majority of skeletal cases that pass through the laboratory for a forensic anthropologist to analyze are usually recovered from a field context. That is, the remains were discovered in a location outside the laboratory. This is in contrast to skeletal material which may occasionally be mailed to the medical examiner's office or even dropped off at the doorstep of the laboratory. In these situations, *where* the bones came from may never be known. However, in the majority of cases, skeletal analysis begins in the field. Sometimes, remains are discovered by citizens who inadvertently stumble upon them while hiking in mountains, hunting, doing construction, and so on. Sometimes, law enforcement personnel discover remains while investigating an area for other reasons. In addition, sometimes, investigators are led to a location by an informant, who specifically points out a location where a decedent is buried. The actual location, or *site*, from where skeletal remains are discovered may contain a wealth of information regarding how they originally got there. This contextual information is critical to ascertain, especially if a crime was committed. Therefore, the proper recovery of skeletal material in the field is of paramount importance to the forensic anthropologist.

While numerous law enforcement personnel and crime scene investigators are trained in processing crime scenes, a forensic anthropologist is needed most when skeletonized (or partially skeletonized) human remains are discovered—and they are especially useful when skeletal remains and evidence are *incorporated into the natural environment*. In these situations, the term **forensic archaeology** is oftentimes used. For many of these crime scenes, skeletal remains and associated material evidence get *buried* by the natural environment, such as under soil, shrubs, grasses, and leaves. In some cases, this occurs naturally (e.g., when a lone hiker dies in the woods and is not discovered until many years later). In other cases, the remains and evidence are covered up on purpose (e.g., when a perpetrator is attempting to conceal a crime). However, in all cases, the incorporation of skeletal remains and material evidence into the natural environment constitutes the **archaeological record**. The record of natural and cultural events, captured above and below ground in that single location, can be very dynamic, because, as the years pass, the record can be transformed by the addition and subtraction of information. Thus, the archaeological record is fragile, and when information needs to be extracted from the archaeological record, a forensic archaeologist is best suited for the task. There are fundamental differences between traditional archaeology and forensic archaeology. Trigger's (1989) goal of archaeology—the study of past human behavior and cultural change—refers more to broader and traditional aspects of the discipline, where the archaeological record of a site (such as an ancient village) is examined to answer research-driven questions. In a more specific sense, forensic archaeology is focused on collecting data from the archaeological record for medicolegal purposes. However, to collect such data correctly, the forensic anthropologist must employ some of the same methods and techniques used in traditional archaeology. Forensic archaeology simply refers to the application of traditional research-oriented methods, theories, and excavation techniques to modern forensic sites and crime scenes.

However, crime scenes come in all shapes and sizes. A crime scene can be in a homeowner's backyard, inside the trunk of a car, on the top of a snowy mountain, in the middle of a corn field, or even in a cemetery. A crime scene can be as small as a shoe box or bigger than a football field. Every crime scene is unique, and some sites where remains are discovered turn out not to be crime scenes at all. A crime scene is any location where a crime may have been committed. If a forensic anthropologist is called to a potential crime scene, it usually means that skeletal remains were discovered at that site. However, not all sites to which a forensic anthropologist is called turn out to be crime scenes, because it is possible that no *crime* was actually committed. For example, it is not uncommon for skeletons to be found in the woods. But, was a crime committed here? Is this a victim of a homicide? Or, was this an accident? Did a lone hiker get attacked by a wild animal? Or, this hiker might just have suffered a heart attack. Here is another example: Perhaps, a forensic anthropologist is called to a scene where a box of bones is found in a private garage. Again, was a crime committed here? Is this a homicide from long ago that someone has tried to cover up? Or, these might be legally acquired bones used for teaching anatomy classes at the local community college. In some places, it was commonplace for families to bury loved ones on their own property rather than in a local cemetery. And, for many locations throughout the world, ancient skeletons that have long slumbered within the ground are rediscovered by intrusion and development of ever-expanding urbanism. Forensic anthropologists are frequently called out to investigate these types of burials.

Modern sanctioned burials deemed by the forensic anthropologist as having little **forensic value** can be simply cataloged and photographed. Ancient archaeological sites should also be cataloged; but do not rush to photograph and dismiss.

As Brothwell points out, "the legislation relating to the excavation, curation, and reburial of human remains has become politically more significant…stimulated by tribal and religious groups who wish to have a say in the process" (Márquez-Grant and Fibiger 2011:xxxiii). Therefore, be sure to notify the local Historic Preservation Office of the discovery of the archaeological finds (and that you did not disturb the site). Sanctioned burials and archaeological sites aside, sites at which a forensic anthropologist is called to a scene by the police require processing as potential crime scenes. For potential crime scenes, it is not necessarily the job of the forensic anthropologist to determine whether or not a crime has actually taken place. That determination is made by the detectives only *after* the scene has been processed and the skeletal remains have been analyzed in the laboratory. Therefore, a good forensic anthropologist responds to all sites with the same amount of professionalism and scientific integrity. The best practice is to process a site as if a crime has been committed, until otherwise determined.

IN THE PAST

Finding human remains and processing a crime scene require an organized, systematic approach. However, crime scenes have not always been treated with the same meticulous attention that they are treated today. Imagine a standard Easter Egg Hunt at a park (a grassy field, some shrubs, and a few trees). A dozen children happily and hastily run off to look for colorful plastic eggs to claim the candy reward inside them. In what direction do the children run off? The answer is: they run in all directions. And, when one child finds an egg in the tall grass, the children who were searching the nearby shrubs abandon their search, run to the tall grass, and begin searching there, instead. In the end, some children find many eggs, while others are not so lucky. Some locations get searched over and over again by the children, hoping that more plastic eggs turn up, while other locations are never searched. The final result is that some eggs are never found. This method of searching describes an **unsystematic** approach. The children worked independently. Their goals were based on personal gain (trying to find and claim the candy reward for themselves before other children could do so). No coordinated search pattern was established, and no single person made sure that all locations were investigated. The search ended not when the area was thoroughly processed but probably when a parent decided it should.

While an Easter Egg Hunt may never reach the same level of significance as searching for human remains at a crime scene, the distinction in the method is paramount. If a plastic egg is left behind at an Easter Egg Hunt, the children's lives will hardly be affected, but if an investigator misses a crucial fragment of human remains or evidence at a crime scene, that error may have significant consequences. It may mean the difference between a mysterious victim being positively identified and him being forever left unknown. It may even mean the difference between solving a homicide and finding the victim's killer or allowing the perpetrator to escape justice.

Unfortunately, the Easter Egg Hunt method is not too dissimilar from how crime scenes used to be processed. In the past, untrained personnel and unsystematic methods were used. Crime scenes were crowded with media reporters, political officials, and curious on-lookers. Evidence was found by looking at random places, rather than by using a thorough and methodical search from one end of the crime scene to the other. Wearing gloves was oftentimes an ignored practice. Evidence was contaminated and sometimes even ignored. Photographs of the crime scene were scarce and/or poor. Documentation of methods used was rare. Occasionally, a victim's body was repositioned, and evidence was discarded (sometimes by law enforcement personnel themselves) for *the sake of decency,*—that is, to avoid potentially corrupting the victim's image and shocking survivors, family members, or the public. The results were poor scientific integrity and lost forensic evidence. While this may seem appalling by today's standards, these deficiencies were simply due to a lack of regulations or standardized practices regarding crime scene processing.

Today, more attention is paid to the science of crime scene investigation, especially in developed nations, where the criminal justice system relies heavily on the validity of strong, credible, and factual evidence. Moreover, with the growing trend of publicizing criminal trials, greater scrutiny is placed on all aspects of a criminal case, including crime scene methods. Scientific evidence is used increasingly in court trials to convict a criminal or to absolve an accused. Over the years, this has significantly improved crime scene processing.

So, the difference between the Easter Egg Hunt and modern crime scene methodology is to process a site in a **systematic** fashion that stands up to the highest level of scrutiny. The Defense Prisoner of War/Missing In Action Accounting Command–Laboratory of the U.S. Department of Defense, defines scientific integrity as a "strict adherence to the scientific method with transparency and accountability for all actions. Scientific integrity requires sound and accepted methods and high ethical standards regardless of the nature of the work" (DPAA Laboratory Manual 2015). In a simplified sense, it means that everyone works together with the common and unbiased goal of processing the crime scene according to set methods and techniques considered to be the best practice to ensure maximum recovery of human remains and forensic evidence and to minimize alteration of the context of the crime scene (see Box 3.1). If you deviate from using systematic methods, it means that you have no system, and if you have no system, you might as well be searching for Easter Eggs.

> **BOX 3.1   THE CSI EFFECT: SCIENCE IN ACTION (BACKWARD AND WRONG)**
>
> The recent popularity of laboratory crime dramas and "whodunits" that focus on the collection and application of evidence, such as *CSI*, *NCIS*, and *Bones*, has exposed viewers to the detailed-oriented world of forensic evidence collection and scientific analysis. An entire generation of viewers has become fascinated with and influenced by the dramatized careers of crime scene investigators, forensic anthropologists, and pathologists, who are fictionalized on these shows, sometimes, as gun-toting, slightly nerdy professionals with model-good looks, who interrogate suspects, serve arrest warrants, and carry out laboratory analysis—all after processing the entire crime scene in the first place. While most viewers ignore the inaccurate portrayal of crime scene investigators and forensic anthropologists in favor of good entertainment, some viewers believe the many Hollywood embellishments as fact. Small inaccuracies (such as the standard laboratory holographic skeleton imaging device) are comical when real forensic anthropologists are asked about them by the public. However, more serious fictionalizations, such as outrageously and inaccurately portrayed timelines to process and analyze evidence (e.g., DNA analysis) can have a detrimental effect on public citizens who may become victims of a real-life crime. Their expectations of the investigation are unrealistic and can be emotionally devastating. In addition, the opinions and expectations that many lay viewers develop about forensic science and how evidence is collected and processed can have detrimental effects on genuine cases, should they be called into court to serve on the jury.
>
> However, the biggest crime of which these television shows are guilty is the presentation of science conducted *backward*. In these shows, the investigators visit a crime scene and, after just moments of examining the scene, they immediately formulate and verbalize their opinions as to *what* crime was committed, *how* it was committed, and even *why* it was committed. A twist half way through the episode, when a laboratory technician discovers something shocking about a piece of evidence, causes a dramatic and sudden change of opinion—such drama! Good forensic science does not work this way, and depicting it as such in these television shows is a gross disservice for those working in these jobs.
>
> In reality, a good forensic scientist never develops a hypothesis about what happened at a crime scene while it is being processed. Instead, the hypothesis and interpretations occur only after *all* the data have been collected and after *all* the facts are presented on the table. Waiting for these facts to surface can take days, weeks, even months. A good forensic scientist begins to create a theory only after all the evidence and facts are gathered. Many television shows, therefore, depict the scientific process backward. If you develop a theory before all the facts are in, you may bias yourself and the inevitable will occur—you will twist the facts to suit your theory—and that makes for poor science.

CONTROLLED DECONSTRUCTION

One of the most fundamental concepts in forensic archaeology is the understanding that to process a crime scene is to tamper with it—even possibly destroy it. Examining the archaeological record through excavation cannot be accomplished without *excavating* the site. The systematic excavation of a site is one of the most intrusive and destructive methods used to examine the archaeological record. In the most basic sense, determining what happened at the crime scene *requires* its alteration. For example, because investigators must physically walk through the crime scene, their footsteps may alter the floor. Because an investigator needs to move a body to the morgue, the position of the victim's arms and legs will change. Because an investigator needs to pick up a weapon and transport it to the laboratory, the weapon will have to be moved from its original position. Even small actions, such as turning on a light and opening a window, can alter the crime scene.

Because of this, it is crucial for an investigator to understand that to process a crime scene, you *have* to alter it—you *have* to walk through it with your footsteps, you *have* to move the body eventually, and you *have* to put the weapon in a bag and take it to the laboratory. The key is to alter the crime scene *correctly*. This is known as controlled destruction or, rather, **controlled deconstruction**, because you are not so much destroying it—you are *taking it apart*. Crime scene methodology is based on a series of steps used to *deconstruct* the crime scene in order to maximize evidence collection, while at the same time minimizing contamination and alteration of the crime scene and preserving the context of the evidence. It is possible that the investigation team that processes the crime scene may be the only people to ever see it. Therefore, the team has a moral obligation to process the site to a high level of scientific integrity, or else, justice may not be served. A forensic anthropologist gets only one chance to process a crime scene or excavate a site. Consequently, a skeleton must be excavated correctly because, if it is not, it cannot be put back in the ground for a second try.

SCENE DOCUMENTATION: PROTECTING THE SCENE INTEGRITY

Because forensic archaeology is a deconstructive science—and because it is possible that the investigation team that processes the crime scene may be the only people to ever see it—it is crucial for the team to document everything that is seen and done there. **Documentation** of a crime scene investigation is the act of using written notes, photography, and digital data collection to preserve the methods used and the contextual data that are recovered and collected at a crime scene. The goal of crime scene documentation is to create a permanent record of what was discovered, how it was discovered, in what condition it was discovered, and the steps taken by the investigators to process the site. This permanent record can be used later by other people, such as law enforcement personnel, laboratory analysts, attorneys, and other forensic anthropologists, to understand the context of everything at the crime scene. Good documentation is crucial, because these other investigators must be able to theoretically (even physically) recreate the crime scene as accurately as possible, weeks, months, or even years after it was initially processed.

Good documentation protects the integrity of not only the investigation but also of the investigator. For cases that result in courtroom testimony, an investigator may be questioned by an attorney years after the initial incident. The time differential can make it difficult for an investigator to remember the details of the case, including questions such as, "Who was with you at the crime scene? What did you do with the evidence once you found it?" Even if an investigator was able to remember these details, he or she may not be able to provide evidence to support the answers, if required in court. Relying on memory for details is hardly an efficient technique. Instead, proper documentation of the crime scene allows an investigator to refer back to the moment of scene processing to answer questions with confidence.

As a baseline, scene documentation begins with the basics of "who-what-where-when-and-how," but ultimately, the documentation should be much more in depth. For example:

- Who called you for help in the first place? (A detective?)
- At what time of day did you arrive at the location?
- Where exactly is the crime scene? (What is the address?)
- If it is an outdoor crime scene, what was the weather like when you arrived? (Was it raining?)
- Who was there when you showed up? (Who might potentially contaminate the scene?)
- Who was part of your investigation team?
- What does the crime scene look like? (Describe this in as much detail as possible.)
- Where were the remains found? (Under a bed? In a tree?)
- What methods did you use to search for more evidence?
- What kind of evidence did you find? (Human remains? Weapons? Clothing?)
- Where did you find the items? (Were evidentiary items touching each other? Were the items far apart?)
- What did you do with the evidence? (How did you contain the items, and where and when were they taken?)
- If you had to excavate, how did you go about doing this?
- How much did you excavate?
- Did you draw a sketch map of the excavated area?

These are only a handful of questions that a good investigator must be sure to answer when documenting a crime scene. It is important to remember that the documentation must be good enough for another person to recreate the crime scene in the future (see Box 3.2). Poor documentation can result in the loss of information and even the loss of crucial evidence.

> **BOX 3.2 DOCUMENTATION EXERCISE: TEST YOUR ABILITY TO DOCUMENT A CRIME SCENE WITH THIS SIMPLE EXERCISE**
>
> You will need two additional helpers and numerous building blocks of different colors and shapes (Figure 3.1), available at your local toy store or craft store. Have one helper build up a tower with the blocks in a random pattern. Once complete, your job is to *deconstruct* the tower, but you must document the process. To test your documentation ability, have the second helper reconstruct the tower *based entirely on your notes*. If your documentation is adequate, the reconstructed tower should be identical to the first.
>
>
>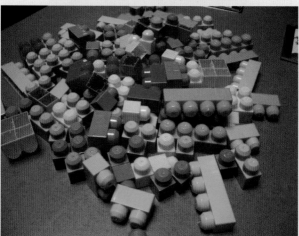
>
> **Figure 3.1** Building blocks of different colors and shapes. (Courtesy of Denise To.)

Photography is paramount in good documentation, because it provides a visual accompaniment for all written descriptions and actions. Context can be much better understood with photographs, and no amount of written description can substitute photographs. Photographs of the activities conducted by the investigation team are important to document the methods used by the team. Activities such as walking a pedestrian survey, clearing vegetation, excavating a burial, screening sediments, and removing and bagging evidence are some of the many actions performed by the team that should be documented photographically. Photographs of evidence and human remains discovered at a scene are vital and should always start with wide shots, in order to capture an item in relation to its surroundings. Progressively closer photographs of the evidence help capture the complexity of a busy site. Finally, detailed, close-up photographs of the evidence while still *in situ* are a must. Good photographs always include a scale (and direction of view for the wider shots—cardinal directions work best).

- **IN-TEXT QUIZ:** You process a crime scene in a wood that has no street address. How would you document that location, so that another investigator can find that exact spot in the future?
- **ANSWER:** For most crime scenes that are *off the beaten path*, you can document the location with a Global Positioning System (GPS) Grid Coordinate. This system uses multiple orbital and/or suborbital satellites to triangulate the location. The coordinates of that location can be plotted with a GPS-receiving device, using one of many different coordinate formats (including Latitude-Longitude, Universal Transverse Mercator, and Military Grid Reference System). Future investigators can input the grid coordinates into another GPS device to lead them to it. Document the distance and bearing of the closest road and town and be sure to document all landmarks in the area in as much detail as possible, so that the crime scene can be relocated. Detailed topographic maps are available for most locations in the United States. In addition, commercial services available online, such as Google Maps®, can provide relatively accurate aerial photography of an area. Be sure to mark the crime scene location on one of these types of maps and include it in your documentation.

PROCESSING THE CRIME SCENE

Some sites are easy; they are small, with simple topography and little vegetation, have very few items to document, and are thus relatively straight forward. Small sites can be quickly processed, with minimal resources and a small number of personnel (sometimes, the team consists of just one person—the forensic anthropologist—and that person does all the work). Other sites are not simple; they are large, complicated, and filled with many items to map and photograph. The site may also be in a heavily wooded area that requires a long, arduous hike. These sites can take many hours, even days, to process and may require the help of other people. The kind of help that other people can provide depends on who are available and what experience they have. However, one must be conservative when choosing help. Community residents, for example, may have good intentions, but their lack of training may lead to accidentally contaminating and disturbing the crime scene. Worse yet, they may unknowingly damage and destroy a crucial piece of evidence. In addition, be aware that the more random volunteers you employ, the greater the risk you take that photographs of the crime scene might end up on social media. Everyone carries personal mobile phones these days—some with high-resolution cameras. Because of this, stick to the professionals, when possible. Available law enforcement personnel can be useful when large searches need to be conducted, but for actual handling and processing of data and evidence, stick to professional crime scene investigators only. However, regardless of how simple or complex the site is, an investigator must follow the same set of principles and methods at all sites.

All investigations begin with a **preliminary assessment**—an initial step where the forensic anthropologist determines what needs to happen at the site. During this initial step, the anthropologist determines if the skeletal remains are human. (To the untrained eye, animal remains are often mistaken for human remains.) If the remains are indeed human, the anthropologist determines if they have forensic value. If so, then the anthropologist makes a logistical assessment of what needs to be done to process the site. Do additional personnel and resources need to be mobilized to assist? Has the medical examiner's office been notified of the discovery of human remains? Does special equipment need to be brought in? How much time will the investigation take? Is there enough daylight to complete the task, or will lights be needed? Will excavation be needed? At this point, processing of the crime scene has begun, and the anthropologist must begin documentation and photography.

The first major step in processing a crime scene is to determine its boundaries. Various survey methods to inspect the site are available and range from simple and inexpensive (e.g., visual pedestrian survey) to more complex and time-consuming (e.g., remote-sensing survey). Some **aerial surveys** of the area might already be available. At the crime scene, a **visual pedestrian survey** is very effective, and its simplicity makes it a method that should always

Figure 3.2 Line of investigators conducting a pedestrian survey. (Courtesy of M. Morrow.)

be utilized first, when possible. This is a systematic and controlled search of the area, wherein a team of people walks across the crime scene, lined up in a row side by side, approximately one arm's length apart from each other, and visually search as much of the crime scene as possible (Figure 3.2). Investigators walk a transect (a horizontal straight line across a parcel of land) across the crime scene, wherein each person visually inspects the area within arm's reach, until the team collectively takes the next step. As possible evidence is discovered, it is marked with a pin flag, numbered, photographed, and left in place, until it can be mapped and documented. The goal of the pedestrian survey is to cover as much of the crime scene as possible, so that no evidence is missed. During the pedestrian survey, the team looks for human remains and items (or even a **cache** of items) that may be related to the case (weapons, clothing, wallets, and so on). They also look for **features** and other clues that may not be so obvious, such as changes in the soil and vegetation (depressions, mounds, odors, stains, and so on), as well as areas with differential insect or animal activity that may indicate decaying organic matter. The number of people needed for a pedestrian survey depends on the size of the area. Figures 3.3 and 3.4 illustrate a typical walking path followed during a pedestrian survey. The survey is conducted first in one direction, followed by another pedestrian survey conducted in the perpendicular direction. This method doubles the area covered and decreases the likelihood that possible evidence is overlooked.

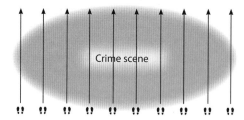

Figure 3.3 Diagram of the first phase of a typical walking path used in a pedestrian survey.

Figure 3.4 Diagram of the second phase of a typical walking path used in a pedestrian survey. The same area is resurveyed from the perpendicular direction to decrease the likelihood of overlooking possible evidence.

UNDERSTANDING SITE FORMATION

As mentioned above, every site is unique, and the distinctive nature of each site dictates the techniques used by the investigator. Some sites may require a mechanical excavator to remove a cement slab overburden, while others may require rappelling equipment to gain access to a remote location. A water hose may be needed at one site to screen muddy sediments, while another site may require machetes to clear tall shrubbery. However, in general, the forensic anthropologist will face two major types of sites: **surface sites**, where the majority of human remains are on top of the ground surface (or relatively near it), and **subsurface sites**, where remains are buried below the ground surface. At surface sites, the human remains are easier to find, but they are often disturbed from the original context because they are more easily moved and affected by other processes. Worse yet, they can disappear from the site, taken away either by animals or by natural processes such as erosion and flooding. As a result, preservation of the remains and the integrity of the site get worse the longer they stay on the surface.

Subsurface sites are those where the majority of human remains are found below the ground surface, such as those buried in a grave or covered by a landslide. These remains can be either shallow or deep and may be very hard to find. These sites can be more time-consuming, as they require more work to recover, since excavation may be required. However, the remains are usually in a better state of preservation, as is the context and integrity of the archaeological record at the site. The general principles of crime scene processing (in terms of documentation, evidence collection, and scientific integrity) are the same, regardless of whether remains are on the surface or below it.

However, subsurface sites pose a new set of challenges for the forensic anthropologist, because the presence of skeletal remains below ground can be confirmed only when remains are actually discovered (and that almost always requires excavation). In other words, the presence of subsurface remains is only hearsay (e.g., when an informant thinks that a body was buried in a specific area). Just because someone *thinks* that there are remains there does not mean that there are *actually* remains there. It is up to the forensic anthropologist to confirm or refute the possibility. The techniques used to investigate the subsurface environment range from minimally invasive and variably effective (such as remote-sensing devices) to highly invasive and very effective (excavation). The unique nature of each site will dictate which technique is most appropriate. For example, the use of a cadaver dog specially trained to identify the scent of decaying human tissue can be very useful, especially when remains are believed to be shallow and relatively fresh (i.e., recently deceased). However, the efficacy of cadaver dogs can decline when the remains have been buried for a significant period of time.

Certain **remote sensing** and geophysical devices can be useful to anthropologists, but these methods do not actually detect skeletal material; they only provide a snapshot of information below the ground and it may or may not be related to human remains. For example, metal detectors and **magnetometers** use a magnetic field that penetrates into the ground surface to detect the presence of anomalies caused by metal items such as firearms, ammunitions, and nails. **Ground-penetrating radar**, a method by which radar pulses are used to image the subsurface, can be useful in detecting anomalies and voids in the soil, such as those similar to ones produced by graves. **Resistivity and conductivity** devices use electricity to view the subsurface, the former by measuring the soil's ability to resist an electrical current and the latter by measuring the soil's ability to conduct an electrical current. Because remote sensing devices do not detect the actual presence of skeletal material, these techniques can be used only as a proxy to determine if human remains may be present. The *images* of the subsurface produced by these machines may not have any relation to human remains. A skeleton with no associated metal items will have little effect on a metal detector. A decaying tree stump detected with ground-penetrating radar can produce an anomaly similar to that of a grave. Moreover, electrical geophysical devices can fluctuate due to soil changes in water and temperature.

Therefore, remote sensing surveys can be useful in determining where human remains *might* be present, rather than actually detecting human remains. In other words, once a specific location is determined to potentially contain human remains, the only way to confirm that possibility is with more intrusive techniques. Simple, small-scale tools, such as soil probes and augers, may provide preliminary samples of what is below ground. In the end, large-scale techniques, such as excavation of test pits and trenches and block excavation, can be the only way to confirm or refute the presence of a burial. Sometimes, *looking* for human remains can take a lot longer than excavating them.

MAPPING

One of the most effective ways of documenting the crime scene is to take spatial control of all items in the area, not just evidence, such as weapons and human remains, but also mundane items such as a desk and chair or a tree and shrub. To maintain spatial control means to understand and document the three-dimensional location of all the items at a crime scene relative to other items. Determining an item's location can be accomplished by superimposing, both theoretically and physically, a grid system onto the crime scene. Sometimes, this is done in real life by using tape measures, string, wooden stakes, and so on. Traditionally, graph paper is used to recreate and document the grid system at the crime scene, and all items are drawn on the graph paper. This constitutes the heart of mapping a crime scene. This type of mapping uses the **Cartesian grid system** (with North and East baselines), but other types of mapping that utilize **triangulation** or **azimuths** are also not uncommon. The tools used to create a sketch map can be simple (graph paper, pencil, tape measure, and compass) or complex (laser-assisted computer modeling **theodolite** systems). The end product should be the same: a highly detailed map of the crime scene illustrating the locations of all the items within the boundaries of the scene (Figure 3.5). If the topography (height or elevation) of important items is deemed a significant influence on the site, more complex maps should be generated (Figure 3.6).

Large-scale maps (such as a state map) and medium-scale maps (such as a city map) already exist and simply require an "X" placed on the map to represent the crime scene. The sketch map drawn by the investigator encompasses the entire crime scene. Additional detailed sketch maps can be used to zoom in on a particular feature. The sketch map also helps guide recovery strategies, as it is drawn concurrently as the crime scene is processed. As an area is searched, it can be documented on the map; as an area is excavated, it can be delineated as such. Smaller areas, such as burials, may require a more detailed map (Figure 3.7).

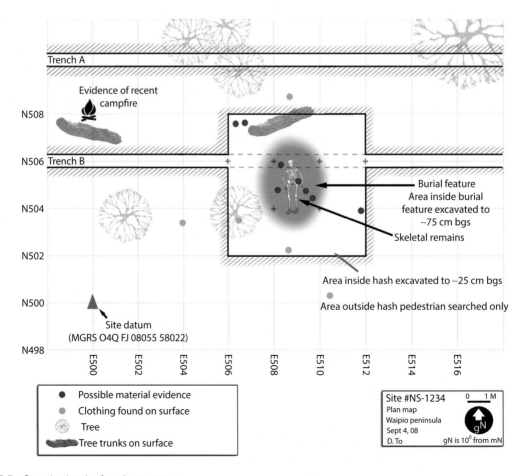

Figure 3.5 Sample sketch of a crime scene.

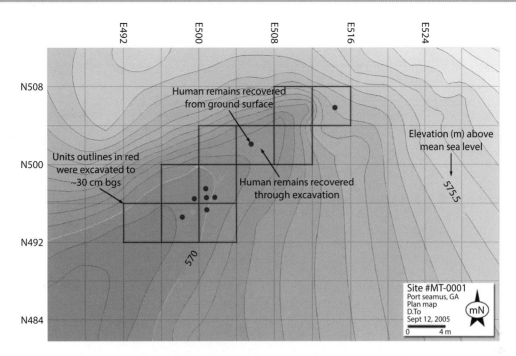

Figure 3.6 Sample sketch of a crime scene illustrating topographic variation.

A sketch map is usually to scale (the horizontal and vertical distances on the graph paper are equal) and includes written information such as the details of the site designations, an indicator pointing North, and descriptions of items. Were the remains found in a box under a bed or buried several feet underground? Were they found at the back of a cave inside a burlap sack or on the ground surface within the grass? This type of information is drawn on the sketch map and is written in the field notes. Sketch maps should always include a **reference datum**—a single point whereby all other items mapped are based on that location. Data taken from the reference datum can include distance measurements, azimuths, and depth.

Between the sketch map, field notes, and photography, a good investigator can obtain two necessary pieces of information to fully understand the crime scene: **provenience** and **context**. *Provenience* refers to the exact location of an item (or a piece of data) in space; it is that item's three-dimensional location at a recovery scene. *Context* refers to the abstract relationship that an item has with other items at the crime scene, in reference to its provenience. While context is the most important information that can be derived from a recovery scene, it is also the most *easily lost* kind of information, especially when best practices are not adhered to. By weaving together contextual information at the crime scene, a forensic archaeologist is able to piece together the archaeological record and better understand the event that occurred in the past and all the subsequent processes that have affected the crime scene since then.

EXCAVATION AND RECOVERY

As the subject matter expert, the forensic anthropologist's main responsibility is the proper recovery of human skeletal remains. Regardless of whether they are on the ground surface or below it, accuracy and care in the steps taken to recover human remains at a crime scene are crucial, in order to maximize data collection and minimize potential damage to the remains. When human remains are discovered at the crime scene, they are often referred to as a **burial**. This term specifically refers to a human body deliberately placed under the ground surface, but sometimes, the term *burial* is used in a general sense to describe all human remains found in the field (even if not entirely accurate). For example, skeletal material found scattered on the surface has been referred to as a *burial* by law enforcement personnel, despite the fact that the remains were not buried. If a single bone is found underground, does a single bone constitute a *burial*? In a generic sense, yes.

When human remains are discovered at a crime scene, it is important to determine the type of skeletal deposition. This is overall description of the remains, as well as a general contextual interpretation. Does this appear to be a grave

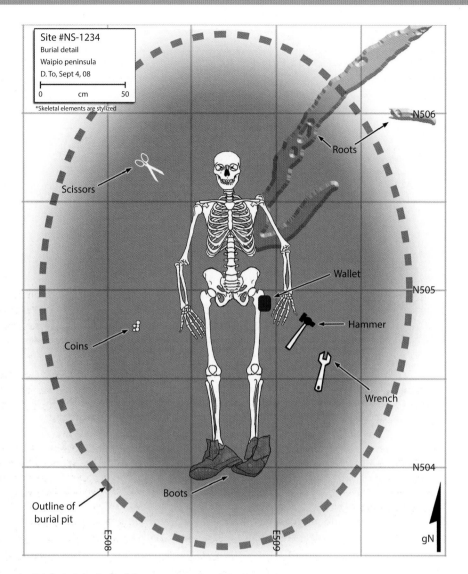

Figure 3.7 Sample detail sketch of a burial.

or some scattered assemblage of bones in a field? Is this a single-person or a multiple-individual burial, or, possibly, even a mass burial? Detailed data collection must be done concomitantly with the recovery of skeletal remains to answer these questions. As much information as possible should be documented before anyone touches the remains, and then, documentation of data should continue regularly during excavation. Where are the remains located (in the context of the crime scene)? Who discovered them? Have they been altered since discovery? In what condition they are? Extensive photographic documentation is crucial at this step.

A detailed description of the remains is vital, since much contextual information can be contained in a description (see Box 3.3 and Figure 3.8). Do the remains appear to be of one individual? Are they articulated? Can you determine the position of the skeleton? Think of a standard American cemetery. Individuals in our society are typically buried **extended** and **supine**. If a skeleton is found in this position in a residential backyard, one might believe the homeowners if they claimed that it was an ancestor who died of natural causes, especially if there was corroborating evidence near it, such as a grave marker. While a residential backyard may not be the standard way to bury the dead in today's American society, the context here might suggest that no major crime was committed. However, take, for example, a similar situation where a skeleton is found in someone's backyard and the remains are **prone** in a **semi-flexed** position. Does this look like a considerate burial of an ancestor, or was foul play involved? And, what if this skeleton was found in the backyard under a concrete slab? A description of skeletal position, therefore, is vital to understanding the context at a crime scene.

BOX 3.3 SKELETAL POSITIONING

One type of data that can be very useful to forensic anthropologists is the description of the skeletal positioning, that is, the position of the skeletal elements when discovered at the crime scene. A skeleton that is *articulated* means that all the elements are still in correct anatomical position as they would normally exist in a living human. This means that the body was whole when it was buried. An articulated skeleton that is *extended* is stretched out to its full height. One that is *semi-flexed* is a skeleton whose limbs are gently bent (see Figure 3.8). One that is *flexed* is a skeleton whose limbs are tightly bent at the joints. When it is *disarticulated*, it means that the elements are positioned in a way that they could not possibly exist in a living human and the body had likely become skeletonized while it was still above ground.

Knowing the position of a skeleton can help determine the context of the burial. Is the positioning of the skeleton, as found at the crime scene, the original position of the body shortly after death? If so, this is called a *primary deposit*. In other words, it has been relatively untouched and undisturbed since the time the body was placed (regardless of whether it was deliberate or accidental). A *secondary deposit* refers to an assemblage of human remains that has been disturbed (either deliberately or naturally). The location and position of the remains at the crime scene, therefore, represent a secondary location after the original primary context.

- **IN-TEXT QUIZ:** What might a scattered, partially articulated skeleton represent?
- **ANSWER:** Partially articulated skeletal remains may indicate that there was soft tissue on the body when the remains ended up there. However, the fact that they are scattered may indicate that the skeleton was disturbed before complete skeletonization. Check for tooth gnawing, which may suggest that an animal has moved the remains, or look for rain erosion channels on the ground, which may indicate that the body was washed down from some other place.

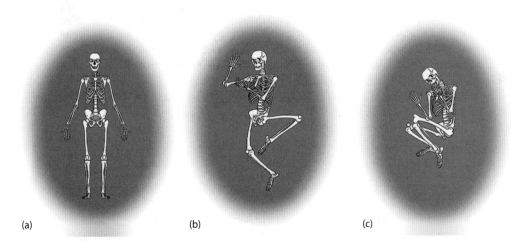

Figure 3.8 Documentation of the position of skeletal remains can assist in understanding the context with which they were buried. The standard positions are: (a) Extended, (b) semi-flexed, and (c) flexed.

The documentation should also include a detailed description of the condition of the remains (skeletal condition), specifically how well they are preserved, color, odor, and degree of fragmentation. Documenting these details at the moment of recovery is essential, because these variables may change once the remains are removed from the archaeological record and transported to the laboratory. The physical movement of remains can greatly affect the condition of the remains, as can the environmental conditions. Remains that are wet can change color as quickly as the car ride to the laboratory.

In addition to a description of the remains, good documentation includes a description of the immediate area surrounding the remains. If they are buried, this documentation should include details on the kinds of soil or sediment that surrounds the remains (whether it is sand, silt, clay, or a combination). Is it muddy? What color is the soil, and what sort of inclusions are in the layer? Each layer of material in the ground is called a **stratum**, and burials can be found between and below many different strata, creating a stratification of the sediments. Interpreting the strata is called **stratigraphy**. Understanding some basic concepts of geology can be crucial in interpreting the stratigraphy of a crime scene. Many of the basic concepts of geology are born out of the idea of **uniformitarianism**. This concept is attributed to the nineteenth century geologist Charles Lyell, who wrote the influential three-volume text *Principles of Geology* between 1830 and 1833. Adams and Valdez Jr. (1997) describe the concept of uniformitarianism as the assumption that there is continuity in the physical processes that form strata on the earth. In other words, processes that act on the earth (such as gravity and water erosion) work in the same way today as they did in the past. This concept may seem simple and obvious, but it functions as the basis to understanding stratigraphy at crime scenes. For example, skeletal remains found within a streambed are likely washed *downstream* rather than *upstream*.

Another important principle in understanding stratification is the **law of superposition**. Adams and Valdez Jr. (1997) describe superposition by explaining that "the oldest layers are on the bottom and the youngest layers are on the top" (236). The stratification of the layers is thus a snapshot of many events (both physical and cultural), with the most recent layer or event represented as the material found on the ground surface. This concept is important for forensic archaeologists, because excavating the sediments in a burial is done from the top-down; therefore, the youngest stratum is excavated first. This law functions in most cases, but specific environments can disrupt it, such as stratigraphy in caves. Sites that have significant animal disturbance (e.g., rodent burrows) may experience mixture of sediments. Likewise, burials result in sediment mixture and disturbed layers, because the burial fill is put back into the grave in a random manner.

Another simple, yet significant concept is the **law of association**. This concept states that the distance of items within a deposit *and* the integrity of the stratum determine the reliability of any observed spatial and chronological relationship between the two objects. For example, if a wallet and body are found buried underground and wrapped in a relatively intact plastic tarp, those two items can be associated with each other by 1) context (i.e., they were found within the same deposit—the tarp) and 2) integrity (i.e., the tarp is relatively intact; therefore, one can assume little disturbance after it was placed in the ground). In fact, if the wallet was found in the pocket of clothing that the individual appears to have been wearing, then the spatial proximity of the wallet strengthens the association. However, it is important to remember that *relatively intact plastic tarps* are hardly the norm in forensic archaeology. Evidence is not normally contained neatly within a plastic bag. Instead, it is usually mixed and disturbed within different deposits of sediment, and items that are spatially associated in the same deposit may, in fact, be chronologically separated by a great deal of time. A vertical sketch of the burial stratigraphy greatly aids the understanding and documentation of the crime scene. This sketch, called a **section drawing**, is drawn much in the same way as a two-dimensional plan map of the site, except that the two dimensions in a section drawing represent depth of the deposits and horizontal distance spanning one specific cross-section (Figure 3.9).

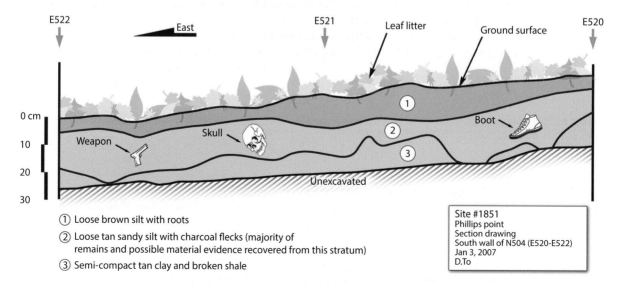

Figure 3.9 Sample section drawing of the sediment stratigraphy at a crime scene.

Finally, any details regarding how the remains got there and any taphonomic **disturbance processes** should be noted. This includes cultural (human-caused) disturbances, such as rubbish and construction material found in association with the remains; natural disturbances, such as a brush fire and rock fall; mechanical disturbances (when remains appear to have been moved by processes such as gravity and water); biological activities from animals, tree roots, or anything organic; and, finally, chemical disturbances, such as adhering corrosion products and oil stains. All these details are fundamental to understanding the context of skeletal remains.

For skeletal materials that are underground, traditional archaeological excavation techniques are necessary to recover them. This process can be very laborious and time-consuming, depending on the depth of the burial, its complexity and preservation, and the density of the sediment around it. However, the labor and time-intensive techniques are necessary to minimize damage to the skeletal remains as well as to maximize preservation of any evidence and clues that the remains might yield in solving the crime and identifying the individual. Excavation of a burial is usually accomplished with trowels, brushes, and other small hand tools. Occasionally, larger tools, such as a shovel, can be used to remove limited overburden. Heavy machinery should be used only to remove heavy or large amounts of overburden and should be employed only by an extremely experienced forensic archaeologist. The potential damage to skeletal remains by heavy machinery such as backhoes is usually not worth the risk, unless it is used to remove a concrete slab or large amounts of sediment to a known burial level, such as the sediment above a coffin during an exhumation. Burial excavation should be conducted only by experienced personnel and is accomplished in three stages: *defining the burial*; *exposing the remains*; and *lifting the remains*.

During the first stage, *defining the burial*, an anthropologist determines how much of the skeleton is present and in what position it is lying. Gentle removal of small amounts of sediment with a trowel and brush exposes small amounts of skeletal material for the anthropologist to identify. Eventually, and without exposing too much bone, the anthropologist can determine where the skeletal elements are lying and in what direction. Removing too much sediment from each element too soon will expose the surface of the bone prematurely. If the remains are exposed to the air and sun longer than they need to be, they stand the risk of being heavily damaged. Ironically, remains are commonly better protected from harsh elements when they are underground and covered with sediment. However, some exposure is necessary in order to define the burial, so it is important to minimize the exposure of the boney surfaces until it is time to move onto the second stage.

The second stage, *exposing the remains*, can take the longest, as this is the stage in which the majority of the sediment is removed from the burial and the skeletal elements are exposed. An experienced anthropologist should direct and closely monitor all work during every stage, but the second stage should be conducted only by the most experienced personnel with tools such as trowels and brushes. Take care not to rush this stage, as written documentation and photography are crucial in capturing the process, and sloppy and fast excavation can scrape and damage bone surfaces. Be mindful of a common mistake called *bone fever*. This happens when skeletal remains are discovered on site (usually found underground) and a person at the crime scene gets excited and rushed with a false sense of urgency and the idea that the remains need to be removed from the ground as fast as possible. When bone fever strikes, a person can get tunnel vision and can get so obsessed with cleaning off and digging out the remains that he or she forgets all the other important steps in processing the crime scene, such as photography, documentation, and observation of other important clues. Bone fever most often affects the inexperienced or the nonanthropologists, but anyone can get it. It can be avoided by taking the necessary steps to process the site in a careful manner. A good forensic anthropologist plans several steps ahead, but never proceeds to the next step until it is time.

The third stage, *lifting the remains*, comes only after the remains have been **pedestaled** and exposed, all data have been collected, sketch maps have been drawn, and photographs have been taken. The remains can then be lifted by hand, one by one. Elements (such as the hand phalanges and ribs) that can be difficult to side in the laboratory and/or elements that are heavily fragmented should be isolated by side or seriation, when possible. The remains should be placed in bags and a padded container, secured, and prepared for immediate transportation to the laboratory.

A final step in excavation is to screen the sediment in the burial in order to maximize the likelihood of recovering small bones, loose teeth, and associated evidence. **Screening** is a process wherein the sediment within a burial fill is placed through a metal mesh to separate larger particles from the sediment matrix. The size of the mesh varies, but a good standard for burial fill is a mesh of at least ¼ inches (but smaller, if possible). The sediment will pass through the mesh, but items larger than ¼ inches will stay in the screen and can be collected. Screening is an important step to catch many of the smaller bones in the body that may not be seen during excavation, such as hand and foot bones. In addition, screening is helpful to find small pieces of evidence that may have been missed, such as buttons, jewelry, and even bullets. It is important not to confuse screening with *sifting,* which is a process wherein the sediment that falls through the mesh is the material that is kept, like when bread baker sifts flour. In archaeology, screening can oftentimes reveal some of the most significant items. All burial fill should be screened, including any disturbed sediment located under the remains. In the end, after the remains and evidence have been removed, the floor of the excavated burial pit should be an undisturbed layer.

EVIDENCE MANAGEMENT

Everything found at a crime scene can be potentially related to the incident. But, should all items be considered evidence? Is everything photographed, placed into a container, and taken to the laboratory? Usually not. Only the items deemed probative are treated as potential evidence; this includes items that are believed to be related to the incident, items that may have come in direct or indirect contact with perpetrators and victims, and items that may contain clues to help solve the riddle of what happened at the crime scene. But, who decides what is evidentiary and what is not? In most situations, the forensic anthropologist works together with crime scene investigators, death investigators, and local detectives to determine that. In many situations, the forensic anthropologist is responsible for the skeletal remains, while fingerprints, weapons, clothing, personal effects, and other types of material items that are considered evidence are handled by the crime scene personnel. However, in reality, the anthropologist who processes the crime scene shares in this responsibility. Therefore, it is crucial to handle all evidence properly. The consequences of mishandling evidence can be significant, including a prosecution that is dismissed in court. It can even lead to an innocent person being wrongly convicted of a crime. Therefore, a forensic anthropologist must be vigilant about handling evidence properly.

One of the biggest mistakes in evidence handling is the **contamination** of evidence or compromising its original integrity from the site. Contamination of remains and evidence is a serious problem, and a good forensic anthropologist takes all the necessary steps to prevent it. The first step to prevent contamination is to minimize the number of people that come in contact with the evidence (see Box 3.4 and Figure 3.10). Only those people who absolutely need to touch something at the crime scene should be allowed to touch it. The second step is for investigators to minimize personal contact with the evidence by wearing gloves, shoe covers, and personal protective equipment, if necessary. The third step is to minimize the amount of handling of an item. Ideally, an item should be picked up only once and then immediately placed in an evidence container. It does not need to be passed around from person to person for examination. When possible, items should be handled only by their edges. Be mindful to handle them in such a way that the fingerprints are not rubbed off. Even an investigator's own skin cells can potentially contaminate evidence. Skin cells can float in the atmosphere and transfer to an item. If that item is subsequently sampled for DNA analysis, the investigator's skin cells may be included in that sample. It is important for all investigators to remember that crime scenes are complicated enough on their own; they do not need to be further complicated through evidence contamination. Collecting the evidence is critical, but evidence can lose much of its value if it is placed in a container with no associated information. Accurate and detailed labeling ensures that once an item is taken to the laboratory, it is not confused with other pieces of evidence from the same case or even mixed up with evidence from other cases. Some cases can contain hundreds—even thousands—of individual items. Without the proper associating information, an item loses all of its potential evidentiary value. At minimum, the following information should be included in or on every container of evidence.

- Unique case number
- Item description
- Provenience
- Date recovered
- Unique evidence/bag number
- Initials of person(s) who collected the item

The final step in collecting evidence is to create a paper trail that permanently follows an item once it is collected at the crime scene and keeps a record of all the people who have had control of the item. This paper trail is called the **chain of custody**. A chain of custody is a simple piece of documentation that contains all the same information that

BOX 3.4 ACCESS CONTROL

One of the most critical steps in processing a crime scene is to control who has access to the site. Some people (such as the forensic anthropologists) need to have access to everything at the scene. But everyone—even law enforcement personnel—get curious about the crime scene and, without controlling access from the beginning, too many people may clutter the area. At worst, having too many people mingling around a crime scene may lead to destruction or contamination of evidence.

There are generally three zones of access for a crime scene. As the lead investigator, it is up to the forensic anthropologist to decide the boundaries of these zones. Marking these areas with "police tape" or "crime scene tape" is sometimes necessary.

- The inner zone (Zone 1) consists of the entire crime scene. This area should be accessed only by the forensic anthropologist and the team processing the scene. Zone 1 should not include government or law enforcement onlookers, some of whom think that their presence is acceptable because they are "allowed to be there." However, be strict and conservative with who is allowed in Zone 1, as anyone in this area runs a high risk of coming in contact with the evidence. In addition, Zone 1 should not include family members related to the crime scene incident when the area is being processed. Some crime scenes are gruesome, and emotional family members can potentially disturb evidence.
- Zone 2 is the area where supporting law enforcement personnel can stage. This zone is where equipment can be staged and where work vehicles are positioned. Local government representatives and additional law enforcement personnel should stage in this zone. Immediate family members related to the crime scene incident may stage here, but they should be kept far away from evidence.
- Beyond the perimeter of Zone 2 is Zone 3—or the rest of the world. Sometimes, law enforcement activity attracts curious bystanders, news and media outlets, or local residents. For best crime scene management and for the safety of these outsiders, it is crucial these persons be kept at a reasonable distance. If necessary, protective barriers (such as tarps) are needed to keep people out. These barriers must also protect the scene from camera lenses that can zoom in from a distance. And, remember that cameras can come from above, too.

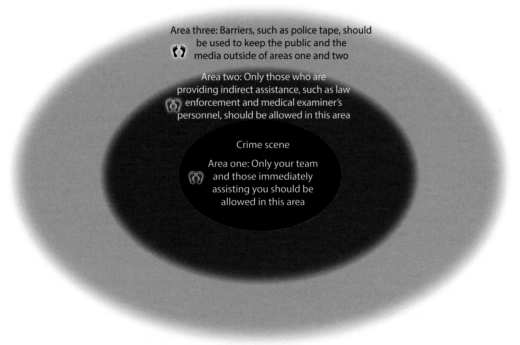

Figure 3.10 Access control of the crime scene should be established early and done so with at least three areas of control.

is on the label of the evidence container. When a person hands control and responsibility of the evidence over to another person, both people sign it until, eventually, it makes its way back to the laboratory, where the evidence manager takes care of it for good. The chain of custody document keeps careful control of who has had access to evidence from the time it was recovered to the time it was checked into the laboratory. If that chain is broken or compromised, it may indicate possible contamination of evidence. With critical evidence from a crime scene, the possibility of contamination can literally be the single reason for which a criminal case gets dismissed from court; therefore, tight control of evidence is top priority.

SUMMARY

When skeletonized (or partially skeletonized) human remains are discovered at a crime scene, a forensic anthropologist can be the most crucial specialist to have, and the more those skeletal remains and their associated evidence have been incorporated into the natural environment, the more important part the principles of forensic archaeology play. The record of natural and cultural events, captured above and below ground in that single location, can be very dynamic, because, as years pass, the record can be transformed by the addition and subtraction of information. Thus, the archaeological record is fragile, and when information (and skeletal remains) needs to be extracted from it, a forensic anthropologist trained in forensic archaeology is best suited for the task. One of the fundamental principles of forensic archaeology is to understand that processing and excavating a recovery scene are to alter the scene itself, so forensic archaeology functions on a controlled deconstruction of steps in order to maximize evidence collection, while at the same time minimizing contamination and alteration. A forensic anthropologist gets only one chance to process a crime scene or excavate a burial. Consequently, a skeleton must be excavated correctly because, if it is not, it cannot be put back in the ground for a second try.

As the subject matter expert, the forensic anthropologist's main responsibility is the proper recovery of human skeletal remains. The type of skeletal deposit, the positioning of the skeletal elements, and the associated taphonomic factors can all be gleaned by proper methods of excavation and recovery. The three traditional archaeological steps of burial excavation (defining the burial, exposing the remains, and lifting the remains) must not be rushed, and bone fever should be avoided at all times.

Documentation of the crime scene investigation through written notes, photography, and digital data is effective only if the documentation is detailed and thorough. In addition, the methods that are used to process the recovery scene must be systematic, should follow the best practices of the discipline, and should stand up to the highest level of scientific scrutiny. Preliminary steps in scene processing, such as aerial and visual pedestrian surveys, can be as important and revealing as the later steps of burial recovery. And, regardless of the size of the recovery scene, maintaining spatial control of all activities and recovered items constitutes the heart of proper mapping and documentation. Through provenience and context, a forensic archaeologist is able to piece together the archaeological record and better understand the event that brought the skeleton to rest at that location, as well as all the subsequent processes that have affected it since then.

Review questions

1. What is forensic archaeology?
2. In your own words, describe the difference between doing something systematically and unsystematically.
3. What is meant by "controlled deconstruction"?
4. How are human remains found on the surface different from those found below ground? How might you approach these situations differently?
5. What is documentation, and why is it so important?
6. What does a forensic anthropologist do during a preliminary survey of a crime scene?
7. What is a visual pedestrian survey? Discuss other methods of searching for human remains. What are the pros and cons of each?
8. What is a cadaver dog?

9. Why is it important to know the position of a body within a burial?
10. Name several processes that can disturb remains.
11. Describe the three stages of excavating remains.
12. Why is it important to prevent contamination of evidence? What are some ways in which this can be accomplished?

Glossary

Aerial survey: an examination of a specific area above the ground; sometimes conducted with manned vehicles (such as airplanes and helicopters) or through digital imagery with unmanned devices (such as drones and satellites).

Archaeological record: an account of natural and cultural events and transformation processes at a single location as documented only by the physical environment.

Archaeological tool kit: the suite of knowledge and resources used by a forensic archaeologist to effectively process a site. Can also refer to the actual assemblage of equipment used by a forensic archaeologist.

Articulated: the positioning of human skeletal elements relative to each other as they would normally exist in a living human.

Azimuth: a method of mapping whereby points are plotted using their distance and clockwise bearing relative to a single reference point, such as the site datum. Azimuth degrees range from 0° to 360°.

Baseline: a specific transect placed at a scene that is used as a reference line for various data collection activities such as measurements, pedestrian surveys, excavation, and so on.

Burial: specifically, a human body that has been deliberately placed subsurface and covered, usually with sediment. In a general sense, any assemblage of human remains found under or near the ground surface.

Cache: a localized deposit or assemblage of something of interest (such as a cache of weapons or a cache of tools).

Cartesian grid system: a method of mapping whereby points are plotted using northings' and eastings' measurements taken from a North and East baseline. The measurements are taken at a perpendicular distance from each baseline (thus, requiring that each baseline is aligned straight, true, and perpendicular to each other).

Chain of custody: a critical step in evidence management whereby the disposition of an item is tracked from initial collection at a crime scene to its accession into a crime laboratory (as well as any subsequent transportation to different locations).

Contamination: any kind of process, activity, or physical contact that can potentially damage, destroy, and/or compromise the forensic significance and context of an item by commingling it with data from elsewhere.

Context: the relationship and association of an item or piece of data relative to another item or piece of data at a recovery scene. An item's location in time and space relative to another item's location in time and space. While context is the most important information that can be derived from a recovery scene, it is also the most *easily lost* kind of information, especially when best practices are not adhered to.

Controlled deconstruction: systematic processing of a site conducted with adequate documentation using transparent and repeatable processes with particular attention to maximizing contextual data collection while minimizing contamination and alteration of the scene.

Diachronic: change over time.

Disarticulated: the positioning of human skeletal elements relative to each other in a way that they could not possibly exist in a living human.

Disturbance processes: all taphonomic activities that can influence human remains after its primary deposition including cultural, mechanical, natural, biological, and chemical activities.

Evidence management: the controlled process and administration of forensically significant items that include identifying, documenting, collecting, protecting, transporting, and curating.

Exhumation: specifically, the act of removing a deliberately buried skeleton from its perceived resting place, usually a cemetery. In a general sense, the removal of a body from any burial location. Also called *disinterment*.

Extended: a skeletal position whereby most of the limbs are stretched out to their maximum length. Modern traditional American burials are extended.

Feature: a specific assemblage of data that appear localized to itself and separate from the surrounding matrix. Examples can include a termite mound, a grass stain, a patch of differential soil disturbance, and a cache of tools. A human burial and associated grave can be considered as a feature.

Flexed: a skeletal position whereby most of the limbs are tightly bent at the joints.

Forensic archaeology: the application of theories and methods used in traditional research-based archaeology to process modern scenes of a medico-legal nature (usually with skeletal remains).

Forensic value: the significance of a site, an item, or an assemblage of skeletal remains with regard to its medico-legal context.

Global positioning system: a process whereby every location on earth is recognized by unique coordinates, as determined by the triangulation of that location using multiple orbital and/or suborbital satellites. The coordinates are obtained with a GPS-receiving device, using one of many available coordinate formats in reference to one of many available reference data.

Ground penetrating radar: a remote sensing method of distinguishing different densities below the ground surface through the measurement of administered radar pulses.

in situ: in its original place. A burial *in situ* means the skeletal remains have not been moved from the exact location where they were found.

Latitude–Longitude: a type of coordinate format system used to recognize a specific location on earth in reference to its place on the latitudinal and longitudinal axes of the globe.

Law of Association: a concept whereby the distance of items within a deposit and the integrity of the strata they are found in determine the reliability of any observed spatial and chronological relationship between the two objects.

Law of superposition: a fundamental principal of stratigraphy whereby the older, more undisturbed strata are generally found on the bottom and the younger, more disturbed strata are generally found on the top. Combined with the Law of Association, an object found in a stratum close to the surface might logically be interpreted as having been deposited more recently than an object found in a stratum far deeper in the ground.

Magnetometer: a remote sensing device capable of recognizing metallic signatures below the ground surface using magnetic forces. A magnetometer focuses on the earth's magnetic field while a metal detector focuses on the magnetic field around a localized object.

Military grid reference system (MGRS): a type of coordinate format system used to recognize a specific location on earth in reference to its place on the MGRS of the globe.

Pedestal: a method of excavation by which a specific item (e.g., a bone or material item) is left in place embedded in its matrix while the surrounding sediment is removed and leveled. This leaves the item slightly raised as the now-level area around it is lower.

Pedestrian survey: a systematic and controlled search of an area wherein a team of people walks across the scene in transects, and visually searches for data and evidence of forensic value.

Preliminary assessment: the initial step in scene processing where the forensic anthropologist makes broad determinations of which methods should be used.

Primary deposit: an undisturbed disposition of forensic evidence whereby the *in situ* location of an item is the same place where that item was originally placed in the ground. For example, a body that is buried in the ground and subsequently discovered by law enforcement in that same location is considered to be a primary deposit.

Prone (face down): a skeletal position where the majority of the body is facing down toward the ground.

Provenience: the exact location of an item (or a piece of data) in space; its three-dimensional location at a recovery scene.

Reference datum: a physical location, usually marked by a fixed point or solid item, whereby all other items mapped are based on that location. Data taken from the reference datum can include distance measurements, azimuths, and depth.

Remote sensing: a suite of methods used to collect data (both on the ground surface and below it) from a distance (i.e. without actually touching it) by means of various devices that *scan* the area in question. Remote sensing methods include metal detection, ground penetrating radar, resistivity, and conductivity.

Resistivity and conductivity: a remote sensing method of distinguishing different densities below the ground surface through the measurement of administered electricity.

Scene documentation: the act of preserving methodological practices of scene recovery as well as preserving the contextual data that are recovered and collected at a crime scene, using written notes, photography, and digital data collection.

Scientific integrity: it is the adherence to the best practices of the scientific method with transparency and accountability for all actions; requires sound and accepted methods and high ethical standards regardless of the nature of the work.
Screening: a process where sediment is placed through a mesh or sieve of a certain size to separate particles larger than that size. In burial recovery, the sediment that makes up the burial fill should be screened through at least a ¼-inch mesh (preferably smaller) to find small bone elements and material items that might be evidentiary.
Secondary deposit: a disturbed disposition of forensic evidence whereby the *in situ* location of an item is *not* the same place where that item was originally placed in the ground. For example, a body that is buried in the ground, years later to be washed away to a different location by flood waters, and subsequently discovered by law enforcement in that different location is considered to be a secondary deposit.
Sediment: a layer of earthly material found on the ground or below the ground placed there by natural or cultural practices. Burials are commonly found within various types of sediment, including soil, sand, and concrete.
Semi-flexed: a skeletal position whereby many of the limbs are loosely bent at the joints.
Shovel test pit: the product of a subsurface testing method whereby a small area (usually the diameter of a shovel) is excavated to variable depths (usually the depth of the shovel) to examine its contents and collect information below the ground surface.
Site formation: the process by which a physical location (usually, but not always, outdoor) is transformed in significance by a cultural event (e.g., when a crime takes place). In reference to forensic anthropology, it is when an assemblage of human skeletal remains is first deposited at a location.
Site transformation: all the natural and cultural processes that happen at a location after the initial site formation. These processes can be slow and long term (such as a trickle of water flowing into a grave) or a one-time, sudden event (such as a coyote digging through a grave). Site transformation processes can significantly alter the contextual data of the primary deposition.
Skeletal position: a description of the physical layout of a human skeleton (or skeletal elements). Can include extended, semi-flexed, flexed, supine, and prone.
Soil: a layer of sediment (such as silt, sand, or clay) usually in the higher strata of the ground that contains heavy organic qualities.
Strata: differing layers of sediment as deposited either by natural or cultural forces (stratum is singular). The strata are distinguished by type of sediment, color, density, moisture, inclusions, texture, and particle size.
Stratigraphy: the analysis and interpretation of differing strata at a particular location relative to each other. Stratigraphy considers when each stratum was deposited relative to each other, as well as how each are related to the formation of the crime scene.
Subsurface sites: scenes where the majority of the data and evidence are below the surface of the ground.
Subsurface testing: a process of collecting data from a specific area below the ground surface.
Supine (face up): a skeletal position where the majority of the body is facing up toward the sky.
Surface sites: scenes where the majority of the data and evidence are on or very near to the surface of the ground.
Systematic methods: transparent and coordinated processes of best practice that are conducted with step-by-step procedures and can be repeated.
Taphonomy: specifically, the "study of the laws of burial." In a general sense, a description and understanding of the processes that can affect human remains after original deposition.
Theodolite: may also be called a "transit." A highly accurate device used to measure the bearing of an object or location from a specified reference point. A transit and theodolite differ in mechanics, but are sometimes used synonymously. While the origins of the device are hundreds of years old, digital and laser-assisted versions are now the norms.
Triangulation: a method of mapping whereby points are plotted using the triangular vertices that are formed relative to each other. For example, mapping an unknown point can be done by measuring the distance of that point from where two other known points are located, as long as the distance between the two known points is also measured.
Uniformitarianism: a fundamental principal of nature whereby properties that affect the earth (such as gravity, erosion, flooding) act in the present day in the same way as they did in the past.

Universal Transverse Mercator (UTM): a type of coordinate format system used to recognize a specific location on earth in reference to its place on the UTM system of the globe.

Unsystematic methods: unrepeatable processes that are conducted largely at random with no semblance of control, structure, or holistic view.

References

Adams REW and Valdez F, Jr. 1997. Stratigraphy. In TR Hester, HJ Shafer, and KL Feder (Eds.), *Field Methods in Archaeology*. 7th edn. Mountain View, CA: Mayfield Publishing Company, pp. 235–252.

DPAA Laboratory Manual: Part I; SOP 1.1; Section 7.0; revised 10 March 2015.

Márquez-Grant N and Fibiger L. 2011. *The Routledge Handbook of Archaeological Human Remains and Legislation*. New York: Routledge, a Taylor and Francis Group, p. xxxiii.

Trigger BG. 1989. *A History of Archaeological Thought*. Cambridge: Cambridge University Press.

CHAPTER 4

Forensic Taphonomy

James T. Pokines and MariaTeresa A. Tersigni-Tarrant

CONTENTS

Introduction	57
Carnivore damage to skeletal remains	60
Rodent damage to skeletal remains	67
Other sources of gnawing	70
Summary	70
Review questions	72
Glossary	72
References	73
Further reading	77

> **LEARNING OBJECTIVES**
> 1. Define *taphonomy*, and provide examples of taphonomic processes that potentially affect forensic anthropological analyses.
> 2. Explain how taphonomic characteristics of a forensic scene can aid in the reconstruction of events that created/formed the scene and assist in interpreting the evidence.
> 3. Describe the characteristic patterns and features of rodent and carnivore scavenging, and how these are distinguished from activity caused by human agents.
> 4. Explain how animal gnawing and dispersal of remains can alter forensic scenes.
> 5. Suggest research designs that would provide an understanding of taphonomic variables that affect forensic analyses.

INTRODUCTION

Efremov (1940) defined taphonomy as the *laws of burial*. As such, taphonomy refers to the depositional history of a particular set of remains, explaining the natural and, in some cases, artificial (Pokines 2015c) processes that have acted upon the remains until the time of their discovery and removal from that depositional environment. Taphonomic research is drawn from multiple academic disciplines, including **ethology**, botany, paleontology, archaeology, and physical anthropology. Taphonomy can illuminate the natural processes that have affected a crime scene or other forensic archaeological sites. Research on taphonomic processes in large part began as a method to interpret osseous data about past environments and ecological relationships in an effort to determine the causative agents of bone destruction and alterations that might bias these interpretations (Behrensmeyer

1991; Weigelt 1989). Taphonomic research is also rooted in differentiation between bone modification caused by hominins (i.e., humans and our close fossil relatives) and that caused by other animals (Binford 1981; Brain 1981; Byard et al. 2002; Kerbis Peterhans 1990; Payne 1983). Current taphonomic research covers a variety of topics, including subaerial weathering (i.e., the effects of exposure to sunlight, precipitation, and so on), damage from plant root growth, gnawing by scavengers, modifications made by fluvial transport or thermal damage, and the effects of acidic soil on buried skeletal remains. See Chapters 15, 17, and 18 of this text for a detailed discussion of other taphonomic processes.

Within forensic anthropology, taphonomic studies provide a wealth of information about the crime scene and the postmortem interval (Ubelaker 1997). If carefully analyzed, the taphonomic characteristics of a forensic scene can aid in the reconstruction of events that occurred at the time of its formation. These characteristics also illuminate the processes that altered the scene between its formation and its processing by law enforcement, assist in interpreting the preservation of osseous and other types of evidence, identify the potential transport or **dispersal** of remains, and may elucidate confusing scene characteristics caused by the taphonomic processes themselves (Haglund and Sorg 1997b; Pokines 2009; Ubelaker 1997). Natural postmortem processes may also mimic perimortem trauma (Chapter 13, this volume) and pathological conditions (Chapter 12, this volume), so forensic anthropologists must be able to distinguish among multiple sources of osseous alteration.

The use of taphonomic research or methodology within forensic anthropology is often referred to as **forensic taphonomy**. In fact, Haglund and Sorg (1997a, p. 3) suggest that forensic taphonomy:

> … refers to the use of taphonomic models, approaches, and analyses in forensic contexts to estimate the time since death, reconstruct the circumstances before and after depositions, and discriminate the products of human behavior from those created by earth's biological, physical, chemical and geological subsystems.

The following sections discuss common types of osseous alteration encountered on the remains from forensic recovery scenes throughout the world: animal scavenging. These processes are largely destructive, removing entire bones or considerable portions of them, and they generally decrease the level of detail of the biological profile obtainable from the skeletal remains. Long bone epiphyses and weaker bones (especially ribs, vertebrae, innominates, and sterna) are especially vulnerable to these postdepositional forces. This information loss often prevents thorough metric assessment of human skeletal remains and removes other key pieces of information, such as the stage of epiphyseal fusion for age determination, fragile portions of the facial area useful for ancestry estimation, and evidence of perimortem trauma directly related to the cause of death. In turn, a thorough analysis of these processes yields other types of information, including estimates of postmortem interval and area of expected dispersal. In many ways, forensic anthropologists must learn to read the postmortem history of bones from these alterations in order to reveal the probable forces that caused them and the sequence in which they occurred. Rarely has only one major taphonomic process affected a set of remains at a forensic scene, and these overlapping taphonomic alterations can be decoded only through careful analysis (Pokines and Symes 2014).

Scavengers also impact forensic scenes through the dispersal of remains: it is very common for many parts of a skeletonized or decomposing human body to have been dispersed throughout the surrounding area (Camarós et al. 2013; Hudson 1993; Kjorlien et al. 2009; Pokines 2014, 2015a; Stiner 1991). While moving water or trampling may cause some of this bone movement, bones tend to get moved most often by the actions of scavenging animals in terrestrial environments. Disarticulation sequences of mammalian remains by carnivores have been studied in multiple environments and have been found to follow clear patterns, maximizing the most efficient consumption of nutrients (Blumenschine 1986, 1988; Hill and Behrensmeyer 1984). Carnivores of all types commonly remove portions of a body for consumption some distance away to reduce the amount of inter- and intra-species competition. Portions of a body may be moved tens or even hundreds of meters, as scavengers pull apart portions, often leaving a central concentration of bones that is very incomplete and partially disarticulated. The loss of bones during the scavenging process, along with the actual consumption of portions of or whole (small) bones, further reduces the biological profile data available to the forensic anthropologist. Therefore, a thorough search of the area around a body for additional remains is essential to any forensic field recovery where scavenging may have occurred (see Chapter 3, this volume).

Notably, scavengers impact indoor and outdoor forensic scenes. Household pets and wild species living in human habitations (i.e., those **commensal** with humans) frequently scavenge remains of individuals who died indoors. Domestic dogs (*Canis familiaris*), in particular, consume soft tissue and gnaw bones (Colard et al. 2015), sometimes leaving little skeletal material remaining (Steadman and Worne 2007). House mice (*Mus musculus*) and rats (*Rattus* spp.) are particularly common commensal rodent species, whose ranges have expanded worldwide through their close association with humans. These rodents can affect any indoor forensic scene by gnawing on bone and consuming soft tissue (Haglund 1992; Tsokos et al. 1999; Tsokos and Schultz 1999).

Multiple other environments, besides terrestrial surface settings, yield human (and often, nonhuman) remains that must be examined by forensic anthropologists, if only to determine that the remains are not of recent origin (Ubelaker 1995) and/or are not of further forensic interest (Duhig 2003; Hughes et al. 2012; Schultz 2012) (Box 4.1). All analysts must be aware of the taphonomic alterations potentially resulting from biological and environmental processes common to these other environments. In addition, terrestrial environments yielding remains with gnawing damage commonly display other taphonomic alterations. These include subaerial weathering (Behrensmeyer 1978; see also Chapter 15, this volume), where the bone becomes bleached and cracked from solar radiation, moisture loss, and expansion and contraction, over the course of many years, splintering into fragments. Bones recovered after extended periods of surface exposure often display dark staining on their surfaces in contact with the soil or decomposing leaf litter and algae, moss, or lichens growth on their surfaces exposed to sunlight (Dupras and Schultz 2014; Junod and Pokines 2014). Thus, the analyst is often presented with a suite of taphonomic characteristics that, when viewed as a whole, often point to a single environment of origin.

> **BOX 4.1 SWGANTH GUIDELINES FOR THE DETERMINATION OF HUMAN OR NONHUMAN REMAINS**
>
> The process of fragmentary remains' identification has become such a concern in recent years, that the Scientific Working Group for Forensic Anthropology (SWGANTH) has dedicated a portion of its "Determination of Medico-legal Significance from Suspected Osseous and Dental Remains" to this topic (www.swganth.org 2013). The purpose of this document is to provide guidance as to the best practices for distinguishing bone from nonbone and human bone from nonhuman bone, and, finally, the determination of whether remains are of medico-legal significance. These guidelines for best practices include suggestions as to when it is appropriate to use microscopic techniques for the analysis of human remains and provide a framework for the discussion of the results of such analyses in terms that can be universally understood.
>
> **Bone versus Non-Bone:** According to the SWGANTH guidelines, microscopic analysis is warranted in the process to determine if material is bone or not. These analyses attempt to identify cellular structures consistent with bone. The terminology used to report the findings of such an analysis are stated as "Consistent with Osseous/Dental Material," "Inconsistent with Osseous/Dental Material," or "Inconclusive." This codified terminology aids the analyst by providing universal terminology by which to account for the microscopic findings for a bone or nonbone analysis.
>
> **Human versus Nonhuman:** The SWGANTH guidelines suggest that microscopic analysis is an appropriate analytic tool for determining human or nonhuman osseous remains in cases of excessive fragmentation. The guidelines suggest looking for micromorphological patterns of plexiform bone or patterned osteons, though they advise extreme caution, as the micromorphology that is consistent with human remains is also found in some large animals. The terminology that they have chosen for this type of analysis reflects this conservative approach by using "Diagnostic of Human," "Diagnostic of Non-Human," and "Inconclusive."
>
> **Unacceptable Practices:** The SWGANTH guidelines are resolute in their discussion of histological techniques and other specialized techniques as important parts of the practice of forensic anthropology. However, they strongly caution against the use of these techniques without proper specialized training. It is imperative, then, that practitioners familiarize themselves with these techniques, through one-on-one tutorial with an expert in the method or through continuing education courses and practice before implementing these methods in medico-legal casework. Alternatively, practitioners who do not feel comfortable with these techniques are encouraged to seek the aid of those who are well versed in each method.

Human remains from terrestrial environments may also be buried before their recovery, perhaps as part of the clandestine disposal of remains (Chapter 3, this volume). Direct contact of bone with soil tends to leave behind its own taphonomic markers, including etching of the bone surfaces (if the soil is acidic), plant root infiltration into the bone, dendritic (i.e., branching) surface etching by plant root contact, and all-over soil staining (Pokines and Baker 2014). Burial types also include marked and unmarked cemeteries, and multiple researchers have examined the particular changes observed in human remains recovered from coffin burials (Berryman et al. 1991, 1997; Nawrocki 1995; Pokines and Baker 2014; Pokines et al. 2016; Rogers 2005; Schultz et al. 2003; Sledzik and Micozzi 1997). Remains from cemeteries, if old enough (>100 years in many jurisdictions), become the responsibility of state historical preservation offices and similar government entities. Therefore, more recent coffin remains are the responsibility of the forensic anthropologist, who must be able to differentiate among the different taphonomic origins of buried remains.

Remains of accident, suicide, and homicide victims are also frequently recovered from marine environments (Haglund and Sorg 2002; Sorg et al. 1997). These remains can be encountered still floating, washed ashore along a beach, or accidentally snagged in fishing nets. Bones from saltwater environments may display multiple taphonomic characteristics that form a unique combination. These characteristics include mineral staining; battering and rounding of surfaces; and adhering sand, barnacles, mollusks, and saltwater species of algae (Higgs and Pokines 2014; Pokines and Higgs 2015).

Multiple cultural practices are also the proximate cause for human skeletal remains coming to the attention of law enforcement. These include former trophy skulls from modern military conflicts (Bass 1983; Harrison 2006; Sledzik and Ousley 1991; Weingartner 1992; Willey and Leach 2009), which were sometimes transported to the United States as souvenirs from WWII and the Vietnam conflict. These skulls are sometimes found among the personal effects of veterans after they have died or are turned in voluntarily by the veterans in an effort to effect their proper burial, decades later. Skeletal remains that were formerly used for anatomical instruction are also brought in for examination (Paolello and Klales 2014; Pokines 2015c), often originally in the possession of anatomy and medical instructors and, sometimes, repurposed as display items. Ritual practices from multiple religions that are spreading in the United States, including Palo Mayombe (Gill et al. 2009; Pokines 2015b; Wetli and Martinez 1983), feature the use of human remains often illegally obtained from cemeteries. Taphonomic alterations, therefore, may reflect more than one stopping point, as bones passed through multiple environments, with each environment leaving some characteristic indicators along the way.

CARNIVORE DAMAGE TO SKELETAL REMAINS

The modification of vertebrate skeletons by other animals is a well-documented and integral aspect of taphonomy. In archaeology, the characteristics of carnivore damage are studied to determine potential effects that these species could have had on human and animal osseous assemblages, both through the complete removal and/or destruction of bone or alterations made to surviving bone (Binford et al. 1988; Blumenschine and Marean 1993; Brain 1981; Haynes 1980, 1983). Within the subdiscipline of paleoanthropology, taphonomic studies have been used to determine whether early hominins (with stone tools) or scavenging animals were responsible for the modification of remains and which species had access to carcasses and in what order (Binford et al. 1988; Blumenschine 1986, 1988; Brain 1981; Kerbis Peterhans 1990; Pobiner 2008; Stiner 1994). In forensic anthropology, the identification of characteristic carnivore damage aids in postmortem interval estimation and trauma analysis, as well as in determining the method of remains' dispersal over a large area.

Consumption of a large vertebrate carcass tends to proceed from the most easily accessible portions first to the least-desirable, lowest-nutrient portions last (Blumenschine 1986, 1988). The sequence often starts with the internal organs (heart, liver, and lungs), given their ease of access through the abdominal wall and high return of nutrients, with little or no bone crushing necessary to consume them. Large muscle masses of the upper hind limbs often come next, leading to the disarticulation of this portion of the carcass. Feeding can then proceed to the muscles of the thoracic cage or the upper front limbs. The ribs, sternum, and vertebrae pose little hindrance to large carnivores consuming a skeleton, and these are often fragmented as a part of overall feeding (Horwitz and Smith 1988; Kerbis Peterhans 1990; Pokines and Kerbis Peterhans 2007). By this time, most portions of the skeleton are disarticulated, allowing easier dispersal of remains over large distances. Movement of bone is a normal consequence of carnivore feeding, as it reduces competition while feeding and enables transportation to a den to feed cubs that are not yet old enough to participate in hunts (i.e., **provisioning**). Haglund et al. (1988, 1989) noted an important difference in the aforementioned sequence of carcass consumption in their research on human remains scavenged by large canids in the Pacific Northwest, U.S. (dogs [*Canis familiaris*] and coyotes [*Canis latrans*]). They found that with human remains, the first portion of the body consumed was often the throat region. This different pattern may be due to the presence of clothing, restricting access by large vertebrates to the portions that are scavenged.

Damage to the skeleton itself follows the same overall pattern—the destruction of easiest portions first and more gradual destruction of the better-protected portions occurring later. The weaker elements of the thoracic cage often undergo some destruction during initial feeding and then undergo more destruction as they are fragmented and consumed while their associated soft tissue is consumed. Fragments and the smaller elements, including those of the hands and feet, can be consumed whole. Carnivores' damage to the skeleton is most notable on the appendicular skeleton, in particular the long bones, since weaker elements (i.e., ribs and vertebrae) are often completely destroyed by gnawing (see Box 4.2). The long bones pose the biggest obstacle to consumption, yet these hold an important food resource: the marrow cavity (in adult mammals) stores large supplies of fat, and their epiphyses also have a high fat

---

BOX 4.2 CASE STUDY: EXAMINATION OF TAPHONOMIC ALTERATIONS ON A COLD CASE

Some forensic anthropologists analyze (or reanalyze) remains as cold cases that have been in storage for multiple years, with no positive identification and few clues as to the identity of the person. A complete skeletal analysis, including taphonomy, is essential in narrowing down the possible identity of the remains in question and understanding the recovery context. DNA testing will also likely be necessary, as advances in technology, funding, and processing speeds have made it possible to complete nuclear DNA testing in months (instead of years) after initial recovery. This information, along with the biological profile of a set of unknown remains, can be uploaded into an online database, such as the National Missing and Unidentified Persons System administered by the U.S. Department of Justice (www.namus.gov).

The case in question was examined 10 years after its initial recovery in eastern Massachusetts in the greater Boston area. The remains consisted of a nearly complete skeleton in excellent condition (Figure 4.1). All portions of the skeleton were represented, with only small elements or portions missing and some desiccated soft tissue (primarily periosteum), and two toe nails were present. The individual was estimated to be a probable Hispanic male who was 71 inches tall and between 27 years and 66 years old (likely at the upper end of that range) at the time of death.

Taphonomically, the bones retained a faint greasy texture and retained organic content. Dark soil staining was present but was patchy throughout (Figure 4.2). Small amounts of sediment or decomposing organic matter adhered externally; additional decomposing organic matter (probable leaf fall breaking down into humus) was present inside the cranium. The right and superior portions of the cranium, in particular, had lain in contact with the soil, as these areas had dark soil stains. Most exterior bone surfaces were almost entirely intact; however, some minor cortical erosion occurred where there was contact with the acidic topsoil. Blowfly larval casings also still adhered to some bones. No subaerial weathering (Behrensmeyer 1978) had developed beyond incipient bleaching in places (especially the right innominate), indicating that the remains either had been in a protected surface location or had not lain on the surface exposed for multiple years. Green fabric residue, likely the remnants of clothing, was present in multiple locations, including vertebrae T11-L5, left rib 2 sternal end, right rib 11 posterior surface, right posterior scapula, and right distal humerus.

As indicated, the remains were nearly complete, with the exception of missing bones of the hands. Could the missing hand bones have been related to the death of the individual, possibly as part of dismemberment, so that the deceased could not be fingerprinted? If it was a natural postmortem process, why were these elements missing and not others of similar size and susceptibility to postmortem destruction? Given the time since initial recovery, further exploration of the area where the skeleton was recovered would not likely yield any of these small elements. Taphonomic examination of the hand bones that were recovered and other skeletal elements supplied the likely scenario. Multiple minor areas of damage fitting the pattern of wet-bone gnawing

(Continued)

> **BOX 4.2 (*Continued*) CASE STUDY: EXAMINATION OF TAPHONOMIC ALTERATIONS ON A COLD CASE**
>
> by omnivorous rodents to obtain bone grease were present throughout the skeleton: distal right humerus, right proximal radius, right distal radius and ulna (Figure 4.3), left proximal ulna and radius, right hamate, two left metacarpals proximally and distally, and three manual phalanges (Figure 4.4). This type of gnawing damage is consistent with *Rattus* spp., including brown rats and black rats. Some areas of gnawing damage were more consistent with small carnivores, as on the distal left fibula (including tooth punctures present; see Figure 4.2) and distal left and right tibia. The overall excellent preservation and completeness of the foot bones and distal leg bones indicate their probable protection within durable footwear during decomposition and scavenging (Pokines and Baker 2014). Thus, the missing hand elements were likely dispersed by scavenging rodents or small carnivores, both of which can transport small skeletal elements. The hands were likely exposed to scavengers, while other portions of the remains were covered and protected (Figures 4.2 and 4.3).

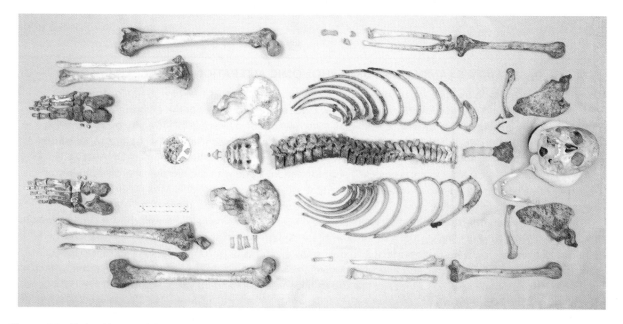

Figure 4.1 Skeletal layout of the case study remains. Note the missing hand elements and completeness of the foot elements. The scale bar is approximately 15 cm.

Figure 4.2 Distal left fibula from the case study. Note the missing cortical bone and tooth punctures (arrow), which are most consistent with carnivore (instead of rodent) gnawing. Variable patches of dark soil staining are also present.

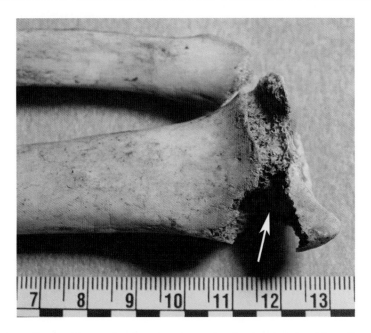

Figure 4.3 Pedestaling (arrow) caused by small rodent gnawing to the distal end of the right radius. The distal ulna was also removed. The bone comes from recent surface terrestrial remains, Massachusetts, U.S.

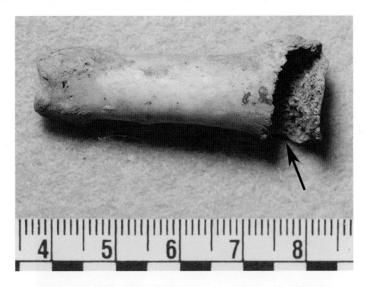

Figure 4.4 Pedestaling (arrow) caused by small rodent gnawing to the proximal end of a first-row manual phalanx. The bone comes from recent surface terrestrial remains, Massachusetts, U.S.

content. Carnivore gnawing on long bones normally starts on the epiphyses (Figures 4.5 and 4.6), as the thin layer of cortical bone covering the ends makes the contents of the bone easier to access. In long bones, where one end tends to be more robust than the other (e.g., the distal humerus being much more robust than the proximal humerus in most mammals), scavengers often skip the more robust end entirely and focus their gnawing efforts on the easier portion (Pokines and Kerbis Peterhans 2007). Then, gnawing proceeds by removal (and frequent consumption) of bone fragments, working inward toward the marrow cavity. This process often leaves a bone **cylinder** (a tube of diaphysis with both ends removed) behind as the final result of carnivore gnawing of a long bone (Kerbis Peterhans 1990). Continued gnawing of these more durable elements frequently takes place away from the kill site or the depositional site, where the carnivore can expend more time, with less chance of food being stolen by a competitor.

The cranium often remains untouched by carnivores owing to the difficulties of gnawing through such a large, rounded object, because it is difficult to fit deep enough within a carnivore's jaws for consumption or to hold onto for

Figure 4.5 Characteristic carnivore gnawing damage on human remains from a terrestrial surface environment in Massachusetts, U.S. Pictured is a proximal human femur (head), with large areas removed by carnivore gnawing; note the areas of missing cortical bone and tooth punctures (arrows). Some desiccated soft tissue still adheres. The species that did the gnawing was likely domestic dog (*Canis familiaris*) or coyote (*Canis latrans*).

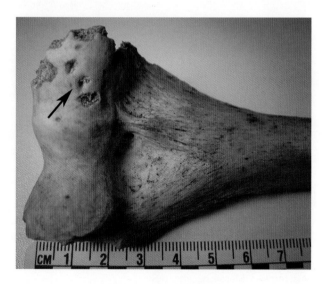

Figure 4.6 Characteristic carnivore gnawing damage on human remains from a terrestrial surface environment in Massachusetts, U.S. Pictured is a proximal human femur (head), with large areas removed by carnivore gnawing; note the areas of missing cortical bone and tooth punctures (arrow). The species that did the gnawing was likely domestic dog (*Canis familiaris*) or coyote (*Canis latrans*).

transport away from the point of initial deposition for later gnawing. An exception to this pattern occurs when larger carnivores attack the crania of smaller species or the juveniles of larger species (Pokines and Kerbis Peterhans 2007). The brain tissue has a high fat content, so gnawing of the cranium by carnivores is largely a way to access this food resource (Stiner 1994). Intact human crania can be transported to several kilometers by scavengers for later consumption. Even a scavenger as small as a striped hyena (*Hyaena hyaena*), a largely solitary species (Horwitz and Smith 1988), has been documented to transport human crania to great distances.

Most gnawing is done by extant carnivores with their **carnassial teeth**, which are the upper and lower teeth that have evolved to shear past each other (like scissors) for cutting meat and crushing bone. Among carnivores, the teeth that evolved for this role are the maxillary fourth premolars and the mandibular first molars. Significant biting force can be applied at this point because of the position of these teeth farther back in the mouth and the design of the mandibular lever system (Pokines 2014:205). The anterior teeth (i.e., the incisors and canines) have the least amount of biting force within the same individual.

The most common characteristics of carnivore damage to skeletal remains are multiple types of tooth marks. These have been defined in the taphonomic literature, primarily in the context of zooarchaeology (Binford 1981; Haglund 1997a; Haynes 1980; Lyman 1994; Pobiner 2008; Pokines and Kerbis Peterhans 2007). Tooth marks are classified into four basic types. **Pits** are small indentations in the bone (less than three times long than wide and usually oval or irregular in shape) that do not penetrate the cortical layer. Pits are typically found in areas with a thicker cortex (Figure 4.7). **Punctures** are similar in form to pits (less than three times long than wide) but are deeper indentations or perforations of the bone that do penetrate the cortical layer; punctures are typically found in thin cortical bone overlying cancellous bone (Figures 4.5 and 4.6). **Scores** are elongated (more than three times long as wide) marks that do not penetrate the cortical layer and are caused by carnivore teeth sliding across the outer surface of the bone (Figure 4.8). **Furrows** are elongated (more than three times long as wide), like scores, but do penetrate into the cancellous bone and are usually

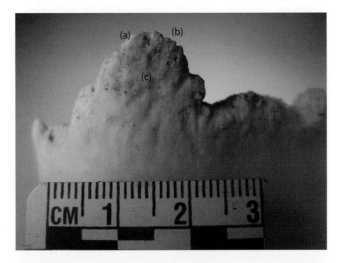

Figure 4.7 Characteristic carnivore gnawing damage to a right proximal humerus shaft of cattle (*Bos taurus*). Note the formation of (a) edge polish on the exposed margin, (b) crenellated margins, and (c) multiple shallow pits left in the cortical bone. The bone was recovered from a surface terrestrial environment in Massachusetts, U.S.

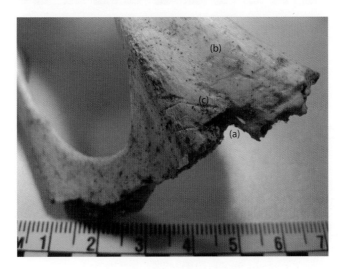

Figure 4.8 Characteristic carnivore gnawing damage to a left innominate of white-tailed deer (*Odocoileus virginianus*). Note the (a) crenellated margin and multiple (b) pits, and (c) striations. The bone was recovered from a surface terrestrial environment in Massachusetts, U.S.

detected in cancellous bone that has been previously exposed by gnawing. In addition to these types of individual tooth marks, **edge polish** may be left behind on exposed cortical bone that has already been fractured but has been worn down by additional gnawing on the freshly exposed surface (Figures 4.7 and 4.9) (Pokines and Kerbis Peterhans 2007). The more general term **gripping marks** refers to unpatterned pits and scores that may be left on long bone shafts as a carnivore shifts the element within its mouth to reposition it for gnawing on the epiphyses or to hold onto the bone for transport. Carnivores often create *crenellated* (or jagged) margins in gnawed bone as they splinter away fragments (Figures 4.7, 4.8, and 4.10). Continued gnawing of any bone tends to destroy earlier marks left behind in weaker areas of the bone. These areas are the most likely to exhibit more diagnostic tooth marks (punctures), yet they are also the areas that are most likely to be destroyed by subsequent gnawing. Similarly, it is rare to find furrows in long bones that have undergone significant gnawing, as continued gnawing will likely result in complete destruction of the epiphyses.

Tooth punctures offer the greatest potential to identify the carnivore taxon that gnawed upon a bone, as individual marks or patterned clusters of marks might conform to a dental shape and size unique to that gnawing taxon (Murmann et al. 2006). These marks are typically formed by the incisors, canines, or carnassial teeth. Canine teeth are not always useful in this regard, as they have a similar cone shape in most carnivores, so smaller canine punctures might be formed by a deeply penetrating smaller carnivore or by a shallowly penetrating larger carnivore.

Large-bodied carnivores consume large amounts of masticated bone fragments and small intact elements (including carpals and phalanges) during their normal feeding activities. Apart from the destructive effects of fragmentation, this ingested bone also undergoes significant modification from gastric acid and digestive enzymes as it passes through the alimentary canal and is deposited as feces or regurgitated (Esteban-Nadal et al. 2010). Carnivore taxa

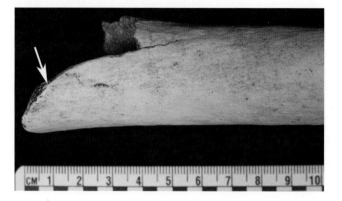

Figure 4.9 Edge polish (arrow) on a proximal human femur formed after the epiphysis was removed by carnivore gnawing; the bone was recovered from a surface terrestrial environment in Massachusetts, U.S.

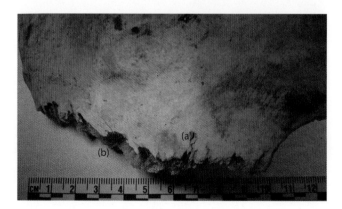

Figure 4.10 Carnivore gnawing, showing (a) tooth punctures and (b) crenellated margin on a human ilium recovered from a surface terrestrial environment in Massachusetts, U.S. The species that did the gnawing was likely domestic dog (*Canis familiaris*) or coyote (*Canis latrans*).

that have been researched in this regard include canids, felids, and hyaenids (Hill 1989; Payne and Munson 1989; Pokines and Kerbis Peterhans 2007; Shipman and Phillips-Conroy 1977; Stiner et al. 2012). This type of alteration is termed **gastric corrosion**. Gastric corrosion causes the formation of small holes in fragments or elements (referred to as **windowing**), thinning of edges, and a characteristic sculpted appearance to bone surfaces (Brain 1981; Kerbis Peterhans 1990; Pokines and Kerbis Peterhans 2007). The size of the fragments is limited by the maximum size of bone that can be swallowed by the specific carnivore species, so larger carnivore taxa tend to leave behind larger and potentially more identifiable digested bone fragments or elements. Payne and Munson (1989) suggested that the corrosive digestive environment is very destructive to bone, but acidity affects weaker parts of consumed bone (e.g., trabecular and thin cortical bone) more than the stronger parts (thick cortical bone). They also indicated that the effects of mastication and digestion tend to be more damaging to the skeletons of smaller vertebrates. Even after exposure to the corrosive effects of carnivore digestion, cortical bone fragments can be identified as human or non-human, provided that the bone is not severely degraded as to damage the inner cortical structure, and the plexiform bone histological structure of the latter is still observable.

While it is usually not difficult to determine if a skeleton has been altered by carnivores, it is more difficult to identify the specific carnivore that caused the damage. In addition to the many known larger species, many species of small carnivores will also gnaw bone and otherwise feed upon or disperse scavenged remains (Rippley et al. 2012; Young et al. 2014). Domínguez-Rodrigo and Piqueras (2003) attempted to develop techniques to determine carnivore gnawing taxa according to tooth mark location and size on the bones. They discovered that the tooth marks by themselves cannot be used to glean taxa information and can only distinguish between large carnivores (such as lions, bears, and large canids) and small carnivores (such as mustelids, small felids, and small canids). Other research into tooth impressions notes that while overall size classes of carnivore may be discernible (Andrés et al. 2012; Coard 2007; Lyver 2000), there is significant overlap between many common gnawing species (Delaney-Rivera et al. 2009; Murmann et al. 2006). Murmann et al. (2006) further noted that the overall curvature of the anterior dental arcade, including the incisors and canines, varies by carnivore family, and the distance between pairs of upper or lower canine tooth impressions (in the rare cases where these are preserved in bone or other materials) may be useful in determining the size class (and hence taxonomic grouping) of the gnawing species.

Many obstacles make studies of this type difficult, including ascertaining if bones recovered from natural settings have been gnawed by only one species. For example, Haynes (1982) discusses the similarities of canids (specifically wolves) and ursids in their modification of more than 125 carcasses and skeletons of bison (*Bison bison*), moose (*Alces alces*), and white-tailed deer (*Odocoileus virginianus*) in North America. He documented the dispersal of bones at kill sites and noted that characteristic gnawing damage occurs at kill sites, scavenge sites, and home sites. Haynes described the patterns of dispersal and accumulation associated with both canids and ursids (bears). Canids and ursids generally do not gnaw on *dry* bones that have been exposed to the elements for longer than six months (Haynes 1982). Carson et al. (2000) attempted to distinguish between canid and bear scavenging upon human remains, based on the disarticulation pattern and specific elemental damage patterns. Bears remove the proximal and distal articular surfaces of limbs and also damage the iliac crest and ischial tuberosity significantly. This pattern of damage, however, is also typical of large canid or other taxa scavenging in multiple settings (Pokines 2014). Subsequent gnawing of an already scavenged bone obscures existing marks potentially definitive as to a carnivore species, family, or size class. Forensic investigators may find it more rewarding to look for other field signs left behind by scavengers in order to identify them, including hair, footprints, and droppings (Einarsen 1956; Elbroch 2003).

RODENT DAMAGE TO SKELETAL REMAINS

In contrast to the typical carnivore characteristics left on skeletal remains, rodent damage can be more subtle. All rodent species (by definition) have paired maxillary and mandibular incisors (for a total of four teeth) that grow continuously (Figure 4.11). These species need to gnaw on relatively hard items to keep these teeth at a usable length and to sharpen the cutting margins (Kibii 2009). Skeletal remains are often an ideal surface on which to wear down these teeth. The incisors, when scraped across a hard surface, produce **parallel grooves** that are unlike the characteristic pitting or striations caused by carnivores (Figures 4.12 and 4.13). The parallel grooves may be found in a single layer or multiple overlying layers if further gnawing has obscured previous tooth striations. If the grooves are singular, then they were likely produced by only one set of teeth dragging across the bone surface; grooves that converge to a point indicate that both the upper and lower incisors were used at the same time (Figure 4.13). Bone margins gnawed

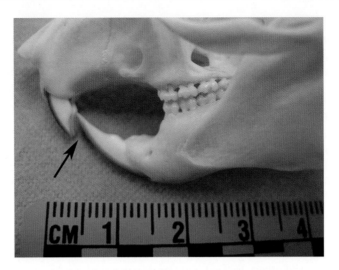

Figure 4.11 Cranium of eastern gray squirrel (*Sciurus carolinensis*), a common rodent for much of eastern and central North America, showing typical rodent incisor morphology (arrow). All rodents have one pair of maxillary and mandibular incisors that grow continuously and must be worn down.

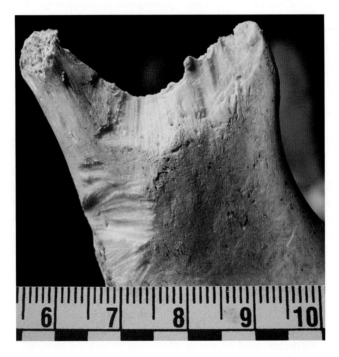

Figure 4.12 Rodent dry bone gnawing to recent surface terrestrial remains (right mandible) from Massachusetts, U.S.; note the many fine parallel incisor grooves.

by rodents tend to have a uniform pitch, since they can slowly wear away bone with a series of short, parallel, or overlapping grooves. The gnawed margins are therefore more uniform than the typically jagged margins left behind by carnivores. Haglund (1997b) suggested that rodents prefer epiphyseal cartilage and adjoining areas of long bones and also noted that rodents tend to gnaw on the cranium where thin, exposed margins such as the eye orbits can fit within their jaws (Gapert and Tsokos 2013).

Rodents also use skeletal remains to obtain nutrients from fat-filled trabecular bone or from the mineral component of the bone. Certain omnivorous rodents, including the widespread species black rat (*Rattus rattus*) and brown rat (*Rattus norvegicus*), are attracted to skeletal remains during the early postmortem interval, when fat is still contained in the trabeculae of the long bones. Brown rat's and other omnivorous rodents' consumption of the trabecular bone

Figure 4.13 Rodent dry bone gnawing to a tibia shaft; note the many fine parallel incisor grooves. The bone originated from a cemetery burial from Massachusetts, U.S., which was later dumped on the ground surface.

in the skeletal epiphyses produces a *pedestaled* appearance, as these rodents gain access to these regions by targeting areas where the cortical bone is the thinnest (typically the epiphyses). They proceed to hollow out the epiphyses, leaving behind the shaft and the articular portion of the epiphysis (Figures 4.3 and 4.4) (Klippel and Synstelien 2007). This gnawing into epiphyses may leave behind areas of adjacent, untouched bone (i.e., pedastaled) (Figure 4.4). Some porcupine species feed upon the mineral content of bone and consume dry bone material while gnawing (Duthie and Skinner 1986). Rodents attracted to bone only after the fat has leached from the bone are more active later in the postmortem interval. Multiple common rodent taxa, including rats, house mice (*Mus musculus*), squirrels (Sciuridae), and voles (including *Microtus* spp.; Pokines, 2015d) gnaw upon dry bone; however, whether the gnawing is confined to incisor-sharpening activity or whether bone minerals are also being consumed is presently undetermined.

Larger rodents produce larger incisor marks. Throughout the United States and Canada, the North American porcupine (*Erethizon dorsatum*) is the most common large rodent known to gnaw on bones (Roze 2009; Woods 1973). This species often consumes additional minerals to make up for deficiencies in its normal plant diet (Roze 2009). The individual marks that the porcupine leaves on bone are visibly larger and deeper than those left by rodents in the rat- to squirrel-size range (Figure 4.14).

Figure 4.14 Extensive North American porcupine's (*Erethizon dorsatum*) gnawing activity on a cattle's (*Bos taurus*) femur cylinder that had been previously machine-butchered (flat surface to the right). The bone was recovered from a surface terrestrial environment in Massachusetts, U.S.

Klippel and Synstelien (2007) attempted to clarify the characteristics of bone gnawing by rodents. Some researchers indicate that rodents prefer dry bone, and others suggest that rodents seek fat-filled bone. Klippel and Synstelien found that brown rats prefer fat-containing cancellous bone, a behavior associated with the acquisition of nutrients; whereas eastern gray squirrels (*Sciurus carolinensis*) choose cortical bone, where fat was no longer present, a behavior associated with acquiring minerals or simply incisor wear. They further suggested that since eastern gray squirrels do not show interest in gnawing bone before at least 30 months of exposure, the presence of eastern gray squirrel gnaw marks on remains in environments similar to that of east Tennessee could provide a minimum time since death estimate. Since rates of bone weathering (i.e., drying out) are variable, based on microenvironment, and the bone-gnawing behavior of a very few rodent species (out of the over 2200 named species) has been studied to date, more research is needed in this area.

OTHER SOURCES OF GNAWING

Other sources of gnawing damage that may appear in forensic cases are more rare. For example, many herbivores have been noted to practice **osteophagia** (i.e., bone consumption) on occasion. The known taxa include deer, sheep, and giraffe, and the reason is likely in response to localized dietary deficiencies that may be alleviated by bone minerals (Cáceres et al. 2013; Hutson et al. 2013). Their broad, multi-cusped cheek teeth (premolars and molars) evolved for grinding large amounts of tough plant material, so the pattern of damage that they leave often takes the form of a "Y-fork" on the ends of long bones, as the bone is ground down, along with general tooth pitting (Pokines 2014). An example of this type of taphonomic alteration from an archaeological context is presented in Figure 4.15. Other sources of bone gnawing include other mammals such as pigs (*Sus scrofa*) and insects such as termites (Pokines 2014); however, termite damage is rarely encountered among forensic cases in the United States.

SUMMARY

The effects of carnivores on bone include the tooth marks that they commonly leave behind (i.e., pits, punctures, scores, and furrows), each based on proportions and degree of penetration into the bone. They may also cause **crenellated margins** to bone due to fragmentation of gnawed portions. Continued gnawing to these exposed margins may wear them down, forming edge polish. Consumption of fragments or whole small bones may cause gastric corrosion, and these remains may be recovered from vomit or feces. Carnivores are also a common source of remains dispersal from forensic scenes, as their feeding activity often includes transport to tens or hundreds of meters away from the point of initial deposition to spend time in feeding in greater isolation.

In contrast, rodent gnawing produces marks that are largely distinguishable from those left behind by carnivores. Rodents often gnaw upon dry bone with their incisors, leaving behind short parallel grooves concentrated on the

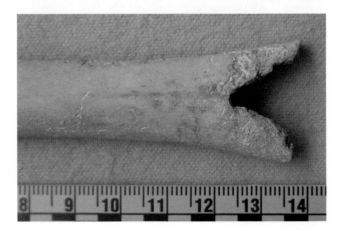

Figure 4.15 Ungulate osteophagia leaving behind the characteristic "Y-fork" pattern on a nonhuman (sheep or goat) proximal metacarpal. Note the overall pitting and battering around the gnawed margin. The bone was recovered from a Bronze Age level of an archaeological site in modern-day Jordan and was likely gnawed upon by other sheep/goats or cattle.

thin margins of bone. The gnawed margins tend to have a uniform pitch, unlike the crenellated margins left behind by carnivores. They favor thinner bone margins, as they can fit their small jaws around the projections. However, some rodents feed on soft tissue and gnaw their way into fresher bones, especially epiphyses, to consume the fat content contained within. This gnawing into epiphyses may leave behind pedestaled areas of adjacent, untouched bone. Other rarer bone-gnawing activity includes osteophagia by herbivores, leaving battered bone portions that are ground down by their cheek teeth. This may result in the classic "Y-fork" pattern to long bone ends.

Gnawing and dispersal by multiple animal species have the potential to alter indoor and outdoor forensic scenes throughout the world. Few natural taphonomic processes are more pervasive, as decomposition at some pace always follows the death of an organism, and many species have evolved to consume these nutrient resources when they are introduced into their environment. Most decomposition occurs due to the actions of microscopic organisms (Box 4.3) or feeding insect larvae (Chapter 15, this volume), but feeding by larger (vertebrate) scavengers greatly reduces the amount of soft tissue present, damages bone, and disperses remains into the surrounding environment. While much of our knowledge of these processes comes from studies of captive animals, research suggests that captive animals may gnaw bones differently than their wild counterparts (Gidna et al. 2013). Therefore, studies of wild behavior are required to document the effects of these species on forensic scenes, whether these studies are experimental in nature or involve gathering data from actual forensic cases.

BOX 4.3 BONE HISTOMORPHOLOGY USED TO IDENTIFY POSTMORTEM TAPHONOMY

Bone histomorphology is used to determine if postmortem taphonomic damage due to biological and nonbiological degradation has occurred to the bone or bone fragment in question. Schultz (2001, 2003) indicated that bone histology could also be used to identify the preservation state of remains. This can aid in determining if further chemical/genetic analysis could be undertaken successfully, which, in turn, saves the needless waste of time and money on the analysis of poorly preserved bone.

Schultz (1997) discussed a case in which differential decomposition was noted only after histological analysis: a buried skeleton was missing a portion of the lower limb, and it was presumed amputated. On further analysis of trabecular bone fragments within the soil, it was determined that the lower limb was present at the time of burial but had subsequently decomposed in a differential manner than the rest of the body.

Numerous taphonomic processes can be identified using thin sections of bone and either light microscopy or scanning electron microscopy. These include root infiltration, fungus or bacterial growth, insect infestation, and fire modification (Jans 2014; Schultz 1997).

Fungus, bacteria, and even algae infiltration can degrade the internal and external structures of bone (Child 1995; Jans 2014; Jans et al. 2004; Schultz 1997). Differential signatures within the bone microstructure suggest the presence of algae, fungus, or bacteria. Schultz (1997) suggested that large tubular spaces in the microstructure are indicative of fungus infiltration, while bacteria cause channels to form within the bone. After the initial damage, these channels are usually filled by fungi, algae, and bacteria that colonize the compromised bone. All of these microorganisms continue to reproduce and damage the bones from the inside for years, until the bone is completely degraded. This degradation can prevent any sort of isotopic or microstructural analysis.

Bones that have been modified by heat or fire are also precluded from isotopic or microscopic analysis, depending on the duration of the heat/fire, as well as the temperature to which the bones were heated. Differential temperatures can lead to diverse morphological changes within the bone microstructure (Absolonova et al. 2013; Schultz 1997; Tersigni 2007). In some cases, the bone is altered minimally at the microscopic level. In other cases, the bone is altered maximally, losing its organic component, thus making it extremely friable and an unlikely candidate for isotopic or microstructural analysis.

At the microscopic level, if the thin section of bone is highly fragmented or is missing large portions of the pertinent microstructure, this sample is excluded from further analysis because of heavy taphonomic damage. In addition, if the section is completely devoid of characteristic microstructure (Haversian systems, osteons, plexiform bone, etc.), then it may be prudent to rule this section inadequate for further analysis.

Review questions

1. Define taphonomy. What kind of processes it might encompass?
2. Why is taphonomy relevant to both forensic anthropology and forensic archaeology?
3. What are the characteristics of rodent gnawing?
4. What are the different types of carnivore tooth marks, and how are they defined? What other kinds of damage do these species typically cause?
5. Which North American species commonly affect outdoor forensic scenes? Which species commonly affect indoor forensic scenes?
6. What is osteophagia, and why does it occur? What traces may it leave on bone?
7. Why is it difficult to determine which carnivore species gnawed on a given set of remains? What evidence could be used to aid in this determination?

Glossary

Carnassial teeth: teeth that have evolved in carnivores and some other species to shear past each other for slicing soft tissue and bone-like scissors.

Commensal: wild species that live among human habitations; examples include mice and rats.

Crenellated margins: the jagged edges left behind on carnivore-gnawed bone as they splinter away fragments.

Cylinders: hollow tubes of bone created as scavengers remove the proximal and distal epiphyses of a long bone.

Dispersal: the spread of remains and related evidence through multiple taphonomic processes, including scavenger action and water transport.

Edge polish: forms on freshly exposed cortical bone margins that become worn down by repeated tooth wear from additional carnivore gnawing.

Ethology: the study of living animal behavior; this research is one aspect of taphonomy.

Furrows: elongated tooth marks that penetrate through the cortical bone into the cancellous bone or marrow cavity; they are at least three times long as they are wide.

Gastric corrosion: found on small bones and fragments consumed by carnivores, whether retrieved from vomit or feces. It is characterized by the thinning of edges and a sculpted appearance to bone surfaces and may include windowing (see below).

Gripping marks: unpatterned tooth pits and scores often left on long bone shafts as a carnivore shifts the element within its mouth to reposition it for gnawing on the epiphyses or for transport.

Osteophagia: consumption of bone by herbivores (including cattle, deer, etc.) likely in response to dietary deficiencies. The grinding of the bone between their broad cheek teeth can leave a characteristic "Y-fork" shape to long bone ends.

Parallel grooves: tooth marks left by rodents as they gnaw with their paired incisors on dry bone.

Pedestaling: taphonomic formation caused by a rodent gnawing into an epiphysis or other area of thin cortical bone, leaving behind untouched areas of bone that therefore have a pedestaled appearance.

Pits: small tooth marks (usually oval or irregular) that do not penetrate the cortical bone; they are less than three times as long as they are wide.

Provisioning: transport of food by adults to their young. As a taphonomic process, it is one reason why remains' dispersal can be so extensive.

Punctures: tooth marks similar in form to pits but with deeper perforations that do penetrate the cortical bone; they are less than three times as long as they are wide.

Scores: elongated tooth marks that do not penetrate through the cortical bone into the cancellous bone or marrow cavity; they are at least three times long as they are wide.

Taphonomy: the study of entire series of changes that organisms go through as portions of their remains return to the inorganic sphere.

Windowing: the creation of small holes (or expansion of existing foramina) in bones by a variety of taphonomic processes, including gastric corrosion.

References

Absolonova K, Veleminsky P, Dobisikova M, Beran M, and Zocova J. 2013. Histological estimation of age at death from compact bone of burned and unburned human ribs. *J Forensic Sci* 58(S1):S135–S145.

Andrés M, Gidna AO, Yravedra J, and Domínguez-Rodrigo M. 2012. A study of dimensional differences of tooth marks (pits and scores) on bones modified by small and large carnivores. *Acs Anthropol Sci* 4:209–219.

Bass WM. 1983. The occurrence of Japanese trophy skulls in the United States. *J Forensic Sci* 28:800–803.

Behrensmeyer AK. 1978. Taphonomic and ecologic information from bone weathering. *Palaeobiology* 4:150–162.

Behrensmeyer AK. 1991. Terrestrial vertebrate accumulations. In PA Allison, and DEG Briggs (Eds.), *Taphonomy: Releasing the Data Locked in the Fossil Record*. Dordrecht, the Netherlands: Springer, pp. 291–335.

Berryman HE, Bass WM, Symes SA, and Smith OC. 1991. Recognition of cemetery remains in the forensic setting. *J Forensic Sci* 36:230–237.

Berryman HE, Bass WM, Symes SA, and Smith OC. 1997. Recognition of cemetery remains in the forensic setting. In WD Haglund, and MH Sorg (eds.), *Forensic Taphonomy: The Postmortem Fate of Human Remains*. Boca Raton, FL: CRC Press, pp. 165–170.

Binford LR. 1981. *Bones: Ancient Men and Modern Myths*. New York: Academic Press.

Binford LR, Mills MGL, and Stone NM. 1988. Hyena scavenging behavior and its implications for the interpretation of faunal assemblages from FLK 22 (the Zinj Floor) at Olduvai Gorge. *J Anthropol Archaeol* 7:99–135.

Blumenschine RJ. 1986. Carcass consumption sequences and the archaeological distinction of scavenging and hunting. *J Hum Evol* 15:639–659.

Blumenschine RJ. 1988. An experimental model of the timing of hominid and carnivore influence on archaeological bone assemblages. *J Archaeol Sci* 15:483–502.

Blumenschine RJ and Marean CW. 1993. A carnivore's view of archaeological bone assemblage. In J Hudson (Ed.), *From Bones to Behavior*. Center for Archaeological Investigations, Occasional Paper No. 21. Carbondale, IL: Southern Illinois University, pp. 273–299.

Brain CK. 1981. *The Hunters or the Hunted? An Introduction to African Cave Taphonomy*. Chicago, IL: University of Chicago Press.

Byard R, James R, and Gilbert J. 2002. Diagnostic problems associated with cadaveric trauma from animal activity. *Am J Foren Med Path* 23:38–44.

Cáceres I, Esteban-Nadal M, Bennasar M, Monfort MDM, Pesquero MD, and Fernandez-Jalvo Y. 2013. Osteophagia and dental wear in herbivores: Actualistic data and archaeological evidence. *J Archaeol Sci* 40:3105–3116.

Camarós E, Cueto M, Teira LC, Tapia J, Cubas M, Blasco R, Rosell J, and Rivals F. 2013. Large carnivores as taphonomic agents of space modification: An experimental approach with archaeological implications. *J Archaeol Sci* 40:1361–1368.

Carson EA, Stefan VH, and Powell JF. 2000. Skeletal manifestations of bear scavenging. *J Forensic Sci* 45:515–526.

Child AM. 1995. Towards an understanding of the microbial decomposition of archaeological bone in the burial environment. *J Archaeol Sci* 22:165–174.

Coard R. 2007. Ascertaining an agent: Using tooth pit data to determine the carnivore/s responsible for predation in cases of suspected big cat kills in an upland area of Britain. *J Archaeol Sci* 34:1677–1684.

Colard T, Delannoy Y, Naji S, Gosset D, Hartnett K, and Bécart A. 2015. Specific patterns of canine scavenging in indoor settings. *J Forensic Sci* 60:495–500.

Delaney-Rivera C, Plummer TW, Hodgson JA, Forrest F, Hertel F, and Oliver JS. 2009. Pits and pitfalls: Taxonomic variability and patterning in tooth mark dimensions. *J Archaeol Sci* 36:2597–2608.

Duhig C. 2003. Non-forensic remains: The use of forensic archaeology, anthropology, and burial taphonomy. *Sci Justice* 43:211–214.

Domínguez-Rodrigo M and Piqueras A. 2003. The use of tooth pits to identify carnivore taxa in tooth-marked archaeofaunas and their relevance to reconstruct hominid carcass processing behaviors. *J Archaeol Sci* 30:1385–1391.

Dupras TL and Schultz JJ. 2014. Taphonomic bone staining and color changes in forensic contexts. In Pokines JT and Symes SA, (eds.), *Manual of Forensic Taphonomy*. Boca Raton, FL: CRC Press, pp. 315–340.

Duthie AG and Skinner JD. 1986. Osteophagia in the Cape porcupine *Hystrix africaeaustralis*. *S Afr J Zool* 21:316–318.

Efremov IA. 1940. Taphonomy, a new branch of paleontology. *Pan-Am Geologist* 74:81–93.

Einarsen AS. 1956. Determination of some predator species by field signs. *Oreg St Monogr Stud Zool* 10:1–34.

Elbroch M. 2003. *Mammal Tracks & Sign: A Guide to North American Species*. Mechanicsburg, PA: Stackpole Books.

Esteban-Nadal M, Cáceres I, and Fosse P. 2010. Characterization of a current coprogenic sample originated by *Canis lupus* as a tool for identifying a taphonomic agent. *J Archaeol Sci* 37:2959–2970.

Gapert R and Tsokos M. 2013. Anthropological analysis of extensive rodent gnaw marks on a human skull using post-mortem multi-slice computed tomography (pmMSCT). *Forensic Sci Med Path* 9:441–445.

Gidna A, Yravedra J, and Domínguez-Rodrigo M. 2013. A cautionary note on the use of captive carnivores to model wild predator behavior: A comparison of bone modification patterns on long bones by captive and wild lions. *J Archaeol Sci* 40:1903–1910.

Gill JR, Rainwater CW, and Adams BJ. 2009. Santeria and Palo Mayombe: Skulls, mercury, and artifacts. *J Forensic Sci* 54:1458–1462.

Haglund WD. 1992. Contribution of rodents to postmortem artifacts of bone and soft tissues. *J Forensic Sci* 37:1459–1465.

Haglund W. 1997a. Dogs and coyotes: Postmortem involvement with human remains. In W Haglund, and M Sorg M (Eds.), *Forensic Taphonomy: The Postmortem Fate of Human Remains*. Boca Raton, FL: CRC Press, pp. 367–381.

Haglund W. 1997b. Rodents and human remains. In W Haglund, and M Sorg (Eds.), *Forensic Taphonomy: The Postmortem Fate of Human Remains*. Boca Raton, FL: CRC Press, pp. 405–414.

Haglund W, Reay DT, and Swindler DR. 1988. Tooth mark artifacts and survival of bones in animal scavenged human skeletons. *J Forensic Sci* 33:985–997.

Haglund W, Reay DT, and Swindler DR. 1989. Canid scavenging/disarticulation sequence of human remains in the Pacific Northwest. *J Forensic Sci* 34:587–606.

Haglund WD and Sorg MH. 1997a. Introduction to forensic taphonomy research. In W Haglund, and M Sorg (eds.), *Forensic Taphonomy: The Postmortem Fate of Human Remains*. Boca Raton, FL: CRC Press, pp. 1–9.

Haglund WD and Sorg MH. 1997b. Method and theory in forensic taphonomy research. In W Haglund, and M Sorg (Eds.), *Forensic Taphonomy: The Postmortem Fate of Human Remains*. Boca Raton, FL: CRC Press, pp. 13–26.

Haglund WD and MH Sorg. 2002. Human remains in water environments. In WD Haglund, and MH Sorg (Eds.), *Advances in Forensic Taphonomy: Method, Theory, and Archaeological Perspectives*. Boca Raton, FL: CRC Press, pp. 201–218.

Harrison S. 2006. Skull trophies of the Pacific War: Transgressive objects of remembrance. *J Roy Anthropol Inst* (new series) 12:817–836.

Haynes G. 1980. Evidence of carnivore gnawing on Pleistocene and recent mammal bones. *Paleobiology* 6:341–351.

Haynes G. 1982. Utilization and skeletal disturbances of North American prey carcasses. *Arctic* 35:266–281.

Haynes G. 1983. A guide for differentiating mammalian carnivore taxa responsible for gnaw damage to herbivore limb bones. *Paleobiology* 9:164–172.

Higgs N and Pokines JT. 2014. Marine environmental alterations to bone. In JT Pokines, and SA Symes (Eds.), *Manual of Forensic Taphonomy*. Boca Raton, FL: CRC Press, pp. 143–179.

Hill AP. 1989. Bone modification by modern spotted hyenas. In R Bonnichsen, and M Sorg (Eds.), *Bone Modification*. Orono, ME: Center for the Study of First Americans, pp. 169–178.

Hill AP and Behrensmeyer AK. 1984. Disarticulation patterns of some modern East African mammals. *Paleobiology* 10:366–376.

Horwitz L and Smith P. 1988. The effects of striped hyaena activity on human remains. *J Archaeol Sci* 15:471–481.

Hudson J. 1993. The impacts of domestic dogs on bone in forager camps; or, the dog-gone bones. In J Hudson (Ed.) *From Bones to Behavior: Ethnoarchaeological and Experimental Contributions to the Interpretation of Faunal Remains*. Center for Archaeological Investigations, Occasional Paper No. 21. Carbondale, IL: Southern Illinois University, pp. 301–323.

Hughes C, Juarez C, Zephro L, Fowler G, Chacon S. 2012. Past or present? Differentiating California prehistoric Native American remains from forensic cases: An empirical approach. *Int J Osteoarchaeol* 22:110–118.

Hutson JM, Burke CC, Haynes G. 2013. Osteophagia and bone modifications by giraffe and other large ungulates. *J Archaeol Sci* 40:4139–4149.

Jans MME. 2014. Microscopic destruction of bone. In JT Pokines, and Symes SA (Eds.) *Manual of Forensic Taphonomy*. Boca Raton, FL: CRC Press, pp. 19–34.

Jans MME, Nielsen-Marsh CM, Smith CI, Collins MJ, Kars H. 2004. Characterisation of microbial attack on archaeological bone. *J Archaeol Sci* 31:87–95.

Junod CA and Pokines JT. 2014. Subaerial weathering. In JT Pokines, and SA Symes (Eds.) *Manual of Forensic Taphonomy*. Boca Raton, FL: CRC Press, pp. 287–314.

Kerbis Peterhans JC. 1990. *The Roles of Porcupines, Leopards, and Hyenas in Ungulate Carcass Dispersal: Implications for Paleoanthropology*. University of Chicago, Chicago, IL: Unpublished PhD dissertation, Department of Anthropology.

Kibii JM. 2009. Taphonomic aspects of African porcupines (*Hystrix cristata*) in the Kenyan Highlands. *J Taphonomy* 7:21–27.

Kjorlien YP, Beattie OB, Peterson AE. 2009. Scavenging activity can produce predictable patterns in surface skeletal remains scattering: observations and comments from two experiments. *Forensic Sci Int* 188:103–106.

Klippel WE and Synstelien JA. 2007. Rodents as taphonomic agents: Bone gnawing by brown rats and gray squirrels. *J Forensic Sci* 52:765–773.

Lyman RL. 1994. *Vertebrate Taphonomy*. Cambridge, U.K.: Cambridge University Press.

Lyver PB. 2000. Identifying mammalian predators from bite marks: A tool for focusing wildlife protection. *Mammal Rev* 30:31–44.

Murmann DC, Brumit PC, Schrader NA, and Senn DR. 2006. A comparison of animal jaws and bite mark patterns. *J Forensic Sci* 51:846–860.

Nawrocki SP. 1995. Taphonomic processes in historic cemeteries. In A Grauer (Ed.), *Bodies of Evidence: Reconstructing History Through Skeletal Analysis*. New York: Wiley-Liss, pp. 49–66.

Paolello JM and Klales AR. 2014. Contemporary cultural alterations to bone. In JT Pokines, and SA Symes (Eds.) *Manual of Forensic Taphonomy*. Boca Raton, FL: CRC Press, pp. 181–199.

Payne S. 1983. Bones from cave sites: Who ate what? In J Clutton-Brock, and C Grigson (Eds.), *Animals and Archaeology: 1. Hunters and Their Prey*. Oxford, U.K.: BAR International Series 163, pp. 149–162.

Payne S and Munson P. 1989. Ruby and how many squirrels? The destruction of bones by dogs. In NRJ Feller, DD Gilbertson, and NGA Ralph (Eds.), *Paleobiological Investigations: Research Design, Methods, and Data Analysis*. Oxford, U.K.: BAR International Series 266, pp. 31–40.

Pobiner BL. 2008. Paleoecological information from predator tooth marks. *J Taphonomy* 6:373–397.

Pokines JT. 2009. Forensic recoveries of U.S. war dead and the effects of taphonomy and other site altering processes. In DW Steadman (Ed.), *Hard Evidence: Case Studies in Forensic Anthropology*, 2nd edn. Upper Saddle River, NJ: Prentice Hall, pp. 141–154.

Pokines JT. 2014. Faunal dispersal, aggregation, and gnawing damage to bone in terrestrial environments. In JT Pokines and SA Symes (Eds.), *Manual of Forensic Taphonomy*. Boca Raton, FL: CRC Press, pp. 201–248.

Pokines JT. 2015a. A procedure for processing outdoor surface forensic scenes yielding skeletal remains among leaf litter. *J Forensic Ident* 65:161–172.

Pokines JT. 2015b. A Santería/Palo Mayombe ritual cauldron containing a human skull and multiple artifacts recovered in western Massachusetts, U.S.A. *Forensic Sci Int* 248:e1-e7.

Pokines JT. 2015c. Taphonomic characteristics of former anatomical teaching specimens received at a Medical Examiner's office. *J Forensic Ident* 65:173–195.

Pokines JT. 2015d. Taphonomic alterations by the rodent species woodland vole (*Microtus pinetorum*) upon human skeletal remains. *Forensic Sci Int* 257:16–9.

Pokines JT and Baker JE. 2014. Effects of burial environment on osseous remains. In JT Pokines, SA Symes (Eds.) *Manual of Forensic Taphonomy*. Boca Raton, FL: CRC Press, pp. 73–114.

Pokines JT and Higgs N. 2015. Macroscopic taphonomic alterations to human bone in marine environments. *J Forensic Ident* 65:953–984.

Pokines JT and Kerbis Peterhans JC. 2007. Spotted hyena (*Crocuta crocuta*) den use and taphonomy in the Masai Mara National Reserve, Kenya. *J Archaeol Sci* 34:1914–1931.

Pokines JT and Symes SA (Eds.). 2014. *Manual of Forensic Taphonomy*. Boca Raton, FL: CRC Press.

Pokines JT, Zinni DP, and Crowley K. 2016. Taphonomic patterning of cemetery remains received at the Office of the Chief Medical Examiner, Boston, Massachusetts. *J Forensic Sci* 61(S1):S71–S81.

Rogers TL. 2005. Recognition of cemetery remains in a forensic context. *J Forensic Sci* 50:5–11.

Rippley A, Larison NC, Moss KE, Kelly JD, and Bytheway JA. 2012. Scavenging behavior of *Lynx rufus* on human remains during the winter months of Southeast Texas. *J Forensic Sci* 57:699–705.

Roze U. 2009. *The North American Porcupine*, 2nd edn., Ithaca, NY: Cornell University Press.

Schultz JJ. 2012 Determining the forensic significance of skeletal remains. In DC Dirkmaat (Ed.), *A Companion to Forensic Anthropology*. Chichester, U.K.: Wiley-Blackwell, pp. 66–84.

Schultz JJ, Williamson M, Nawrocki SP, Falsetti A, and Warren M. 2003. A taphonomic profile to aid in the recognition of human remains from historic and/or cemetery contexts. *Fl Anthropol* 56:141-147.

Schultz M. 1997. Microscopic investigation of excavated skeletal remains: A contribution to paleopathology and forensic medicine. In WD Haglund and MH Sorg (Eds.), *Forensic Taphonomy: The Postmortem Fate of Human Remains*. Boca Raton, FL: CRC Press, pp. 201–222.

Schultz M. 2001. Paleohistopathology of bone: A new approach to the study of ancient diseases. *Yearb Phys Anthropol* 44:106–147.

Schultz M. 2003. Light microscopic analysis in skeletal paleopathology. In DJ Ortner (Ed.), *Identification of Pathological Conditions in Human Skeletal Remains,* 2nd edn. Amsterdam: Academic Press, pp. 73–109.

Shipman P and Phillips-Conroy JE. 1977. Hominid tool-making versus carnivore scavenging. *Am J Phys Anthropol* 46:77–86.

Sledzik P and Micozzi MS. 1997. Autopsied, embalmed, and preserved human remains: Distinguishing features in forensic and historic contexts. In WD Haglund and MH Sorg (Eds.), *Forensic Taphonomy: The Postmortem Fate of Human Remains*. Boca Raton, FL: CRC Press, pp. 483–495.

Sledzik PS and Ousley S. 1991. Analysis of six Vietnamese trophy skulls. *J Forensic Sci* 36:520–530.

Sorg MH, Dearborn JH, Monahan EI, Ryan HF, Sweeney KG, and David E. 1997. Forensic taphonomy in marine contexts. In WD Haglund and MH Sorg (Eds.), *Forensic Taphonomy: The Postmortem Fate of Human Remains*. Boca Raton, FL: CRC Press, pp. 567–604.

Steadman DW and Worne H. 2007. Canine scavenging of human remains in an indoor setting. *Forensic Sci Int* 173:78–82.

Stiner MC. 1991. Food procurement and transport by human and non-human predators. *J Archaeol Sci* 18:455–482.

Stiner MC. 1994. *Honor Among Thieves*. Princeton, NJ: Princeton University Press.

Stiner MC, Munro ND, and Sanz M. 2012. Carcass damage and digested bone from mountain lions (*Felis concolor*): Implications for carcass persistence on landscapes as a function of prey age. *J Archaeol Sci* 39:886–907.

Tersigni MA. 2007. Frozen human bone: A microscopic investigation. *J Forensic Sci* 51:16–20.

Tsokos M, Matschke J, Gehl A, Koops E, Püschel K. 1999. Skin and soft tissue artifacts due to postmortem damage caused by rodents. *Forensic Sci Int* 104:47–57.

Tsokos M, Schultz F. 1999. Indoor postmortem animal interference by carnivores and rodents: report of two cases and a review of the literature. *Int J Legal Med* 112:115–119.

Ubelaker DH. 1995. Historic cemetery analysis: practical considerations. In A Grauer (Ed.), *Bodies of Evidence: Reconstructing History Through Skeletal Analysis*. New York: Wiley-Liss, pp. 37–48.

Ubelaker D. 1997. Taphonomic applications in forensic anthropology. In W Haglund and M Sorg (Eds.), *Forensic Taphonomy: The Postmortem Fate of Human Remains*. Boca Raton, FL: CRC Press, pp. 77–90.

Weingartner JJ. 1992. Trophies of war: U.S. troops and the mutilation of Japanese war dead, 1941–1945. *Pac Hist Rev* 61:53–67.

Weigelt J. 1989. *Recent Vertebrate Carcasses and Their Paleobiological Implications*, Translation of 1927 edn. Chicago, IL: University of Chicago Press.

Wetli CV and Martinez R. 1983. Brujeria: Manifestations of Palo Mayombe in South Florida. *J Fla Med Assoc* 70:629–634, 634a.

Willey P and Leach P. 2009. The skull on the lawn: Trophies, taphonomy, and forensic anthropology. In DW Steadman (Ed.), *Hard Evidence: Case Studies in Forensic Anthropology*, 2nd edn. Upper Saddle River, NJ: Prentice Hall, pp. 179–189.

Woods CA. 1973. *Erethizon dorsatum*. *Mammalian Species* 29:1–6.

www.namus.gov. 2015. NamUs: National Missing and Unidentified Persons System. Online Document.

www.swganth.org. 2013. Determination of Medico-legal Significance from Suspected Osseous and Dental Remains. Online Document.

Young A, Stillman R, Smith MJ, and Korstjens AH. 2014. An experimental study of vertebrate scavenging behavior in a northwest European woodland context. *J Forensic Sci* 59:1333–1342.

Further reading

Andrews P. 1990. *Owls, Caves, and Fossils: Predation, Preservation, and Accumulation of Small Mammal Bones in Caves, with an Analysis of the Pleistocene Cave Faunas from Westbury-Sub-Mendip, Somerset, U.K.* Chicago: University of Chicago Press.

Bonnichsen R and Sorg MH (Eds.). 1989. *Bone Modification*. Orono, ME: Center for the Study of the First Americans.

Buikstra JE and Ubelaker DH. 1994. *Standards for Data Collection from Human Skeletal Remains*. Fayetteville, AR: Arkansas Archeological Survey Research Report No. 44.

Haglund WD and Sorg MH (Eds.). 1997. *Forensic Taphonomy: The Postmortem Fate of Human Remains*. Boca Raton, FL: CRC Press.

Haglund WD and Sorg MH (Eds.). 2002. *Advances in Forensic Taphonomy: Method, Theory, and Archaeological Perspectives*. Boca Raton, FL: CRC Press.

Hart DL and Sussman RW. 2008. *Man the Hunted: Primates, Predators, and Human Evolution* (rev. Ed.). Cambridge, MA: Westview Press.

Kruuk H. 2002. *Hunter and Hunted: Relationships between Carnivores and People*. Cambridge: Cambridge University Press.

Miller GJ. 1969. A study of cuts, grooves, and other marks on recent and fossil bones. I. Animal tooth marks. *Tebiwa* 12:9–19.

Miller GJ. 1975. A study of cuts, grooves, and other marks on recent and fossil bones: II. Weathering cracks, fractures, splinters, and other similar natural phenomena. In EH Swanson (Ed.) *Lithic Technol*. The Hague: Mouton, pp. 212–226.

Morlan RE. 1984. Toward the definition of criteria for the recognition of artificial bone alterations. *Quaternary Res* 22:160–171.

Section II

The Skeleton and Skeletal Documentation

5 Human Osteology
6 Human Odontology and Dentition in Forensic Anthropology
7 Skeletal Examination and Documentation

CHAPTER 5

Human Osteology

MariaTeresa A. Tersigni-Tarrant and Natalie R. Langley

CONTENTS

Bone as a connective tissue	82
Bone structure	84
Bone types and joints	86
Bone growth	86
Body orientations and anatomical terminology	88
Cranium and mandible	88
Terminology	89
Bones of the skull	91
Frontal bone	91
Parietals	92
Temporals	92
Auditory ossicles	93
Occipital	93
Ethmoid	93
Sphenoid	93
Lacrimals	93
Nasals	94
Vomer	94
Inferior nasal conchae	94
Palatines	94
Zygomatics	94
Maxilla	94
Mandible	95
Postcranial skeleton	95
Axial skeletal elements: hyoid, vertebrae, sacrum, sternum, and ribs	95
Hyoid	95
Vertebrae	95
Sacrum	99
Sternum	99
Ribs	99

Shoulder girdle: Clavicle and scapula	100
Clavicle	100
Scapula	100
Upper limb: Humerus, radius, ulna, carpals, metacarpals, and phalanges	101
Humerus	101
Radius	101
Ulna	101
Carpals	102
Metacarpals	103
Phalanges	103
Pelvic girdle	103
Os coxae (innominates)	103
Lower limb: Femur, patella, tibia, fibula, tarsals, metatarsals, and phalanges	104
Femur	104
Patella	104
Tibia	105
Fibula	105
Tarsals	105
Metatarsals	107
Phalanges	107
Summary	108
Review questions	108
Glossary	108
References	109
Supplementary reading and reference materials	109

LEARNING OBJECTIVES

1. Define human osteology, and explain why it is essential to the practice of forensic anthropology.
2. Discuss the macroscopic and microscopic structures of bone, and identify the main features on a diagram or illustration.
3. Describe the function of each of the three major types of bone cells.
4. Explain the mechanism and cells involved in bone formation or modeling and bone remodeling.
5. Compare and contrast the three structural categories of joints.
6. Discuss the anatomical locations of bones and bone features by using proper anatomical terminology.
7. Identify the cranial bones and the cranial landmarks associated with each bone.
8. Identify side (right or left) and pertinent anatomical features of each skeletal element.

Human osteology is the study of the bones of the human skeleton ("osteo" means "bone," and "ology" means "the study of"). Forensic anthropologists are trained extensively in human osteology. A well-trained osteologist has a thorough understanding of microscopic and macroscopic bone anatomies, bone growth and development, normal and pathological variation, and comparative osteology. To develop skills in comparative osteology, many forensic anthropologists take courses in nonhuman osteology, or **zooarchaeology**. Proper knowledge of human and nonhuman osteology is necessary, because many forensic anthropology cases contain both human and nonhuman skeletal remains. This chapter provides a brief overview of human osteology for study in an introductory level course. It is not an exhaustive treatise on human osteology. Students desiring more detail or advanced study should refer to the list of supplementary reading materials.

BONE AS A CONNECTIVE TISSUE

Bone is a calcified **connective tissue** that supports and protects the soft tissues of the body; provides attachment sites for muscles; produces blood cells; and stores calcium, nutrients, and lipids. Bone tissue is composed of an inorganic component that gives bone its rigidity and an organic component that provides flexibility. The inorganic constituent of bone is a carbonated hydroxyapatite crystalline matrix ($Ca_{10}[PO_4]_6[OH]_2$), and the organic component is primarily **Type I collagen**. Bone is laid down, maintained, and remodeled throughout life by several different types of bone maintenance cells. These cells include **osteoprogenitor cells** (which eventually give rise to osteoblasts), osteoclasts, osteocytes, and bone-lining cells.

Osteoclasts are cells that dissolve (or resorb) bone. Osteoclasts are formed by the fusion of **monocytes** in the red marrow, which is found in spongy bone. This fusion is initiated by the release of parathyroid hormone when the serum calcium concentration in the blood is low. Fusion of the single-nucleated monocytes forms a multinuclear and functionally polarized large cell (i.e., the osteoclast). Osteoclasts have a specialized ruffled acidic border that provides additional surface area for bone resorption. First, the acidic border demineralizes the bone tissue, and then, enzymes dissolve the collagen. Osteoclasts can move quickly and erode as much as tens of micrometers of bone away per day. Osteoclasts live in resorptive bays, or spaces, called Howship's lacunae (Figure 5.1).

Osteoblasts are **mononucleic** bone-building cells that are formed from differentiated mesenchymal osteoprogenitor cells. Mesenchymal cells are embryonic precursor cells, or stem cells, that are capable of differentiating into a variety of cell types, including bone cells, cartilage cells, and fat cells. The formation of osteoblasts is a two- to three-day process, triggered by mechanical or metabolic stress (i.e., heightened serum calcium concentration triggers mesenchymal cells to differentiate into osteoblasts). Osteoblasts produce *osteoid*, which is the unmineralized organic portion of bone matrix, and are responsible for laying down new bones. Osteoid is composed of collagen, noncollagenous proteins, proteoglycans, and water, making it a gel-like substance when deposited. The water is replaced by minerals (hydroxyapatite) as the osteoid mineralizes. Osteoblasts can lay down osteoid at a rate of 1 µm per day during **skeletogenesis** and fracture repair.

Osteocytes are actually osteoblasts that become surrounded by the bone matrix secreted by the osteoblasts. Once the osteoblasts are encased in the bone matrix, they become mature bone cells (osteocytes), and their function is converted from bone production to bone maintenance and communication. Osteocytes reside in pores, or spaces, called lacunae

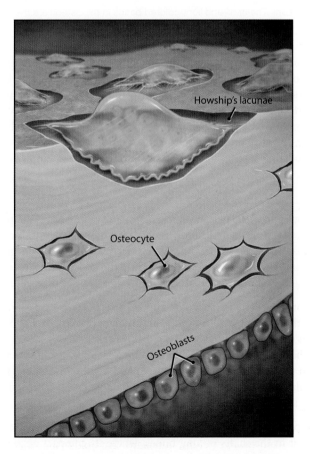

Figure 5.1 Multinucleated osteoclasts in Howship's lacunae. (Courtesy of Jennifer Stowe.)

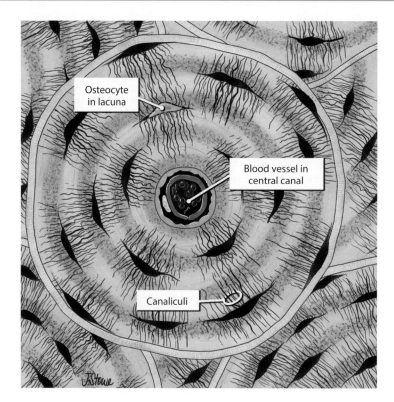

Figure 5.2 Haversian system (osteon). The dark- and light-colored bands in the osteon are the circumferential lamellae. (Courtesy of Jennifer Stowe.)

and communicate with other osteocytes through tentacle-like projections that are housed in canaliculi (little canals) (Figure 5.2). The projections communicate through ion transfer at gap junctions; that is, the tentacles come close enough to each other to pass ions and small nutrients to one another through and positive/negative flow. In this manner, the inside of the bone is connected to the outside of the bone, and the top of the bone is connected to the bottom of the bone, thereby allowing for communication (biochemical reciprocity) throughout the bone. Bone-lining cells are also converted osteoblasts; however, they live on the bone surface and do not become embedded in the osteoid. These are the cells that initiate bone remodeling (resorption and deposition) in response to mechanical stress such as fracturing.

BONE STRUCTURE

Two layers frame the microscopic structure of bone: periosteum and endosteum. The periosteum is an external fibro-cellular sheath that contains collagen fibers and fibroblasts; it forms an **osteogenic**/nourishing layer on the compact bone surface. The periosteum is connected to the underlying bone by bundles of collagen fibers (Sharpey's fibers) that protrude from the periosteum into the outer layers of bone tissue. The endosteum lines the surface of the inner marrow cavity, as well as all Haversian and Volkmann's canals. Endosteum consists of osteogenic cells and thin reticular fibers.

Adult bone consists of two forms of bone: trabecular and compact (Figure 5.3). Trabecular bone, also known as cancellous or spongy bone, is the internal porous bone found in irregular bones, flat bones, and the articular ends of long bones. Trabecular bone is the latticework of bony spicules, or trabeculae, that are found within the shaft and the epiphyses (Ross et al. 1989). The bone matrix is laid down in the form of trabeculae that are usually less than 200 μm thick. In life, the pores within trabecular bone are connected and filled with red marrow. The arrangement of these trabeculae is somewhat irregular, although often in an orthogonal formation. Trabeculae can be visualized as beams of bone that form inner to the scaffolding of bones.

Compact bone, also known as cortical bone, is the dense bone that forms the outer cortex covering the trabecular bone. Cortical bone is most abundant in the shafts of long bones, where its compact arrangement provides much-needed strength (Figure 5.3). The microorganization of this bone consists of many small formations that allow the inside of the

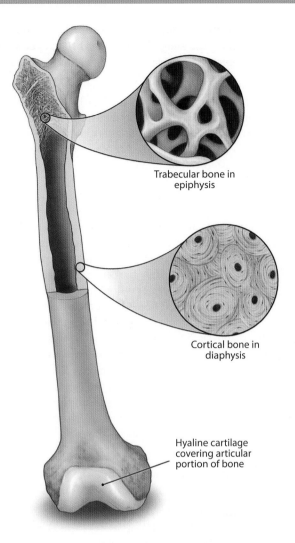

Figure 5.3 Locations of trabecular and cortical bones in a typical long bone. (Courtesy of Jennifer Stowe.)

bone (endosteum) to communicate with the outside of the bone (periosteum). These formations include Haversian canals, Volkmann's canals, and resorptive bays (Howship's lacunae) (Figure 5.2). Haversian canals are narrow trenches aligned with the long axis of the bone that contain nerves and capillaries that allow nutrients to move throughout the entire bone matrix. Volkmann's canals are short transverse canals connecting Haversian canals to one another and allowing communication to the outer bone surface through blood vessels. Howship's lacunae are scoured-out areas in which osteoclasts live during the initiation of bone remodeling (discussed in the previous section). The density/porosity of both types of bone is dynamic and can be affected by pathologies or by mechanical or metabolic stimuli. Thus, trabecular bone can become more compact, and compact bone can become more porous, depending on functional or metabolic demands. Any change in the original porosity of the bone affects the mechanical/structural properties of the bone.

Compact bone can be divided further into primary and secondary bones. Primary bone is the first type of bone to be laid down during development and fracture repair. There are two types of primary bone found within different species: circumferential lamellar bone and plexiform bone. Circumferential lamellar bone (Haversian bone) has lamellae, or layers, that are laid down in a parallel orientation to the bone's surface. Blood vessels are encompassed by these circular lamellae, creating Haversian canals that run longitudinally throughout the bone. These blood vessels are used to move minerals, nutrients, and remodeling cells to the section of bone that needs attention. The entire unit of a circular set of lamellae surrounding a single Haversian canal is called a Haversian system, or osteon (see Figure 5.2). This type of primary bone is found in humans. Plexiform bone is a mixture of both woven and lamellar bones found in some animals. Plexiform bone is often described as having a brick wall pattern, because the vascular structure is rectilinear, thus creating the image of individual bricks of bone separated by the vasculature, as the brick's mortar. This type of bone is typical in large fast-growing animals, such as horses, cows, sheep, and goats.

Secondary bone is formed by the resorption of existing bone and the laying down of new lamellar bone in its place. This process is known as remodeling. Secondary osteons usually consist of a number of consecutive circumferential lamellae that have formed surrounding a Haversian canal that houses blood vessels and nerves. Secondary osteons are surrounded by a cement line, which demarcates where the osteoclasts (bone resorption cells) ended their resorption process. Most of the compact bone in adult humans consists of secondary bone, leaving a characteristic morphology of complete osteons and pieces of old osteons (interstitial lamellae) within a thin section of bone.

BONE TYPES AND JOINTS

The bones of the body are classified into four categories, based on their shape: long bones, short bones, flat bones, and irregular bones. Long bones are longer than they are wide and are found in the arms and legs. Long bones have a diaphysis (shaft) and two epiphyses (articular portions), one at either end of the diaphysis (see Figure 5.3). The diaphysis is composed of dense compact bone covered by periosteum. The center of diaphysis is open and is called the medullary cavity, or intramedullary canal. The medullary cavity contains yellow bone marrow, which is primarily fat. The epiphyses are the ends of the bone and are composed of trabecular bone that contains red marrow. Articular cartilage, composed of **hyaline cartilage**, covers the external surface of each epiphysis and aids in decreasing joint friction. Long bones also have a nutrient foramen located on the diaphysis. The nutrient foramen serves as a site for the passage of blood vessels into the medullary canal. Frequently, the orientation of the nutrient foramen is useful in siding bone fragments. Short bones are composed of trabecular bone covered by a thin layer of compact bone. Short bones are found in the wrists and the ankles (e.g., the carpals and tarsals). Flat bones are composed of two layers of compact bone covering a thin trabecular bone layer (usually referred to as diploë). Examples of flat bones are the bones of the cranial vault and ribs. Irregular bones are bones of various sizes and do not fit into the other categories. They are composed of spongy bone with a thin compact bone covering. The vertebrae and the bones of the pelvic girdle are examples of irregular bones.

Joints are the junction (i.e., articulation) points between two bones. Joints are classified on the basis of structure (primary tissue components) and function (range of movement). The skeleton contains three structural categories of joints: fibrous, cartilaginous, and synovial joints (Figure 5.4). Fibrous joints are joints where the bones are bound together by fibrous connective tissue. Fibrous joints have no joint cavity and do not permit much movement. Examples of fibrous joints include the cranial sutures and the specialized syndesmosis joints between the shafts of the radius and ulna and between the tibia and fibula.

Cartilaginous joints are joints in which the bones are bound together with cartilage. Cartilaginous joints do not have a joint cavity. Cartilaginous joints are divided into primary cartilaginous joints (synchondroses) and secondary cartilaginous joints (symphyses), based on the amount of movement permitted at the joint. The bones united by synchondroses are joined with hyaline cartilage and allow for no movement, except for bone growth, at the site of the joint. The epiphyseal plates of the long bones (also known as growth plates) and the spheno-occipital synchondrosis in the base of the skull are examples of synchondroses. The bones in a symphysis are united by fibrocartilage and allow for a small amount of movement at the joint surface. These include the intervertebral disks and the pubic symphysis.

Synovial (diarthrodial) joints permit a great deal of movement. Synovial joint components include a joint cavity containing a membrane that secretes synovial fluid and articular (hyaline) cartilage covering the bone surfaces invested in the joint. The articular cartilage coupled with the synovial fluid facilitates smooth movement of the bones within the joint. There are six different classifications of synovial joints, based on the type of articulation or the type of movement produced at the articulation: plane, or gliding, joints; hinge joints; pivot joints; condylar joints; saddle joints; and ball-and-socket joints.

BONE GROWTH

The adult body has 206 bones; however, occasionally, an extra vertebra, rib, or additional sesamoid bones are present. However, the subadult skeleton has more than twice as many bony elements (around 450 ossification centers are present at birth). *Growth* is a term used to describe the changes in size and shape, or morphology, that occur as an organism develops and ages. Bones develop *in utero* from cartilage models (i.e., endochondral ossification) and/or vascular membranous

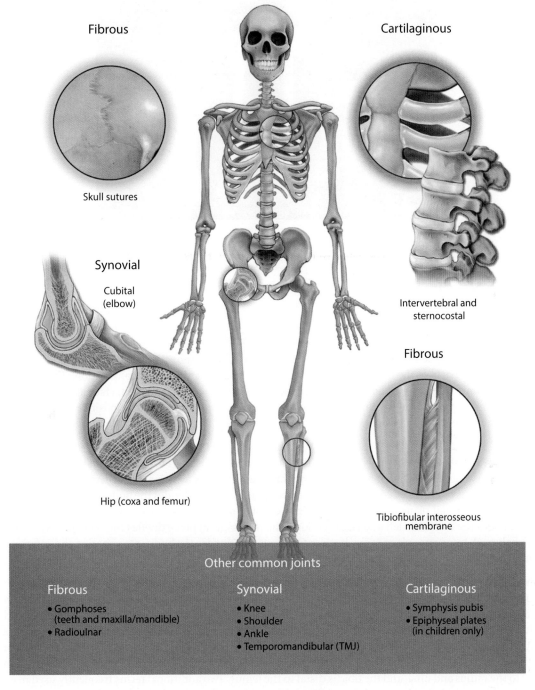

Figure 5.4 Fibrous, cartilaginous, and synovial joints. (Courtesy of Jennifer Stowe.)

templates (i.e., intramembranous ossification). During this developmental process, which is referred to as osteogenesis or ossification, the embryonic precursor tissues are replaced by bones at sites called primary ossification centers. In general, intramembranous ossification produces cortical and diploic bones, and endochondral ossification produces cancellous/trabecular bones (Scheuer and Black 2000). Secondary ossification centers or epiphyses appear after birth and are separated from the primary ossification centers by a layer of cartilage referred to as the epiphyseal plate. The epiphyseal plate attributes to the lengthening of the growing bone. Once bone growth is complete, the primary and secondary ossification centers fuse, the epiphyseal plate disappears, and the bone assumes its adult size and shape (Figure 5.5).

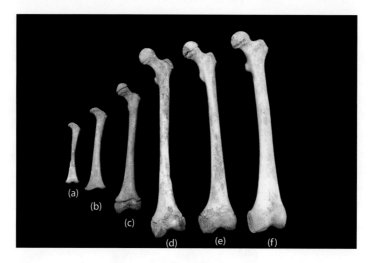

Figure 5.5 Growth and development of human femur: (a and b) Diaphysis without secondary ossification centers (epiphyses). (c) Secondary ossification centers for femoral head and distal end present; (d and e) secondary ossification center for femoral head, distal end, and greater trochanter present; and (f) all epiphyses completely united to diaphysis (growth is complete and bone has assumed adult morphology). (Courtesy of Sandra Cridlin.)

BODY ORIENTATIONS AND ANATOMICAL TERMINOLOGY

Because the human body assumes a number of positions and changes position relative to its environment, references to the body or parts of the body are made with respect to standard anatomical position. The human body is in standard anatomical position when standing erect, with arms by the sides and palms facing forward (Figure 5.6). Standard anatomical position of the skull is referred to as *Frankfurt horizontal*. In this position, the lower margin of the orbit and the upper margin of the external acoustic meatus (ear hole) lie in the same horizontal plane, and that plane is parallel to the ground.

Now that we have established a standard position for the body, let us discuss basic anatomical orientation and directional terminology. Figure 5.6 illustrates three primary reference planes, which are imaginary planes used to divide the body into sections or halves. A coronal plane is any vertical plane that divides the body into anterior and posterior portions; the mid-coronal plane divides the body into equal front and back halves. A sagittal plane is any vertical plane that divides the body into left and right sections; the mid-sagittal plane divides the body into symmetrical left and right halves. A transverse plane is any horizontal plane that divides the body into superior and inferior sections and is perpendicular to the coronal and sagittal planes.

The terms *anterior* and *posterior* refer to front and back, respectively. Sometimes, the terms *ventral* and *dorsal* are also used to refer to front and back (or belly side and back side). *Superior* and *inferior* mean "above/toward the head" and "below/away from the head" and are used to describe the trunk of the body. Similarly, the terms *cranial* and *caudal* mean "toward the head" and "toward the tail," but in human osteology, the terms *superior* and *inferior* are used more frequently than *cranial* and *caudal*. When referring to limb bones, the terms *distal* and *proximal* are used preferentially. *Distal* means "farthest from the trunk," and *proximal* means "nearest the trunk." For example, the humerus is proximal to the radius and ulna, and the tibia is distal to the femur. Two additional terms frequently used in osteology are *medial* and *lateral*. Medial is toward the midline of the body, and lateral is away from the midline of the body.

CRANIUM AND MANDIBLE

The cranium and mandible are collectively referred to as the skull. Humans have always had a certain fascination with the skull, and, perhaps for this reason, it is the single skeletal element that has received the most attention in the research literature. Certainly, the skull is an important element in forensic anthropology, as it can be used to determine sex, ancestry, and age. However, many postcranial elements provide superior estimates of sex and age compared with those offered by the cranium. In addition, some postcranial elements are useful in determining ancestry

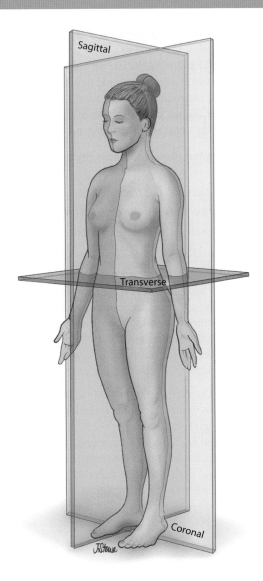

Figure 5.6 Anatomical position and planes. (Courtesy of Jennifer Stowe.)

(specifically the os coxa, femur, and tibia). Typically, the element(s) used to estimate the various parameters of the biological profile are determined simply by the bones that are present and complete for analysis.

As we begin to explore the bones of the skull, it is important to make certain that we understand the common terminology used to discuss its main components. Therefore, the first part of this section will focus on some general terminology. Next, the bones of the skull will be discussed in brief detail. As this is not an osteology text, the discussion of each bone is limited to key anatomical landmarks, including the main sutures, metric, and nonmetric landmarks. Students interested in pursuing a degree in physical anthropology should take a human osteology course that covers in detail various anatomical landmarks, as well as growth and development of each bone of the skull.

Terminology

The skull refers to the complete osseous frame of the head, including the mandible. The mandible refers to the lower jaw. The portion of the skull that contains the brain is often referred to as the neurocranium, while the bony structure of the face is the splanchnocranium. The skull without the mandible is called the cranium. The cranium without the bones of the face is referred to as the calvaria (this refers to all the bones that surround the brain). If the base is absent from the calvaria, the bony portion that remains is referred to as the calotte.

Cranial bones are unique in terms of structure and composition. Particularly, the bones of the cranial vault, or neurocranium, are composed of two layers of cortical bone and a layer of dense spongy bone, referred to as diploe. The diploe is sandwiched between the endocranial and ectocranial cortical bone layers. Its relative density, compared with spongy bone in other areas of the skeleton, is due to the functional demands of the cranial vault (i.e., to provide a rigid, protective structure to house the brain). The neurocranial elements are flat bones, and the joints between these bones are unlike the joints in any other portion of the skeleton. Because of their jagged appearance, these joints are called *sutures*. The major cranial vault sutures are the sagittal, coronal, lambdoidal, and squamosal sutures (Figure 5.9a):

- *Sagittal suture*: It runs along the mid-sagittal plane and forms the articulation between the right and left parietal bones.
- *Coronal suture*: It forms the articulation between the frontal bone and the parietal bones.
- *Lambdoidal suture*: It forms the articulation between the occipital bone and the parietal bones.
- *Squamosal suture*: There are two squamosal sutures (right and left); they form the articulation between the temporal and parietal bones.

The remaining cranial bones are more structurally complex than the neurocranial elements and vary widely in terms of thickness and composition. These bones also articulate at suture sites, many of which derive their names from each of the two articulating elements (i.e., the zygomaticomaxillary suture). In the infant cranium, the gaps between the sutures are large, in order to allow for cranial growth. The sutural junction sites are composed of cartilaginous membranes called fontanelles ("soft spots") that eventually ossify (Figure 5.7). As sutures represent naturally weak portions of the skull, they are frequently the site of fractures in cranial trauma cases; consequently, knowing the names and locations of these features is important for the forensic anthropology practitioner.

Another aspect of cranial terminology that is vital to the forensic anthropology practitioner is cranial landmarks. Cranial landmarks are points on the skeleton that are defined by the intersection of sutures, maximum or minimum curvature, or biological significance. These landmarks serve as the sites from which cranial measurements are taken. These measurements are used in conjunction with discriminant functions or regression equations or with software packages such as the Fordisc program (see Chapter 14) to estimate sex and ancestry. Figure 5.8 shows some of the standard cranial landmarks used in forensic anthropology. Refer to the publications of Langley et al. (2016) and Buikstra and Ubelaker (1994) in the Supplementary Reading and Reference Materials section for detailed descriptions of cranial landmarks and measurements.

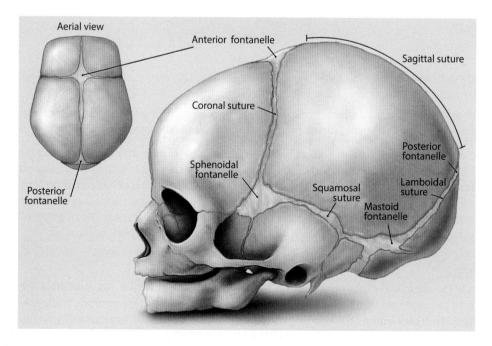

Figure 5.7 Fontanelles in infant cranium. (Courtesy of Jennifer Stowe.)

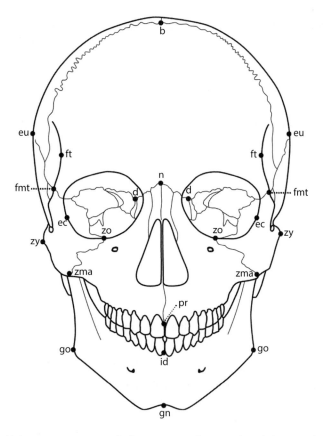

Figure 5.8 Cranial landmarks. b, bregma; eu, euryon; ft, frontotemporale; n, nasion; d, dacryon; fmt, frontomalare temporale; ec, ectoconchion; zy, zygion; al, alare; ns, nasospinale; go, gonion; and gn, gnathion.

Bones of the skull

This section provides a brief description of each of the bones of the skull, including the anatomical location of each bone, the cranial bones with which it articulates, key bony features, and pertinent metric landmarks that lie on the bone or at its borders. Chapter 13 provides an explanation of how cranial landmarks are used in forensic anthropology.

Frontal bone

The frontal bone comprises the anterior portion of the face, including the forehead and superior portion of the eye orbits (represented in pink in Figure 5.9b). The frontal bone articulates with the following bones:

- Right and left parietals (posteriorly) via the coronal suture
- Nasals (inferiorly) via the frontonasal suture
- Maxillae (inferiorly) via the frontomaxillary sutures
- Ethmoid (inferiorly) via the frontoethmoidal sutures
- Lacrimals (inferiorly) via the frontolacrimal sutures
- Zygomatic (inferio-laterally) via the zygomaticofrontal sutures
- Sphenoid (inferio-posteriorly)
- Right and left halves of the frontal fuse in the midline at the metopic suture (occasionally a visible remnant of this suture persists into adulthood)

Some key features of the frontal bone include the supraorbital margins, superciliary arches, temporal lines, and the frontal eminences or bosses. Cranial landmarks used for metric analysis that lie on the frontal or on its borders include bregma, dacryon, frontomalare temporale, frontotemporale, glabella, and nasion.

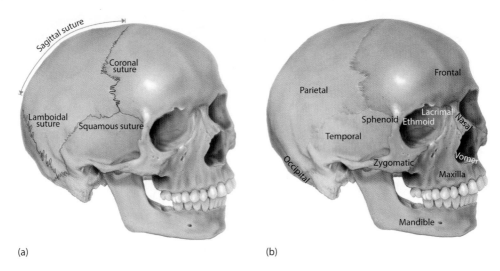

Figure 5.9 (a) Skull sutures and (b) bones. (Courtesy of Jennifer Stowe.)

Parietals

The parietal bones are paired bones (right and left) that make up the sides and top of the skull (depicted purple in Figure 5.9b). The right and left parietal bones articulate with the following bones:

- Both articulate with the frontal bone (anteriorly) via the coronal suture.
- Both articulate with the occipital bone (posteriorly) via the lambdoidal suture.
- Right and left parietals articulate with each other in the midline of the skull at the sagittal suture.
- The right parietal articulates inferiorly with the right temporal via the squamosal suture and the right greater wing of the sphenoid at the sphenoidal angle.
- The left parietal articulates inferiorly with the left temporal via the squamosal suture and the left greater wing of the sphenoid at the sphenoidal angle.

Some key features of the parietal include the temporal lines, parietal foramina, meningeal grooves (endocranially), and parietal bosses. Cranial landmarks used for metric analysis that lie on the parietals or on the parietal borders include bregma, euryon, asterion, and lambda.

Temporals

The right and left temporal bones make up the most inferio-lateral portions of the cranial vault (depicted in light blue in Figure 5.9b). They are the lateral connection of the cranial vault to the cranial base. The right and left temporal bones house the delicate bones of the inner ear, including the incus, malleus, stapes, and the **vestibulocochlear complex**. The right and left temporals articulate with the following bones of the skull:

- Both temporals articulate with the occipital (posteriorly) via the occipitomastoid sutures
- The right temporal articulates with the right parietal (superiorly) via the squamosal suture and posterio-inferiorly via the parietomastoid suture, with the right zygomatic (anterio-laterally) via the zygomaticotemporal suture, and with the right greater wing of the sphenoid (anteriorly) via the sphenotemporal suture
- The left temporal articulates with the left parietal (superiorly) via the squamosal suture and posterio-inferiorly via the parietomastoid suture, with the left zygomatic (anterio-laterally) via the zygomaticotemporal suture, and with the left greater wing of the sphenoid (anteriorly) via the sphenotemporal suture
- The inferior surface of the right and left temporal bones articulate with the corresponding condyle of the mandible at the temporomandibular joint (TMJ)

Some key features of the temporal bones are the temporal squama, external acoustic/auditory meatus (EAM), suprameatal crest, supramastoid crest, zygomatic process, mastoid process, styloid process, temporomandibular articular surface, and petrous pyramid (endocranially). Cranial landmarks that lie on the temporals or on the temporal borders include euryon, mastoidale, porion, and radiculare.

Auditory ossicles

The auditory ossicles are the three small bones found within the petrous portion of the right and left temporals. The auditory ossicles are the incus (anvil), malleus (hammer), and stapes (stirrup). The ossicles are the smallest bones in the body and are typically not recovered in a forensic context.

Occipital

The occipital is the most posterior bone of the skull. A portion of the occipital is visible posterior to the parietal in Figure 5.9b. The occipital articulates with the following bones:

- Right and left parietals (superio-anteriorly) via the lambdoidal suture
- Sphenoid (inferio-anteriorly) via the spheno-occipital synchondrosis (or basilar suture)
- Right and left temporals (antero-laterally) via the occipitomastoid sutures
- First cervical vertebra, or atlas (inferiorly), via the occipital condyles

Key features of the occipital are the foramen magnum, external occipital protuberance, superior and inferior nuchal lines, hypoglossal canals, cruciform eminence (endocranially), and the occipital condyles. Cranial landmarks that lie on the occipital or on its borders include asterion, basion, inion, lambda, opisthocranion, and opisthion.

Ethmoid

The majority of the ethmoid bone is hidden from view when looking at the external skull. The ethmoid is situated between the bony orbits inferior to the frontal in the midline of the skull. The ethmoid articulates with the following bones:

- Frontal (superiorly) via the frontoethmoidal suture
- Sphenoid (posteriorly) via the sphenoethmoidal suture
- Nasals, lacrimals, inferior nasal conchae, and vomer (anteriorly)
- Maxilla and palatines (inferiorly)

Some of the key features of the ethmoid include the cribriform plate, crista galli, and perpendicular plate.

Sphenoid

The sphenoid spans the anterior width of the cranial base and comprises the posterio-inferior surface of the eye orbits (shown in yellow in Figure 5.9b). It is often referred to as the "bat-wing" or "butterfly"-shaped bone. The sphenoid articulates with the following bones:

- Frontal (antero-superiorly) via the sphenofrontal suture
- Occipital (postero-inferiorly) via the spheno-occipital synchondrosis (basilar suture)
- Right and left parietals (posteriorly) at the sphenoidal angle
- Right and left temporals (posteriorly) at the greater wings of the sphenoid via the sphenotemporal suture
- Right and left zygomatics on the anterior surface of the right and left greater wings
- Vomer, ethmoid, palatines (anteriorly), and, sometimes, maxilla

Some key features of the sphenoid include the greater wings, lesser wings, medial and lateral pterygoid plates, and the sella turcica.

Lacrimals

The right and left lacrimals are extremely fragile, thin bones that surround the right and left lacrimal glands (the glands that produce tears), respectively, in the anteromedial portion of each eye orbit. The lacrimal articulates with the following bones:

- Frontal (superiorly)
- Ethmoid (posteriorly)
- Inferior nasal conchae (inferiorly)
- Maxilla (anteriorly)

Nasals

The right and left nasal bones are small, rectangular bones that meet in the midsagittal plane (depicted in purple in Figure 5.9b). The nasal bones form the nasal bridge, and the inferior border of the nasals makes the roof of the bony nasal aperture. The right and left nasals articulate with the following bones:

- Frontal (superiorly) via the frontonasal suture
- Right and left nasals articulate with each other (medially)
- Maxillae (laterally) at the frontal processes of the maxilla

Vomer

The vomer is a thin plate-like bone that articulates inferiorly with the maxillae, creating the postero-inferior portion of the nasal septum that divides the nasal cavity into two halves (visible in orange in the nasal opening; Figure 5.9b). The vomer articulates with the following bones:

- Sphenoid (superiorly)
- Ethmoid (anterosuperiorly)
- Maxillae and palatines (inferiorly)

Key features of the vomer include the perpendicular plate and wings.

Inferior nasal conchae

The right and left inferior nasal conchae are thin-walled bones that lie along the lateral edge of the inferior nasal aperture. The inferior nasal conchae articulate with the following bones:

- Ethmoid and lacrimals (superiorly)
- Maxillae and palatines (inferiorly)

Palatines

The right and left palatines are thin, plate-like bones in the shape of an "L" that articulate with the posterior portion of the maxillae. The palatine bones articulate with the following bones:

- Vomer, inferior conchae, and ethmoid (superiorly)
- Right and left palatines articulate with each other (medially)

Key features of the palatines include the posterior nasal spine and horizontal and perpendicular plates.

Zygomatics

The right and left zygomatics comprise the lateral portion of the anterior face and the inferior portion of the eye orbits (shown in green in Figure 5.9b). The zygomatics are often called the "cheek bones." The zygomatics articulate with the following bones:

- Both articulate with the frontal (supero-laterally) via the zygomaticofrontal sutures
- Sphenoid (posteriorly)
- The right zygomatic articulates with the right temporal (posteriorly) via the zygomaticotemporal suture and the right maxilla (medially) via the zygomaticomaxillary suture
- The left zygomatic articulates with the left temporal (posteriorly) via the zygomaticotemporal suture and the left maxilla (medially) via the zygomaticomaxillary suture

Some key features of the zygomatics include the frontal, temporal, and maxillary processes. Cranial landmarks that lie on the zygomatics or on zygomatic borders include ectoconchion, frontomalare temporale, zygion, zygomaxillare anterior, and zygoorbitale.

Maxilla

The right and left maxillae are a set of paired bones that make up a large portion of the anterior face and hold the maxillary dentition (depicted in blue in Figure 5.9b). The maxilla makes up the medial portion of the eye orbits, the lateral and inferior portions of the nasal aperture, and the anterior part of the hard palate. The right and the left maxillae articulate with the following bones:

- Both articulate with the frontal (superiorly) at the frontomaxillary sutures, with the sphenoid and ethmoid (posteriorly), and with the vomer (anteriorly)
- The right maxilla articulates with the right palatine (posteriorly), the right nasal bone (supero-medially), with the right lacrimal (supero-laterally), with the right inferior nasal concha (infero-medially), and with the right zygomatic (laterally)
- The left maxilla articulates with the left palatine (posteriorly), with the left nasal bone (supero-medially), with the left lacrimal (supero-laterally), with the left inferior nasal concha (infero-medially), and with the left zygomatic (laterally)
- The right and left maxillae articulate with one another medially

Key features of the maxilla are the anterior nasal spine; infraorbital foramen; and the alveolar, zygomatic, frontal, and palatine processes. Cranial landmarks that lie on the maxilla or on its borders include alveolon, ectomolare, prosthion, zygomaxillare anterior, and zygoorbitale.

Mandible

The mandible is also known as the lower jaw and is shown in orange in Figure 5.9b. It articulates with the skull via the right and left temporal bones at the TMJ. The mandible serves as an attachment site for the muscles of mastication and houses the lower dentition. Like the frontal bone, the mandible develops as two separate halves and fuses in the midline (the symphysis).

Some key features of the mandible include the body (corpus), mental protuberance/eminence, ascending ramus, mandibular condyles, coronoid processes, gonial angles, and mental foramina. Cranial landmarks that lie on the mandible include condylion, gnathion, gonion, and infradentale.

POSTCRANIAL SKELETON

The postcranial skeleton is divided into axial and appendicular elements (Figure 5.10). The axial skeleton contains the bones of the trunk, including the ribs, sternum, vertebrae, sacrum, and hyoid. The appendicular skeleton consists of the bones of the limbs and limb girdles. Figure 5.11 illustrates the location and orientation of each postcranial element in the skeleton. More detailed descriptions and additional photographs of osteological features can be found in the reference texts suggested in "Supplementary Reading and Reference Materials."

Axial skeletal elements: hyoid, vertebrae, sacrum, sternum, and ribs

Hyoid

The hyoid is a U-shaped bone found anterior to the cervical vertebrae in the neck. It is one of the few bones in the body that does not have a bony articulation with any other bone. The hyoid has three distinct parts: the body, greater horns, and lesser horns. The body of the hyoid is in the center of the bone. It has multiple sites for muscle attachments, many of which are used during speech. The greater horns project postero-laterally from the body. The lesser horns are found on the superior-lateral aspect of the body. The fusion of the three parts of the hyoid is variable and is not known to be age- or sex-related (Scheuer and Black 2000). The hyoid can be important in a forensic context, as it can show evidence of neck trauma that may occur during strangulation or hanging.

Vertebrae

The *vertebral column* is the term used to identify the 33 bones that house the spinal cord, spanning from the base of the skull to the pelvic girdle. The vertebral column supports the body weight of an erect human, and its unique S shape aligns the trunk over our center of gravity (located at the pelvis). The vertebral column is composed of five different types of vertebrae: cervical, thoracic, lumbar, sacral, and coccygeal vertebrae. The cervical, thoracic, and lumbar vertebrae are found in the posterior midline of the body, and these vertebrae are normally moveable; however, they may fuse in older or pathological individuals and become considerably less mobile (Figure 5.12). The sacral vertebrae are found within the pelvic girdle and are normally fused to one another in an adult, as are the coccygeal vertebrae.

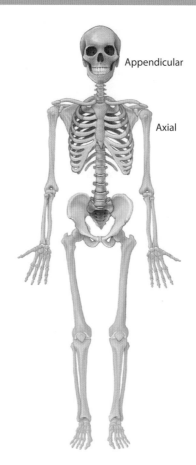

Figure 5.10 Axial and appendicular skeleton. (Courtesy of Jennifer Stowe.)

The cervical, thoracic, and lumbar vertebrae have several common features, including the fact that each of the 24 vertebrae that fall within these groups is separated by intervertebral discs. These fibrocartilage discs aid in shock absorption and allow for movement among the vertebrae. Other components shared by all vertebrae include the vertebral body, superior and inferior articular facets, spinous process, transverse processes, vertebral arch, vertebral foramen, lamina, and pedicle. The vertebral body is the ovoid-shaped portion of the vertebra that lies anterior to the pedicle and provides most of the weight-bearing support of the vertebral column. The vertebral bodies increase in size from superior to inferior, with the cervical bodies being the smallest and the lumbar bodies being the largest. The lateral portion of the thoracic vertebral bodies contains articular facets for rib-head attachment. The superior articular facets are located on the superior surface of each vertebra and articulate with the inferior articular facets from the vertebra situated above. The spinous process is on the posterior surface of each vertebra in the midline of the bone. It serves as the site of ligamentous and muscular attachments and aids the movement of the vertebral column. The transverse processes are located on the lateral surfaces of each vertebra. The transverse processes are attachment sites for muscles and ligaments and aid in movement of the vertebral column. The transverse processes of the thoracic vertebrae have articular facets that serve as the sites of rib-neck attachment. The vertebral arch is the ring of bone that surrounds the spinal cord. The vertebral arch is anterior to the spinous process and posterior to the vertebral body. The vertebral foramen is the space inside the vertebral arch, through which the spinal cord passes. The lamina is a flattened portion of the vertebral arch that attaches the spinous process (posterior) to the pedicle (anterior). The pedicle is the portion of the vertebral arch. It connects the vertebral body (anterior) to the lamina (posterior).

Typically, seven cervical vertebrae sit directly below the skull in the posterior neck region. These cervical vertebrae are designated "C" (for cervical), followed by a number 1–7. They are numbered from superior to inferior, with the most superior cervical vertebra numbered as C1 and the most inferior cervical vertebra numbered as C7. The first and second cervical vertebrae have a slightly different morphology than C3–C7. The atlas (C1) is the cervical vertebra that articulates with the inferior portion of the skull. It is named after the Greek Titan who was sentenced by Zeus

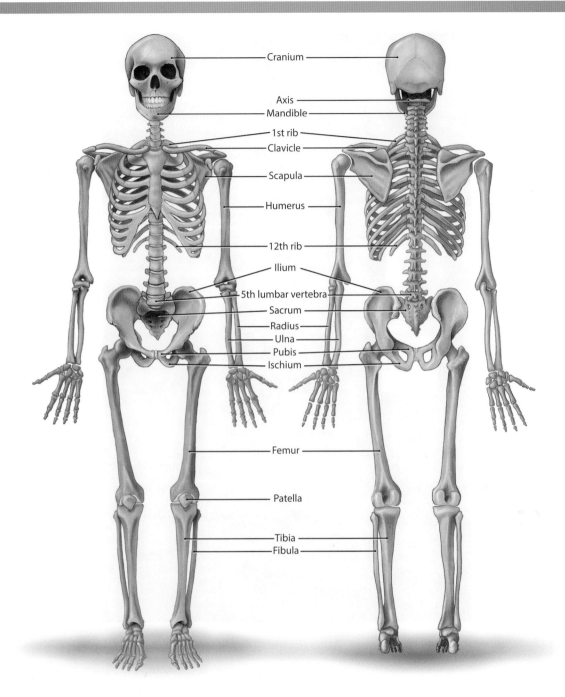

Figure 5.11 Major skeletal elements. (Courtesy of Jennifer Stowe.)

to hold up the sky. The atlas has no body, spinous process, or intervertebral discs on the superior or inferior surfaces. The axis (C2) is the cervical vertebra that articulates inferior to the atlas and superior to C3. The axis does not have a vertebral body, yet its odontoid process (dens) projects superiorly and articulates with the atlas, creating a pivot point for the atlas. When nodding your head "yes," the cranial base pivots on the superior articular surfaces of the atlas. When shaking your head "no," the inferior articular surfaces of the atlas pivot on the superior articular surfaces of the axis. A unique feature of cervical vertebra is that each has transverse foramina located on vertebral arch lateral to the vertebral bodies. In life, the vertebral arteries travel through these foramina into the base of the brain. Cervical transverse processes are small, and the spinous processes are orientated horizontally and are frequently bifurcated at the tip.

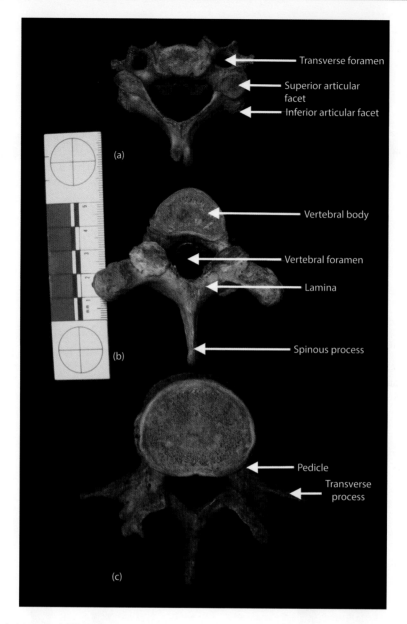

Figure 5.12 Vertebral elements: (a) Typical cervical, (b) thoracic, and (c) lumbar vertebrae.

Differentiating: C1 and C2 are easy to distinguish from the other cervical vertebrae, based on the characteristics mentioned previously. C7 can be distinguished, because it is considered a transition vertebra (i.e., transitional in form between cervical and thoracic vertebrae, as it articulates with the first thoracic vertebra inferiorly). C7 has the largest body of all the cervical vertebrae and a spinous process similar to the thoracic vertebrae.

There are usually 12 thoracic vertebrae in the posterior thorax, though some individuals may have 13. These vertebrae are designated "T" (thoracic), followed by a number 1–12, with the most superior thoracic vertebra designated as T1 and the most inferior designated as T12. Thoracic vertebrae have characteristic costal facets or demifacets (half facets) located on the lateral surfaces of the vertebral bodies and the transverse processes. These allow for the articulation of the head of the rib and the neck of the rib, respectively. The spinous processes of the thoracic vertebrae are long and project inferiorly. When a thoracic vertebra is in proper anatomical position, with the spinous process facing you, the overall morphology is suggested to resemble that of a giraffe.

Differentiating: Several of the thoracic vertebrae can be distinguished from the others based on their costal facet morphology. The first thoracic vertebra has a whole costal facet on the superior aspect of its body and a demifacet on the inferior aspect. T1 also has a spinous process and body similar to that of the cervical vertebrae. T10 has a single costal facet on the superior surface of both sides of the vertebral body and a facet on each transverse process. T11 has a single costal facet on the superior aspect of either side of the vertebral body but no articulation on the transverse processes. T12 resembles T11, but its inferior articular facets are oriented in an antero-lateral position, much like the articular facets of the lumbar vertebrae. Since T12 is a transitional vertebra, it resembles a lumbar vertebra.

The lumbar vertebrae are the largest of the 24 trunk vertebrae. There are usually five lumbar vertebrae, though this can vary, with some people having six lumbar vertebrae, while others having a sacralized fifth lumbar vertebra (L5 is fused to the sacrum). The lumbar vertebrae have large vertebral bodies and do not exhibit costal facets or transverse foramina. The spinous process is large and flat, and the transverse processes are smaller than the transverse processes on thoracic vertebrae. When the lumbar vertebra is in anatomical position, with the spinous process facing you, the overall morphology is said to resemble that of a moose.

Differentiating: The first lumbar vertebra (L1) is the smallest of the lumbar vertebrae, while the fifth lumbar vertebra is the largest, with widely flaring inferior articular facets for articulation with the sacrum.

Sacrum

The sacrum is composed of five sacral vertebrae that begin fusing together in early adolescence but do not complete fusion until postadolescence (S1–S2 are the last segments to complete fusion). The sacrum articulates superiorly with the inferior surface of the fifth lumbar vertebra, laterally with the iliac blades, and inferiorly with the coccyx. The major morphological characteristics of the sacrum include the alae, sacral auricular surfaces (sacroiliac joints), anterior sacral foramina, superior articular facets, and the median spine. The alae, or wings, project laterally and articulate with the iliac blades at the sacral auricular surfaces. The sacral nerves pass through the anterior sacral foramina on the anterior surface of the sacrum. The superior articular facets articulate with the inferior articular facets of the fifth lumbar vertebra. The median spine is a median projection on the posterior surface of the sacrum.

The most inferior portion of the vertebral column is called the coccyx. The coccyx is composed of three to five coccygeal segments that are variably fused. Often, portions of the coccyx are fused to the inferior surface of the sacrum. The *cornua* are the large tubercles on either side of the superior coccygeal body.

Sternum

The sternum, or breast bone, is a midline structure located on the anterior portion of the thorax. It articulates superiorly with the clavicles and laterally with the ribs. The sternum is made up of three components: manubrium, sternal body, and xiphoid process. The manubrium is the thick and square-shaped superior portion of the sternum. The manubrium articulates with the clavicle via the clavicular notches on its superior surface. Between the clavicular notches, in the midline of the superior surface of the manubrium, is the suprasternal (jugular) notch. The sternal body (corpus sterni) is the middle portion of the sternum that is long and thin. The corpus sterni is formed by the fusion of several sternal segments, or sternebrae. The lateral portions of both the manubrium and the sternal body have structures called the costal notches, by which the ribs articulate with the sternum. The xiphoid process is the variably shaped structure at the inferior surface of the sternum. The three portions of the sternum fuse variably, and fusion (or lack thereof) is not known to be related to age or sex. In a small percentage of adults, the sternum has a midline foramen called a sternal foramen. To the untrained eye, this foramen can be mistaken for trauma (Figure 5.13).

Ribs

There are 24 ribs in the human skeleton—12 right ribs and 12 left ribs. Occasionally, there may be 11 or 13 ribs on a side, with the extra ribs typically occurring in the cervical or lumbar area (Scheuer and Black 2000). Ribs 1–10 articulate anteriorly with the sternum and posteriorly with the thoracic vertebrae, while ribs 11 and 12 articulate only with the vertebrae. Ribs 1–7 are referred to as true ribs, because they articulate directly with the sternum via individual cartilaginous attachments. Ribs 8–10 are called false ribs, because they articulate with the sternum via a common cartilaginous attachment. Ribs 11 and 12 are called *floating ribs*, because their medial ends do not articulate with the sternum.

Siding ribs 2–12: The sternal end of the rib is flat, with a U-shaped excavation, and the vertebral end is rounded (this rounded feature is referred to as the rib head); the cranial or superior border of the rib is blunt, and the caudal or

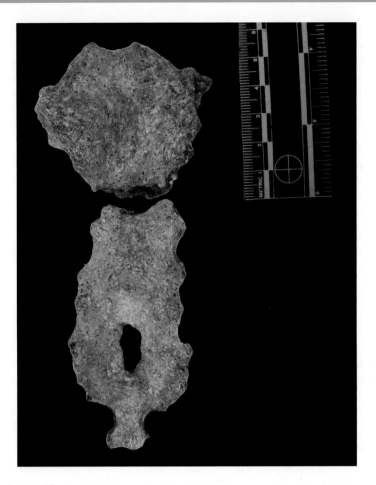

Figure 5.13 Sternum with sternal foramen.

inferior surface is sharp. Rib 1 is flat and unusual, compared with the rest of the ribs. To side the first rib, the head should point inferiorly when the sternal and vertebral ends are properly oriented. Ribs should be X-rayed and examined closely in forensic cases for evidence of sharp trauma and gunshot trauma. Furthermore, the subadult rib cage is frequently a site of healed fractures in child abuse cases.

Shoulder girdle: Clavicle and scapula

Clavicle

The clavicle is a tubular, S-shaped bone that is commonly called the "collar bone." The medial (sternal) end of the clavicle articulates with the sternum, and the lateral (acromial) end articulates with the acromion of the scapula. The clavicle is the first bone to ossify *in utero* and the last bone to fuse (the medial epiphysis completes fusion in the mid-1920s). As such, the clavicle is important in a forensic context in that it provides accurate age estimates into the mid-1920s to late 20s (Black and Scheuer 1996; Langley-Shirley and Jantz 2010; McKern and Stewart 1957; Todd and D'Errico 1928; Webb and Suchey 1985).

Siding: The medial or sternal end is round; the lateral or acromial end is flat; the superior surface is smooth; and the inferior surface is irregular and rough. The medial half of the bone bows anteriorly; the lateral half bows posteriorly; and the acromial tip curves anteriorly.

Scapula

The scapula is a mostly flat, triangular-shaped bone that is located posterior to the rib cage and is commonly referred to as the shoulder blade. The scapula provides muscular attachment sites for the rotator cuff muscles, arm muscles, and many back muscles. The scapula articulates laterally with the head of the humerus and the acromial end of the clavicle. The glenoid cavity or glenoid fossa is the tear-drop-shaped surface on the lateral aspect of the bone, where the humeral head articulates. The scapular spine runs medio-laterally along the posterior surface of the bone and terminates at the lateral end into the enlarged process known as the *acromion* (the articulation site for the lateral aspect of the clavicle). The process protruding opposite of and anterior to the acromion is the coracoid process.

Siding: The smooth featureless surface of the scapula is anterior; the scapular spine is posterior; and the glenoid fossa is lateral. Much like the underlying ribs, the scapula should be inspected carefully for sharp force trauma. However, linear cracking of the thin bone can occur as the bone dries and weathers, and these taphonomic signatures should not be confused with sharp trauma.

Upper limb: Humerus, radius, ulna, carpals, metacarpals, and phalanges

Humerus

The humerus is the largest bone of the upper limb. It articulates proximally with the glenoid fossa of the scapula and distally with the radius and ulna. The humeral head is the rounded proximal feature of the bone that articulates with the glenoid fossa of the scapula in the gleno-humeral joint (a synovial joint). The rugose, roughened area on the lateral portion of the shaft is the deltoid tuberosity—an attachment site for the deltoid muscle of the shoulder. The distal end of the bone has an impression on the posterior surface, called the olecranon fossa. The olecranon fossa accommodates the olecranon of the ulna during full extension of the arm and may be perforated, forming a foramen called a septal aperture. This foramen represents natural variation and should not be confused with perimortem trauma. The distal portion of the humeral shaft flares medially and laterally to form the medial and lateral epicondyles. The medial epicondyle is larger and more prominent than the lateral epicondyle; the epicondyles are attachment sites for many forearm muscles. The distal articular surface of the humerus is spool-shaped. The capitulum is the rounded lateral eminence of the spool that articulates with the head of the radius, and the trochlea is the medial grooved portion of the spool that articulates with the trochlear notch of the ulna.

Siding: The head is medial; the olecranon fossa is posterior; the deltoid tuberosity is lateral; and the nutrient foramen opens distally toward the elbow. The humerus is an important contributor to the biological profile, because several measurements from humerus can be used to estimate sex with a high degree of accuracy (>90%) (Mall et al. 2001).

Radius

The radius is the lateral bone of the forearm. It articulates proximally with the capitulum of the humerus and the radial notch of the ulna and distally with the ulna, lunate, and scaphoid. The proximal end of the radius contains the rounded radial head. Immediately distal to the head is a rounded protuberance called the radial tuberosity or bicipital tuberosity, where the biceps brachii muscle of the arm inserts. The ulnar notch is the concave hollow on the distal and medial surface, where the radius articulates with the neighboring ulna. The styloid process is the pointed projection on the distal and lateral portion of the bone. The dorsal tubercle is the large tuberosity on the posterior aspect of the distal end. The interosseous crest is the sharp margin along the medial portion of the bone shaft, where a tough fibrous membrane (a fibrous syndesmosis joint) attaches to the bone to divide the forearm into anterior and posterior compartments.

Siding: The head is proximal; the styloid process is distal and lateral; the ulnar notch is distal and medial; the interosseous crest and radial tuberosity are medial; the dorsal tubercle is posterior; and the nutrient foramen opens toward the elbow.

Ulna

The ulna is the medial bone of the forearm. It articulates proximally with the trochlea of the humerus and the head of the radius and distally with the ulnar notch of the radius. The proximal articular end of the ulna contains the U-shaped trochlear notch. The most proximal portion of the bone is the large, prominent olecranon process that serves as the insertion site of the triceps brachii muscle. You can feel the olecranon process as your bony elbow. The coronoid process is located opposite of and distal to the olecranon process. The radial notch is the small oval surface medial to the trochlear notch; it receives the head of the radius. The interosseous crest is the sharp medial edge of the shaft. The styloid process is the sharp, most distal projection of the bone.

Siding: The olecranon process is proximal; the coronoid process is proximal and anterior; the styloid process is distal and posterior; the interosseous membrane and radial notch are lateral; and the nutrient foramen opens toward the elbow.

Carpals

The wrist is composed of eight carpal bones. They are oriented in two rows, a proximal row and a distal row. When the bones are in anatomical position, the carpals in the proximal row (lateral to medial) are the scaphoid (navicular), lunate, triquetral, and pisiform. The carpals in the distal row (medial to lateral) are the hamate, capitate, trapezoid (lesser multangular), and trapezium (greater multangular) (Figure 5.14). The saying "so long to pinky, here comes the thumb" is used often to remember the order of the carpals.

- *Scaphoid (navicular)*: It articulates with the lunate and the distal radius, as well as with the capitate, trapezium, and trapezoid. The scaphoid has a *tubercle*, which is the blunt nonarticular portion of the bone. *Siding*: With the concave surface facing toward you and the tubercle curving toward the ground, the tubercle will be on the side it is from.
- *Lunate*: It articulates with the distal radius, scaphoid, capitate, and triquetral, as well as indirectly with the ulna. *Siding*: If the crescent is facing down and the rounded facet points away from you, the flat facet facing you leans toward the side of origin.
- *Triquetral*: It articulates with the lunate, capitate, and pisiform. *Siding*: With the point of the triquetral facing up, the large facet is on the side of origin.
- *Pisiform*: It articulates with only the triquetral and forms within the flexor carpi ulnaris muscle. *Siding*: Place the bulging side of the pisiform away from you, with the round facet for the triquetral pointing down. The protrusion will point toward the side from which the bone comes.
- *Trapezium (greater multangular)*: It articulates with the scaphoid, trapezoid (lesser multangular), and the first metacarpal. *Siding*: With the saddle facet facing upward, the ridge will pass down from the facet and point toward the side of origin.
- *Trapezoid (lesser multangular)*: It articulates with the trapezium (greater multangular), scaphoid, capitate, and second metacarpal. The trapezoid is shaped like a small boot. *Siding*: With the V-shaped ridge of the boot facing you and the sole of the boot on the table, the toe will point toward the side from which it comes.

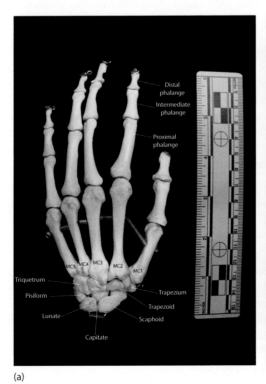
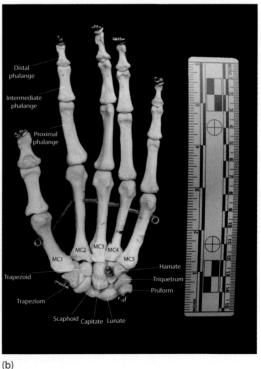

Figure 5.14 (a) Dorsal view of the hand and (b) palmar view of the hand.

- *Capitate*: It articulates with the trapezoid (lesser multangular), scaphoid, lunate, hamate, and the third metacarpal. The *head* of the capitate is the rounded end of the bone, while the *base* is the squared portion of the bone. *Siding*: With the base pointed downward, the head up, and the narrow articular surface facing you, the articular surface is on the side of origin.
- *Hamate*: It articulates with the triquetral, capitates, and the fourth and fifth metacarpals. The *hamulus* of the hamate is the arch-shaped projection that does not articulate with any other bone. *Siding*: With the nonarticular flat surface pointing down, the large articular surface pointing toward you, and the hamulus pointing away from you, the hamulus leans toward the side from which the bone comes.

Metacarpals

The palm of the hand has five metacarpals. These metacarpals are designated MC 1 (thumb) through MC5 (the little finger), each corresponding with the developmental rays of the hand (Figure 5.14). The metacarpals are tube-shaped bones, each with a single distal articular surface that is rounded (head) and a proximal square base. It is this base that allows for differentiating and siding the metacarpals. Metacarpals 2–5 articulate with one another and with the distal row of carpals and the proximal row of phalanges.

- The first metacarpal does not articulate with any of the other metacarpals but does articulate proximally with the trapezium (greater multangular) at a rather large articular facet and distally with the first proximal phalanx. *Siding*: With the dorsal surface facing you, the base of MC1 projects farther on the side from which the bone comes.
- The second metacarpal is usually the longest; MC2 articulates proximally with the trapezoid, capitate, and trapezium; medially with MC3; and distally with the second proximal phalanx. The base has two processes, one pointed and the other more broad. *Siding*: With the dorsal (palmar) surface facing you, the longer process on the base of MC2 points toward MC3. The butterfly-shaped facet is medial and articulates with MC3.
- The third metacarpal has a single, sharp projection (styloid process) at the base. The base of MC3 articulates with the capitate; MC3 articulates laterally with MC2, medially with MC4, and distally with the third proximal phalange. *Siding*: With the dorsal side facing you, the styloid process points toward MC2 and the butterfly-shaped facet points toward and articulates medially with MC4.
- The fourth metacarpal is shorter than the second and third metacarpals. Its very square base articulates proximally with the hamate and sometimes with the capitate. The base also articulates laterally and medially with MC3 and MC5, respectively. The head of MC4 articulates distally with the fourth proximal phalange. *Siding*: With the dorsal side facing you, the single, butterfly-shaped facet is medial and articulates with MC5, while the lateral surface has two small rounded facets that articulate with MC3.
- The fifth metacarpal is the shortest of MC2–MC5. Its base articulates proximally with the hamate and laterally with MC4. The head of MC5 articulates distally with the fifth proximal phalange. *Siding*: The side of the base that does not have an articular surface is medial. The articular surface of the base points laterally toward MC4.

Phalanges

The hand has 14 phalanges: three phalanges on each of the second to fifth fingers and two phalanges on the thumb. Each finger has proximal, middle, and distal phalanges, except for the thumb, which has only proximal and distal phalanges (Figure 5.14). Proximal phalanges have a single concave proximal articular facet for articulation with the head of the metacarpals. Intermediate phalanges have a double concave proximal articular facet. Distal phalanges have a double concave proximal articular facet and a flat nonarticular distal end. The finger phalanges are anteroposteriorly flattened and semicircular in crosssection. The proximal and distal thumb phalanges are readily distinguishable from the other hand phalanges, as the proximal and distal thumb phalanges are shorter and thicker than the other proximal and distal phalanges of the hand. Siding phalanges is nearly impossible, unless the whole hand is present or antemortem radiographs are obtained. Therefore, they are inventoried as the number of proximal, intermediate, and distal phalanges that are present.

Pelvic girdle

Os coxae (innominates)

The hip bones are referred to as the right and left os coxae (plural), meaning "bone of the tail region." They are irregularly shaped and, as such, are also called the *innominates*, meaning "bones with no name." Each os coxa (singular) articulates posteriorly and medially with the sacrum, laterally with the head of the femur, and anteriorly with its counterpart at the pubic symphysis. The os coxa consists of three separate elements in subadults: the ilium, ischium, and pubis. These elements fuse to form a single bone in the adult skeleton. The ilium is the large, laterally situated, blade-like portion of the os coxa. The postero-inferior portion of the ilium contains the U-shaped greater sciatic notch. The posterior portion contains the ear-shaped auricular surface, where the ilium articulates with the sacrum. The ischium is the blunt inferior portion of the os coxa, known as the "sit bones." The hamstring muscles attach to the rugose ischial tuberosity. The pubis is the anterior portion of the os coxa. The pubic symphysis is the oval-shaped joint surface located medially. The superior pubic ramus is the strut of bone that connects the pubis to the ilium, and the ischiopubic ramus is the strut of bone that connects the pubis to the ischium. The obturator foramen is the large foramen formed by the superior pubic ramus and ischiopubic ramus. The acetabulum is the round socket on the lateral surface of the os coxa, where the head of the femur articulates with the pelvis. The acetabulum is formed from portions of the ilium, ischium, and pubis.

Siding: The acetabulum is lateral; the ilium is lateral and superior; the auricular surface is posterior; the pubic symphysis is anterior; and the ischial tuberosity is inferior. Forensically, the os coxa is one of the most important skeletal elements for constructing a biological profile. Many pelvic features are used to determine sex, including the greater sciatic notch, ischiopubic ramus, and obturator foramen, as well as the overall shape of the bone (see Chapter 8). In addition, the pubic symphysis and auricular surface are important features for age estimation (see Chapter 10).

Lower limb: Femur, patella, tibia, fibula, tarsals, metatarsals, and phalanges

Femur

The femur is the longest and heaviest bone in the body. It is commonly called the "thigh bone," and it articulates proximally with the os coxa at the acetabulum and distally with the tibia and patella. The head of the femur lies at the proximal end of the bone and is the rounded portion of the bone that articulates with the acetabulum of the os coxa via a ball-and-socket joint (a synovial diarthrotic joint). The head of the femur can be distinguished easily from the head of the humerus, as the head of the femur is more sphere-shaped, while the head of the humerus is typically hemispheric. The fovea capitis is the small notch found on the medial surface of the femoral head, where the ligamentum teres attaches the femoral head firmly to the acetabulum of the os coxa. The neck of the femur is the strut of bone that attaches the head to the shaft of the femur. The greater trochanter is the large prominence on the lateral surface of the proximal femur that is the insertion site for multiple muscles of the hip and thigh that aid in bipedal locomotion and stabilization. The lesser trochanter is a smaller prominence on the postero-medial surface of the proximal femur near the femoral neck; it is the site of hip flexor muscle attachments. The femoral shaft is the term used to refer to the long tube of bone that extends from the proximal to distal ends of the femur. The linea aspera is an elevated ridge running most of the length of the posterior femoral shaft; it serves as a site of muscle attachment for the powerful adductor muscle group, as well as some of the quadriceps and hamstring muscles. The nutrient foramen is found on the posterior surface of the midshaft near the linea aspera. The distal end of the bone has both condyles and epicondyles. The medial condyle is the large articular surface on the medial side of the distal femur. The medial condyle typically projects more distally than the lateral condyle. The lateral condyle, then, is the articular surface on the lateral side of the distal femur. The medial epicondyle is the projection on the medial surface of the medial condyle and is the site of ligamentous attachment. The lateral epicondyle is a projection on the lateral surface of the lateral condyle and is the site of a ligamentous and muscle attachments. The patellar articular surface is a smooth articular surface on the anterior portion of the distal femur that allows for the movement of the patella over the distal femur during flexion and extension. The lateral portion of this articular surface is raised and projects more anteriorly than the medial aspect to guard against lateral dislocation of the patella during extension.

Siding: The head is proximal and medial, and the condyles are distal. The greater trochanter is lateral. The nutrient foramen opens distally or away from the knee ("flees the knee"). The patellar articular surface is anterior, and the lesser trochanter and linea aspera are posterior. The femur contributes important metric information to help determine the sex, ancestry, and stature components of the biological profile.

Patella

The patella is a large sesamoid bone (a bone that develops and is encased in a tendon). The patella articulates with the distal femur and is often called the knee cap. It lies within the quadriceps tendon and glides smoothly over the knee joint. The patella has three major components: the apex, the medial articular facet, and the lateral articular facet.

The apex of the patella is the pointy nonarticular portion of the patella that points distally. The medial articular facet is on the posterior surface and is smaller than the lateral articular facet. The lateral articular facet is on the posterior surface of the patella and is the larger of the two facets.

Siding: Place the patella on a flat surface, with the articular processes touching the surface and with the apex pointed away from you; the patella will fall to the side from which it comes (since the larger articular surface is lateral).

Tibia

The tibia, or shin bone, is the larger of the two bones in the leg, and it supports the majority of the weight placed on the leg. It lies in the medial position and articulates superiorly with the femur, laterally with the fibula (at the proximal and distal ends), and distally with the talus. The tibial plateau is the proximal surface with which the distal femur articulates. The tibial plateau is divided into the medial condyle and the lateral condyle. The medial condyle is the medial portion of the tibial plateau, while the lateral condyle is the lateral portion of the tibial plateau. The superior fibular articular facet is on the postero-inferior portion of the lateral condyle. The tibial tuberosity is on the anterior surface of the proximal tibia, just inferior to the tibial plateau; the patellar ligament attaches to the tibial tuberosity. The nutrient foramen is on the posterior surface of the bone, proximal to the midshaft. The anterior crest is the sharp portion of the anterior shaft that is commonly referred to as the shin. The interosseous crest is on the lateral portion of the shaft and serves to attach the interosseous membrane that binds the shafts of the tibia and fibula. The medial malleolus is a bulbous projection on the disto-medial surface of the tibia. It forms the medial bulge of the ankle and articulates with the talus. The fibular notch is the most disto-lateral edge of the tibia. This is the site of the ligamentous attachment, creating the distal tibio-fibular syndesmosis. This syndesmosis binds the distal ends of both bones in a tight U-shaped attachment that surrounds the trochlea of the talus. The inferior fibular articular surface is the articular surface of the distal tibia that lies on the inferior portion of the fibular notch.

Siding: The tibial plateau is proximal. The medial malleolus is distal and medial. The fibular notch is distal and lateral. The tibial tuberosity and anterior crest are anterior. The nutrient foramen opens away from the knee ("flees the knee"). The tibia provides the metric information important for estimating stature and sex.

Fibula

The fibula is the lateral bone in the leg. It is a long and skinny bone that is easily broken with laterally applied force or **eversion** of the foot. The fibula articulates proximally with the tibia and distally with the tibia and talus. The head, styloid process, shaft, interosseous crest, malleolar articular surface, malleolar fossa, and the nutrient foramen are the important features of the fibula. The head is the bulbous proximal end of the fibula. It is larger and more rounded than the distal end of the fibula. The head has a proximal projection on the posterior surface, called the styloid process. The shaft of the fibula is a thin tube of bone attaching the proximal and distal ends of the fibula. The interosseous crest is a raised crest of bone that runs the medial length of the shaft. It is the site of the interosseous ligament attachment that connects the shafts of the tibia and fibula and provides surface area for muscle attachment. The malleolar articular surface is the distal articular surface of the fibula. It is a flattened triangular surface that points medially and articulates with the talus. The malleolar fossa lies posterior to the malleolar articular surface and is the site of ligamentous attachments for the ankle joint. The nutrient foramen is found on the posterior surface of the fibula near midshaft.

Siding: The head is proximal, and the malleolar fossa is distal and posterior. The nutrient foramen opens away from the knee ("flees the knee"). If you position the head superiorly, turn the malleolar fossa toward you, and place your thumb in the fossa, your thumb nail will point in the direction from which the bone comes. The fibula can provide accurate stature estimates.

Tarsals

The tarsals form the posterior portion of the foot and the anterior portion of the longitudinal arch. The foot has seven tarsals: calcaneus, talus, navicular, cuboid, medial (first) cuneiform, intermediate (second) cuneiform, and lateral (third) cuneiform (Figure 5.15).

- The calcaneus is the largest tarsal and forms the heel of the foot. The calcaneal or Achilles tendon inserts on the most posterior portion of the calcaneus (the *calcaneal tuber*). The calcaneus articulates superiorly with the talus and anteriorly (distally) with the cuboid. The medial shelf-like projection that supports the head of the talus is

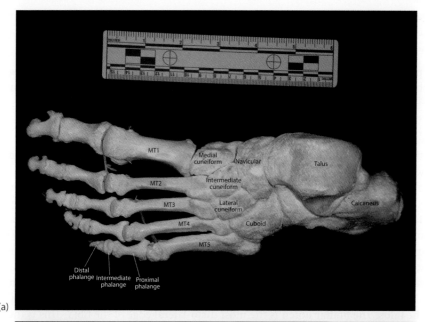

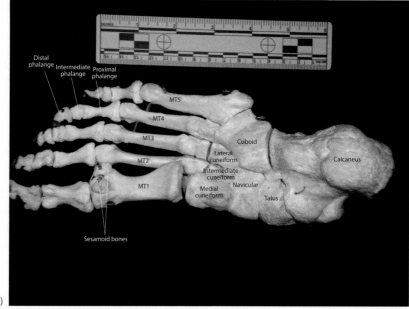

Figure 5.15 (a) Dorsal view of the foot and (b) plantar view of the foot.

referred to as the *sustentaculum tali*. *Siding*: The calcaneal tuber is posterior; the largest articular facet is superior; and the *sustentaculum tali* is medial. When held in anatomical position, the larger superior facet is on the side from which the bone comes.

- The talus is the second largest tarsal bone. It articulates superiorly and medially with the tibia, laterally with the fibula, inferiorly with the calcaneus, and anteriorly with the navicular. The most obvious features of the talus are the saddle-shaped *trochlea* that articulates with the tibia and fibula and the round *head* that articulates with the navicular. *Siding*: The trochlea is superior, and the head is anterior. When viewed from above, the head is on the medial side of the bone and the articular surface for the fibula is on the side from which the bone comes.
- The navicular is a bowl-shaped bone because of the concave proximal surface, where it articulates with the head of the talus. Distally, the navicular articulates with the medial, intermediate, and lateral cuneiforms

via a three-part articular surface. *Siding*: The concave surface is proximal; the three-part surface is distal; and the projecting tubercle is medial.

- The cuboid is located on the lateral side of the foot; it articulates proximally with the calcaneus, medially with the navicular and lateral cuneiform, and distally with the fourth and fifth metatarsals. The cuboid has a large tuberosity on the infero-lateral surface of the bone and is called the *cuboid tuberosity*. *Siding*: Look at the flat nonarticular surface of the bone, with the cuboid tuberosity facing you. The bone will resemble a ballerina's foot, and the toe (the pointed tip) is on the side from which the bone comes.
- The medial (first) cuneiform is the largest cuneiform. It articulates proximally with the navicular, distally with the first metatarsal, and laterally with the intermediate (second) cuneiform. *Siding*: Hold the bone with the point up and the L-shaped articular facet facing you (the L will be upside-down). The tip points to the side from which the bone comes.
- The intermediate (second) cuneiform is the smallest cuneiform. It articulates proximally with the navicular, distally with the second metatarsal, medially with the medial cuneiform, and laterally with the lateral cuneiform. *Siding*: Hold the bone with the pistol-shaped facet facing you. The barrel of the gun points to the side from which the bone comes.
- The lateral (third) cuneiform articulates proximally with the navicular, distally with the third metatarsal, medially with the intermediate cuneiform, and laterally with the cuboid. *Siding*: Hold the bone with the double facet facing you (this facet resembles a butterfly's wing). The narrower side of the bone points to the side from which the bone comes.

Metatarsals

The five metatarsals form the anterior portion of the longitudinal arch and are situated between the tarsals proximally and the phalanges distally (Figure 5.15). Metatarsals have a rounded head distally and a concave or flat base proximally.

- The first metatarsal is the shortest and most robust metatarsal. It articulates distally with the first proximal phalange, proximally with the medial (first) cuneiform, and laterally with the second metatarsal. *Siding*: The concave base is proximal; the round head is distal; the single round portion of the head is superior; the grooved portion of the head is inferior; the flat nonarticular surface is medial; and the convex lateral side with an articular facet for the second metatarsal is lateral.
- The second metatarsal is usually the longest metatarsal. It articulates distally with the second proximal phalange, proximally with the intermediate cuneiform, medially with the first metatarsal, and laterally with the third metatarsal. The base of the second metatarsal is triangular in shape, and the distal shaft has a single medial facet for articulation with the first metatarsal and a double lateral facet for articulation with the third metatarsal.
- The third metatarsal articulates proximally with the lateral cuneiform, distally with the third proximal phalange, medially with the second metatarsal, and laterally with the fourth metatarsal. The base is triangular in shape, and the distal shaft has a double medial facet for articulation with the second metatarsal and a rather large, single lateral facet for articulation with the fourth metatarsal.
- The fourth metatarsal articulates proximally with the cuboid, distally with the fourth proximal phalange, medially with the third metatarsal, and laterally with the fifth metatarsal. The base of the fourth metatarsal is distinctly square and resembles a slice of bread. The distal shaft has single medial and lateral articular facets for the adjacent third and fifth metatarsals.
- The fifth metatarsal articulates proximally with the cuboid, distally with the fifth proximal phalange, and medially with the fourth metatarsal. The distal shaft has a single large medial articular facet for the adjacent fourth metatarsal and a long pointed nonarticular lateral surface.
- *Siding metatarsals 2–5*: The base is proximal; the round head is distal; the single rounded portion of the head is dorsal (superior); and the double rounded portion of the head is plantar (inferior). Held in anatomical position, the base of the metatarsal leans to the side from which the bone comes.

Phalanges

Like the hand, the foot has 14 phalanges: three phalanges on each of the second to fifth toes and two phalanges on the great toe. Each toe consists of proximal, middle, and distal phalanges, except for the great toe, which has only proximal and distal phalanges (Figure 5.15). Proximal phalanges have a single concave proximal articular facet and a head with two rounded protrusions. Intermediate phalanges have a double concave proximal articular facet and a head with two rounded protrusions. Distal phalanges have a double concave articular facet and a flat nonarticular distal end. Proximal toe phalanges can be distinguished from proximal finger phalanges by examining the shape of the shaft: finger phalanges are anteroposteriorly flattened and ovoid in crosssection, while toe phalanges are narrower in the medio-lateral dimension and rounder in crosssection. Intermediate and distal toe phalanges are not likely to be confused with their finger counterparts, because toe phalanges are much smaller. Many times, the small intermediate and distal phalanges of the lateral digits fuse together. Siding phalanges is a nearly impossible task unless the entire foot is present, and even then, it can be extremely challenging. Consequently, phalanges are usually inventoried only as proximal, intermediate, and distal when writing up a forensic case report. Interestingly, foot bones can tell a lot about a person, including preferred shoe type, athletic activities, habitual activities and postures, and health information (i.e., gout and congenital defects).

SUMMARY

This chapter briefly introduces human osteology and provides an overview of histology, joints, and bone morphology. The chapter covers governing principles and overarching concepts, including bone growth and development, bone classifications, bone shapes categories, and general terminology. The descriptions provide a cursory review of the bones of the human body, including key features and bony landmarks, anatomical location relative to other skeletal elements, and tricks for siding skeletal elements. Cranial landmark information and other relevant correlations with forensic anthropology practice are also presented. Forensic anthropologists take advanced courses in human osteology, human growth and development, human gross anatomy, bone histology, zooarchaeology, biomechanics, and bone pathology. Students who wish to enhance their osteology skillset should consider coursework in these areas and are encouraged to consult the Supplementary Reading and Reference Materials section of this chapter.

Review questions

1. What is osteology? Why is it important in the practice of forensic anthropology?
2. Describe the three major types of bone cells. Differentiate between the functions of each.
3. Explain the structure of bone on macroscopic and microscopic levels. Draw and label a picture to accompany your description.
4. Differentiate between periosteum and endosteum.
5. Compare and contrast compact bone and trabecular bone. Give an example of where each type of bone is found in the human skeleton.
6. Explain the difference between primary and secondary bones.
7. Identify the four shape categories of bone, and give one example of a human bone from each category.
8. Compare and contrast the three different structural categories of joints. Give an example of each.
9. Describe the differences between endochondral and intramembranous ossifications.
10. Discuss the relationship of the following structures by using the proper anatomic terminology: skull/pelvis, frontal/occipital, hand phalanges/humerus, femur/tarsals, tarsals/foot phalanges, parietals/sphenoid, cervical vertebrae/hyoid, ulna/radius, fibula/tibia, and pubis/sacrum.
11. Differentiate between the following terms: cranial/postcranial, axial/appendicular, tarsal/carpal, and metatarsal/metacarpal.

Glossary

Connective tissue: one of the four types of tissues in the human bodies. Connective tissues support or connect other tissues. Connective tissues are comprised of specialized cells embedded in an extracellular matrix.

Eversion: a movement of the foot that rotates the plantar surface of the foot laterally, away from the midline of the body.

Hyaline cartilage: transparent cartilage found on many joint surfaces, in the nose, ears, trachea, and larynx. Hyaline is the most abundant of the three types of cartilage in the human body.
Monocytes: the largest of the three major types of white blood cells.
Mononucleic: a cell that has a single nucleus.
Osteogenic: concerned with bone production, growth, or repair.
Osteoprogenitor cells: cells that arise from mesenchymal stem cells in the bone marrow. Osteoprogenitor cells give rise to osteoblasts.
Skeletogenesis: the embryological process of skeleton formation.
Type I collagen: the most abundant form of collagen in body. Type I collagen is found in bone, where it is the structural protein in the extracellular space of this connective tissue.
Vestibulocochlear complex: a series of specialized structures in the inner ear that regulate balance/equilibrium and hearing. The vestibulocochlear complex is innervated by the vestibulocochlear, or auditory, nerve (cranial nerve VIII).
Zooarchaeology: the study of faunal remains, or the remains left by animals after they die.

References

Black S and Scheuer L. 1996. Age changes in the clavicle from the early neonatal period to skeletal maturity. *Int J Osteoarchaeol* 6:425–434.

Langley-Shirley NR and Jantz RL. 2010. A Bayesian approach to age estimation in modern Americans from the clavicle. *J Forensic Sci* 55(3):571–583.

Mall G, Hubig M, Buttner A, Kuznik J, Penning R, and Graw M. 2001. Sex determination and estimation of stature from the longbones of the arm. *Forensic Sci Int* 117(1–2):23–30.

McKern T and Stewart T. 1957. *Skeletal Age Changes in Young American Males Analysed from the Standpoint of Age Identification.* Natick, MA: Quartermaster Research and Development Center, Environmental Protection Research Division. pp. 89–97.

Ross MH, Romrell LJ, and Gordon IK. 1989. Bone In *Histology A Text and Atlas*, 3rd ed., Baltimore, MD: Williams and Wilkins. pp. 150–187.

Scheuer L and Black S. 2000. *Developmental Juvenile Osteology.* New York: Academic Press.

Todd T and D'Errico J. 1928. The clavicular epiphyses. *Am J Anat* 41:25–50.

Webb P and Suchey J. 1985. Epiphyseal union of the anterior iliac crest and medial clavicle in a modern multiracial sample of American males and females. *Am J Phys Anthropol* 68(4):457–466.

Supplementary reading and reference materials

Bass WM. 2005. *Human Osteology: A Laboratory and Field Manual*, 5th edn. Colombia, MO: Missouri Archaeological Society.

Buikstra JE and Ubelaker DH. 1994. *Standards for Data Collection from Human Skeletal Remains.* Fayetteville, AR: Arkansas Archeological Survey.

Langley NR, Meadows Jantz L, Ousley SD, Jantz RL, and Milner GR. 2016. *Data Collection Procedures for Forensic Skeletal Material 2.0.* Knoxville, TN: University of Tennessee Anthropology Department.

Scheuer L and Black S. 2000. *Developmental Juvenile Osteology.* New York: Academic Press.

Scheuer L, Black S, and Schaefer MC. 2008. *Juvenile Osteology: A Laboratory and Field Manual.* New York: Academic Press.

White TD and Folkens PA. 2005. *The Human Bone Manual.* New York: Academic Press.

CHAPTER 6

Human Odontology and Dentition in Forensic Anthropology

Debra Prince Zinni and Kate M. Crowley

CONTENTS

Introduction	111
General overview of dentition in forensic anthropology	112
Tooth anatomy and dental morphology	112
Tooth anatomy	113
Dental numbering systems	115
Dental morphology	116
Common dental anomalies and pathologies	122
Genetic dental anomalies	122
Environmental dental anomalies	122
Summary	123
Review questions	123
Glossary	123
References	124
Suggested readings	124

> **LEARNING OBJECTIVES**
> 1. Explain how forensic anthropologists use the dentition in their practice.
> 2. Describe the nature and location of the dental tissues in the dental-facial complex.
> 3. Construct a dental chart by using the Universal Numbering System.
> 4. Name the types of teeth in the human dentition, and discuss the differences between the permanent and deciduous dentitions.
> 5. Discuss some of the common genetic and environmental anomalies.

INTRODUCTION

This chapter provides a general overview of the role of the dentition in forensic anthropology and a synthesis of tooth anatomy and dental morphology, dental numbering systems, and common dental anomalies and pathologies. Estimating the biological profile for an unknown individual is a crucial part of forensic anthropology, bioarchaeology, and paleodemography. Teeth are often recovered in forensic cases, mass disasters, armed conflicts, and mass graves associated with human rights violations, all of which may produce mass casualties or commingled remains. Owing to its vast postmortem longevity, dentition is an important tool in determining the biological profile and identity of an individual. Please refer to Appendix A for applications of dentition in forensic anthropology (i.e., determining age, sex, and ancestry from dental features).

GENERAL OVERVIEW OF DENTITION IN FORENSIC ANTHROPOLOGY

Dentition offers immense information in the forensic setting, providing important clues about the identity of an unknown individual. Owing to their mineralized composition, teeth are highly resistant to physical and chemical influences and therefore have a vast postmortem longevity. Teeth are the most durable structures in the human body; they are more resilient than bones and thus are often the only human remains recovered from forensic scenes and archaeological sites (Maples 1978; Marcsik et al. 1992; Ohtani 1995; Prince 2004; Prince and Konigsberg 2008).

Physical anthropologists utilize dental features in three main areas: paleontology, skeletal biology, and forensic anthropology. Within paleontology, dental morphology is often utilized to reconstruct phylogenetic relationships (i.e., evolutionary relationships among organisms). Skeletal biologists use dental characteristics to surmise population adaptations to environmental factors, as well as to assess an individual's ability to survive in particular environments. Forensic anthropologists use dentition as a means of positive identification. This chapter focuses on the latter use of dental information and provides an overview of dental morphology and how dentition is utilized in the identification process.

Forensic anthropologists work closely with forensic odontologists, who are licensed dentists applying their knowledge of dentition, dental appliances, and dental work to medico-legal matters, particularly in the realm of positive identification and bite-mark analysis. Forensic odontology is the scientific discipline that utilizes the principles of dentistry to positively identify human remains. Assistance of forensic odontologists is often requested when human remains are decomposed, burned, skeletonized, or beyond the point of clear facial recognition due to severe disfigurement. The use of the human dentition as a means to identify unknown individuals is related to three basic facts about teeth: the survivability of dental evidence, infinite points of dental comparison, and dynamic and accessible patient dental databases (Kessler 1997).

Teeth have an outer enamel layer that makes them the hardest substance in the human body. High temperatures, decomposition, desiccation, or long-term submersion in water does not easily destroy them, which is particularly beneficial in incidences of mass disasters or fragmentation (Kessler 1997). In addition, the materials that dentists use during the restorative phase of treatment, including amalgam, porcelain, gold, and resins, are resilient to exposure to extreme environments (Kessler 1997). Specifically, gold alloys, fused porcelain, synthetic porcelain, and porcelain denture teeth can withstand temperatures greater than 1600°F (Kessler 1997). Consequently, dental evidence has a high survivability and is sometimes the only evidence remaining at a scene.

Another aspect of dentition that makes it useful in a forensic setting is that, much like fingerprints, each person's dentition is unique. The human permanent dentition is typically composed of 32 teeth. Each tooth can be virgin, restored, rotated, fractured, and/or missing. When restored or fractured, any of five different tooth surfaces may be involved for each of the 32 teeth. In addition to restored and fractured surfaces, forensic odontologists can analyze the size, depth, and anatomy of altered tooth surfaces, providing infinite points of dental comparison between antemortem and postmortem dental radiographs (Kessler 1997).

Much of the general population seeks dental care on a regular basis, therefore providing criteria for positive identification in the form of patient records and databases. Dentists commonly update dental radiographs every 12–16 months, creating a dynamic database of individuals' unique and changing dental features (Kessler 1997). These radiographs depict the unique pattern of virgin, restored, or damaged teeth and document the tooth roots, thereby providing an additional data set with which to compare antemortem and postmortem dental radiographs (Kessler 1997).

TOOTH ANATOMY AND DENTAL MORPHOLOGY

The dental-facial complex, often referred to as the *splanchnocranium*, is not a static entity, as it changes with the development of the mandible, face, and dentition. A dynamic relationship exists between the hard and soft palates of the oral cavity. Several factors affect the dental-facial complex: genetic composition, environmental factors (e.g., nutrition and healthcare), and functional factors (e.g., muscle movements). The main function of the dental-facial complex is mastication (chewing); secondary functions include breathing, speech, vision, olfaction, and hearing.

Tooth anatomy

Three groups of tissues comprise the dental-facial complex: skeletal tissues, soft tissues, and dental tissues. Skeletal tissues in the splanchnocranium include bone and cartilage, while soft tissue refers to muscles, glands, nervous, and vascular structures. The five dental tissues are *enamel, dentin, cementum, pulp,* and *gingiva* (Figure 6.1). Teeth are highly mineralized, with the majority of the tooth composed of hydroxyapatite $(Ca_{10}[PO_4]_6[OH])_2$. Enamel is the hardest element in the human body and is composed of approximately 97% hydroxyapatite and 3% water and organic material (Schroeder 1991). Dentin is less mineralized and is composed of approximately 70% hydroxyapatite, calcium, and phosphate; approximately 20% organic materials (primarily proteins); and approximately 10% water (Schroeder 1991). Cementum is composed of roughly 45% hydroxyapatite, 33% proteins (mainly collagen), and 22% water (Schroeder 1991).

The crown and root comprise the gross anatomy of the tooth. The *crown* is covered by enamel and is the portion of the tooth that is exposed in the oral cavity. The *root* is covered by *cementum* and is the portion of the tooth anchored to the alveolar bone in a special joint called a gomphosis. Within the gomphosis, the periodontal ligament anchors the tooth to the alveolar bone of the maxilla and mandible. The alveolar bone between two teeth is called the *interdental septum*, while the alveolar bone between two roots of a single tooth is called the *interradicular septum*. The neck of the tooth refers to the region of the root immediately adjacent to the crown, while the *cementoenamel junction* (*CEJ*) is the point of junction of the crown and the root. The apical foramen is located at the apex of the tooth root and is the opening through which the nerve, blood, and lymphatics supply each tooth.

Enamel is the hardest, most mineralized, and most brittle (i.e., least pliable or ductile) material in the human body and is composed mainly of calcium and phosphorus (Schroeder 1991). Enamel is an acellular material that covers the crown of the tooth (Schroeder 1991) and does not have the capacity to remodel.

Dentin forms the bulk of the tooth and is not normally exposed in the oral cavity. In the crown, enamel covers the coronal dentin, and in the root, a hard substance called cementum covers the root dentin. Attrition, carious lesions, age, and fracture can expose dentin. Microscopically, dentin is composed of intertubular dentin, which is formed during *odontogenesis*, and of dentin tubules, which contain specialized odontoblastic processes that secrete dentin during tooth formation (Schroeder 1991). Peritubular dentin, which lines the inner walls of the dentin tubules, is deposited gradually throughout life (Schroeder 1991). Peritubular dentin has a higher mineralized content than

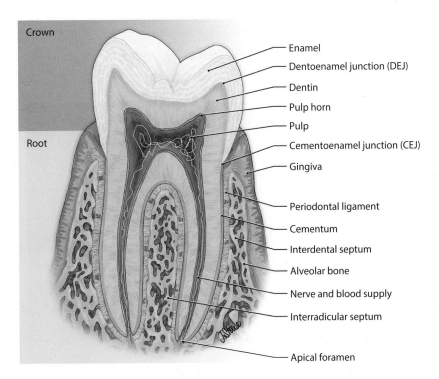

Figure 6.1 Diagram of tooth cross-sectional anatomy.

intertubular dentin, owing to differences in the organic matrices of these compounds (the former is composed of mucopolysaccharides, while the latter is composed of collagen) (Schroeder 1991). Since peritubular dentin is deposited throughout life, the pulp cavity is gradually reduced, as is the diameter of the dentin tubules. Primary dentin consists of all dentin formed until completion of the root, while secondary dentin refers to dentin deposited after the completion of the root (Schroeder 1991).

Cementum is a mineralized, avascular connective tissue that coats the tooth root between the dentin and the periodontal ligament (Schroeder 1991). The primary function of cementum is to anchor the collagen fibers of the periodontal ligament to the tooth, thus anchoring the tooth firmly to the alveolar bone (Schroeder 1991). Cementum also performs adaptive and reparative processes in response to orthopedic forces and trauma to the root (Schroeder 1991). Cementum is laid down in two primary forms, acellular and cellular, and five types of cementum can be distinguished in human teeth: (1) acellular, afibrillar cementum, which does not contain cementocytes or collagen fibers; (2) acellular extrinsic fiber cementum, which is almost exclusively composed of Sharpey's fibers; (3) cellular mixed fiber cementum, which is composed of cementocytes, Sharpey's fibers, and intrinsic bundles of collagen fibers; (4) cellular intrinsic fiber cementum, which is composed of cementocytes and intrinsic bundles of collagen fibers; and (5) intermediate cementum (Kagerer and Grupe 2001; Schroeder 1991). Intermediate cementum is acellular cementum that forms a thin sheath around the entire root. An additional, thicker layer of acellular cementum is located at the cervical (neck) region of the tooth, while cellular cementum is laid down on the remaining one-half to two-thirds of the tooth root (Schroeder 1991).

The pulp, or pulp organ, is the nerve center of the tooth and is centrally located in the pulp chamber of the tooth (Schroeder 1991). The pulp has sensory, reparative, and nutritive properties (Schroeder 1991). Odontoblasts, which are specialized cells that secrete dentin, line the inside of the pulp chamber. As a result, the pulp chamber changes shape with age, becoming smaller as secondary dentin is deposited in the pulp chamber.

The gingivae (gums) are part of the oral mucosa and mirror the alveolar bone and cementoenamel junction. The gingivae anchor the teeth and secure them in the alveolar bone (Schroeder 1991). In addition, the high turnover rate of epithelial and connective tissue components of the gingivae offers a peripheral defense against infection (Schroeder 1991).

Standard dental nomenclature can be found in Figure 6.2, including directional terms, such as *mesial* (toward the midline), *distal* (away from the midline), *lingual* (the tooth surface that is tongue-side), *labial* (lip-side), *buccal* (cheek-side), *incisal* (biting surface of the anterior teeth), and *occlusal* (the biting surface of the posterior teeth).

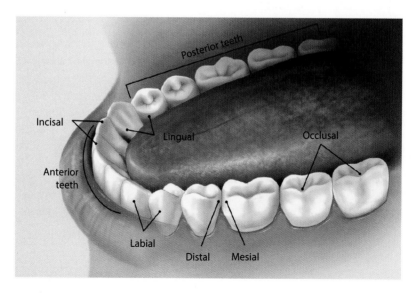

Figure 6.2 Diagram of standard dental nomenclature for tooth surfaces and directional terminology.

DENTAL NUMBERING SYSTEMS

Several notational methods for dental charting have been devised as shorthand methods to quickly identify a tooth without writing out the entire cumbersome anatomic description (Alt and Türp 1998; Hillson 1996; Sopher 1976). Over 30 systems exist today for charting teeth (Clark 1992); however, this chapter presents the two most widely used systems in forensic contexts: the Universal Numbering System (UNS) and the FDI (Fédération Dentaire Internationale) System.*

In 1882, the German Dentist Gustav Julius Parreidt proposed the Universal Numbering System (UNS), which is frequently used today throughout the United States (Figure 6.3). For the permanent (adult) dentition, the UNS assigns a unique number (1–32) to each tooth, starting with the maxillary right third molar (#1) and continuing anteriorly and medially across the maxillary dentition to the right central incisor (#8) and then distally from the left central incisor (#9) to the maxillary left third molar (#16). This numbering system continues with the mandibular left third molar (#17) and continues mesially and anteriorly across the mandibular dentition to the left central incisor (#24), then distally from the right central incisor (#25), and ends with the mandibular right third molar (#32). The deciduous (baby or primary) dentition (Figure 6.4) is labeled similarly, using the letters A–T, beginning with the maxillary right second deciduous molar (A), continuing across the maxillary arcade to the maxillary left second deciduous molar (J), dropping down to the mandibular left second deciduous molar (K), and ending with the mandibular right second deciduous molar (T). Notation for supernumerary teeth

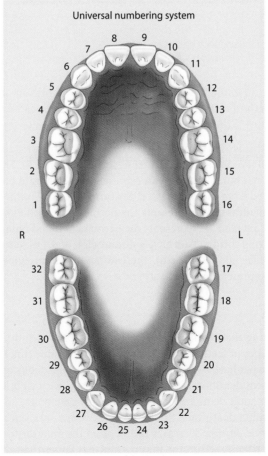

Figure 6.3 Diagram of the UNS for the permanent dentition.

---

* For a discussion on other dental notation systems, please see Hillson (1996) or Alt and Türp (1998).

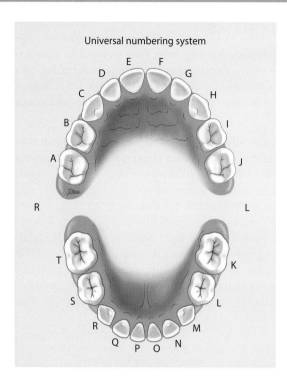

Figure 6.4 Diagram of the UNS for the deciduous dentition.

(additional teeth other than the 32 found in the normal permanent dentition or the 20 found in the deciduous dentition) varies. A common method is to note where the extra tooth is located (e.g., "a supernumerary tooth is located between #8 and #9").

In 1971, the FDI devised a system that is used throughout the world by organizations such as Interpol, the World Health Organization, and the International Association of Dental Research (Figure 6.5). This system provides a unique two-digit number for each tooth. The first number in the pair represents the quadrant of the mouth, and the second number delineates the tooth, numbered from mesial to distal per quadrant. Any number beginning with "1" represents the permanent maxillary right quadrant, "2" represents the permanent maxillary left quadrant, "3" represents the permanent mandibular left quadrant, and "4" represents the permanent mandibular right quadrant. All central incisors have a second digit of 1, while all third molars have a second digit of 8. Deciduous quadrants are delineated with the first numbers 5–8 in the same fashion (Figure 6.6). These notation systems allow for quick entry into a computer database, with a unique number representing each tooth.

DENTAL MORPHOLOGY

Humans are heterodontic, meaning that we have different types of teeth with different functions. Homodonts, on the other hand, have the same types of teeth, all of which perform the same function; many reptiles and bony fish are homodonts. In addition, humans have two separate sets of teeth (deciduous and permanent) and are therefore *diphyodonts*. The human permanent dentition is composed of 32 teeth, with eight teeth per quadrant of the mouth. There are four tooth types in the adult dentition: incisors, canines, premolars, and molars. The dental code, then, for each quadrant of the normal adult dentition is 2:1:2:3, that is, two incisors, one canine, two premolars, and three molars (Figure 6.7). The maxillary dentition decreases in size from the anterior to the posterior teeth, such that the central incisor is larger than the lateral incisor, the first premolar is larger than the second premolar, and the first molar is larger than the second molar, which, in turn, is larger than the third molar. The mandibular dentition exhibits the opposite form, in that the central incisor is smaller than the lateral incisor and the first premolar is smaller than the second premolar; however, the molars follow the same sizing pattern of the maxillary dentition (the first molar is larger than the second, which, in turn, is larger than the third).

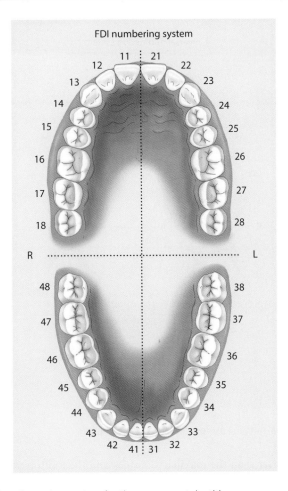

Figure 6.5 Diagram of the FDI dental notation system for the permanent dentition.

The incisors are categorized as central and lateral incisors, according to their location in the dental arcade. Incisors are single-rooted, spatulate-shaped teeth, and their main functions are cutting, tearing, and ripping. The permanent maxillary central incisors (teeth #8 and #9) are the longest and widest anterior teeth. The crown of the maxillary central incisor is wider mesiodistally than labiolingually. To side the maxillary central incisors, the mesioincisal edge exhibits a straight edge, while the distoincisal aspect is rounded. The lingual surface exhibits a concavity, called the lingual fossa, and a convex aspect, called the cingulum. The maxillary lateral incisors (teeth #7 and #10) resemble the maxillary central incisors, but the crowns are smaller in all dimensions. The crown of the maxillary lateral incisor is wider mesiodistally than labiolingually. Siding the maxillary lateral incisors is similar to the maxillary central incisors, in that the mesioincisal edge is straighter than the distoincisal edge.

The mandibular central incisors (teeth #24 and #25) are the smallest teeth in the dental arcade. Each mandibular central incisor is wider labiolingually than mesiodistally and has a symmetric crown. Owing to the symmetry of the crown, siding the mandibular central incisors is often difficult; however, the cingulum is slightly offset to the distal aspect of the tooth. The mandibular lateral incisors (teeth #23 and #26) are slightly larger than the mandibular central incisors and are less symmetric. Therefore, the distoincisal edge is often more rounded than the mesioincisal edge. The cingulum of the mandibular lateral incisor is offset slightly toward the distal aspect of the tooth.

Canines exhibit a somewhat pointed shape that is transitional between the incisors and the premolars. Their main functions are cutting, tearing, and ripping. The maxillary canines (teeth #6 and #11) have a large and centered cingulum and are wider labiolingually than mesiodistally. Typically, the maxillary canines have the longest roots of all dentition. The distal aspect of the maxillary canine is more rounded than the mesial aspect, which aids in siding. In addition, the root often deviates distally. The mandibular canines (teeth #22 and #27) have a longer crown than the maxillary canines;

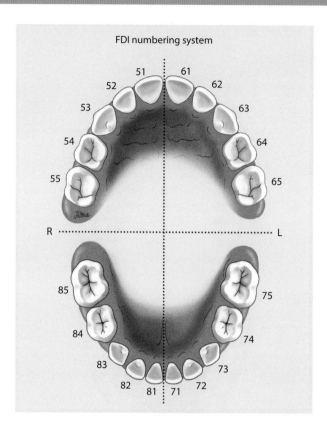

Figure 6.6 Diagram of the FDI dental notation system for deciduous dentition.

however, the root is typically shorter than the maxillary canines and the tooth is smaller mesiodistally and labiolingually. Similar to the maxillary canine, the mandibular canine is rounder distally, and the root often deviates distally.

Premolars are transitional in form and function between the canines and molars. The main functions of the premolars are tearing and grinding. The maxillary first premolars (teeth #5 and #12) have a buccal cusp and a lingual cusp, with the former being the larger of the two. Owing to the buccal and lingual cusps, this tooth is larger buccolingually than mesiodistally. The root of the maxillary first permanent premolar is often bifurcated. The maxillary first premolar also has a prominent mesial marginal groove extending onto the mesial aspect of the tooth (Figure 6.8). In addition, the buccal cusp is offset toward the distal surface. The maxillary second premolars (#4 and #13) are slightly smaller than the maxillary first premolars. The maxillary second premolars also have more symmetrically sized buccal and lingual cusps. As with the maxillary first premolar, the maxillary second premolar is larger buccolingually than mesiodistally.

The mandibular first premolars (#21 and #28) resemble the canines more than the molars in form and function. The buccal cusp is much larger than the lingual cusp in the mandibular first premolar; both cusps are offset to the mesial aspect of the tooth. The mandibular second premolars (#20 and #29) are larger than the mandibular first premolars, and the cusps are more symmetric in size. The mandibular second premolar has one buccal cusp and either one or two lingual cusps. In cases where there are two lingual cusps, the mesiolingual cusp is typically larger; therefore, the lingual groove is offset to the distal aspect. In cases of one lingual cusp, the buccal cusp is larger than the lingual cusp (see Figure 6.9).

The main function of the molars is grinding. The typical morphology of the crown and root of the molars is presented; however, additional cusps and roots are exhibited in some specimens. The maxillary molars have three roots, two of which are set buccally (one mesial and one distal) and the third is lingual. The lingual root is typically the longest root, followed by the mesiobuccal root. The crowns of the molars are wider buccolingually than mesiodistally. The maxillary first molars (#3 and #14) are typically the largest teeth in the dental arcade. The maxillary first molar usually consists of four cusps: mesiobuccal, distobuccal, mesiolingual, and distolingual. The mesiolingual cusp is typically the largest cusp; the mesiobuccal cusp is the next largest; and the distolingual cusp is the smallest. Although

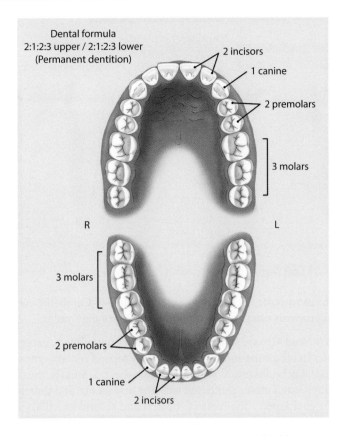

Figure 6.7 Diagram of the human dental formula (by quadrant) for the permanent dentition.

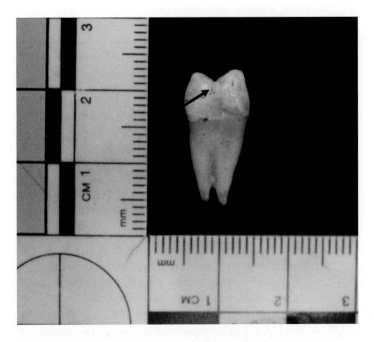

Figure 6.8 Mesial surface of tooth #5 depicting the mesial marginal groove and bifurcated root.

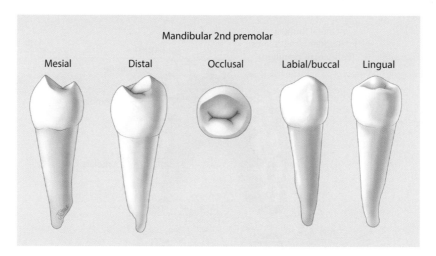

Figure 6.9   Diagram of mandibular right second premolar (tooth #29) viewed from each of the five dental surfaces.

the maxillary first molar is the most static, an accessory cusp (the cusp of Carabelli) may be located on the mesiolingual cusp. A mesial marginal groove is usually present on the maxillary first molar.

The maxillary second molars (#2 and #15) are similar to the maxillary first molars, except that the crowns are smaller. The distobuccal cusp is smaller in all dimensions in the maxillary second molar compared with the first molar, and therefore, the distolingual cusp may be observed from the buccal aspect. Similarly, the distolingual cusp is smaller in all dimensions in the maxillary second molar compared with the first molar, and therefore, the distobuccal cusp can often be observed from the lingual aspect. The roots of the first and second maxillary molars are similar in length. The buccal roots of the second maxillary molar are closer together and equal in length compared with the roots of the maxillary first molar, so it is not uncommon for the buccal roots to be fused.

The maxillary third molars (#1 and #16) are similar to the maxillary second molars, except that they are smaller in all dimensions. The roots are shorter and often fused together. The distolingual cusp is small and frequently missing. Maxillary and mandibular third molars are often congenitally absent or impacted and are the most variable of all teeth.

The mandibular molars have two roots, a mesial root and a distal root. On the crown, the mandibular molars are wider mesiodistally than buccolingually. Whereas maxillary first molars have four cusps, mandibular first molars (#19 and #30) typically exhibit five cusps: mesiobuccal, distobuccal, mesiolingual, distolingual, and distal cusps. The mesiobuccal cusp is the largest of the cusps, while the distal cusp is the smallest. The mesiolingual and distolingual cusps are similar in size, with the former exhibiting slightly larger proportions. The distobuccal cusp is the second smallest cusp. The roots of the mandibular first molar deviate distally.

The mandibular second molars (#18 and #31) are the most symmetric of all the molars and have four cusps arranged in a "+" pattern: mesiobuccal, distobuccal, mesiolingual, and distolingual. The cusps of the mandibular second molar are more equal in size than those of the mandibular first molar; however, the mesiobuccal cusp is normally the largest, while the distolingual cusp is the smallest. The roots of the mandibular second molar angle to the distal aspect and are sometimes fused. The mandibular third molars (#17 and #32) can have five cusps, such as in the mandibular first molars, or four cusps, such as in the mandibular second molars. The roots of the mandibular third molars are often fused and typically shorter than the roots of the first and second molars.

There are 20 deciduous teeth in the human dentition, with five teeth per quadrant: two incisors, one canine, and two molars. Therefore, the dental code for each quadrant of deciduous teeth is 2:1:0:2, as there are no premolars in the deciduous dentition (Figure 6.10). The deciduous teeth are smaller in all respects compared with their permanent counterparts. Since the deciduous dentition is replaced by the permanent dentition approximately by the age of 12 years, less developmental emphasis is placed on building highly mineralized deciduous teeth. As such, the deciduous dentition is not as strong as the permanent dentition, and the enamel is neither as mineralized nor as thick, owing to differences in diet and facial musculature. On account of these mineralization differences, the deciduous dentition exhibits a grayish hue. The mixed dentition period (the stage during which deciduous and permanent teeth are present in the oral cavity

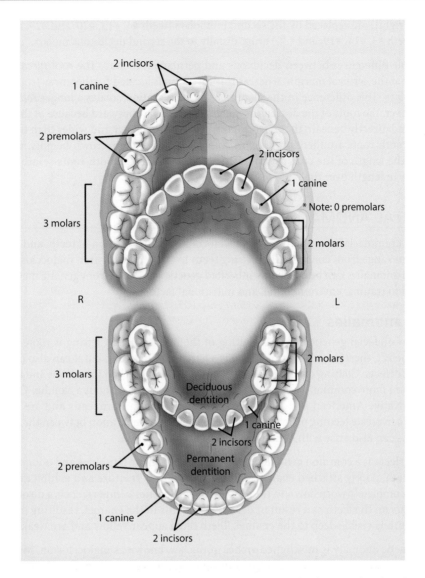

Figure 6.10 Diagram of the human dental formula (by quadrant) of the deciduous dentition overlayed on the permanent dentition.

at the same time) typically occurs between six years and 12 years of age. The permanent incisors replace the deciduous incisors; the permanent canines replace the deciduous canines; and the permanent premolars replace the deciduous molars. The permanent molars erupt distal to the deciduous second molars. The first permanent molar erupts approximately at the age of 6 years, the second molar at the age of 12 years, and the third molar at the age of 18 years.

The deciduous maxillary central incisors (teeth E [right] and F [left]) are located on either side of the midline in the upper dental arcade and are replaced by the permanent maxillary central incisors (teeth #8 and #9, respectively). The deciduous mandibular central incisors (teeth O [left] and P [right]) are located on either side of the midline in the lower dental arcade and are replaced by the permanent mandibular central incisors (teeth #24 and #25, respectively). The deciduous maxillary lateral incisors (teeth D [right] and G [left]) and the mandibular lateral incisors (teeth N [left] and Q [right]) are located distal to the deciduous central incisors and are replaced by the permanent lateral incisors (teeth #7, #10, #23, and #26, respectively). Moving distally in the dental arcade, the deciduous maxillary canines (tooth C [right] and H [left]) and deciduous mandibular canines (M [left] and R [right] are replaced by the permanent canines (teeth #6, #11, #22, and #27, respectively). The first deciduous maxillary molars (teeth B [right] and I [left]) and mandibular molars (teeth L [left] and S [right]) are replaced by the permanent first premolars (teeth #5, #12, #21, and #28, respectively). The deciduous second maxillary molars (teeth A [right] and J [left]) and mandibular molars (teeth K [left] and T [right]) are replaced by the second premolars (teeth #4, #13, #20, and #29, respectively). The first permanent molars (teeth #3, #14, #19, and #30) erupt distally to the second deciduous molars.

There are a few notable differences between deciduous and permanent molars. The root *furcation* of the deciduous molars occurs almost at the cervical margin, whereas this furcation is located further away from the cervical margin in the permanent molars. This difference in the placement of the furcation creates a longer *root trunk* in the permanent dentition. Moreover, the roots of the deciduous molars flare or splay outward because of the forming permanent premolar crown located directly beneath them. Deciduous crowns are shorter and wider than their permanent counterparts. Deciduous tooth roots are therefore long in comparison with their crown heights, with roots measuring approximately twice the length of the crowns; however, because deciduous tooth roots resorb during the shedding process, they decrease in length over time.

COMMON DENTAL ANOMALIES AND PATHOLOGIES

Dental anomalies are craniofacial abnormalities of facial form, function, position of teeth, and tissues of the jaw and mouth. Anomalies can range from congenitally missing teeth to stained teeth, arch malocclusions, and cleft palate. The etiology of dental anomalies can be caused by inherited genetic defects or from genetic mutations. Dental deformities are also related to trauma, environmental, and nutritional factors.

Genetic dental anomalies

The most common craniofacial genetic defect is clefting of the lip or palate. Clefting is more prevalent in families with histories of the defect, therefore suggesting that it is a hereditary disorder, but it can also occur as part of a syndrome or sequence of defects. Clefting of the lip is more common in males than in females and is typically unilateral, whereas palate clefts are more common in females and may occur with or without a cleft lip. Cleft palate and lip are twice as common in Native American and Asian groups compared with Europeans and are most rare in African populations. Clefting can create feeding problems because of the lack of separation between the nasopharynx and the oropharynx. Problems can also arise with dental development and eruption.

Dentinogenesis imperfecta is a common genetic defect involving the dentin tubules and odontoblasts that results in severely discolored teeth, poorly attached enamel, and teeth that easily fracture and exhibit extreme attrition, with the crown becoming completely worn down to the gum line. Amelogenesis imperfecta is a disorder that is correlated with thin enamel layers on the teeth as a result of defective proteins in the enamel, resulting in crown translucency so severe that the dentin is visible deep to the enamel. Teeth often appear yellow and are weak and cavity-prone.

A less debilitating genetic anomaly is misaligned arches of the jaw, known as malocclusion. Most malocclusions are inherited; however, they can be caused by accidents (i.e., traumatic loss of teeth) and individual habits, such as prolonged thumb sucking. Malocclusion and tooth crowding are increasing because of evolutionary factors affecting the modern human dentition. Teeth are not downsizing at the same pace as the soft tissues. Therefore, rates of malocclusion and crowding are on the rise, along with the number of corrective orthodontic procedures performed to change appearance or alleviate associated physical discomfort. Other dental defects include microdontia, an inherited condition that causes smaller than normal teeth, and anodontia, a dental condition in which one or more permanent teeth fail to develop and are missing in the adult dentition.

Environmental dental anomalies

Although teeth are very durable after death, they are susceptible to environmental factors during life. Environmental factors that may cause dental anomalies include the effects of diet/nutrition and agents such as fluoride and antibiotics. For example, a lack of vitamin C can cause scurvy, which affects the periodontal and connective tissues of the individual, resulting in red, inflamed, and bleeding gums. A lack of iron, such as in anemia, can cause fiery red gums, whereas fluorosis, defined as an overabundance of fluorine in the diet, creates white or "mottled" spots on teeth. Caries (i.e., cavities), the most common dental pathology, are the result of bacterial plaque coming in contact with the tooth surface. Acid produced from the bacteria attacks the surface of the tooth and penetrates the enamel or dentin, causing destruction to the tooth and, in severe cases, infection of the alveolar bone in the form of an abscess. Likewise, habits such as tobacco use and eating caustic foods (e.g., acidic fruits and sugary foods) can affect the condition of the dentition and provide clues about lifestyle.

SUMMARY

This chapter provides a general overview of dentition in forensic anthropology, tooth anatomy, dental morphology, dental numbering systems, and common dental anomalies and pathologies. Dentition offers immense information that can be utilized by forensic anthropologists and odontologists presented with unknown remains. Parameters of the biological profile such as age, sex, and ancestry can be obtained through analysis of dentition, as well as information pertaining to the individual's life history, such as diet, disease, habits, and nutrition. Most importantly, forensic odontologists use dentition to positively identify unknown individuals through comparison of antemortem and postmortem dental records. Owing to the survivability of the dental evidence, the infinite number of dental comparisons, and dynamic and accessible patient dental databases, the human dentition is an essential investigative tool in the medico-legal setting.

Review questions

1. What is the hardest element in the human body? Describe a developmental anomaly involving this element or substance.
2. What are the five dental tissues found in the dental-facial complex? Explain the location and/or distribution of these tissues around the surfaces and features of a tooth. (You may choose to draw the answer to this question.)
3. Which dental numbering system is most frequently used in the United States? Draw this dental system for the permanent and deciduous dentitions, labeling each tooth with its full name and its abbreviation.
4. Name the four tooth types in the permanent dentition. Explain the differences between the permanent and deciduous dentitions.
5. What tooth type is not represented in the deciduous dentition? Describe the eruption sequence of the permanent dentition.
6. What is the dental code or formula for the permanent dentition?
7. What is the most common craniofacial genetic defect, and what problems are inherent with this defect?
8. Which are the two features found in the deciduous molars that distinguish them from the permanent molars?

Glossary

Buccal: next to or toward the cheek.
Cementoenamel junction (CEJ): the junction between the enamel and cementum on a tooth; also known as the cervical line. The CEJ separates the enamel of the crown from the cementum of the tooth root.
Cementum: connective tissue forming the external layer of the tooth root.
Cervix: a narrow or constricted portion of a tooth in the region of the junction of the crown and root; also known as the "neck" of the tooth.
Crown: the portion of the tooth that is covered with enamel and normally is visible in the oral cavity.
Dentin: hard, yellowish tissue underlying the enamel and cementum; dentin constitutes the majority of the tooth crown and root.
Diphyodonts: an animal with two successive sets of teeth (deciduous and permanent).
Distal: away from the midline.
Enamel: white, external surface of the crown of the tooth; hardest substance in the human body.
Furcation: location on multirooted teeth where the root trunk divides into separate roots.
Gingiva: soft tissue that covers the alveolar process of the maxilla and mandible; surrounds the cervical portion of the tooth.
Incisal: the biting surface of the incisors (equates to the occlusal surface of the posterior teeth).
Interdental septum: the alveolar bone between two teeth.
Interradicular septum: the alveolar bone between two or more roots of a single tooth.
Labial: next to or toward the lip; also referred to as the facial surface.
Lingual: next to or toward the tongue.

Mesial: towards the midline.
Occlusal: the biting surface of a posterior tooth.
Odonotogenesis: the formation or development of teeth.
Pulp: soft tissue located in the center of the tooth crown and root.
Root: the portion of the tooth embedded in the alveolar process (bone) and covered with cementum.
Root trunk: the portion of a multirooted tooth between the cervical line and the bifurcation or trifurcation of the separate roots.
Splanchnocranium: the facial skeleton.

References

Alt KW and Türp JC. 1998. Roll Call: Thirty-two white horses on a red field. The advantages of the FDI two-digit system of designating teeth. In KW Alt, FW Rösing, and M Teschler-Nicola (Eds.), *Dental Anthropology: Fundamentals, Limits, and Prospects*. New York: Springer, pp. 41–55.

Clark DH (Ed.) 1992. *Practical Forensic Odontology*. Oxford, U.K.: Butterworth-Heinemann.

Fédération Dentaire Internationale (FDI). 1971. Two-digit system of designating teeth. *Int Dent J* 21:104–106.

Hillson S. 1996. *Dental Anthropology*. Cambridge, U.K.: Cambridge University Press.

Kagerer P and Grupe G. 2001. Age-at-death diagnosis and determination of life-history parameters by incremental lines in human dental cementum as an identification aid. *Forensic Science International*, 118:75–82.

Kessler HP. 1997. An overview of forensic dentistry. In HP Kessler and CW Pemble (Eds.), *33rd Annual Course Forensic Dentistry*, Bethesda, MD, March 3–8.

Maples WR. 1978. An improved technique using dental histology for estimation of adult age. *J Forensic Sci* 23:764–770.

Marcsik A, Kósa F, and Kocsis G. 1992. The possibility of age determination on the basis of dental transparency in historical anthropology. In P Smith and E Tchernov (Eds.), *Structure, Function and Evolution of Teeth*. London, U.K.: Freud Publishing House, pp. 527–538.

Ohtani S. 1995. Studies on age estimation using racemization of aspartic acid in cementum. *J Forensic Sci* 40:805–807.

Prince DA. 2004. Estimation of adult skeletal age-at-death from dental root translucency. PhD dissertation, The University of Tennessee, Knoxville, TN.

Prince DA and Konigsberg LW. 2008. New formulae for estimating age-at-death in the Balkans utilizing Lamendin's dental technique and Bayesian analysis. *J Forensic Sci* 53(3):578–587.

Schroeder HE. 1991. *Oral Structural Biology: Embryology, Structure, and Function of Normal Hard and Soft Tissues of the Oral Cavity and Temporomandibular Joints*. Stuttgart, Germany: Georg Thieme Verlag.

Sopher IM. 1976. *Forensic Dentistry*. American Lecture Series No. 990. Springfield, IL: Charles C. Thomas.

Suggested readings

Bowers CM. 2004. *Forensic Dental Evidence: An Investigators Handbook*. San Diego, CA: Elsevier Academic Press.

Bowers CM and Bell GL (Eds.) 1997. *Manual of Forensic Odontology*, 3rd edn. Ontario, Canada: Manticore Publishers.

Cawson RA and Eveson JW. 1987. *Oral Pathology and Diagnosis*. Philadelphia, PA: W.B. Saunders Company.

Fuller JL and Denehy GE. 1999. *Concise Dental Anatomy and Morphology*, 3rd edn. Chicago, IL: Year Book Medical Publishers.

Hillson S. 1996. *Dental Anthropology*. Cambridge, U.K.: Cambridge University Press.

Luntz LL and Luntz P. 1973. *Handbook for Dental Identification: Techniques in Forensic Dentistry*. Philadelphia, PA: J.B. Lippincott.

Whittaker DK and MacDonald DG. 1989. *Color Atlas of Forensic Dentistry*. London, U.K.: Wolfe Medical.

CHAPTER 7

Skeletal Examination and Documentation

Lee Meadows Jantz

CONTENTS

Skeletal preparation	126
Soft tissue removal	126
Dry bone	128
Post-processing care and handling	128
Preparing a case report	130
Summary	132
Review questions	132
Appendix 7A	132
Description of scene and field methods	133
Biological profile and skeletal analysis	134
Sex: Male	134
Age: 20–30 years	134
Ancestry: White	134
Stature	134
Trauma	134
Pathology	134
Taphonomy	134
Summary	135
Appendix 7B	135
Introduction	135
Biological profile	136
Sex: Female	136
Age: 50+ years	136
Ancestry: European American or white	136
Stature: 5′2″–5′6″	136
Trauma	136
Identification	138
Summary	138
Glossary	139
References	139

> **LEARNING OBJECTIVES**
> 1. Explain the proper sequence of preparing remains for a forensic anthropological analysis and case report.
> 2. Describe the procedure used to remove tissue from remains for skeletal analysis.
> 3. Describe the essential elements of a forensic anthropology case report and explain how these elements may differ from one case to another.

In order to make skeletal material and features more accessible for forensic anthropological analyses, and in order to procure material that may be have additional investigative value, it is sometimes necessary to sample and/or process skeletal material or other tissues associated with skeletal material. The condition of the remains at arrival and throughout the process of preparation and sampling should be documented. Any alterations caused by the process of sampling and preparation should also be documented.

—Scientific Working Group for Forensic Anthropology, 2011

SKELETAL PREPARATION

This chapter emphasizes the methods and techniques useful for the preparation of remains, in order to enable the anthropological examination and prepare a case report. The idea is to minimize any contamination or unnecessary destruction during the preparation of the skeletal material. Ultimately, the goal of skeletal preparation is to remove any adhering soft tissue or soil, in order to examine the bone morphology or surface features more closely, while being mindful of the potential evidentiary nature of all the material recovered. The recovered remains may be relatively fresh material taken at autopsy, fragmentary or burned remains from an explosion or plane crash, decomposing soft tissue remains, or dry skeletal remains. Each of these conditions requires unique processing techniques.

Before any preparation, a case number or identification number should be assigned and maintained. This allows for the tracking of the remains throughout cleaning and analysis. It is important to obtain *radiographic images* of the recovered materials before any cleaning or maceration, in order to identify any potential nonbone evidence such as bullets or bullet fragments, as well as to identify any possible accessory bones such as sesamoids in the hands or feet. If this step is omitted, this vital evidence may be lost or washed down the drain. Examination of the radiographs will allow the examiner to identify non-osseous items, as well as to identify the approximate location of these materials (see Appendix 7B: example from Scott county case). The recovery of this type of evidence before cleaning may prevent loss or further damage or destruction of the evidence. Proper handling and bagging of this evidence, as well as notification of the investigating agency, are necessary. In addition, all remains and associated evidence should be thoroughly photographed before cleaning.

Soft tissue removal

In situations where remains are presented or recovered with any soft tissue adhering to the bone, the goal is to remove all of this soft tissue in order to facilitate a detailed osteological examination. Personal protective equipment (PPE) such as gloves, masks, and goggles or eye shields should be worn when processing and handling remains. You should also be familiar with the Occupational Safety and Health Administration's (OSHA) guidelines for dealing with *biohazardous materials* and obtain any training required by your laboratory or institution. Tools that permit grasping, such as forceps, pick-ups, and hemostats, will pull most of the tissue away from the bone. If necessary, scissors may be used to cut portions away from the bone. A scalpel or knife should be used only if a large amount of tissue removal is required. These instruments, known as *sharps*, not only require special handling protocols but may also cause small cuts in the bone during the removal process. Special handling of sharps is necessary, in order to prevent a cut or puncture by a potentially contaminated tool. Most laboratories have protocols in place regarding the use of these tools as well as the safe disposal of sharps in designated containers (see Figures 7.1 and 7.2).

Any tool marks made on the bone during processing, such as scratches and nicks, must be recorded carefully. If not, then these marks made during the cleaning of the bones may be misinterpreted as trauma relating to the individual's death. Once the majority of the soft tissue has been removed, it is typically necessary to heat the bone with water. A slow cooker, pot, or steam-jacketed kettle may be used. It is important to put the bones in the water before the water reaches a hot temperature, in order to prevent cracking of the bones or teeth. The bones may be soaked in a very warm

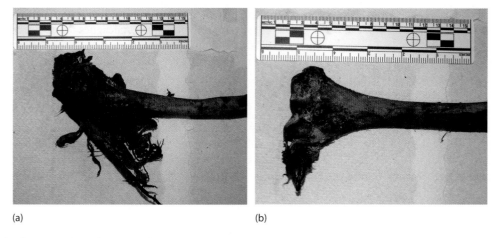

Figure 7.1 (a) Posterior view of a distal left radius with adhering mummified tissue and (b) anterior view of distal right humerus with adhering tissue.

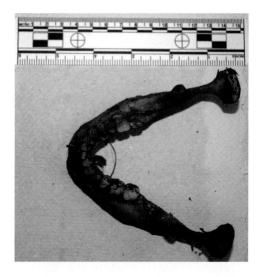

Figure 7.2 Occlusal view of mandible with adhering soil debris.

to simmering water bath for a period of 1–2 hours. The water should not be brought to a boil, as this may destroy the integrity of the bone or, at the very least, damage the DNA in the bone. Circumstances may indicate that the use of a mild soap or detergent is warranted, but it is not usually necessary. The current best practice is to alter the bone as little as possible during cleaning. Once the bones have *cooked* for a period, they should be removed from the water and allowed to reach room temperature naturally. Do not immediately place the bones in cool or cold water, as the risk of cracking increases with any rapid change in temperature. Once cool enough to handle, the bones may be rinsed with water and brushed with a soft- to medium-bristle brush to remove any remaining adhering tissue. Metal brushes should be avoided, as they can be fairly destructive. Again, forceps or other grasping or scraping tools may be used to remove any remaining tissue. The bones should then be allowed to dry at room temperature. This may be repeated as many times as necessary to remove the soft tissue and degrease the bone, as long as the bone surface remains in good condition. If the bone is already compromised due to bone thinning or disease, it is best to simply allow some of the tissue to dry and remain. If *adipocere* or other tissue is very difficult to remove, then the bones should be allowed to dry, followed by brushing with a firmer brush while dry. This will often successfully remove this stubborn material.

Finally, it is imperative to take care when cleaning the areas of the skeleton that are examined for the biological profile evaluation. For example, the areas of the pubic symphysis, auricular surface, sternal rib ends, or any epiphyses should be treated very carefully. It is possible to scrub these areas too hard and remove some of the bony surface. This should be avoided. Also, severely burned or excessively fragile remains require expert handling. Practitioners with little or no experience in handling these types of cases should consult more experienced colleagues for advice to avoid inadvertently damaging the remains. Forensic anthropologists should also communicate with law enforcement in these instances, so that they can be present at the scene to ensure that remains are properly handled, documented, and transported to the laboratory for processing.

Dry bone

If skeletal material has been recovered in a relatively dry or skeletal condition, it is still essential to radiograph it before any brushing of dirt or leaf litter, as there may be adhering evidence. On completion of the imaging, the bone may be brushed with a soft- to medium-bristled brush to remove any loose material. If the bone is covered in dried mud, water may be used to rinse it from the bone surface.

POST-PROCESSING CARE AND HANDLING

Once the remains are clean and ready for analysis, they should be placed either on trays or on a table. Delicate bones (e.g., the skull or elements with trauma) should be placed on soft protective pads, so that they are not damaged. Care should be taken that the case number or identification number is associated with the remains either by tags or labels on the trays or table. The bones should be placed in anatomical position in order to determine if any elements are missing, repeated, or duplicated. A careful and detailed inventory should be recorded to identify the skeletal elements present, as well as to identify if duplicated elements are present, which may represent multiple individuals. The inventory may be in the form of a diagram or a list (Figure 7.3). A diagram allows a layman or an investigator to see straightforwardly what elements are missing.

During this inventory, skeletal elements that are not human are identified as nonhuman and sorted separately from the human remains. The species of nonhuman bones should be identified, if possible, and examined for any evidence

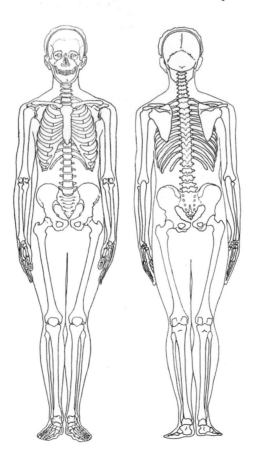

Figure 7.3 Inventory diagram.

of association with the human remains. It is common to have nonhuman skeletal remains gathered at the same time as the human, particularly if nonexperts are involved. This is to be commended and recommended. Because many of the investigators who search and recover skeletal remains have little or no training in human osteology, they should be encouraged to collect all bones that they discover. Ultimately, the amount of detail provided for nonhuman remains in the final report should be based on the potential involvement in the case. It may be that a simple species list is sufficient, or it may be determined that an inventory for each species and detailed description is necessary. This is something that is decided between the anthropologist and the investigators.

If the human skeletal remains must be *curated* or retained for further analysis, each element should be labeled with the identifying case number. A fine-tipped permanent ink pen is recommended. The location of the label may be determined by the anthropologist. Figure 7.4 illustrates the locations used for labeling by the Smithsonian Institution Natural History Museum (David Hunt 2011, pers. comm.) and the Forensic Anthropology Center, University of Tennessee. Because the cortex of the bone is porous, it will often absorb the ink along fine cracks in the bone surface.

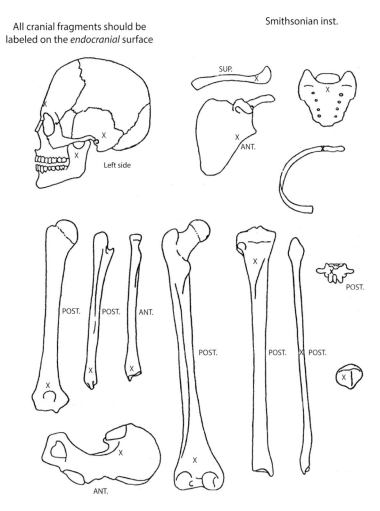

Figure 7.4 Bone-labeling diagram.

Do not put the label on or near an area of the bone examined for biological profile evaluations or observations of trauma or disease. The labels should be small but legible. Loose teeth, small bones, and fragments may be difficult to label; however, these elements must be treated carefully in the event that they become disassociated with the other material. Material that is too small or greasy may be put into separate cloth or paper bags, with a paper label marked in pencil. Pencil will not fade with light or contact with the grease, whereas ink may fade.

Once the analysis is complete, Forensic Data Bank (FDB) forms should be completed and submitted to the curators of the FDB for inclusion in this important database. These forms are available as downloads within FORDISC 3 (Jantz and Ousley 2005) and in an appendix of the *Data Collection Procedures for Forensic Skeletal Material 2.0* (Langley et al. 2016). The latest version of the *Data Collection Procedures for Forensic Skeletal Material 2.0 (DCP 2.0)* manual can be accessed and downloaded for free at http://fac.utk.edu/pdf/DCP20_webversion.pdf. The FDB inventory page also allows for a quick and easy documentation of the inventory. Circumstances of the case, biological profile parameters, metrics, and dental information are also easily recorded on the FDB forms. Refer to Chapter 14 for a discussion of the FDB and FORDISC software program.

PREPARING A CASE REPORT

A number of factors may be included in a forensic anthropology case report. The contents of the report should be directed at the intended audience (i.e., medical examiner or coroner and law enforcement agency). Each case is unique and requires thought about the appropriate amount of detail and the sections that must be included in the report. Regardless, *every* report should be written with the idea that it may ultimately be presented in a court of law. If the primary audience is a law enforcement officer or agency, then excessive technical jargon should be avoided. However, this does not mean that proper terminology should be avoided. If the forensic anthropology report is for a forensic pathologist, the same recommendation holds. Why? Because it is possible that your report will be used at a trial and thus read by a jury of laypeople. Furthermore, some medical examiners may prefer that you do not make a statement about the cause or manner of death, as this is the primary task of the pathologist's report, and having an additional statement about this may unnecessarily complicate things in a court of law. A report's summary section on the first or last page of the report, with less technical jargon, is useful for nonexperts who need to glean important details from the report.

Formats for forensic reports may be dictated by the agency employer such as the Federal Bureau of Investigation, the Defense POW/MIA Accounting Agency (DPAA), and a large medical examiner office. It may also be the case that, as an independent practitioner or consultant, you determine the format. Formats may range from a single-page fill-in-the-blank form to extensive prose. Regardless of the format, the same general content must be conveyed, including:

- The inventory and condition of elements present for examination
- The factors in a biological profile (age, sex, ancestry, stature, disease, and so on), if the remains are not positively identified
- Trauma analysis, if applicable
- Estimation of time since death (if necessary)
- A dental assessment or dental chart (this may be provided by a forensic odontologist)
- Supporting diagrams and/or photographs of these elements (all photographs should include a scale and a label with the case number)

If possible and applicable, a discussion of the recovery of the elements, including a map of the location of the remains, may be warranted. A description of how you received the case (e.g., recovered at the scene or received from a law enforcement or medical examiner's office) is useful. A summary of the findings is also helpful. All of these topics may play a role in a forensic anthropology analysis. Because all these are parts of an analysis, you will want to keep detailed *bench notes* on these, which are considered supporting documents. If the format of the final report is the brief one-page style, the bench notes provide the information to support your findings and may be subject to examination in a court of law. These notes serve as the references for your conclusions and document all stages of the case, including correspondences with law enforcement and medical examiners. Currently, both the American Board of Forensic Anthropology and the SWGANTH (2011) recommend/require supporting documents for case reports.

The University of Tennessee Forensic Anthropology Center (FAC) sends reports to the investigator(s), the county medical examiner, and the state medical examiner. The state medical examiner serves as the representative for the state's department of health. Local prosecutors may also require a copy of the report. The case number, date of report, and subject are included in the beginning of the report (see Appendices 7A and 7B for case report examples). The body of the report begins with a brief introduction, explaining how the case began and proceeded. This is followed by the biological profile findings; descriptions of trauma, taphonomy, and pathology; and the dental chart and odontological findings. If the investigators are interested in an estimation of the time since death, the report should also include your analysis and interpretations. Photographs are an excellent means to illustrate points made in a report. Figure 7.5 shows the *in situ* position of the victim's arms in the burial. A map or drawing can also be employed (Figure 7.6); however, the photograph has the stronger impression. If you are new to writing case reports, it is a good idea to have a colleague peer review your report and make suggestions before submitting the official report.

Figure 7.5 Photograph of a skeleton *in situ* in a burial, showing the position of the body as it was placed in the burial. Note the arm extended above the head.

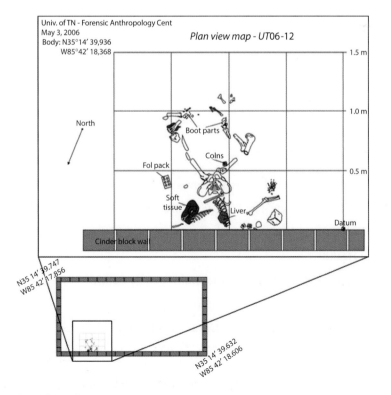

Figure 7.6 Map of skeletal remains before recovery.

SUMMARY

Before completing an analysis and case report, a forensic anthropologist must prepare the remains for skeletal examination. This process may differ, depending on the state of the remains (burned, fresh, decomposed, or skeletal). After thoroughly documenting the condition of the remains with photographs and X-raying them to inspect the body for foreign objects or evidence, the anthropologist prepares the bones for analysis. Soft tissue can be removed with scissors, forceps, scalpels, and brushes. If the tissue is particularly difficult to remove, the bones may have to be simmered in a warm water bath to soften up and release the soft tissue. Care should be taken when removing tissue with sharp instruments, and any inadvertent damage to the bones should be recorded immediately, so that it is not confused with trauma during the skeletal analysis.

Once the remains have been cleaned, they are ready for analysis. The anthropologist takes a skeletal inventory and begins to assemble the elements of the report: biological profile, trauma analysis, taphonomy and pathology assessments, dental chart, and time since death. Some reports contain only a few of these elements, whereas others contain them all. The sections included in the report are determined by the request of the medical examiner and by the material available for analysis. All steps of the analysis should be documented in bench notes, so that the forensic anthropologist has material that can be consulted in the event that questions arise later (e.g., after the remains have been returned to the family). Bench notes are not part of the formal report but may be examined in court. Occasionally, all or part of the remains must be retained for long-term curation. In this case, it is imperative to label all elements with a permanent ink pen and store them appropriately. It is a good idea to have a colleague peer review case reports occasionally and offer suggestions for improvement, especially for challenging cases.

REVIEW QUESTIONS

1. Why do forensic anthropologists X-ray remains before processing and examination?
2. You receive a set of remains, with the skull and most of the abdomen and thorax skeletonized, but a fair amount of decomposing soft tissue is present on the limbs. Explain how you would prepare these remains for skeletal analysis.
3. What is the recommended procedure for handling nonhuman remains?
4. Describe the procedures for preparing remains that must be kept as evidence.
5. Why would a forensic anthropologist refrain from stating cause of death in the report?
6. What are the essential elements of forensic anthropology case reports?
7. What is the purpose of bench notes?

APPENDIX 7A

To: Larry B. Davis, Special Agent, Tennessee Bureau of Investigation

William Barker, Bomb and Arson Section, State Fire Marshall's Office

Byron Harbolt, D.O., Grundy County Medical Examiner

Tennessee Department of Health

From: Lee Meadows Jantz, PhD, Coordinator, Forensic Anthropology Center

Joseph T. Hefner, MA, Asst. Coordinator, Forensic Anthropology Center

Joanne Devlin, PhD, Forensic Anthropology Center, University of Tennessee

Re.: Recovery and analysis of burned human remains from fatal house fire in Grundy County, Tennessee

Case No.: UT06-12F

Date: March 27, 2012

Figure 7.7 Grundy County, Tennessee, and the approximate location of the structure fire.

On Tuesday, May 2, 2006, TBI Special Agent Davis contacted the Forensic Anthropology Center (FAC), requesting assistance in the recovery and analysis of burned human remains discovered in the burned remnants of a small residential structure located on White City Circle in Tracy City, Tennessee, a small community located within Grundy County (Figure 7.7). According to investigators, the structure had burned on April 27, six days before the arrival of the FAC recovery team. The human remains were discovered on May 2, at which time the area was secured and our laboratory was contacted. The Forensic Anthropology Center Recovery Team, composed of Lee Meadows Jantz, Joseph Hefner, Rebecca Wilson, Donna McCarthy, and Kevin Hufnagl, responded to the scene on Wednesday, May 3, 2006, to assist investigators with the recovery of all associated skeletal material.

Description of scene and field methods

An initial search of the secured area by the forensic recovery team revealed human skeletal elements within an isolated area in the northeast quadrant of the remaining block foundation. Although partially obscured by debris from the fire, the remains of a single individual were visible. The skull was located next to the block foundation, with the postcranial elements extending in the general direction of the center of the structure. The position, condition, and associated fire debris indicate that the remains were burned during the fire that occurred at the structure.

A 2m × 1.5m grid was established over the area to control for spatial distribution and facilitate a thorough excavation and recovery. The grid was divided into units of 50 cm size, which were processed using standard archaeological procedures. Using hand trowels, the top 2–3 cm of ash and debris were removed to expose skeletal elements. At this stage of excavation, a plan-view map was produced, which suggests that all of the skeletal elements are in approximate anatomical position. A photographic record was produced using a Nikon D70 digital camera. Photographs were taken before, during, and after processing. Field notes were collected throughout the process.

Skeletal elements were identified and collected. Following recovery of all bones and fragments, any remaining debris was screened through wire mesh to enable identification of small osseous or evidentiary material.

All recovered elements were transported to the Forensic Anthropology Center at the University of Tennessee for analysis.

Biological profile and skeletal analysis

A nearly complete skeleton was recovered from the scene. There is no irregular duplication of elements, and all skeletal elements are consistent in morphology, age, and taphonomy. All elements display evidence of exposure to fire. Minimal soft tissues remained, all of which exhibit charring and heat damage. Exposure to heat dehydrates and can subsequently destroy the organic component of bone, resulting in fragile and distorted material. Heat-affected bone also demonstrates differential coloration, fragmentation, and shrinkage. The skeletal elements recovered from Grundy County exhibit characteristics consistent with thermal alteration.

Sex: Male

Although heat-induced shrinkage occurred, the robust condition of the remains is consistent with male morphology. A visual inspection of several regions of the skeleton confirms this assessment. A prominent external occipital protuberance and blunted orbital margins are present on the skull. Pelvic elements display a moderately narrow sciatic notch, absence of a ventral arc, and a narrow subpubic concavity.

Age: 20–30 years

Age at death was determined using multiple indicators of the postcranial skeleton. The pubic symphyseal morphology is consistent with Phase III of the Suchey–Brooks method, suggesting an estimated age of 30 years (±10 years). The sternal ends of the ribs are between Phases 3 and Phases 5 (range of 20–30 years), determined using the Iscan and Loth technique for age estimation for males. The anterior border of the first sacral vertebra is in the process of fusing, but has not yet completely fused to S2, indicating an age <30 years. The superior aspect of the S1 centrum shows marked lipping; however, spondylolisthesis present on L5 may have resulted in increased mechanical loading and osteophytosis on the first sacral vertebra. Slight osteophytic activity is present on the vertebral elements. The presence of osteophytosis to this degree is generally indicative of early middle age, that is, 20s to 30s.

Ancestry: White

Although the skeletal material is extremely fragmented, several nonmetric traits indicative of European ancestry (e.g., white) are noted. These include the following: the presence of a nasal sill, a marked anterior nasal spine, a jagged transverse palatine suture, and a *curved* zygomaticomaxillary suture.

Stature

The degree of shrinkage observed in the postcranial skeleton of UT06-12 may result in error for stature estimation; therefore, living stature was not calculated.

Trauma

Although gross, radiographic, and macroscopic analyses did not reveal evidence of perimortem trauma, the degree of thermal alteration exhibited may obscure or destroy such signatures.

Pathology

Skeletal pathology includes slight osteoarthritis and osteophytosis on several vertebrae in the form of arthritic lipping. While arthritic lipping is often associated with age-related changes, it may also be the result of stress-related activity. A *pars defect*, or spondylolisthesis, is noted on the lamina of the fifth lumbar vertebra. The etiology and significance of this pathological condition is uncertain in the clinical literature; however, it is noted in approximately 6%–12% of the population.

Taphonomy

Coloration and preservation is grossly visualized and appears to be resultant from thermal destruction.

Summary

The skeletal remains presented to this laboratory for analysis represent a single individual. Analysis indicates that the remains are those of a 20- to 30-year-old White male. All remains show evidence of burning and concomitant thermal damage. Coloration, fragmentation, and distortion of the bone are consistent with exposure to high temperatures (in excess of 800°F).

If we can be of further assistance in this or any other matter, please do not hesitate to contact us.

Lee Meadows Jantz, PhD

Coordinator Forensic Anthropology Center

Joseph T. Hefner, MA

Forensic Anthropology Center

Joanne Devlin, PhD

Forensic Anthropology Center

APPENDIX 7B

To: Steve Vinsant, Special Agent, Tennessee Bureau of Investigation

Bobby Ellis, Sheriff, Scott County Sheriff's Office

Maxwell Huff, MD, Scott County Medical Examiner

Tennessee Department of Health

From: Joanne B. Devlin, PhD, Forensic Anthropology Center

Lee Meadows Jantz, PhD, Coordinator, Forensic Anthropology Center

Re: Recovery, analysis, and positive identification of human remains recovered from burial in Scott County, Tennessee

Case No: UT10-06

Date: July 6, 2010

Introduction

On May 11, 2010, Meadows Jantz received a call, requesting assistance in the excavation and recovery of a burial identified by an informant. Arrangements were made for Meadows Jantz, Devlin, and graduate assistant Sandra Cridlin to travel to the Chitwood Mountain area of Scott County that afternoon. Once at the suspected burial site, examination revealed an area of relatively large flat rocks arranged in a semicircle near the edge of the creek bank. Following minimal probing, this area was cleared of leaf litter and overburden, revealing a small zone of discoloration, more yellow brown than the surrounding soil. Preliminary trowel scraping revealed loose soil and a continuation of the brown-yellow-colored soil. While no grave boundary was apparent at this time, a trench was placed in the N–S direction and excavated to a depth of approximately 2 ft below the surface; bone was encountered. At this time, the excavation was expanded in the E–W direction to reveal the full extent of the burial. Ultimately, a full set of skeletal remains was exposed, with the individual in a supine position, with the arms extended over the head and flexed at the elbow. These remains were recovered and taken back to Knoxville for further analysis.

Immediately after recovery, the anthropologists traveled to the Scott County Sheriff's Office to collect dental records of *Jean Johnson*, a woman reported missing in February 2007. In addition, TBI SA Vinsant provided several radiographs of Jean Johnson.

The remains were predominantly skeletonized, although residual soft tissues maintained the posterior portion of the torso in articulation. Several elements were in a fragmentary form at the time of recovery. Before submitting the remains for a radiographic analysis, a metallic object was discovered within the cranial vault. This artifact was released to SA Vinsant. All of the skeletal elements were radiographed, which were all negative for radiopacities; however, the brain matter did exhibit tiny radiopacities.

Biological profile

Sex: Female

Based on nonmetric skeletal traits exhibited by the os coxae, including wide pubic bones, presence of a ventral arc, subpubic concavity, pronounced pre-auricular sulcus, and a ridge on the medial aspect, the remains are estimated to represent a female (Phenice 1969). The skull exhibits prominent frontal bossing, reduced mastoid processes, and moderately sharp supraorbital margins. These attributes, in conjunction with the quality of the mental eminence, indicate that the remains are those of a female.

Age: 50+ years

The skeletal remains are those of an adult, as evidenced by closed epiphyses and the condition of the medial clavical. Standard techniques for estimating the degree of degenerative change to the os coxa and the right fourth rib were used to estimate the age. The pubic symphysis is characterized by erosion along the ventral margin, with irregular ossification and crenulations across the depressed surface; late phase 5/early phase 6 on the Suchey–Brooks scale, mean ages of 48.1 years (95% range: 25–83 years) and 60 years (95% range: 42–87 years), respectively (Suchey and Katz 1998). Following the Lovejoy approach, the auricular surfaces are irregular; slight microporosity and dense patches are visible. Although there is some residual transverse organization, the apex and superior demiface exhibit some lipping, osteophytes are visible in the retroauricular area, and the inferior face extends beyond the body of the innominate. These characteristics are consistent with age ranges of 50–60 years (Lovejoy et al. 1985). Assessment of the rib following the approach of Iscan et al. (1985) reveals a slightly irregular U-shaped rim, a prominent plaque-like deposit, and small windows suggestive of phase 6 mean age of 50.7 years (95% confidence interval: 43.3–58.1 years). The synthesis of these techniques indicates that the remains represent those of an individual >50 years of age.

Ancestry: European American or white

Ancestry of this individual is estimated as White or European. Morphological features of the cranium include angled orbits, pinched nasals, and a relatively narrow nasal aperture. All of these features support the estimate as white (Bass 2005). Metric analysis of the skull was not performed due to trauma.

Stature: 5′2″–5′6″

FORDISC 3.0 (Jantz and Ousley 2005) was used to estimate stature, using the maximum lengths of the clavicle (132 mm), femur (437 mm), and fibula (350 mm); all elements are from the right side of the body. Estimated stature with a 90% prediction interval is 61.8–66 in.

Trauma

The skeletal remains of this individual exhibit evidence of perimortem trauma. The cranium shows evidence of ballistic trauma, while the humeri exhibit characteristics consistent with blunt force trauma. The skeleton also exhibits evidence of injuries that occurred during the antemortem period. The left femur demonstrates evidence of surgical repair by the presence of bone growth around an orthopedic rod and associated screws. The left shoulder also indicates bone growth around the humeral head and the glenoid fossa of the scapula, evidence of an antemortem injury that affected the bones of this joint. Further, healed fractures are visible on the left ribs (4–8) at the midpoint of each shaft.

The cranial trauma consists of a network of fractures across the craniofacial skeleton and vault, with a projectile traveling anterior to posterior. The entrance is located just left of midline, impacting the left nasal and maxilla, producing separation of the palatine process, fragmentation of the sphenoid, and separation of both zygomatics. The mandible is fractured vertically through the left mental foramen, separating the ramus and resulting in two mandibular fragments. A corresponding fracture is visible across the mandibular dental plate (Figure 7.8). The projectile impacted the endocranial surface of the occipital squama approximately 1 cm left of midline, resulting in radiating and concentric fractures, which displaced three fragments of bone (Figure 7.9). These fragments were

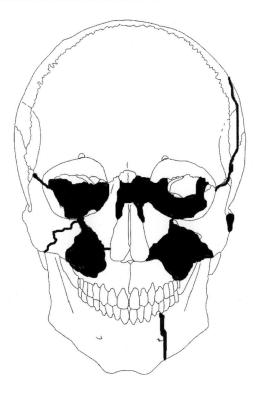

Figure 7.8 Anterior view, with highly fragmented and unreconstructable portions colored black.

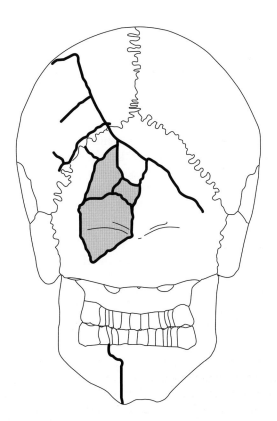

Figure 7.9 Posterior view. Fracture lines noted and displaced fragments shaded.

maintained in anatomical position in the vault, given the presence of soft tissues. Removal of these three fragments revealed external beveling commonly associated with exit wounds. As noted, a metallic object was recovered from within the cranial vault. The skull was reconstructed to facilitate radiographic comparison against antemortem records.

Perimortem trauma, specifically blunt force trauma, is present in both the left and right humeri. The left humeral diaphysis exhibits a butterfly fracture slightly distal of midshaft. The position of the wedge segment indicates that the force was applied to the anteriormedial aspect of the shaft superior to the medial edge of the trochlea. The right humeral diaphysis exhibits two fractures: a butterfly fracture slightly proximal to midshaft and a transverse fracture approximately 15 mm inferior to the surgical neck. The nature of these intersecting fractures is such that the transverse fracture occurred before the butterfly fracture. The latter consists of a large wedge of bone, which indicates that a force impacted the lateroanterior aspect of the shaft.

Identification

The identity of the remains was established via several complimentary sources. Radiographs of the skull and lower torso, written dental records, and notations of injuries and treatment were provided to the FAC by SA Vinsant.

Antemortem radiographs of Jean Johnson were available for evaluation and comparison. Specifically, an anteroposterior view of the frontal sinus patterns of the skull and the frontal sinus pattern, as reflected in the antemortem radiograph of Johnson (dated September 30, 2006), were compared. While the most superior portion of the frontal sinus was not captured in the antemortem film, the inferior patterns of the sinus margins corresponded with those exhibited in the postmortem radiograph. No inconsistencies were identified, and the following specific regions of correspondence were noted: (1) Left lateral aspect of the frontal sinus from the supraorbital notch extending across the orbital margin and continuing in a superiorly arcing path; (2) the shape of the frontal sinus, as observed within the superiomedial area of the right orbit; (3) the frontal sinus shape, as observed within the superiomedial area of the left orbit; (4) the undulating inferior margin of the sinuses across the interorbital region; and (5) the shape and location of two centrally located sinus pouches extending from the anatomical landmarks nasion to glabella.

Additional radiographs dated September 30, 2006, further supported the identity of the remains. The proximal portion of the left humerus showing what appears to be an old, healed injury resulting in an unusual morphology is recorded. This is also noted in the medical records provided to the FAC. The left humerus recovered from the burial site also exhibits the same unusual morphology involving the head, neck, and tubercle.

Another area of consistency concerns the dentition and the written dental record with a last entry of June 10, 1991. The written dental record of Johnson notes the presence of only six mandibular teeth; all incisors and the both canines. The record notes that Johnson was edentulous in the maxillae. The dental record further notes the presence of a partial mandibular plate. The recovered remains consist of an edentulous maxilla and a mandible with the crowns of five single-rooted teeth (the incisors and the right canine), with only the root present for the left canine. A partial dental plate for the mandible is recovered. The teeth represented in the dental arcade of the remains (and the presence of the plate) are consistent with the written dental record of Johnson.

A final point of correspondence is the documentation of a well-healed fracture of the left femur that was reportedly treated with an orthopedic rod insertion into the bone. The left femur recovered from the burial site exhibits a well-healed fracture of the distal third of the shaft, with an orthopedic rod inserted into the medullary cavity of the bone.

In conjunction, these points of correspondence are such that the remains represent Jean Johnson.

Summary

The remains recovered from a burial in the Chitwood Mountain area of Scott County represent a white female >50 years of age. The individual reported to have been in this location is Jean Johnson, a white female 69 years of age.

Perimortem trauma is evident by means of a gunshot to the head and blunt force to both upper arms. Comparison of the antemortem radiographs to the skeletal remains resulted in no inconsistencies, and the reported dental condition, old femur trauma, and orthopedic treatment are consistent. Based on all of these points, we believe that the recovered skeletal remains are of Jean Johnson.

These remains were returned to the family for burial on June 3, 2010. If we can be of further assistance, please feel free to contact us.

Joanne B. Devlin, PhD

Forensic Anthropology Center

Lee Meadows Jantz, PhD

Coordinator, Forensic Anthropology Center

Glossary

Adipocere: a waxy substance formed by the anaerobic hydrolysis of fat during the decomposition process; adipocere is sometimes called "grave wax."
Bench notes: notes that accompany a forensic case report. They are not included as part of the official report document but are subject to subpoena and examination in court.
Biohazardous materials: biological substances that pose a threat to living organisms or to the environment.
Curate: the process of organizing materials (skeletal remains) for later use or observation.
Radiographic images: medical images such as radiographs (X-rays) and computed tomography (CT) scans that can be used to visualize the inside of a body, identify radiopaque foreign objects, and facilitate positive identification.

References

Bass WM. 2005. *Human Osteology: A Laboratory and Field Manual*, 5th edn. Columbia, MO: Missouri Archaeological Society.

Iscan MY, Loth SR, and Wright RK. 1985. Age estimation from the rib by phase analysis: White females. *J Forensic Sci* 30:853–863.

Jantz RL and Ousley SD. 2005. *FORDISC 3: Computerized Forensic Discriminant Functions. Version 3.0.* Knoxville, TN: The University of Tennessee.

Langley NR, Meadows Jantz L, Jantz RL, Ousley SD, and Milner G. 2016. *Data Collection Procedures for Forensic Skeletal Material 2.0.*

Lovejoy CO, Meindl RS, Pryzbeck TR, and Mensforth RP. 1985. Chronological metamorphosis of the auricular surface of the ilium: A new method for the determination of adult skeletal age at death. *Am J Phys Anthropol* 68:15–28.

Phenice TW. 1969. A newly developed visual method of sexing the os pubis. *Am J Phys Anthropol* 30:297–301.

Scientific Working Group for Forensic Anthropology (SWGANTH). 2011. Skeletal sampling and preparation. Issue Date: May 25, 2011.

Suchey JM and Katz D. 1998. Applications of pubic age determination in a forensic setting. In KJ Reichs (Ed.), *Forensic Osteology: Advances in the Identification of Human Remains*, 2nd edn. Springfield, MA: Charles C. Thomas, pp. 204–236.

Section III

Skeletal Individuation and Analyses

8 Sex Estimation of Unknown Human Skeletal Remains
9 Ancestry Estimation: The Importance, The History, and The Practice
10 Age Estimation Methods
11 Stature Estimation
12 Pathological Conditions as Individuating Traits in a Forensic Context
13 Analysis of Skeletal Trauma
14 Introduction to Fordisc 3 and Human Variation Statistics

CHAPTER 8

Sex Estimation of Unknown Human Skeletal Remains

Gregory E. Berg

CONTENTS

Introduction	144
Sex versus gender	144
Sexual dimorphism	144
DNA	145
Documentation	146
Morphological approaches to sex determination	146
Pelvis	146
Skull	148
Other postcranial bones	152
Metric approaches to sex determination	152
Population-specific standards	156
When methods disagree	157
Conclusion	157
Summary	158
Review questions	158
Acknowledgments	158
Glossary	159
References	159

> **LEARNING OBJECTIVES**
> 1. Explain the difference between sex and gender.
> 2. Explain the difference between metric and morphological/nonmetric techniques for determining sex from skeletal remains.
> 3. Define and describe human sexual dimorphism (i.e., when it develops and why it is relevant to sex determination).
> 4. Discuss best practices in sex determination.
> 5. List the bony elements that forensic anthropologists frequently use to determine sex in the order of accuracy (most to least accurate), and name several methods for each element.
> 6. Explain the concept of a sectioning point, and discuss factors that influence sectioning points.
> 7. Name the variables that should be taken into account when determining sex from the skeleton, and explain how these variables might affect one's assessment.

INTRODUCTION

This chapter explores the approaches and methods used to determine sex from the human skeleton. Skeletal sex determination in medico-legal or archaeological context is a key component of personal identification and for establishing the remainder of the biological profile. Sexing methods are typically divided into two categories: morphological (shape) and metric (size). Morphological aspects of the human skeleton that are sex-specific are found primarily in the pelvis and, less so, in the other bones of the body. Metric differences between the sexes are mainly found in the skull but also in the postcranial skeleton. Both categories have varying accuracy rates, with the pelvis being the best indicator overall; however, some metric methods are accurate >90% of the time. This chapter details several morphological techniques and provides a discussion of the statistical aspects of metric sex determination. As with any rigorous science, a solid foundation in basic techniques is necessary, and that foundation is the goal of this chapter.

SEX VERSUS GENDER

Two specific terms are frequently confused in common usage: sex and gender. This is a major issue in science, because the difference between these words is a distinction between a biological reality and a social category. For scientists, sex is a biological fact, whereas gender is a socially ascribed and perceived identity. The confusion may stem from several sources: it could be because the general public lacks the awareness of the distinction between the two terms, and perhaps, from the common usage of *gender* on governmental and business forms when, in fact, *sex* is meant. Overall, the confusion likely lies in the fact that most people believe these words to be synonyms, when they are not. As anthropologists, we can only determine the biological sex of an individual from skeletal remains, not that individual's gender, character, or role. In some cases, gender might be implied from examining the material culture found in conjunction with skeletal remains such as clothing, jewelry, weaponry, and class symbols, but this determination can be very difficult and is fraught with uncertainty. We must remember that when a police investigator asks for the *gender* of an unknown decedent, it is not a question of how that individual dressed, acted, or performed various roles in society but rather a simple question of biologically *male* or *female*. We should be mindful of how we apply these terms in everyday professional discussions, casework, or publications, so that we do not perpetuate this misconception.

SEXUAL DIMORPHISM

When an anthropologist is asked to determine the sex of an individual, he or she always starts with 50/50 odds of being correct, as there are only two biological states: male and female. In order to increase the odds of accuracy, the anthropologist needs to understand sexual dimorphism. Sexual dimorphism generally refers to size and shape differences between males and females of a given species, although other differences in anatomy, function, behavior, and psychology are present. In many ways, this statement oversimplifies the definition; however, for purposes of this chapter, it functions as a basis for the discussion on sexing skeletal remains. Easily seen soft tissue differences between the sexes are not the only physical manifestation of sexual dimorphism; the hard tissues of the skeleton are also affected. In some species, the differences are extremely marked, but in others, only subtle differences exist. For the skeletal system, size is often the most dramatic and easily observed disparity between males and females. For instance, given a selection of 25 male and 25 female gorilla skulls mixed together, an untrained observer could place the skulls into a contiguous line, from males to females, solely based on size and would likely be 100% correct. Given the same scenario for human skulls, the untrained observer may achieve a 60% correct classification based on size, if lucky, and where the males and females meet in the middle of the chain would be very hard to determine. This is because humans are not as sexually dimorphic as gorillas—human sexual dimorphism is less pronounced and often population-specific.

Skeletally, sexual dimorphism develops with the onset of sexual hormone production around the time of puberty; these hormones control and regulate the expression of the secondary sexual characteristics. As would be expected in most human populations, these changes are better defined nearing the age of 17 years; however, they can be present much earlier in developmentally precocious individuals. Therefore, estimation of sex from the skeleton is easier in adults and late adolescents and is much harder in subadults. Most anthropologists would argue that sexing individuals younger than 15 years results in an educated guess; however, techniques are available. The accuracy of these techniques depends on the preservation of the skeletal elements, the population from which they are originated, and whether known-sex comparative specimens from that same population are available. The Scientific Working Group for Forensic Anthropology (SWGANTH) has specifically weighed in on this topic. The SWGANTH is a joint venture

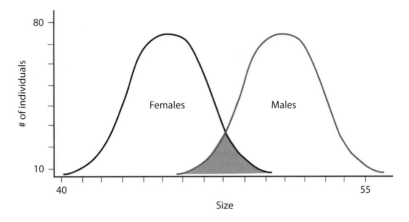

Figure 8.1 Two even 20% overlapping bell curves. The gray-shaded area where the curves overlap is an ambiguous zone in which misidentifications occur. (Courtesy of J. Hefner.)

between the Federal Bureau of Investigation and the Department of Defense Central Identification Laboratory that develops consensus best-practice forensic anthropology guidelines and establishes minimum standards. In addition, the committee disseminates the guidelines, studies, and findings from their meetings at http://www.swganth.org/index.html. For subadult sex determination, the SWGANTH states that sexing individuals younger than 12 years is an unacceptable practice; however, results may be achieved if the pelvis is fusing and methods for adults are applied to the remains (that are older than 14 years) (SWGANTH 2010, p. 3). Other good reviews of the challenges, techniques, and pitfalls for subadult sex estimation can be found in Komar and Buikstra (2007) and Scheuer and Black (2000).

For the most part, sexual dimorphism in the human skeleton can be stated as "males are larger and more robust than females, or females are smaller and more gracile than males." A favorite saying is that males are usually bigger than females. Males can be upward of 20%–30% larger than females in certain skeletal dimensions (e.g., the size of the femoral and humeral heads). However, individual variation can make this general rule hard to follow—as with any bell curve, some individuals are much larger or smaller than the average. Given two even, 20% overlapping bell curves, the area where the curves overlap is where misidentification happens (Figure 8.1). Said differently, small, gracile males can be easily confused with large, robust females. Complicating the picture, human populations vary across time and geographic space. As a simple example, populations with life ways focused on hunting and gathering frequently tend to produce large and rugose individuals, whereas individuals in sedentary populations are often less rugose and potentially smaller owing to less stress placed on skeletal tissues. If a hunter/gatherer female is compared with a sedentary male, it is quite possible to confuse the sexes of these two individuals based on size and rugosity, and vice versa. This example is not meant to factor in all of the biological complexities that surround this issue; rather, it is a simple explanation of a complex problem. For these reasons, anthropologists must examine as much skeletal data as possible before assigning sex to an unknown individual, and population- and period-specific standards should be used for each case. Familiarity with the individual variation within a population can alleviate some problems. If one is not familiar with the range of variation, seriating the population (creating a series from largest to smallest) can be a good practical exercise to become accustomed to a population's variation.

DNA

Scientific advances may someday make sex determination easier for the anthropologist. The state of molecular technology (DNA testing) is such that a simple test for the presence or absence of a Y chromosome can determine sex from skeletal remains. This is particularly helpful in cases involving juvenile remains. Unfortunately (or fortunately, depending on your view), DNA testing is not cheap, and for large skeletal assemblages, it is cost-prohibitive. The time required for DNA tests is frequently several weeks to several months, versus a determination by the anthropologist, which can be made in a matter of minutes. Furthermore, DNA analysis requires destruction of bone, and acquiring the necessary permissions for testing in archaeological contexts can be very difficult. These limiting factors will dictate the need of anthropologists to determine sex of most unidentified individuals for the foreseeable future. That stated, it behooves the cautious anthropologist to seek out DNA testing in medico-legal cases involving juvenile remains, where this is a feasible option.

DOCUMENTATION

Perhaps, the most important aspect of an anthropologist's work is the documentation of the procedures, methods, and interpretations made in the course of a skeletal analysis. Proper scientific documentation is the backbone of a thorough analysis. Each SWGANTH document expressly indicates that relevant tests, observations, exemplars, and decision-making processes are documented as part of a case file. A standardized notes packet is helpful in this regard. For sex determination, good documentation includes a comprehensive skeletal inventory, notes on the condition of the remains, the age of the remains (both skeletally and temporally), details of any pathological conditions or deleterious processes that have affected the remains, and any contrary or disparate sex indicators. The notes should state the observations complied, measurements taken, methods used, and the results of these analyses. An overall sex assignment is annotated, and the reasons for the assignment are explicitly stated. Any departures from normal methods or processes used should be communicated clearly and effectively. When a case is approached in this manner, anthropologists can be confident of a positive outcome for their portion of the case. Consistently and completely documenting the entirety of an analysis is best practice, while cutting corners risks a negative professional outcome.

There are two main approaches to sex determination for the human skeleton: morphological and metric. Each approach has advantages and detractors, and each technique has associated error rates. Understanding the limitations to each method is crucial to the correct diagnosis of sex. In addition, the approach employed will depend on the elements present for analysis. Archaeological preservation, taphonomic change, scavenger activity, intentional destruction, and other factors affect the skeletal elements that may be present for analysis. The competent anthropologist evaluates each case; uses the best approach, given the remains present; and frequently employs both morphological and metric methods. Techniques used in both approaches will be examined in detail later, focusing on various regions of the skeleton.

MORPHOLOGICAL APPROACHES TO SEX DETERMINATION

Traditional morphological approaches to sex determination from the human skeleton focus on the pelvis and the skull. As one would expect, based on human reproductive differences, the pelvis is highly sexually dimorphic, and even on gross observation, females are usually quite easy to separate from males. Similarly, multiple morphological traits on the human skull are sexually dimorphic; however, the accuracy rates tend to be lower than those of the pelvis. Anthropologists can expect a sexing accuracy rate of 70% to high 80% from the cranium and 90%–95% from the pelvis. The most common morphological traits for both areas are defined and scoring systems are provided in *Standards for Data Collection from Human Skeletal Remains* (Buikstra and Ubelaker 1994). The scoring systems are employed widely across the biological anthropology field and have been improved and quantified by later research (e.g., Garvin et al. 2014; Klales et al. 2012; Walker 2005, 2008). The pelvis should be used as the primary indicator of sex, when present.

Pelvis

Several distinctive traits of the pelvis are used for sex determination. In gross appearance, the female pelvis (to include the sacrum) is much broader and has a round pelvic inlet compared with the narrower male counterpart, with a heart-shaped inlet. The pelvic outlet is also narrower and more constricted in males than in females. The pelvic blades are broader and more flared in females and smaller and narrower in males. The male sacrum is narrower and more anteriorly curved than the female sacrum. Frequently, the female sacroiliac joint forms a low plateau above the surrounding bone, while in males, the normal form is an even contour that forms a smooth surface. As is generally true with the rest of the skeleton, the male pelvis is larger and more muscle marked than the female pelvis. Figure 8.2 shows a male pelvis and a female pelvis for comparison.

Arguably, the most accurate morphological sexing method was developed in the late 1960s by Phenice (1969). In his study of 275 males and females from the Terry Collection, he developed a visual method of sexing the pelvis that has an accuracy rate of 95% or greater. The Phenice method is based on three traits of the subpubic region: the ventral arc, the degree of subpubic concavity, and the shape of the medial aspect of the ischiopubic ramus. The pubic bones meet anteriorly in the midline of the body at a joint called the pubic symphysis; the area inferior to this joint is the subpubic region. The lateral and ventral portions of the joint complex are the pubic body, or the pubic bones. The ventral arc is a slightly elevated ridge of bone that runs inferiorly and laterally across the ventral surface of the pubis, merging with the medial border of the ischiopubic ramus (Figure 8.3a and b). In females, this ridge of bone is nearly always present; in males, it is usually absent. Subpubic concavity is observed inferior to the pubic symphysis and is the shape of the ischiopubic ramus,

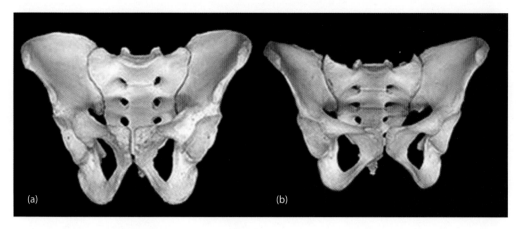

Figure 8.2 (a) Male pelvis versus (b) female pelvis.

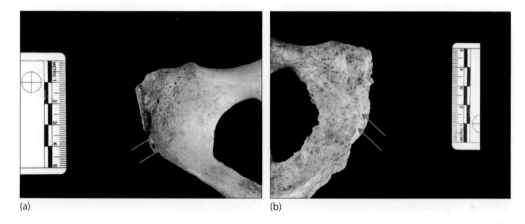

Figure 8.3 (a) Female pubis (ventral aspect): arrows indicate the ventral arc. (b) Male pubis (ventral): arrows show location where ventral arc would be located; this feature is absent in males.

when examined in an antero-posterior orientation. In females, the ischiopubic ramus exhibits a large concavity in this location, whereas in males, there is little to no concavity (Figure 8.4a and b). For the final trait, the pelvis is oriented such that the pubic symphysis is directly facing the observer. Females exhibit a narrow region of bone, sometimes containing a sharp ridge directly inferior to the pubic symphysis, while males exhibit a wide and flat medial aspect to the ischiopubic ramus in this location (Figure 8.5a and b). Collectively, these three traits are now referred to as the *traits of Phenice*. If there is no ventral arc or subpubic concavity, and the medial aspect of the ischiopubic ramus is wide and flat, the pelvis is a male's, and this determination is 96% accurate. Conversely, the pelvis that exhibits a ventral arc, subpubic concavity, and a sharp medial aspect of the ischiopubic ramus is female's. The traits can also be used singly, or in different combinations, but the accuracy rates decline when all three are not used together (Phenice 1969).

Recently, Klales and colleagues (2012) revised the Phenice method to clarify the morphological states and add statistical surety to the method. They modified the original scoring method for the three traits and expanded the definitions of each trait into five distinct morphological states, instead of the original binary scheme. They produced new scoring definitions, augmented with line drawings and photographic examples. To test these revisions, they employed a random sample of individuals (primarily American whites and blacks) from the William Bass Donated Skeletal Collection and the Hamann-Todd Human Osteology Collection ($n = 310$ pelves). As in the original study, the ventral arc produces the greatest correct sex classifications (88.5%). They provide a logistic regression equation that employs all three features. The equation has a 94.5% accuracy rate for experienced observers and a slightly lower rate for less-experienced practitioners. Inter- and intraobserver error tests show very good agreement between observers, regardless of the experience level, indicating high reliability of the method for scoring each morphological feature. The authors also provide a website to assist with the revised method (http://nonmetricpelvissexing.weebly.com/).

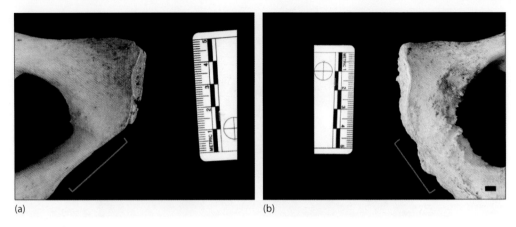

Figure 8.4 (a) Female pubis (dorsal view): bracket showing area of ischiopubic concavity. (b) Male pubis (dorsal): bracket indicating absence of concavity.

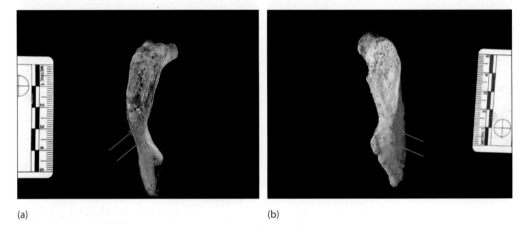

Figure 8.5 (a) Female pubis (medial view): arrows pointing to narrow ischiopubic ramus, with a ridge on the medial aspect. (b) Males pubis (medial view): arrows indicating wide, flat ischiopubic ramus.

Two other pelvic traits are examined using a standardized scoring system: the greater sciatic notch and preauricular sulcus (Buikstra and Ubelaker 1994). The most important of these, the greater sciatic notch is usually a wide, open notch in its female form and a narrow, U-shaped notch in the male expression. The notch is scored on an ordinal scale from 1 (female) to 5 (male) (Figure 8.6a through e). A study by Walker (2005) using an American/British white sample and an American black sample ($n = 296$) found that sciatic notch scores of 3 or greater were 91% likely to be male and those with a score of 1 were 88% likely to be female. The indeterminate category, a score of 2, was still more likely to be from a male (66% chance) than from a female. The preauricular sulcus is scored as 0 (absent) or in a range of 1–4 (depending on the character states of the sulcus). This sulcus is much more variable; a female expression tends to be a large, wide sulcus with deep pits, while the male expression is either absent or a small, narrow, and shallow groove (Figure 8.7a and b). Due to its variability, this trait is not as powerful as other pelvic sex indicators, as males and females frequently exhibit contradictory morphology.

Skull

Unlike the pelvis, which has sexually dimorphic features due to the necessary reproductive differences, the cranial shape differences are due primarily to size. These differences manifest with the onset of puberty. Many experienced anthropologists consider the skull the second best area in the body to use for morphological sex estimation. Relative to male crania, female crania are smaller and more gracile. Several tendencies are present: male crania have narrower and less bossed parietal and frontal bones; the posterior root of the male zygomatic process typically extends as a crest beyond the external auditory meatus; and the frontal sinuses are larger. Most authors do not recommend sexing

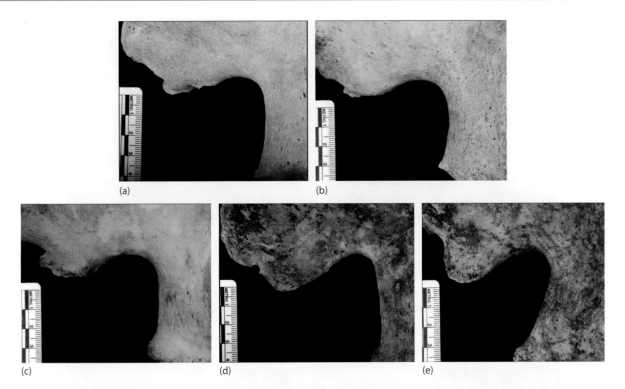

Figure 8.6 (a) Wide sciatic notch (score = 1; female). (b) Relatively wide sciatic notch (score = 2; probably female). (c) Score = 3 (ambiguous). (d) Relatively narrow sciatic notch (score = 4; probable male). (e) Narrow sciatic notch (score = 5; male).

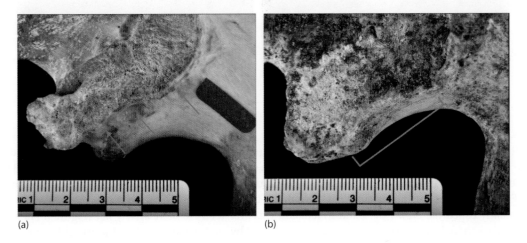

Figure 8.7 (a) Female sacroauricular joint with preauricular sulcus. Arrows showing pits in the sulcus. (b) Male sacroauricular joint. Bracketed area demonstrating absence of preauricular sulcus (score = 0).

the skull without other individuals from the population from which it was drawn from as comparatives because morphological traits are not without intermediate forms, and absolute differences seldom exist (Bass 2005; Braz 2009; White and Folkens 2005). Overall, accuracy rates for sexing skulls range from 70% to the high 80%, but the higher accuracy rates are attributable to more experienced anthropologists (Stewart 1979).

Scoring systems for skull morphology have been codified for several important features (see Buikstra and Ubelaker 1994). Five features are typically scored: nuchal crest, mastoid process, supraorbital margin, glabellar region, and mental eminence (chin). Scores range from 1 (female) to 5 (male). On the posterior aspect of the skull, a large nuchal crest, to include having an inion hook, is a very male character state, whereas a smooth nuchal region without

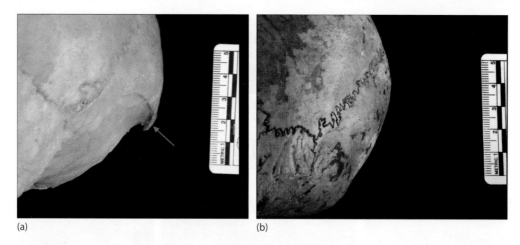

Figure 8.8 (a) Male nuchal crest with an inion hook (score = 5). (b) Female nuchal region—note the absence of rugosity (score = 1).

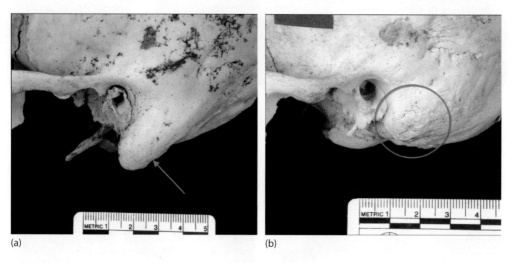

Figure 8.9 (a) Male mastoid (score = 5). (b) Female mastoid (score = 1).

prominent ridging is a very female expression (Figure 8.8a and b). The mastoid process in males is wider and longer in relation to the external auditory meatus, as compared with a smaller and shorter process in females (Figure 8.9a and b). The supraorbital margin in males is rounded and blunt, and in females, it is more superiorly angled, creating a sharp contour (Figure 8.10a and b). The glabellar region in males is more robust and pronounced, showing a heavy ridge of bone (frequently called the *brow ridge*), and in females, this area tends to be relatively flat (Figure 8.11a and b). The mental eminence in males has a massive bony expression, and by contrast, in females, this area is less prominent and continuous with the contour of the mandibular body (Figure 8.12a and b) (but see caveats below).

Building upon these trait categorizations, Rogers (2005) set out to examine a suite of 17 morphological traits of the skull and determine their effectiveness in sex determination. She identified four traits as particularly useful for visual sex assessment: nasal aperture, zygomatic extension, malar size/rugosity, and supraorbital ridge (glabella). Williams and Rogers (2006) then examined 21 cranial traits for accuracy and precision. Using a modern sample of 50 European white crania, they identified six skull features that achieved high precision and accuracy: mastoid, supraorbital ridge, skull size and architecture, zygomatic extension, nasal aperture, and gonial angle of the mandible. Each trait scored over 80% accurate for sex estimation, with ≤10% intraobserver error. The most accurate feature was the mastoid, with an accuracy rate of 92%.

Similar to the morphological traits of the pelvis, Walker (2008) examined the likelihood of correct sex determination by using five skull features. His study sample was composed of 304 modern American and English white individuals

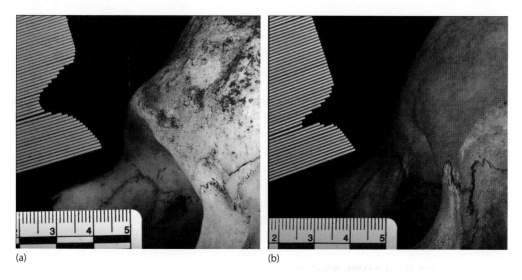

Figure 8.10 (a) Male supraorbital margin (score = 5). (b) Female supraorbital margin (score = 1). The contour gauge in upper left corner is showing the profile of the supraorbital margin. This instrument was pressed onto the supraorbital margin in order to obtain a profile view of the sharpness/thickness of the margin.

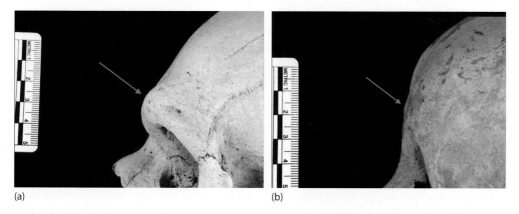

Figure 8.11 (a) Male glabellar region (score = 5). Note the prominent glabellar protuberance indicated by the arrow and the sloping frontal bone. (b) Female glabellar region. Note the absence of a glabellar ridge and the vertical frontal bone.

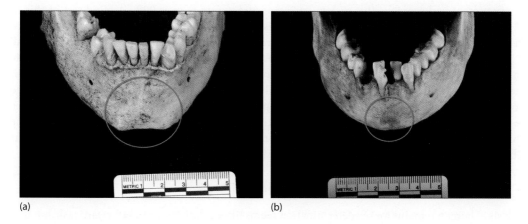

Figure 8.12 (a) Male mandible showing prominent mental eminence (score = 5). The male mental region is square compared with females. (b) Female mandible (score = 1/2) with small mental eminence and a more V-shaped mental region.

and a Native American test sample. Using a statistical approach, he provided empirical probabilities of being male for a given score for each cranial trait, as well as logistic discriminant equations that can be used to evaluate sex from multiple traits. Walker scored each trait on a scale of 1–5, with "1" being the most female form of expression and "5" being the most male form of expression. If a single trait is evaluated and scored as a 4 on the scale of 1–5, the probability of the individual being male is 0.9, or 90%. If multiple traits are evaluated, discriminant function equations are used to estimate sex. For example, one equation that scores the glabella, mental eminence, and mastoid process can predict sex with an approximate 87% accuracy. Walker's work provides anthropologists with an estimate of the error rate for sex determination from morphological traits of the cranium. This popular five-trait method is referred to as the *Walker method*.

Garvin et al. (2014) argued that a surprising lack of formal validation studies has tested the Walker method and therefore the method on six different populations. They explored additional variables such as body size, age, and population with regard to cranial morphology to determine if these factors influence correct sex estimates. Their sample of nearly 500 males and females included US blacks and whites, as well as Native Americans and medieval Nubians. In general, their research supports Walker's (2008) early findings, with a pooled sample of all populations reaching a correct sex classification of 85%. Several other findings were part of their work. First, Garvin et al. (2014) found that age and body size do not affect sex estimates significantly enough to warrant their inclusion in sex estimation procedures, but cranial trait scores do vary significantly by population. To this end, they highly recommended that sex estimation of an unknown skull should employ a statistical approach (e.g., discriminate function analysis) that is population-specific. Second, they re-affirmed the notion that some cranial traits are more dimorphic than others—glabella and the mastoid process were the most reliable traits, while nuchal crest and mental eminence fared the poorest (see Walker 2008; Williams and Rogers 2006).

Multiple authors have tested sex determination from the mandible, with varied results. Loth and Henneberg (1996) produced a method using the posterior border of the mandible and reported a sexing accuracy rate of over 90%. Blind tests of this method by Donnelly et al. (1998) and Haun (2000) showed much lower accuracy rates, between 62% and 68%. Berg (2008, 2014) explored seven mandibular morphological traits for sex determination; nearly all of the traits showed some amount of sexual dimorphism, but the traits were population-dependent. Discriminant functions based on up to five traits achieved cross-validated accuracy rates between 63% and 80%, but the magnitude and direction of the misclassification rates were such that they are not recommended as stand-alone traits to determine sex, without accompanying mandibular metric variables. Thus, while the mandible offers utility for sex determination, using morphological traits, accuracy rates may not warrant sole reliance on them.

Other postcranial bones

Other morphological methods for the postcranium focus on the distal humerus and the medial clavicle. Rogers (1999) devised a system for scoring sexually dimorphic characters of the distal humerus by using a predominantly American white skeletal sample. She examined the shape of the olecranon fossa, the angle of the medial epicondyle, the degree of trochlear constriction, and the amount of trochlear asymmetry. Females have a more oval olecranon fossa, an angled medial epicondyle, and a more symmetrical and constricted trochlea than males. When all four characters are present and scorable, this system has an accuracy rate of over 90%. A blind test of this method by Falys et al. (2005) found that, indeed, the distal humerus is sexually dimorphic and is recommended for use. For the medial clavicle, Rogers et al. (2000) determined that the presence or absence of a rhomboid fossa is a sexually dimorphic trait. Rogers et al. (2000) examined the rhomboid fossa on 344 clavicle pairs from the William F. McCormick Collection (231 males and 113 females). A deeply pitted rhomboid fossa (an excavation on the inferior medial clavicle at the insertion of the costoclavicular ligament) is typically associated with male individuals. Using a grading scale that employs alpha designators for the various states of expression, a rhomboid fossa scoring at least a "C" on the left side of the body indicates a male individual 92.2% of the time, and if present on the right side, 81.7% of the time. As with any morphological sex determination method, best practice stipulates that comparison photographs and/or line drawings be consulted in the original publications to classify and assign scores to visually assessed traits.

METRIC APPROACHES TO SEX DETERMINATION

Metric analysis of human remains for sex determination seems like a relatively straightforward task, but it can be very complex. Metric evaluations have been used in anthropology since the inception of the discipline and even before this time, when anthropology was principally incorporated into the field of anatomy. Innumerable publications on metric sex determination exist on nearly every bone in the human body, using a wide variety of statistical approaches. These methods typically use some of the 34 skull and 43 postcranial measurements that are standard for human skeletal documentation (Buikstra and Ubelaker 1994; Moore-Jansen et al. 1994). Commonly, cranial measurements are employed, followed by various diameters and lengths of the postcranial skeleton; however, other measurements have been used. This section is primarily dedicated to providing the reader with an understanding of the principles of several commonly encountered statistical methods (e.g., sectioning points and discriminant functions) used in metric sex determination methods, so the reader can use any number of published sources easily and effectively. In addition, a brief overview of the computer program FORDISC 3.1 (Jantz and Ousley 2005) is highlighted. (For a more detailed description of this program, see Chapter 14.) Other statistical methods, such as simple and multiple linear regression analyses and principal component analysis, are beyond the scope of this chapter.

Examples of sex determination using quantitative measures are proliferative in the anthropological literature, and because of this abundance, only a few examples are discussed in this introductory text. Cranial measurements, initially explored by Giles and Elliot (1963) in the infancy of multivariate metric sex determination, yield relatively high accuracy rates (typically 80%–90%). Their work was built upon by other researchers, who surveyed sexual dimorphism from world populations (cf Giles 1970; Howells 1973, 1989). When a complete skull is not present, fragments from the cranial base (Holland 1986), as well as complete or partial mandibles (Berg 2008, 2014), can be used to determine sex. Postcranial elements are more effective metric sex estimators than the skull (Spradley and Jantz 2011). Frequently, measurements from the humerus, femur, tibia, or other postcranial elements yield correct sex determinations, with 80%–95% accuracy rates (Dittrick and Suchey 1986; France 1998; Holland 1991; Holman and Bennett 1991; Iscan and Miller-Shaivitz 1984; King et al. 1998; Spradley and Jantz 2011; Spradley et al. 2015; Stewart 1979; Tise et al. 2013). Even the metacarpals and metatarsals produce similarly high accuracy rates (Case and Ross 2007; Lazenby 1994; Robling and Ubelaker 1997). These are just a few studies that anthropologists have to choose from when determining sex from human remains. But, how do most of these methods work?

A sectioning point is a simple statistical approach to metric sexing of the skeleton. A sectioning point for a data set is the mathematical point that maximizes the differences between two groups (e.g., males and females) in one dimension. It is a univariate analysis method (an analysis using one variable instead of many variables) and is common in older publications, although it also appears in recent research articles (Berrizbeitia 1989; MacLaughlin and Bruce 1985; McCormick et al. 1991). A sectioning point can be thought of as the point where two data distributions, frequently expressed as curves, overlap (Figure 8.13). This approach generally uses one variable, such as femur head diameter; however, multiple variables expressed as a ratio (which reduces the variables to a single variable) can be incorporated. The accuracy of a sectioning point depends on the amount of sexual dimorphism present in the data spread of two populations. The data spread is the total value range for a given measurement in an analysis, regardless of population. For example, the femur head diameter might have a data spread for white males and females ranging from 34 mm (extremely female) to 57 mm (extremely male). Highly sexually dimorphic measurements have a large data spread, with little overlap between males and females, and therefore, they produce excellent prediction

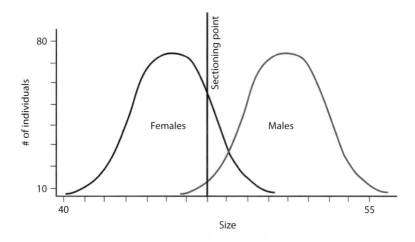

Figure 8.13 The sectioning point for a single variable from two data distributions (male and female). (Courtesy of J. Hefner.)

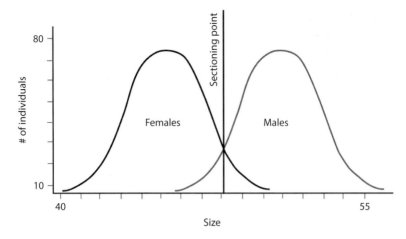

Figure 8.14 An adjusted sectioning point for a single variable from two data distributions (male and female). As this sectioning point has been shifted toward the male distribution, it will classify males correctly at the expense of misclassifying females. (Courtesy of J. Hefner.)

accuracies. Conversely, a narrow data range, with a large amount of data overlap between the sexes, produces lower accuracy rates. While this sounds simple, complex sectioning points can be created that minimize error in one direction, while maximizing it in another direction. For example, a sectioning point can be created that is 90% accurate for male sex determination but misclassifies females 30% of the time. This occurs when the sectioning point is moved along the data curve until 90% of one group is to the right of the point (Figure 8.14). Such accuracy trade-offs can be valuable but should be used judiciously.

In the anthropological literature, sectioning point methods are expressed as a single value, with males having higher values than females—remember, males are typically larger than females. For example, a sectioning point for the maximum diameter of the femur head might be expressed in a publication as

$$\text{Female} = <45.0 \text{ mm}> = \text{Male}$$

Interpreting this equation is straightforward: all measurements <45.0 mm are classified as female, and all measurements >45.0 mm are classified as male. Section point methods can have an indeterminate category built into the equation, as well. An example of a sectioning point with an indeterminate category can be found in Stewart's (1979) sex determination method for the maximum diameter of the femur head. His analysis of 100 white and black males and females in the Terry Collection generated the following sectioning system:

$$\text{Female} = <43.0 \text{ mm}, 44.0–46.0 \text{ mm} = \text{Indeterminate}, >45.0 \text{ mm} = \text{Male}$$

For this study, all individuals with femur head measurements of 44.0–46.0 mm cannot be sexed accurately, while those with measurements of 43.0 mm and below are females and those with measurements of 47.0 mm and above are males. In applications such as the previous example, an accuracy level could be established for a predetermined value, perhaps, 90%. Therefore, the high and low sectioning points would be 90% accurate and the values in between would be classified as indeterminate for sex. Since there are so many different ways of generating a sectioning point, it is imperative to know how each was created, in order to avoid misapplication.

Sexing the human skeleton by using metric measurements greatly moved forward with the adoption of the discriminant function statistic developed by Fischer in the 1940s. For many years, discriminant function analysis was unpractical and unwieldy, requiring many hand calculations and long formulae to develop each classification function. The personal computer changed that situation; today, calculations are automated and the time needed to generate a discriminant function has become minutes or seconds, instead of days and weeks. The application of these computer-generated functions is easily conducted with a hand calculator in a matter of minutes. In forensic anthropology today, discriminant function analysis is relied upon heavily for sex assessment.

The discriminant function has allowed anthropologists to move away from univariate and simpler statistical approaches into complex multidimensional space. The objective of a discriminant function is to maximize the differences between two or more groups. The function accomplishes this objective by weighting and combining two or more discriminating variables in a linear manner, so that the intercorrelations of the variables are considered and the ratio of between-group variance to within-group variance is maximized (Pietrusewsky 2000). These transformed variables are used to place or assign individuals or groups into multidimensional space. The distances between these groups then can be visualized or measured. Each individual in the analysis is classified into one of the original groups based on his or her discriminant score, and the probability of group membership is calculated. In this manner, unknown individuals can be placed into the analysis and their probability of group membership to one of the reference populations can be computed (see Pietrusewsky 2000 for a detailed explanation of the computations). Leave-one-out analyses (also called cross-validation analyses) or boot-strapping analyses of the function are useful tools to determine the unbiased classification accuracy of a discriminant function. These analyses are computed within the computer program generating the function, which outputs the cross-validated or boot-strapped accuracy rates as part of the results. A good discriminant function produces a cross-validated accuracy rate between 85% and 95%.

In a more simplified explanation, the discriminant procedure takes the analytical variables and rank-orders them based on the amount of information they contain about differences between the reference populations. A centroid is computed for each reference population. (The centroid is the average value of all variables entering the analysis for a population in multidimensional space.) Once these positions are determined, an unknown individual can be mathematically mapped into the same multidimensional space, and the distance from each reference population centroid can be measured. These distances are used to classify the unknown; the smaller the distance, the higher the classification score. A large assumption made when using discriminant function analysis is that the unknown individual to be classified is actually *from* one of the reference populations. If the unknown does not originate from any one of the reference populations, the individual will classify, but the probability of group membership will likely be extremely low.

A discriminant function is presented as a formula containing the variables, each multiplied by a function weight, added, or subtracted together and added to a constant. The constant is a special number that is computed for the function and does not vary (hence the name). The resulting value is compared to a sectioning point (frequently set to 0), and the groups (in this case, the sexes) are on either side of zero. A three-variable cranial measurement sex function may look like this:

$$X = (0.245) + BPL(0.123) - WFB(0.754) + 0.232$$

In this example, ZYB (bizygomatic breadth) is multiplied by 0.245 and added to the product of BPL (basion-prosthion length) and 0.123; the resulting value has the product of WFB (minimum frontal breadth) and 0.754 subtracted from it, and finally, 0.232 is added to the total. If X is greater than 0, the skull is male, if it is less than 0, then it is female. Functions typically compare two groups at a time; for more than two groups, the variables have to be added into a second, third, or fourth function, and those results will determine the final classification. If more than two groups are in the analysis, for example, black and white males and females (a four-way discriminant function analysis), the function will be computed in a computer and not by hand, as the analysis is too complex.

Discriminant function analysis sex determination of unknown human remains has become fairly commonplace, particularly due to the introduction of the computer program FORDISC, developed at the University of Tennessee by Richard Jantz and Steve Ousley. As of 2015, the program is in its third version (FORDISC 3.1). This program is used widely in forensic anthropology for constructing custom discriminant functions to classify unknown remains into populations. In its most simplistic use, the program allows the user to enter the standard cranial and postcranial measurements of an unknown skeleton into a user interface. The user selects either cranial or postcranial measurements, as well as the reference populations against which to compare the unknown (e.g., white males and females). Owing to limitations of the program, cranial and postcranial measurements cannot be analyzed simultaneously, and the program in its present version cannot be used to determine the sex of unknown remains solely; instead, sex determination must be made in the context of ancestry. Therefore, prior knowledge of the unknown's ancestry is helpful but not necessary—large analyses that include multiple ancestral reference populations can be undertaken easily. After the data are entered, FORDISC creates a series of custom discriminant functions that classifies the unknown remains into one of the chosen reference populations. The cross-validated classification accuracy for the overall model is provided in the final output. In addition, the program calculates posterior and typicality probabilities (additional analyses that inform the user about the likelihood of a correct assignment to a given population, and how typical [i.e., metrically similar] the unknown is for that population). Therefore, the practitioner can formulate a detailed and informed opinion, based on the output generated from the program.

In FORDISC, the available reference data vary depending on the region of the skeleton. The software contains more reference populations for cranial data than for the postcranial bones. Standard cranial reference populations include white, black, Native American, Hispanic, and Japanese males and females, as well as male samples of Vietnamese, Guatemalan, and Chinese populations. For the postcranial measurements, only black and white males and females have representative populations. A separate module in the program contains additional cranial data sets, including Howell's sample of worldwide populations and the University of Tennessee data sets of the nineteenth and twentieth century black and white groups. This module has additional, nonstandard measurements that can be used in the analyses. In more advanced applications of the program, users can import their own data, create discriminant functions, and classify unknown remains by using custom functions. In addition, the program allows the user to select specific statistical procedures to be used in the function generation and permits the use of shape transformations for cranial analyses. Since this program can conduct analyses containing only a few to a large number of variables, it is extremely flexible and is usually a principal and valuable component in the anthropologist's toolbox for estimating sex and ancestry from unknown human remains.

With any metric analysis, the user must understand the limitations of the method. In the case of FORDISC, the user must be aware that the program will classify the unknown remains into one of the provided reference populations, *regardless* of whether the unknown is actually from *any* of the reference populations (refer to the above discussion on discriminant analysis). The practitioner must interpret the data and determine if the statistical probabilities support classification of the unknown remains into one of those groups. Often, this issue is more complicated in the assignment of ancestry and less so for the determination of sex. If the output statistics indicate only male classifications with high accuracy rates, then in all probability, the unknown remains are male. Having stated this point, if the analyzed remains are *not* from the reference populations provided in the program, the probability of a sex misclassification will increase, and the user should be aware of this possibility.

POPULATION-SPECIFIC STANDARDS

Throughout this chapter, various allusions have been made to using appropriate reference populations when conducting a sex estimation analysis. This issue has gained attention over the last decade, and it is increasingly apparent that we, as forensic practitioners, need to be aware of the potential benefits and pitfalls associated with skeletal analyses using limited populations as reference groups. Researchers are attempting to develop new standards (metric and morphological) for populations across the globe that are deemed forensically important (e.g., Saini et al. 2012; Tise et al. 2013; to name a few). Much effort is being spent on characterizing populations that have received little focus in the foundational literature, such as Hispanics, Indians (subcontinent), Thai, Japanese, and Koreans. The impact of these studies on the practice of modern forensic anthropology is invaluable, because before their inception, little reference data were available to classify unknowns with metric and morphological characteristics dissimilar to the existing reference databases (i.e., largely American white, black, and Native American individuals). Increasing diversity in the American population demands updated standards and reference data for all possible unknown forensic cases. But, why must forensic anthropologists use population-specific formulae for sex estimation?

A short perusal of the research reported in Spradley and Jantz (2011) provides the answer. For postcranial elements, the most accurate sex estimator (skeletal element) varies based on ancestry, or population group. Degrees of sexual dimorphism and body proportions may vary between populations; therefore, the most accurate sex estimator may be different between groups. For example, the femur head is considered a classic measurement to use for sex estimation in white groups, but the length of the clavicle is more accurate for sex estimation in Hispanic individuals (Tise et al. 2013).

When we consider the possibilities in a forensic case, the potential groups from which a person may be represented are largely context-specific. For example, in the United States, an unknown decedent found in the desert between Tucson, Arizona, and Mexico most likely is not of Chinese descent. Methods that have the highest likelihood of a correct classification for sex estimation should be based on contextual information. In this example, a selection of reference groups from Hispanic, white, and black ancestries is arguably the logical first step. If results conflict, then widening the comparison population groups is an advisable second step. If we consider another scenario, say, a partial skeleton found in a garbage dump in Sacramento, California, what reference populations might we use in our analysis? Perhaps, in this instance, using the greatest number of possible groups in a single analysis will be the correct starting point, and narrowing the search based on the initial results would be an appropriate way forward. Alternately, conducting analyses to determine the ancestry of the unknown may be another way forward when the pool of possibilities is extremely large.

WHEN METHODS DISAGREE

One of the trickiest problems in sex determination is when conflicting results are obtained from the same skeleton. Each anthropologist has his/her own way of dealing with this particular problem. The solution is based on experience, the types of analyses conducted, and familiarity with and confidence in the available methods; all of these solidify to form a sort of *gestalt* in the mind of the anthropologist. A gestalt is the unwritten criteria needed to assign a sex category or other determination to a set of unknown remains. A novice anthropologist may not be willing to commit to a sex assignment for a particular set of remains based solely on midrange cranial morphology, whereas a more experienced practitioner may definitively state sex for the same remains due to trust in the methods and level of experience. As experience increases, the use of a gestalt approach yields extremely good results, often better than 80% accurate, even for decisions that are made quickly (Berg and Tersigni-Tarrant 2014). Regardless, what is absolutely necessary for sex determination is the proper documentation of the tests and observations that form the basis of the opinion. Without appropriate and extensive documentation, the anthropologist's opinion can be challenged, potentially jeopardizing the outcome of an analysis.

Experience level aside, anthropologists must have a method for weighting sex assessments from various skeletal elements. In this chapter, sex determination from the pelvis is emphasized. The pelvic girdle should be consulted first because of the distinct *biological* reasons for differences between males and females in this region of the skeleton. According to the guidelines produced by the SWGANTH (2010, p. 2), "Methods based on the shape and size, and the presence or absence of features of the pelvis are preferred." However, if the markers are missing, or the pelvis is incomplete, then on which methods does the anthropologist place the most emphasis for sex assignment, particularly if the results disagree?

Until recently, few statistical methods were available to determine error associated with morphological sex assessments from skull and pelvic features. However, methods are available now that quantify the amount of error associated with morphological sex determination from the pelvis and skull or assign a probability that the answer is correct (Garvin et al. 2014; Klales et al. 2012; Walker 2005, 2008). In addition, morphological sex determination from several other skeletal elements has calculable error rates. Metric assessments of skeletal remains using discriminant functions or sectioning points also provide error rates, and in many cases, postcranial metrics are more accurate sex estimators than cranial measurements (Spradley and Jantz 2011). In the end, it is a matter of weighing the error and determining the method(s) with the smallest chance of a misclassification.

A final set of options for sex determination is available to the anthropologist, which is to use qualifying language to describe the results of a particularly troublesome analysis. Since there are only two biological states, male and female, there can then be up to *three* additional qualified states: probable male, probable female, and indeterminate. A *possible* sex estimate cannot exist, because if it is *possibly* female, then it could be *possibly* male as well. The *indeterminate* state is used in those cases that have a poor representation of skeletal elements, in which a sound scientific opinion cannot be rendered. The *probable* categories are used in situations where the majority of the evidence indicates one sex, but the associated method's error rates do not inspire high confidence in a definitive sex assignment. It is up to the practitioner to decide how much error is allowed before placing a set of remains into a *probable* category. Conversely, a *probable* category can be used to *upgrade* an assignment from the *indeterminate* category, if sufficient justification exists. As a practitioner becomes more familiar with various methods and examines cases more frequently, the judicious use of the *probable* category will undoubtedly come into the anthropologist's toolkit.

CONCLUSION

Sex determination of human skeletal remains is a primary anthropological undertaking, whether in cases of medico-legal interest or in research on ancient populations. Determining sex is foundational first step for developing the rest of the biological profile. An accurate determination is necessary, even if the most accurate assignment is *indeterminate*. Morphological sex determination from the pelvis is the best indicator for a complete set of remains. In cases of partial remains, the skeletal elements available will dictate the most suitable approach. Frequently, both morphological and metric methods are utilized to assess sex. Proper documentation of the tests and findings is imperative. Increasing practice and familiarity with sex estimation methods make the task of sexing a set of unknown remains more accurate and straightforward.

SUMMARY

- Forensic anthropologists determine the sex of an individual, not the gender. Sex is a biological condition or state, and gender is a social construct.
- Sexual dimorphism can be characterized as metric (size) and morphological (shape or form) differences in the human skeleton.
- Proper and extensively written and photographic documentation of all observations, calculations, and determinations should be generated for any skeletal analysis.
- Sex determination should not be attempted on individuals younger than 12 years.
- The most accurate skeletal element from which to determine sex is the pelvis.
- While many morphological methods for sex determination from the pelvis are available, those utilizing the Phenice traits are arguably the most popular and produce the best results.
- Simple metric sex determination methods utilize sectioning points. More complex methods are constructed by using discriminant function analysis or regression models.
- Postcranial metric methods can achieve better accuracy rates than cranial metric and morphological methods.
- Population-specific methods should be utilized, when possible.
- When methods disagree, error rates, probabilities, and experience play a role in the ultimate determination.
- Categories utilizing *probable* designators or an *indeterminate* category are acceptable.

Review questions

1. What is the difference between sex and gender? What types of evidence lends itself to one or the other?
2. What are the necessary components for documentation when conducting skeletal analysis?
3. What are the two broad types of variables used in sex determination methods? When and why is one preferred over the other?
4. When present, what is the primary skeletal element that should be used for sex determination? Why?
5. Describe the three traits of Phenice, and where are they located? How did Klales et al. (2012) modify the Phenice method?
6. What is a sectioning point? Can it be changed or modified? Explain.
7. Can a skull be classified using a program such as FORDISC if the true population is not present in the reference populations? What are the effects on the classification when the reference population is not present?
8. Why is ancestry an important aspect of sex determination?
9. What biological categories should be considered before conducting a sex determination on an unknown case?
10. What should the forensic anthropologist do when conflicting results are obtained in a case?
11. What are the three qualified states or categories that can be used in sex determination? Why are they used?

Acknowledgments

The author and editors would like to thank the C.A. Pound Human Identification Laboratory at the University of Florida and the University of Tennessee Department of Anthropology for use of the skeletal elements pictured in Figures 8.2 through 8.12.

Glossary

Discriminant function: a mathematical function that maximizes the differences between two or more groups in multivariate space.

Gender: the social role or characterization of biological sex to which an individual ascribes or wishes to be perceived.

Gestalt: an impression of whole form that is derived from a summation of its parts but cannot be teased apart. The visual assessment of sex or ancestry from a skeletal Gestalt is the product of many years of experience and is difficult to quantify.

Morphology: the shape of a skeletal feature/trait or skeletal element.

Scientific Working Group for forensic ANTHropology (SWGANTH): a governmental working group assigned to determine best practice for the field of forensic anthropology. The working group was funded primarily through the joint POW/MIA Accounting Command and the Federal Bureau of Investigation. This working group has been superseded by the working group for anthropology through NIST (the National Institute of Standards and Technology).

Scoring system: a system by which morphological states of a skeletal feature are segregated using concise definitions. Scores of sexually dimorphic traits are devised as graded expressions (e.g., score = 1, 2, 3, …) starting from a gracile expression and progressing to a robust expression of the feature.

Sectioning point: a mathematical point that maximizes the differences between two groups in one dimension for a given feature or skeletal measurement.

Sexual dimorphism: differences in size and shape between males and females of a given species. Forensic anthropologists are concerned with sexual dimorphism of skeletal features.

References

Bass WM. 2005. *Human Osteology: A Laboratory and Field Manual*, 5th edn, Special Publication No. 2. Columbia, MO: Missouri Archaeological Society.

Berg GE. 2008. Biological affinity and sex determination using morphometric and morphoscopic variables from the human mandible. PhD dissertation, University of Tennessee, Knoxville, TN.

Berg GE. 2014. Biological affinity and sex from the mandible utilizing multiple world populations. In G Berg and SC Taala (Eds.), *Biological Affinity in Forensic Identification of Human Skeletal Remains: Beyond Black and White*. Boca Raton, FL: CRC Press, pp. 43–82.

Berg GE and Tersigni-Tarrant MT. 2014. Sex and ancestry determination: Assessing the "gestalt." *Pro Acad Forensic Sci* XX:414–415.

Berrizbeitia EL. 1989. Sex determination with the head of the radius. *J Forensic Sci* 34:1206–1213.

Buikstra JE and Ubelaker DH (Eds.). 1994. *Standards for Data Collection from Human Skeletal Remains*. Fayetteville, AR: Arkansas Archeological Survey Research Series No. 44.

Braz VS. 2009. Anthropological estimation of sex. In S Blau and DH Ubelaker (Eds.), *Handbook of Forensic Anthropology and Archaeology*. Walnut Creek, CA: Left Coast Press.

Case DT and Ross AH. 2007. Sex determination from hand and foot bone lengths. *J Forensic Sci* 52:264–270.

Dittrick J and Suchey JM. 1986. Sex determination of prehistoric Central California skeletal remains using discriminant analysis of the femur and humerus. *Am J Phys Anthropol* 70:3–9.

Donnelly SM, Hens SM, Rogers NL, and Schneider KL. 1998. Technical note: A blind test of mandibular ramus flexure as a morphologic indicator of sexual dimorphism in the human skeleton. *Am J Phys Anthropol* 107:363–366.

Falys CG, Schutkowski H, and Weston DA. 2005. The distal humerus—A blind test of Rogers' sexing technique using a documented skeletal collection. *J Forensic Sci* 50:1–5.

France DL. 1998. Observational and metric analysis of sex in the skeleton. In KJ Reichs (Ed.), *Forensic Osteology: Advances in the Identification of Human Remains*, 2nd edn. Springfield IL: Charles C Thomas, pp. 163–186.

Garvin HM, Sholts SB, and Mosca LA. 2014. Sexual dimorphism in human cranial trait scores: Effects of population, age, and body size. *Am J Phys Anthropol* 154:259–269.

Giles E. 1970. Discriminant function sexing of the human skeleton. In TD Stewart (Ed.), *Personal Identification in Mass Disasters*. Washington, DC: National Museum of Natural History, Smithsonian Institution, pp. 99–109.

Giles E and Elliot O. 1963. Sex determination by discriminant function analysis of crania. *Am J Phys Anthropol* 21:53–68.

Haun SJ. 2000. Brief communication: A study of the predictive accuracy of mandibular ramus flexure as a single morphologic indicator of sex in an archaeological sample. *Am J Phys Anthropol* 111:429–432.

Holland TD. 1986. Sex determination of fragmentary crania by analysis of the cranial base. *Am J Phys Anthropol* 70:203–208.

Holland TD. 1991. Sex assessment using the proximal tibia. *Am J Phys Anthropol* 85:221–227.

Holman DJ and Bennett KA. 1991. Determination of sex from arm bone measurements. *Am J Phys Anthropol* 84:421–426.

Howells WW. 1973. *Cranial Variation in Man: A study by Multivariate Analysis of Patterns of Difference among Recent Human Populations*, Papers of the Peabody Museum of Archaeology and Ethnology, Harvard University Press, Cambridge, MA, Vol. 67.

Howells WW. 1989. *Skull Shapes and the Map. Craniometric Analyses in the Dispersion of Modern Homo*, Papers of the Peabody Museum of Archaeology and Ethnology, Harvard University Press, Cambridge, MA, Vol. 79.

Iscan MY and Miller-Shaivitz P. 1984. Discriminant function sexing of the tibia. *J Forensic Sci* 29:1087–1093.

Jantz RL and Ousley SD. 2005. *FORDISC 3.0*. Knoxville, TN: University of Tennessee.

Klales AR, Ousley SD, and Vollner JM. 2012. A revised method of sexing the human innominate using Phenice's nonmetric traits and statistical methods. *Am J Phys Anthropol* 149:104–114.

King CA, Iscan MY, and Loth SR. 1998. Metric and comparative analysis of sexual dimorphism in the Thai femur. *J Forensic Sci* 43:954–958.

Komar DA and Buikstra JE. 2007. *Forensic Anthropology. Contemporary Theory and Practice*. Oxford, U.K.: Oxford University Press.

Lazenby RA. 1994. Identification of sex from metacarpals: Effect of side asymmetry. *J Forensic Sci* 39:1188–1194.

Loth SR and Henneberg M. 1996. Mandibular ramus flexure: A new morphologic indicator of sexual dimorphism in the human skeleton. *Am J Phys Anthropol* 99:473–485.

MacLaughlin SM and Bruce MF. 1985. A simple univariate technique for determining sex from fragmentary femora: Its application to a Scottish Short Cist population. *Am J Phys Anthropol* 67:413–417.

McCormick WF, Stewart JH, and Greene H. 1991. Sexing of human clavicles using length and circumference measurements. *Am J Forensic Med Pathol* 12:175–181.

Moore-Jansen PM, Ousley SD, and Jantz RL. 1994. Data collection procedures for forensic skeletal material. Report of Investigations No. 48, The University of Tennessee, Knoxville, TN.

Pietrusewsky M. 2000. Metric analysis of skeletal remains: Methods and applications. In MA Katzenberg and SR Saunders (Eds.), *Biological Anthropology of the Human Skeleton*, New York: Wiley-Liss, pp. 375–416.

Phenice TW. 1969. A newly developed visual method of sexing the os pubis. *Am J Phys Anthropol* 30:297–301.

Robling AG and Ubelaker DH. 1997. Sex estimation from the metatarsals. *J Forensic Sci* 42:1062–1069.

Rogers NL. 1999. A visual method of determining the sex of skeletal remains using the distal humerus. *J Forensic Sci* 44:57–60.

Rogers NL, Flournoy LE, and McCormick WF. 2000. The rhomboid fossa of the clavicle as a sex and age estimator. *J Forensic Sci* 45:61–67.

Rogers TL. 2005. Determining the sex of human remains through cranial morphology. *J Forensic Sci* 50:493–500.

Saini V, Srivastava R, Rai RK, Shamal SN, Singh TB, and Tripathi SK. 2012. Sex estimation from the mastoid process among north Indians. *J Forensic Sci* 57:434–439.

Scheuer L and Black S. 2000. *Developmental Juvenile Osteology*. San Diego, CA: Academic Press.

Spradley MK and Jantz RL. 2011. Sex estimation in forensic anthropology: Skull versus postcranial elements. *J Forensic Sci* 56:289–296.

Spradley MK, Anderson BE, and Tise ML. 2015. Postcranial sex estimation criterion for Mexican Hispanics. *J Forensic Sci* 60(S1):S27–S30.

Stewart TD. 1979. *Essentials of Forensic Anthropology*. Springfield, IL: Charles C Thomas.

SWGANTH (Scientific Working Group for Forensic Anthropology). 2010. Sex assessment. Issue date 06/03/2010. Electronic document available at http://swganth.startlogic.com/Sex%20Rev0.pdf.

Tise ML, Spradley MK, and BE Anderson. 2013. Postcranial sex estimation of individuals considered Hispanic. *J Forensic Sci* 58(S1):S9–S14.

Walker PL. 2005. Greater sciatic notch morphology: Sex, age, and population differences. *Am J Phys Anthropol* 127:385–391.

Walker PL. 2008. Sexing skulls using discriminant function analysis of visually assessed traits. *Am J Phys Anthropol* 136:39–50.

White TD and Folkens PA. 2005. *The Human Bone Manual*. San Diego, CA: Elsevier/Academic Press.

Williams BA and Rogers TL. 2006. Evaluating the accuracy and precision of cranial morphological traits for sex determination. *J Forensic Sci* 51:729–735.

CHAPTER 9

Ancestry Estimation
The Importance, The History, and The Practice

M. Katherine Spradley and Katherine Weisensee

CONTENTS

Introduction	163
The importance of ancestry estimation in the biological profile	164
Searching for missing individuals	164
Population-specific data	164
History of race and ancestry in forensic anthropology	165
The practice of ancestry estimation	166
How are forensic anthropologists able to estimate ancestry?	166
Why are craniometric data used to estimate ancestry?	166
Estimation of ancestry	167
Problems with ancestry estimation in forensic anthropology	168
Summary	169
Review questions	170
Glossary	171
References	171
Further reading suggestions	174

> **LEARNING OBJECTIVES**
> 1. Explain the importance of ancestry estimation in developing the other elements of the biological profile.
> 2. Discuss the history of the race concept in anthropology, and explain how Boas' immigrant study impacted the typological view of race.
> 3. Explain the theory and methodology behind the use of craniometric data to estimate ancestry.
> 4. Define "secular change," discuss the causes, and clarify its effect on skeletal morphology.
> 5. Explain why ancestry estimation is possible in the United States, and how forensic anthropologists are able to do so reliably.
> 6. Discuss the pros and cons of ancestry estimation by using nonmetric traits.

INTRODUCTION

The estimation of the biological profile (sex, ancestry, age, and stature) from skeletal remains is a central pillar of forensic anthropology. Despite the many changes in the field of forensic anthropology over the past several decades (see Chapter 1), an accurate biological profile is a key component for matching unidentified individuals, with missing person records ultimately leading to a positive identification. Of all the components of the biological profile, ancestry estimation is the most controversial. Some critics suggest that because biological races do not exist, estimating race is not useful and is a practice that reverts back to the nineteenth century typology, thus reinforcing racial typology (Armelagos and van Gerven 2003; Armelagos and Goodman 1998; Smay and Armelagos 2000). Conversely, other practitioners in the field suggest that there is a congruency between social race categories and skeletal biology. Ancestry estimation increases the likelihood of personal identification of unidentified individuals. In order to address this controversy, this chapter will focus on three key concepts:

1. The importance of ancestry estimation in the biological profile
2. History of race and ancestry in forensic anthropology
3. The practice of ancestry estimation

THE IMPORTANCE OF ANCESTRY ESTIMATION IN THE BIOLOGICAL PROFILE

Searching for missing individuals

Early practitioners of forensic anthropology drew from their general knowledge of human skeletal variation in past and present populations to provide law enforcement with information about the biological profile. This practice continues today, when unidentified human remains are found. The forensic anthropologist creates a biological profile that includes information on ancestry, sex, age and stature—the same categories that are presented in a missing person's report. This information is used to search through missing person's databases to find potential matches. The biological profile is important because it helps narrow the search of an unidentified individual. For example, when a forensic anthropologist examines a case for law enforcement, a biological profile, along with any individualizing characteristics (e.g., description of dental restorations, healed antemortem trauma, and surgical appliances), is provided to begin a search through missing persons' records. When an individual goes missing, a family files a missing person's report that includes information regarding age, sex, weight, height, and race. The more accurate the biological profile, the more likely a match will be found in a timely manner, which may ultimately lead to a positive identification. Moreover, the individual components of the biological profile also inform the appropriate reference samples to be used for other components of the biological profile. As an example, the relationship between ancestry and sex estimation is discussed below.

Population-specific data

Ancestry estimation is important because other aspects of the biological profile hinge on using correct *reference data* used for comparison of the unidentified individual. For example, different population groups have varying degrees of sexual dimorphism, so an accurate determination of ancestry will increase the likelihood that sex is estimated correctly. Sexual dimorphism refers to differences in size and shape between males and females. Males, on average, are larger than females in most skeletal elements. The pelvis in males and females also differs in shape due to the fact that the pelvis in females must accommodate childbirth. Sex estimation from pelvic gross morphology provides accurate results (Phenice 1969; Volk and Ubelaker 2002). (Refer to Chapter 8 for a discussion of sex estimation.) If, however, the pelvis is missing due to an incomplete recovery in the field or from taphonomic processes such as scavenging, postcranial metric sex estimation is the next best method of sex estimation and provides more accurate estimates compared with metric and nonmetric analyses of the skull (France 1998; Konigsberg and Hens 1998; Spradley and Jantz 2011). When estimating sex from postcranial remains, it is important to use a method derived from an appropriate reference population that is similar to the skeletal remains that you are analyzing. Postcranial methods of sex estimation provided in most osteology or forensic anthropology textbooks are population-specific and represent individuals born before the twentieth century and therefore may not be appropriate to use on a recent forensic case (Bass 2005; Byers 2002; Krogman and Iscan 1986; Schwartz 2007). For example, until recently (Spradley et al. 2015; Tise et al. 2013), there were no postcranial sex estimation methods for individuals considered Hispanic. As a result, methods derived from American Whites were typically applied to this group, resulting in considerably greater error than attained with Hispanic reference standards. When applying American White criteria for humeral head diameter on Hispanic individuals, 100% of females were classified as females, but only 47% of males were correctly classified, for an overall classification rate of 73.5% (Spradley et al. 2008). When a population-specific method was developed for a Hispanic sample derived from US/Mexico border crossers, using humeral head diameter, 88% of females and males were correctly classified (Spradley et al. 2015). Thus, it is important to estimate the ancestry of an unknown skeleton before estimation of sex. If the sex estimate is incorrect in a forensic anthropologist's report to law enforcement, then the individual will likely remain unidentified.

A practical application of using biological profile to narrow the search for missing individuals is the use of this information in conjunction with automated databases such as Combined DNA Index System (CODIS, www.fbi.gov/about-us/lab/codis/codis). The University of North Texas Center for Human Identification receives funding for processing DNA samples from unknown skeletal remains and uploading the information directly into CODIS, maintained by the Federal Bureau of Investigation (FBI). CODIS contains DNA profiles of missing persons (if available), profiles of unidentified human remains, and profiles from family members (also known as Family Reference Samples or FRS). The CODIS system can compare the unidentified profiles to FRS or the missing person's profile to assist with identifications. However, CODIS operates by first filtering through metadata (age, sex, race, date last missing, etc.) in order to narrow down a list of DNA profiles for comparison. Thus, the more accurate the parameters of the biological profile (especially ancestry) for unidentified skeletal remains, the better the chance may be for making an identification. For example, if you search for a missing male, this narrows your search by half. However, if you include age and race, your initial search will include fewer possibilities, increasing the likelihood of a timely match.

HISTORY OF RACE AND ANCESTRY IN FORENSIC ANTHROPOLOGY

The topic of ancestry remains under close scrutiny largely due to the typological approach of the nineteenth century. The typological approach refers to the existence of fixed racial types with nonoverlapping patterns of variation across geographic regions. Advocates for this approach took measurements and scored *morphoscopic traits* from multiple groups and used these data to rank each group from superior to inferior. Because Europeans were making these observations and because many of these observations were made during colonization, Europeans were ranked *superior* over all other groups. Nott and Gliddon's (1854) depiction of an African adjacent to a chimpanzee was meant to suggest that Africans did not rank close to Europeans, rather suggesting that Africans are subhuman. This work had a profound effect on anthropology by generating data on skull shapes or brain weights and by classifying and ranking different groups, with European populations on the top of the chain and African populations on the bottom of the chain (Gould 1981). This work was used to justify the social conditions of different racial groups, such as differences in mortality rates or educational achievement, as a natural product of the evolutionary process. Specifically, it was thought that the social conditions of different racial groups could not be improved through policy interventions, because different racial groups were fixed and unchanging and would therefore not be affected by changes in the environment.

In 1911, Boas published a significant study on foreign-born versus American-born immigrants. The significance of this study cannot be overstated, owing to its profound impact on the nineteenth century typology. Boas (1911) found that anthropometric dimensions, specifically the cephalic index,* can change within a single generation due to a new environment and diet alone. The major philosophical implication of the potential for biological change was that human potential is not fixed (Mielke et al. 2011). Boas examined the U.S. population at the turn of the twentieth century and found that cephalic index changed for American-born children compared with Europe-born children, indicating that if the cephalic index can change, racial types are not fixed, and therefore, social rank is mobile, not fixed. Critics of ancestry estimation in forensic anthropological casework continue to cite Boas' study as evidence that craniofacial morphology is not fixed and thus cannot be used for ancestry estimation (Armelagos and van Gerven 2003; Smay and Armelagos 2000; Thomas 2000; Williams et al. 2005).

Recent reanalysis of these data does suggest that there was change in the cephalic index between foreign-born and America-born children (Gravlee et al. 2003a, 2003b; Sparks and Jantz 2002, 2003). However, Sparks and Jantz (2002, 2003) point out that this difference in the cephalic index is negligible and that the majority of variation is found *among* ethnic groups used in the study rather than *between* America-born and foreign-born individuals. This would suggest that the majority of variation in Boas' original samples is due to genetic variation rather than due to environmental influence (Sparks and Jantz 2002, 2003).

---

* Cephalic index = (maximum cranial breadth/maximum cranial length) × 100

More recently, Jantz and Logan (2010) found that the abandonment of cradle boarding caused the biggest difference in the cephalic index of Hebrews, the group with the most change in cephalic index in the Boas study. So, rather than a new geographic environment, it was the abandonment of a common cultural practice that caused the dramatic change. The changes observed in the immigrant study can be referred to as *secular changes*. However, these secular changes do not obscure the underlying genetic structure of the population groups, meaning that American-born children are most similar (in terms of cranial morphology) to the foreign-born children, despite the change in the cephalic index (Sparks and Jantz 2003). In other words, while the environment may influence the phenotype, the underlying genetic structure is not obscured (Relethford 2004:384).

Throughout the early twentieth century, explanations for biological variation shifted away from the typological studies and focused on variation in human populations in the context of evolutionary explanations. This shift culminated in Sherwood Washburn's call for a "new physical anthropology" (Washburn 1951). Subsequent analysis of *quantitative traits* in human populations focused on evolutionary processes of selection, genetic drift, gene flow, and assortative mating for analyzing human variation. Analyses of worldwide human biological variation indicate that 85%–90% of variation is found within groups, while the remaining 10%–15% variation exists between groups. For example, if you derive an F_{st} estimate* from multiple quantitative traits for 20 population groups, the majority of variation is found within all 20 groups, while only 10%–15% of the variation is found among the 20 population groups (Mielke et al. 2011). Some interpret these data as suggesting that so little variation between groups makes it difficult to morphologically distinguish between population groups at a level smaller than geographically separated continents (Williams et al. 2005). However, despite the fact that only 10%–15% of variation exists among groups in various quantitative traits, this variation is structured in a way that is informative to geographic ancestry (Bamshad et al. 2004). In addition, as previously stated, secular changes in craniofacial morphology do not erase the underlying genetic structure of population groups (Relethford 2004). Ancestry estimation, as utilized by forensic anthropologists today, focuses on applying these evolutionary models for explaining human population variation to assist in the effort to identify unknown remains.

THE PRACTICE OF ANCESTRY ESTIMATION

How are forensic anthropologists able to estimate ancestry?

The chapter on FORDISC 3.1 (Jantz and Ousley 2005) of this text (see Chapter 14) instructs the reader on the specific applications of estimating ancestry by using metric data. This section addresses the underlying theory of ancestry estimation, with an emphasis on osteometric data from the cranium. *Osteometric data* refers to measurements that archive the overall morphology (i.e., size and shape) of the skeleton. Osteometric data, particularly *craniometric data*, are the most widely used data source for the estimation of ancestry (Dirkmaat et al. 2008; Jantz and Ousley 2005). When the skull is not recovered, select postcranial remains may provide an accurate estimate of ancestry (Holliday and Falsetti 1999; Jantz and Ousley 2005; Wescott 2005). Holliday and Falsetti (1999) found that long bone lengths, trunk height, and bi-iliac breadth provide good results for ancestry estimation for American Whites and Blacks. Wescott (2005) found the platymeric index ([subtrochanteric anterio-posterior diameter/subtrochanteric medio-lateral diameter] *100) a useful indicator to distinguish between Native Americans and American Blacks and Whites. However, postcranial studies of ancestry are not as abundant as studies using the cranium. Currently, FORDISC 3.1 provides postcranial discriminant function analysis (DFA) for American Blacks and Whites by using standard postcranial measurements.

Why are craniometric data used to estimate ancestry?

Craniometric data are considered complex or quantitative traits (polygenic), meaning that they are influenced by a number of genes, as well as the environment, and provide an archive of overall craniofacial morphology (phenotype). Narrow sense heritability (h^2), also referred to as *heritability*, "measures the degree to which, in a given population, the offspring phenotype is explained by the parental phenotypes" (Abney et al. 2000:629). Thus, pedigreed data sets are utilized to estimate heritability. In quantitative genetic analyses, heritability is typically estimated for each linear measurement and then the heritabilities for all measurements are averaged. Heritability estimates range from 0.0 (meaning that genes do not contribute to the phenotypic expression of traits) to 1.0 (genes are the sole reason for the

---

* F_{st} is a measure of subpopulation variation relative to total population variation. For more information on F_{st}, see Edwards (2003).

phenotypic expression). Heritability estimates of craniometric data suggest that, on average, craniofacial morphology is moderate to highly heritable, with an average h^2 of 0.55 (Devor 1987).

Quantitative traits have a genetic component and an environmental component. Secular changes (i.e., short-term generational changes or microevolutionary changes) have been documented in the cranium. These changes are due to improvements or declines in environmental conditions and can be either positive or negative (Cameron et al. 1990). Secular changes impact the generation born during these dramatic shifts in environmental conditions. For example, studies of craniofacial secular change in the United States in both American Whites and Blacks by Jantz (2001), Jantz and Meadows Jantz (2000), and Wescott and Jantz (2005) found that the most significant secular changes are in the cranial vault. They also suggested that the cranial vault is positively and significantly correlated with birth year. Furthermore, although the observed changes are moving in the same direction, significant morphological differences in these two population groups remain. In other words, even though vault height is increasing in the American population as a whole, it is still possible to distinguish ancestral groups for purposes of ancestry estimation in a forensic case report.

These studies suggest that secular changes are likely due to a combination of factors, including gene flow (admixture), selection, and environmental factors, rather than solely due to environmental factors. Furthermore, in a study of worldwide craniometric variation, Relethford (2004) suggested that short-term secular changes in the cranium resulting from the environment, selection, and/or gene flow do not alter the long-term population structure and population history reflected in overall craniofacial morphology. Population structure refers to factors that affect mate choice and genetic distance between population groups (Mielke et al. 2011).

Estimation of ancestry

In the United States, ancestry estimation is possible because there is a concordance (agreement) between the social race categories *White* and *Black* and the population structure and population history of these two groups (Ousley et al. 2009; Sauer 1992). Sauer (1992) posed the seminal question, "if races don't exist, why are forensic anthropologists so good at identifying them?" (107). As Ousley et al. (2009) pointed out, Sauer suggested that forensic anthropologists are good at ancestry estimation because there is a concordance between social race and skeletal biology. Ousley et al. (2009) further suggested that this concordance is due to population history and *assortative mating* practices. The majority of Black individuals in the United States have ancestors from West Africa, while the majority of White individuals have ancestors from Europe. Furthermore, when enslaved West Africans were brought to the American colonies, there was limited gene flow between these two groups. Today, assortative mating practices still maintain differences in population structure between these two groups (Ousley et al. 2009). While admixture* is often thought of as a force that may obscure ancestry estimation, only 2.9% of the U.S. population in 2010 was considered admixed according to the U.S. Census Bureau.

In sum, ancestry estimation is possible in the United States because of population history and assortative mating practices. Multiple population groups live in the United States, including American Black, American White, and Hispanic. Although there are more population groups in the United States, these groups are the three largest groups and, consequently, the groups for which forensic anthropologists have the most data. Each group has different ancestral origin. For example, American Blacks derive a large number of genes from Africa, particularly West Africa, with some European admixture (Parra et al. 1995, 1998). American Whites derive most of their genes from Europe. The population group considered Hispanic derives genes from multiple sources, depending on the nationality (geographic origin) of the individuals that make up this group. According to the U.S. Census Bureau's American Community Survey (U.S. Census Bureau 2006) of the individuals who report their ethnicity as Hispanic, 64% report Mexican origin. Individuals from Mexico have varying amounts of Native American- and European-derived genes and, in some areas, African-derived genes (Bonilla et al. 2005; Cerda-Flores et al. 2002; Rangel-Villalobos et al. 2008; Lisker et al. 1986, 1990, 1996; Rubi-Castellanos et al. 2009). Thus, depending on the mixture of genes, all these groups have different population histories that have influenced their genetic structure. Because of their different population histories, genetic structure, and assortative mating practices, it is possible for forensic anthropologists to use *multivariate statistical methods* to reliably estimate ancestry from osteometric data (Ross et al. 2004; Spradley et al. 2008). Figure 9.1 shows skulls of individuals from five different ancestral groups. Notice the variation in the craniofacial morphology of each of these skulls.

---

* In this case, admixture refers to individuals that self-identified as belonging to two or more race categories in the 2010 U.S. Census.

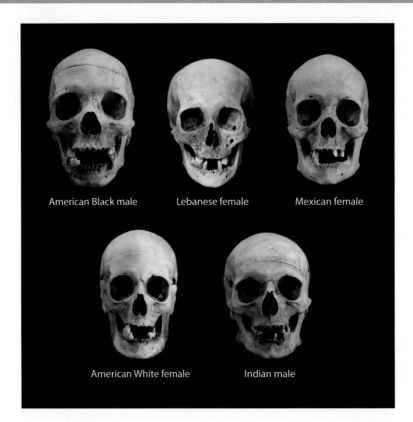

Figure 9.1 Skulls of individuals from five different ancestral groups. (Courtesy of Texas State Donated Skeletal Collection, San Marcos, TX.)

Problems with ancestry estimation in forensic anthropology

When a student interested in forensic anthropology sits down with a textbook, usually only three groups are portrayed: African, Asian, and European (Bass 2005; Burns 2006; Byers 2002, 2008; Klepinger 2006), often referred to as Negroid, Mongoloid, and Caucasoid, respectively. (See Brace [1996] for a discussion on the origin of these terms.) Forensic anthropology texts that use the three-group model fail to present discussions on human variation and do not acknowledge that ancestry estimations can be accomplished using geographic origin (within a continent or between continental regions), tribal affinity, and/or temporal groups (Jantz 1977, 2001; Jantz and Meadows Jantz 2000; Jantz and Willey 1983; Ousley et al. 2009, 2005; Relethford 1994, 2001, 2004; Roseman 2004; Ross et al. 2004, 2008; Spradley et al. 2008). It should be made clear that these texts are not advocating the nineteenth century typology and are in no way ranking groups of people. However, these texts typically instruct a student how to *determine* race without providing the background on human variation that is necessary to understanding why and how forensic anthropologists estimate race. Furthermore, lists of nonmetric or morphoscopic traits (e.g., orbit shape and postbregmatic depression) are usually provided to help the student estimate ancestry.

These trait lists can be traced to Earnest Hooton (1887–1954). As presented, these lists serve to reinforce descriptive typological analysis and lack statistical rigor (Hefner 2009). Hefner (2009) demonstrated that the distribution of each trait should be taken into consideration rather than forcing the trait into one of the three racial categories (i.e., African, Asian, and European). More recently, machine learning techniques have been used for cranial morphoscopic data, with good classification accuracy (Hefner et al. 2015; Hefner and Ousley 2014). These machine learning techniques (e.g., Random Forest Models and Support Vector Machines) are better suited for cranial morphoscopic data, as these data do not meet statistical assumptions (e.g., a normal distribution) when applied to DFA. Furthermore, machine learning techniques can be combined with continuous data (craniometrics) (Hefner and Ousley 2014).

Despite the advances in cranial morphoscopic data analysis, the problem remains that many anthropologists still use observational data and personal expertise by observing the overall cranial gestalt (overall craniofacial morphology) (Hefner and Ousley 2006; Hefner 2009). In this case, an estimation by observation is only as good as the experience of the observer. Wheat (2009) found that forensic anthropology practitioners in the Western United States were better at estimating ancestry for individuals considered Hispanic, while practitioners from the Southeastern United States were better at estimating ancestry for American Black individuals. Wheat's findings are in agreement with the population demographics of the United States and suggest that a forensic anthropologist's familiarity with his or her caseload influences his or her ability to accurately estimate ancestry from nonmetric traits. Individuals considered Hispanic have higher population densities in the Western United States, and American Black individuals have higher population densities in the Southern United States (www.census.gov). However, if a Polynesian skull is brought to the laboratory of any forensic anthropologist in the continental United States, an observation of the gestalt will only likely perplex the observer. Nonetheless, in some cases, osteometric analyses are not possible (e.g., remains that are highly fragmentary or burned), and anthropologists must rely on nonmetric trait analyses to determine ancestry.

SUMMARY

This chapter provides a review of the importance of ancestry estimation, the history of race and ancestry in forensic anthropology, and the practice of ancestry estimation. Ancestry and sex estimation provide key information for missing individual's search, especially when filtering through missing person's automated databases and websites, such as CODIS and NamUs. Because sex estimation is population-specific, ancestry estimation is needed for accurate sex estimation in the absence of the pelvis (Spradley et al. 2008). Furthermore, even though many identifications are made through the use of DNA, it is the components of the biological profile that help forensic practitioners arrive at a potential match for DNA comparison (See Box 9.1 and 9.2).

BOX 9.1   OFFICE OF MANAGEMENT AND BUDGET STATISTICAL DIRECTIVE NUMBER 15

The Office of Management and Budget (OMB) is a component of the Executive Office of the President of the United States that helps implement policies set by the president. Such policies include affirmative action plans, desegregation plans, and access to the Equal Credit Opportunity Act (http://www.whitehouse.gov/omb/fedreg_notice_15). In order to monitor these policies, racial and ethnic categories were created, in order to statistically analyze data. These categories are also used by the U.S. Census Bureau to keep track of our constantly changing population. The OMB recognizes that these categories are not scientific or anthropological in nature (http://www.whitehouse.gov/omb/fedreg_notice_15); rather, they are used to track trends in population growth and monitor presidential policies.

The OMB Statistical Directive Number 15 includes the following racial and ethnic categories:

Ethnicity: Spanish, Hispanic, or Latino (Mexican, Mexican American, Puerto Rican, Cuban, or other). If a person is Hispanic, he or she must check "yes" and then choose a race.

Race: White, Black, American Indian, or Alaska native. Other categories include Asian Indian, Chinese, Filipino, Japanese, Korean, Vietnamese, Native Hawaiian, Guamanian, Samoan, and other Pacific Islander (write in), or some other race (write in).

Forensic Anthropology and the OMB Statistical Directive Number 15

Forensic anthropologists are not bound by the OMB definitions of race and ethnicity. However, a missing person's report is usually reported with OMB terminology, and this terminology is widely used in our society. Forensic anthropologists cannot estimate ethnicity from skeletal remains; rather, they can estimate ancestry or the most likely geographic origin of an individual. Thus, in an analysis using FORDISC 3.1, an individual may exhibit craniofacial morphology most similar to the Hispanic group. The Hispanic reference group is largely composed of individuals who died crossing the U.S.-Mexico Border. Therefore, the forensic anthropologist would tell law enforcement that the skeletal remains represent someone who is Hispanic. In this sense, Hispanic is viewed as a population group of people who share common genetic structure rather than an ethnicity.

For more information on the OMB, visit: http://www.whitehouse.gov/omb/fedreg_notice_15.

For more information on the racial and ethnic categories used by the U.S. Census Bureau, visit: http://www.census.gov/prod/2001pubs/c2kbr01-1.pdf.

> **BOX 9.2 THE FORENSIC ANTHROPOLOGY DATA BANK**
>
> The Forensic Anthropology Data Bank (FDB) was started in 1986 by Dr. Richard L. Jantz, with a grant from the National Institute of Justice. The concept for the FDB was discussed in a meeting among members of the Physical Anthropology Section of the American Academy of Forensic Sciences. In attendance was Dr. Clyde Snow, who discussed the fact that when forensic anthropologists positively identify an individual, the remains are given back to the family. Dr. Snow and others suggested that a system should be in place for forensic anthropologists to collect and submit data to a central repository, so that information could be archived and used for research.
>
> It was Dr. Richard L. Jantz who applied for and received funding from the National Institute of Justice to start the FDB. In the early years of the FDB, Dr. Jantz and his graduate students traveled to skeletal collections in the United States to collect data for the FDB. Dr. Lawrence Angel's forensic case reports were also obtained and input into the FDB. Currently, over 100 laboratories submit data to the FDB. However, the University of Tennessee still remains the largest contributor, owing to the efforts of Dr. Jantz training his students and sending them out to collect data. In addition, the data obtained from the William M. Bass Donated Skeletal Collection, curated at The University of Tennessee, now numbering over 800 individuals, are also entered into the FDB.
>
> The importance of the FDB cannot be overstated. It is an archive of the cases seen around the country by forensic anthropologists. It also generates reference samples for methods of sex, age, ancestry, and stature estimation. It serves as the forensic reference sample in the program FORDISC 3.1. The FDB is unique because the majority of cases in it are derived from forensic anthropology cases, and then, it is applied to forensic cases (e.g., analyzing skeletal remains in FORDISC 3.1). Furthermore, the FDB is reflective of the U.S. population demographics. According to the U.S. Census Bureau, the three largest population groups in the United States are White, Hispanic, and Black, the three largest population groups in the FDB.
>
> The FDB depends on submissions from forensic anthropologists around the country in order to continue growth. According to Dr. Jantz, "the population is a moving target, and like daily weather reporting requires continued updating" (Jantz 2007, pers. comm.). Fortunately, submissions to the FDB allow forensic anthropologists access to data from recent skeletal remains.
>
> To download a copy of the FDB, visit the Inter University Consortium for Political and Social Research (ICPSR) and search for forensic anthropology. http://www.icpsr.umich.edu/icpsrweb/ICPSR/ and search for forensic anthropology.
>
> For more information on the FDB, visit: http://web.utk.edu/~fac/databank.html.

Osteometric data are utilized to estimate ancestry, particularly craniometric data. Because craniometric data are considered quantitative and heritable phenotypic traits, they are particularly useful for ancestry estimation. Although secular changes have been documented in the American population, and demographic population structures are changing, the population history of the United States, along with assortative mating practices, enables forensic practitioners to reliably estimate ancestry. Forensic anthropologists do not "determine" race or any other feature of the biological profile, they provide an estimation. While critics claim that ancestry estimation reverts back to the nineteenth century typology, forensic anthropologists do not, in practice or in publications, rank populations from superior to inferior.

Review questions

1. Explain why it is best practice to use appropriate reference samples to estimate parameters of the biological profile. Give an example of what might happen if an incorrect reference sample is used.
2. Describe the relationship between the various parameters of the biological profile. Why is it important to know ancestry when trying to estimate sex from postcranial remains?
3. What two major types of skeletal data do forensic anthropologists use to estimate ancestry? Describe the differences between these types of data.
4. Explain the concept of heritability. How does the heritability of quantitative traits affect the practice of ancestry estimation?
5. What sources of bias may affect ancestry determination from nonmetric traits?

Glossary

Assortative mating: a form of mating (sexual selection) in which individuals select mates who are phenotypically (and therefore genotypically) more similar as opposed to a random mating pattern in which no selection is involved.

Craniometric data: skeletal data consisting of measurements of the cranium.

Morphoscopic traits: skeletal features with ordinal grades of expression used for estimating ancestry or sex.

Multivariate statistical methods: statistical methods that analyze many variables simultaneously to determine how the variables are related to one another and test their combined predictive power to separate and classify groups.

Osteometric data: skeletal data consisting of bone measurements.

Quantitative traits: measurable phenotypes (traits) that are contributed to polygenic effects, or the effects of multiple genes.

Reference data: a dataset consisting of known values that is used to determine unknown parameters.

Secular changes: changes in human biology that occur over a relatively short period of time primarily due to environmental variables (e.g., increased stature in populations throughout the world).

References

Abney M, McPeek MS, and Ober C. 2000. Estimation of variance components of quantitative traits in inbred populations. *Am J Hum Genet* 66:629–650).

Armelagos G and van Gerven DP. 2003. A century of skeletal biology and paleopathology: Contrasts, contradictions, and conflicts. *Am Anthropol* 105(1):51–62.

Armelagos GJ and Goodman AH. 1998. Race, racism, and anthropology. In AH Goodman and TL Leatherman (Eds.), *Building a New Biocultural Synthesis*. Ann Arbor, MI: The University of Michigan Press.

Bamshad M, Wooding S, Salisbury BA, and Stephens JC. 2004. Deconstructing the relationship between genetics and race. *Nature Rev Genet* 5:598–609.

Bass WM. 2005. *Human Osteology: A Laboratory and Field Manual*. Columbia: Missouri Archaeological Society.

Boas F. 1911. Changes in bodily form of descendants of immigrants. Final report. Vol. 38, Washington, DC: Government Printing Office.

Bonilla C, Gutiérrez G, Parra EJ, Kline C, and Shriver MD. 2005. Admixture analysis of a rural population of the state of Guerrero, Mexico. *Am J Phys Anthropol* 128(4):861–869.

Brace C. 1996. A four-letter word called "Race". In L Lieberman and L Reynolds (Eds.), *Race and Other Misadventures: Essays in Honor of Ashley Montagu in his Ninetieth Year*. Dix Hills, NY: General Hall, pp. 106–141.

Burns KR. 2006. Forensic Anthropology Training Manual. Upper Saddle River, NJ: Prentice Hall.

Byers SN. 2002. *Introduction to Forensic Anthropology: A Textbook*. Boston, MA: Allyn and Bacon.

Byers SN. 2008. *Forensic Anthropology Laboratory Manual*. Boston, MA: Pearson

Cameron N, Tobias PV, Fraser WJ, and Nagdee M. 1990. Search for secular trends in calvarial diameters, cranial base height, indices and capacity in South African Negro crania. *Am J Hum Biol* 2:53–61.

Cerda-Flores RM, Budowle B, Jin L, Barton SA, Deka R, and Chakraborty R. 2002. Maximum likelihood estimates of admixture in northeastern Mexico using 13 short tandem repeat loci. *Am J Hum Biol* 14(4):429–439.

Devor EJ. 1987. Transmission of human craniofacial dimensions. *J Craniofac Genet Dev Biol* 7:95–106.

Dirkmaat DC, Cabo LL, Ousley SD, and Symes SA. 2008. New perspectives in forensic anthropology. *Am J Phys Anthropol* 137(S47):33–52.

France DL. 1998. Observational and metric analysis of sex in the skeleton. In KJ Reichs (Ed.), *Forensic Osteology: Advances in the Identification of Human Remains*, 2nd edn. Springfield, IL: Charles C. Thomas.

Gould SJ. 1981. *The Mismeasure of Man*. New York: W.W. Norton & Company.

Gravlee CC, Bernard HR, and Leonard WR. 2003a. Heredity, environment and cranial form: A Reanalyss of Boas' immigrant data. *Am Anthropol* 105(1):125–138.

Gravlee CC, Bernard HR, and Leonard WR. 2003b. Boas' *Changes in Bodily Form*: The immigrant study, cranial plasticity, and Boas' physical anthropology. *Am Anthropol* 105(2):326–332.

Rangel-Villalobos H, Muñoz-Valle J, González-Martín A, Gorostiza A, Magaña M, and Páez-Riberos L. 2008. Genetic admixture, relatedness, and structure patterns among Mexican populations revealed by the Y-chromosome. *Am J Phys Anthropol* 135:448–461.

Hefner JT. 2009. Cranial nonmetric variation and estimating ancestry. *J Forensic Sci* 54:985–995.

Hefner, JT and Ousley SD. 2006. Morphoscopic traits and the statistical determination of ancestry II. In *Proceedings of the 58th Annual Meeting of the American Academy of Forensic Sciences*, pp. 20–25.

Hefner JT and Ousley SD. 2014. Statistical classification methods for estimating ancestry using morphoscopic traits. *J Forensic Sci* 59:883–890.

Hefner JT, Pilloud MA, Black CJ, and Anderson BE. 2015. Morphoscopic trait expression in "Hispanic" populations. *J Forensic Sci* 60:1135–1139.

Holliday TW and Falsetti AB. 1999. A new method for discriminating African-American from European-American skeletons using postcranial osteometrics reflective of body shape. *J Forensic Sci* 44(5):926–930.

Jantz RL. 1977. Craniometric relationships of Plains populations: historical and evolutionary implications. *Plains Anthropol* 114:146–155.

Jantz RL. 2001. Cranial Change in Americans: 1850–1975. *J Forensic Sci* 46(4):784–787.

Jantz RL and Logan MH. 2010. Why Does head form change in children of immigrants? A reappraisal. *Am J Hum Biol* 22(5):702–707.

Jantz RL and Meadows Jantz L. 2000. Secular change in craniofacial morphology. *Am J Hum Biol* 12:327–338.

Jantz RL and Ousley SD. 2005. *FORDISC 3.1: Personal Computer Forensic Discriminant Functions*. Knoxville, TN: The University of Tennessee.

Jantz RL and Willey P. 1983. Temporal and geographic patterning of relative head height in the Central Plains and Middle Missouri areas. *Plains Anthropol* 28:59–67.

Klepinger LL. 2006. Fundamental of Forensic Anthropology. Cartmill M, and Brown K, editors. Hoboken: John Wiley & Sons.

Konigsberg LW and Hens SM. 1998. Use of ordinal categorical variables in skeletal assessment of sex from the cranium. *Am J Phys Anthropol* 107:97–112.

Krogman WM and Iscan MY. 1986. *The Human Skeleton in Forensic Medicine*. Springfield, IL: Charles C. Thomas.

Lisker R, Perez-Briceno R, Granados J, Babinsky V, de Rubens J, Armendares S, and Buentello L. 1986. Gene frequencies and admixture estimates in a Mexico City population. *Am J Phys Anthropol* 71(2):203–207.

Lisker R, Ramirez E, and Babinsky V. 1996. Genetic structure of autochthonous populations of Meso-America: Mexico. *Hum Biol* 68(3):395–404.

Lisker R, Ramirez E, Briceno RP, Granados J, and Babinsky V. 1990. Gene frequencies and admixture extimates in four Mexican urban centers. *Hum Biol* 62(6):791–801.

Mielke JH, Konigsberg LW, and Relethford JH. 2011. *Human Biological Variation*. New York: Oxford University Press.

Nott JC and Gliddon GR. 1854. *Types of Mankind*. Philadelphia, PA: Lippincott, Grambo, & Co.

Ousley S, Jantz R, and Freid D. 2009. Understanding race and human variation: Why forensic anthropologists are good at identifying race. *Am J Phys Anthropol* 139(1):68–76.

Ousley SD, Billeck WT, and Hollinger RE. 2005. Federal repatriation legislation and the role of physical anthropology in repatriation. *Am J Phys Anthropol* 128(S41):2–32.

Parra EJ, Marcini A, Akey J, Martinson J, Batzer MA, Cooper R, Allison DB, Deka R, Ferrell RE, and Shriver MD. 1998. Estimating African American admixture proportions by use of population-specific alleles. *Am J Hum Genet* 63:1839–1851.

Parra EJ, Teixeira Ribeiro JC, Caeiro JLB, and Riveiro A. 1995. Genetic structure of the populations of Cabo Verde (West Africa): Evidence of substantial European admixture. *Am J Phys Anthropol* 91:381–389.

Phenice T. 1969. A newly developed visual method of sexing the os pubis. *Am J Phys Anthropol* 30(2):297–301.

Relethford JH. 1994. Craniometric variation among modern human populations. *Am J Phys Anthropol* 95:53–62.

Relethford JH. 2001. Global analysis of regional differences in craniometric diversity and population substructure. *Hum Biol* 73(5):629–636.

Relethford JH. 2004. Boas and beyond: Migration and craniometric variation. *Am J Hum Biol* 16:379–386.

Roseman CC. 2004. Detecting interregionally diversifying natural selection on modern human cranial form by using matched molecular and morphometric data. *Proc Natl Acad Sci USA* 101:12824–12829.

Ross AH, Slice DE, Ubelaker DH, and Falsetti AB. 2004. Population affinities of 19th century Cuban craia: Implications for identification criteria in south Florida Cuban Americans. *J Forensic Sci* 49:1–6.

Ross AH, Ubelaker DH, and Guillen S. 2008. Craniometric patterning within ancient Peru. *Latin Am Antiq* 19(2):158–166.

Rubi-Castellanos R, Martínez-Cortés G, Muñoz-Valle JF, González-Martín A, Cerda-Flores RM, Anaya-Palafox M, and Rangel-Villalobos H. 2009. Pre-Hispanic Mesoamerican demography approximates the present-day ancestry of Mestizos throughout the territory of Mexico. *Am J Phys Anthropol* 139(3):284–294.

Sauer NJ. 1992. Forensic anthropology and the concept of race: If races don't exist, why are forensic anthropologists so good at identifying them? *Soc Sci Med* 34:107–111.

Schwartz JH. 2007. *Skeleton Keys*. New York: Oxford University Press.

Smay D and Armelagos GJ. 2000. Galileo wept: A critical assessment of the use of race in forensic anthropology. *Transform Anthropol* 9(2):19–29.

Sparks CS and Jantz RL. 2002. A reassessment of human cranial plasticity: Boas revisited. *Proc Natl Acad Sci USA* 99(23):14636–14639.

Sparks CS and Jantz RL. 2003. Changing times, changing faces: Franz Boas's immigrant study in modern perspective. *Am Anthropol* 105:333–337.

Spradley MK and Jantz RL. 2011. Sex estimation in forensic anthropology: Skull vs. postcranial elements. *J Forensic Sci* 56(2):289–296.

Spradley MK, Jantz RL, Robinson A, and Peccerelli F. 2008. Demographic change and forensic identification: Problems in metric identification of Hispanic skeletons. *J Forensic Sci* 53(1):21–28.

Spradley MK, Anderson BE, and Tise ML. 2015. Postcranial sex estimation criteria for Mexican Hispanics. *J Forensic Sci* 60:S27–S31.

Tise ML, Spradley MK, and Anderson BE. 2013. Postcranial sex estimation of individuals considered Hispanic. *J Forensic Sci* 58:S9–S14.

Thomas DH. 2000. *Skull Wars: Kennewick Man, Archaeology, and the Battle for Native American Identity*. New York: Basic Books.

US Census Bureau. 2006. American Community Survey.

Volk C and Ubelaker DH. 2002. A test of the Phenice method for the estimation of sex. *J Forensic Sci* 47(1):19–24.

Washburn SL. 1951. Section of anthropology: the new physical anthropology. *Transactions of the New York Academy of Sciences* 13(7 Series II): 298–304.

Wescott DJ. 2005. Population variation in femur subtrochanteric shape. *J Forensic Sci* 50(2):231–245.

Wescott DJ and Jantz RL. 2005. Assessing craniofacial secular change in American Blacks and Whites using geometric morphometry. In DE Slice (Ed.), *Modern Morphometrics in Physical Anthropology*. New York: Kluwer Academic/Plenum Publishers, pp. 231–245.

Wheat AD. 2009. Assessing ancestry through nonmetric traits of the skull: A test of education and experience. Master's Thesis, Texas State University San Marcos.

Williams FLE, Belcher RL, and Armelagos GJ. 2005. Forensic misclassification of ancient Nubian crania: Implications for assumptions about human variation. *Curr Anthropol* 46(2):340–346.

Further reading suggestions

Ousley S, Jantz R, and Freid D. 2009. Understanding race and human variation: Why forensic anthropologists are good at identifying race. *Am J Phys Anthropol* 139(1):68–76.

Bamshad M, Wooding S, Salisbury BA, and Stephens JC. 2004. Deconstructing the relationship between genetics and race. *Nature Rev Genet* 5:598–609.

Relethford JH. 2004. Boas and beyond: Migration and craniometric variation. *Am J Hum Biol* 16:379–386.

Edwards AW. 2003. Human genetic diversity: Lewontin's fallacy. *BioEssays* 25(8):798–801.

CHAPTER 10

Age Estimation Methods

Natalie R. Langley, Alice F. Gooding, and MariaTeresa A. Tersigni-Tarrant

CONTENTS

Subadult age-at-death estimation	176
Dental development, calcification, and eruption	177
Fetal age estimation: Diaphyseal length of long bones	177
Appearance and union of epiphyses	178
Adult age-at-death estimation	178
Age estimation in young adults	180
Cranial and palate sutures	181
Gross morphology of the auricular surface	183
Gross morphology of the pubic symphysis	185
Gross morphology of the fourth sternal rib end	187
Other age estimation methods	188
Other age indicators	189
Summary	190
Review questions	190
Glossary	191
References	191

> **LEARNING OBJECTIVES**
> 1. List the four criteria that the SWGANTH Age Estimation Committee recommends for acceptable age indicators.
> 2. Explain the difference between chronological age at death and skeletal age at death.
> 3. Describe the methods that forensic anthropologists use to estimate age in subadults, and compare and contrast these methods in terms of accuracy.
> 4. Discuss the age indicators used to determine age in young adults.
> 5. Describe the basic premise behind adult age estimation methods, and list several commonly used age indicators.
> 6. Recognize the features used to estimate age from the pubic symphysis, auricular surface, and sternal end of the right fourth rib.
> 7. Determine the appropriate method(s) to use for a forensic case, and apply these methods to derive an age estimate.

Forensic anthropologists employ a number of methods to estimate age at death from skeletal remains. These methods are based on skeletal morphology related to growth and developmental changes or to the aging process and degenerative changes. Variables include fusion of skeletal epiphyses, dental mineralization and eruption timing, fusion and obliteration of cranial sutures, **subchondral bone** destruction at joint surfaces, and pathological changes related to aging (e.g., **osteoarthritis**, enamel wear, and declining bone density). The resulting age estimate is an expression of skeletal age at death, which is at best a rough approximation of chronological age at death.

Consider a hypothetical individual who was born in 1970 and died in 2012. The individual was a professional athlete and led an active life until seven years before her death, when she was diagnosed with cancer and lost a tremendous amount of weight during many rounds of harsh chemotherapy. While this person is chronologically aged 42 years, chances are that she may have appeared older skeletally because of the repetitive wear and tear to her joints during her life time as an athlete and because of the rapid decline in bone density and bone health during her illness. Therefore, the most reliable and accurate age estimation methods are based on skeletal indicators, with a strong correlation to chronological age. These methods are not easily influenced by pathological and taphonomic influences.

Forensic anthropologists use a variety of techniques and skeletal features to arrive at an age estimate. The method used depends on the age category of the skeleton (adult or subadult) and on the elements available for examination. According to the Scientific Working Group for Forensic Anthropology:

> Acceptable age indicators fulfill four criteria: (1) morphological changes proceed unidirectionally with age, (2) features have a high correlation with chronological age, (3) changes occur roughly at the same age in all individuals (at least within a distinguishable subgroup), and (4) the characteristics are measured or classified with known intra- and interobserver error rates (SWGANTH 2010) (www.swganth.org).

This chapter provides a cursory overview of age estimation in adult and subadult remains. We describe some of the features most commonly used to estimate age and review some of the methods developed on these skeletal features. Brief attention is given to method limitations; however, the advanced reader is encouraged to refer to Appendix B for a detailed discussion of methodological, statistical, and practical considerations of age estimation research and practice.

SUBADULT AGE-AT-DEATH ESTIMATION

This chapter focuses mainly on adult skeletal aging methods, but an understanding of the methods used to age subadult remains is imperative. For our purposes, *subadult* refers to any individual whose skeleton has not completed skeletal or dental development, that is, a skeleton that is skeletally or dentally immature. Skeletal maturity is reached when all of the epiphyses in the skeleton have fused to their respective diaphyses. Similarly, dental maturity is reached when all of the permanent dentition has mineralized and erupted into the oral cavity, completely replacing the deciduous dentition. Although these two maturation events can occur at different times within a single skeleton and are variable throughout populations of skeletons, the term *subadult* usually refers to an individual younger than 18 years.

The processes of skeletal and dental development and maturation occur at known rates in humans. What follows is a summary of the three most commonly used subadult aging techniques: dental development, calcification, and eruption; long bone/diaphyseal length; and the appearance and union of epiphyses. Although some populations may reach skeletal and dental maturity more precociously than others, most methods for aging immature remains are applied across ancestral groups, as well as for both sexes, mainly because of the dearth of methods available for subadult age estimation. It is important to note that adult age estimation methods are not applicable to subadult remains, as most of the adult aging methods do not include subadults in the reference sample populations. In addition, most of the methods used for adult age-at-death estimation assess the degradation of skeletal morphology over time, a process that does not begin until development is complete.

Age estimates of subadult remains should be derived from multiple lines of evidence, if present. The dentition is most strongly correlated with chronological age and is therefore considered the most accurate age estimator. Dental calcification rates are more accurate than tooth eruption sequences. When dealing with fetal remains, if tooth crowns are present, the age estimate from the dentition is usually given more weight than the estimate from the diaphyseal length of long bones. Likewise, postnatal subadult age estimates from dental calcification and eruption are considered more strongly correlated with chronological age than estimates derived from epiphyseal union. As a rule, dental calcification is under stricter genetic control than dental eruption, and dental development is less susceptible to environmental influences than skeletal maturation (assessed by diaphyseal length and epiphyseal union). Thus, a childhood illness is more likely to affect the growth rate of long bones than the mineralization of tooth crowns. Regardless, all age indicators should be assessed and inconsistencies should be evaluated carefully for the possibility of growth and developmental issues. Ideally, the multitude of age indicators paints a consistent picture, and a concise age estimate can be derived. In cases where inconsistencies arise, best practice dictates using the age estimate from the dentition, unless the practitioner suspects that developmental or pathological conditions are affecting the teeth.

A strong working knowledge of adult and fetal osteology, deciduous and permanent dentitions, skeletal growth and development, skeletal variation, and pathology is imperative before attempting age estimation from immature skeletal remains. The authors recommend hands-on work with fetal and immature skeletons in addition to coursework in human growth and development and/or developmental human osteology to prepare students interested in undertaking forensic casework involving immature skeletal remains.

Dental development, calcification, and eruption

The development, calcification, and eruption of human dentition occur at a regular and predictable rate from fetal development to adulthood (refer to Chapter 5 for a thorough overview of dental development, including eruption sequences). This predictability facilitates the construction of an unambiguous method for age-at-death estimation, based on the absence, partial presence, or complete presence of tooth features (crown, root, etc.) and the timing of their eruption into the oral cavity. (Refer to Appendix A for an overview of age-at-death estimation methods by using dentition.)

The most widely used standards for the aging of immature remains based on tooth mineralization and eruption sequences are those described by Moorees et al. (1963a, 1963b). The Moorees et al. (1963b) method uses an alphanumeric scoring system that describes the level of formation of the crown, root, and apex of deciduous and permanent teeth. As much of this method requires the assessment of dental development occurring within the alveolar bone (e.g., crown development, root length, and apical closure), postmortem radiographs of the skull and mandible are necessary to assess age accurately. This method can be used to assess age in remains from four months' gestation to ~25 years of age. Buikstra and Ubelaker (1994) simplified this scoring system with a 14-stage coding system that can be used to score teeth with single or multiple roots, as well as deciduous or permanent dentition. Ubelaker (1989) published a sequence diagram for the formation and eruption of the dentition, detailing the deciduous and permanent dentitions at various ages. This diagram, although based on data for American Indians, clearly depicts the periods of deciduous eruption, mixed dentition, and full permanent dental eruption, along with suggested age ranges.

Recently, AlQahtani and colleagues (2010) developed an atlas to estimate age from tooth development and alveolar eruption for individuals between 28 weeks *in utero* and 23 years. Tooth development was determined according to Moorees et al. (1963b), and eruption was assessed relative to the alveolar bone. Dry bone and radiographs were used to develop the atlas. A separate validation study found that the London Atlas performs better than two other popular dental development charts (Schour and Massler 1941; Ubelaker 1989; AlQahtani et al. 2014). The London Atlas has diagrams and a free downloadable interactive software app. A pdf of the atlas is available online in 20 languages at https://atlas.dentistry.qmul.ac.uk/?lang=english.

Fetal age estimation: Diaphyseal length of long bones

Just as dental development begins during the fetal period, long bone diaphyses (bone shafts) begin forming before birth and grow at a predictable rate, until the epiphyses fuse to the diaphysis, preventing further growth (Figure 10.1). Diaphyseal measurements are correlated to developmental age and to crown-rump length (and crown-heel length) and, therefore, can be used to estimate age from fetal skeletal remains. Fazekas and Kosa (1978) provide regression formulae derived from diaphyseal lengths to determine skeletal age at death of fetal remains. Buikstra and Ubelaker (1994) summarize these measurements and provide a concise diagram of each pertinent measurement and a recording form to aid in determining age at death from long bone metrics. Scheuer and Black (2000) also incorporate the Fazekas and Kosa (1978) standards into their text *Developmental Juvenile Osteology*. The reference sample for the Fazekas and Kosa (1978) regression equations consists of dry bone measurements from 138 spontaneously aborted fetuses of European origin. Warren (1999) provides similar standards for age estimation from radiographic measurements of fetal diaphyses.

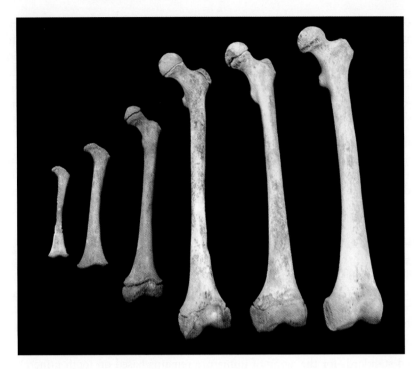

Figure 10.1 This series of photographs of the human femur shows the process of growth and maturation from infancy (far left) to adulthood (far right). (Courtesy of Sandra Cridlin.)

Appearance and union of epiphyses

The appearance and union of epiphyses are used frequently to assess skeletal age at death in immature skeletons. Careful documentation of all skeletal epiphyses is an essential part of the skeletal inventory of a subadult forensic case, as their mere presence will aid in the age estimation. The epiphyses of the human appendicular skeleton form early in life. In contrast with the primary ossification centers, which are located in bone diaphyses and appear *in utero*, epiphyses are secondary ossification centers that facilitate bone growth, beginning at the time of their appearance and proceeding until they fuse to the diaphysis, typically around puberty. In long bones, the epiphyses are located at the proximal and distal ends of the long bone. Between the epiphysis and the diaphysis lies the metaphysis, which is the region containing the epiphyseal plate, or growth plate. The epiphyseal plate is a **hyaline cartilage** plate that facilitates endochondral ossification. Once ossification is complete, the epiphyseal plate is replaced with an epiphyseal line, which is visible on radiographs.

Epiphyses fuse to the diaphysis in a regular sequence and at known age intervals. Axial skeletal epiphyses, such as those of the vertebrae and ribs, develop in a similar manner. Age at death is estimated by assessing which epiphyses are present and scoring the degree of fusion (none, partial, or complete) (Figure 10.2). The most widely used reference text for assessing age from epiphyseal union is Scheuer and Black's (2000) volume *Developmental Juvenile Osteology*. Scheuer and Black (2000) used combination of early work on skeletal development (e.g., Fazekas and Kosa 1978) and more recent research on skeletal maturation to assemble a text detailing the development of the immature skeleton. This text describes each skeletal element from the appearance of primary ossification centers to the fusion of secondary ossification centers. The authors provide summary charts, illustrations, and a detailed developmental chronology for each bone. A condensed laboratory manual version is also available (Schaefer et al. 2009); however, the authors caution using this manual without prior study of the Scheuer and Black volume.

ADULT AGE-AT-DEATH ESTIMATION

Age estimation of subadult remains is based on skeletal development and maturation, while adult skeletal age estimation is based on degenerative changes at various joint surfaces (i.e., the breakdown, or wear and tear, of the skeleton).

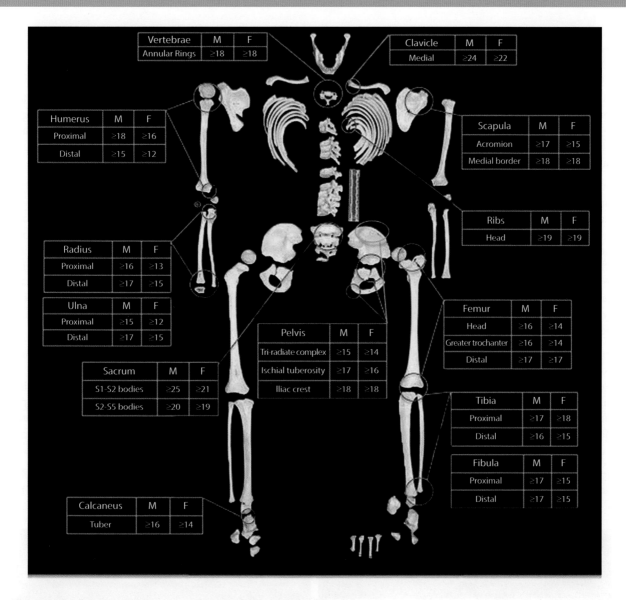

Figure 10.2 Epiphyseal union timing of some of the major skeletal epiphyses.

While degenerative changes take place at a relatively predictable rate, these changes are not the same in everyone because of individual and population variations in activity levels, lifestyle, and personal habits. Wide age ranges are necessary to encompass this variation, so adult age estimates are not as precise as subadult age estimates. In other words, precision (a narrow age range) is sacrificed for accuracy (a correct estimate).

The remainder of this chapter briefly reviews the most frequently used skeletal indicators and presents several popular methods to estimate adult skeletal age at death. Appendix B explains the problems with some of these methods and gives an advanced overview of the ways in which the forensic anthropology community is striving to address these shortcomings.

The most commonly employed methods for aging adult remains focus on gross morphological changes of the pubic symphysis, auricular surface, and sternal end of the right fourth rib. It is precisely because of their frequency of use that these methods should be subjected to continued validation and accuracy of use assessments. Methods using the dentition are also available; these are covered in Appendix A. In addition, cranial and maxillary palate suture closure may be used for age estimation; however, these methods are less favored, largely due to their imprecision.

The following sections provide a description of the gross morphology of each age indicator and a brief literature review of several of the most common methods of age estimation.

Age estimation in young adults

Fusions of the medial clavicle epiphysis and S1–S2 sacral segments provide precise and accurate age estimates for young adults. The eruption of the third molar around the age of 18 years is another age indicator that offers precise age-at-death estimates for individuals in their late teens. The epiphyseal development of pubic symphysis also continues into early adulthood, and a new method is available to estimate age in young adults from this joint surface (Dudzik and Langley 2015). This method will be discussed with pubic symphyseal aging methods, later in this chapter.

The epiphysis at the sternal end of the clavicle is the last skeletal growth plate to fuse. The epiphysis unites to the shaft of the clavicle during the second decade of life, making this epiphysis informative of age in young adults. When examining a skeleton with fused epiphyses, one should examine the medial clavicle (if present) to ascertain whether or not the individual is likely older than 25–30 years. Medial epiphyseal ossification begins at the onset of puberty, but the epiphysis does not fuse to the shaft completely until some 10 years after its initial appearance (Kreitner et al. 1998; Scheuer and Black 2000). The epiphysis appears initially as a small speck of bone in the central area of the sternal end and spreads until it nearly covers the entire medial surface. Scheuer and Black (2000) offer the following timeline for clavicular maturation: a well-defined medial flake appears between 16 years and 21 years; the flake covers the majority of the medial surface between 24 years and 29 years; and complete fusion occurs between 22 years and 30 years. Figure 10.3 shows the various stages of epiphyseal union of the medial clavicle.

Fusion ages of the medial clavicle have been documented in a number of sources from skeletal collections, ranging from historic European populations to modern Americans (Todd and D'Errico 1928; McKern and Stewart 1957; Webb and Suchey 1985; Black and Scheuer 1996; Langley-Shirley and Jantz 2010). Many of these researchers used different scoring systems to record the stage of epiphyseal union. Langley-Shirley and Jantz (2010) found that a simple three-phase system (not fused, fusing, and fused) provides more accurate age estimates and lower observer error rates than more complicated scoring methodologies. The Langley-Shirley and Jantz (2010) standards were derived from a modern autopsy sample by using probability-based statistical methods.

Another late-fusing maturation site is the segment between the first and second sacral vertebral bodies (S1–S2). The sacral segments fuse in a caudal to cranial direction, with S4–S5 and S3–S4 fusing around 12 years, S2–S3 fusing

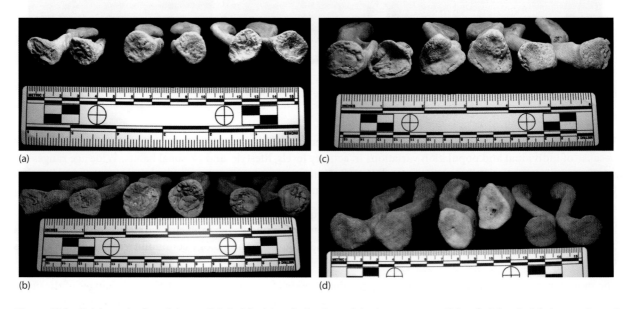

Figure 10.3 Epiphyseal union of the medial clavicle. (a) no fusion (no epiphysis present; medial end of the clavicle is vascular and coral-like in appearance), (b, c) fusing (epiphysis is in the process of uniting to the diaphysis, ranging from a small speck of bone to near-complete coverage of the medial end), and (d) complete fusion (medial end appears quiescent, with no remnants of the fusion event).

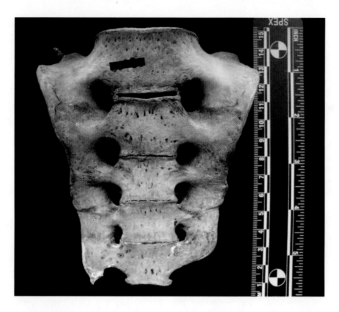

Figure 10.4 Fusion of the sacral vertebral bodies. Note the pattern of fusion, beginning caudally and proceeding cranially.

around puberty, and S1–S2 fusing during early adulthood (25+ years) (Scheuer and Black 2000). The sacrum in Figure 10.4 illustrates the caudal–cranial fusion of the sacrum. Note that S1–S2 is unfused, S2–S3 is not completely fused, and S3–S4 and S4–S5 are in the process of fusing. One might infer from the information that this individual could be around the age of puberty and probably younger than 20 years.

Once all teeth are erupted and the late-fusing epiphyses are completely fused, the only methods available for age estimation are based on degenerative changes in the joints of the skeleton. Since the process of wear and tear at joint **articulations** is highly variable, age ranges derived from degenerative skeletal changes are necessarily wide to incorporate this large degree of variation.

Cranial and palate sutures

In the 1920s, Todd and Lyon (1924) identified specific points on the cranium to assess the degree of suture closure. Later, using research from Baker (1984) and Mann et al. (1987), Meindl and Lovejoy (1985) suggested a revised method in which 17 landmarks of the cranium and palate are scored. The degree of closure of a suture at a particular point is scored as one of the four categories: open, minimal closure, significant closure, and complete obliteration. Figure 10.5 illustrates each of the four categories. A composite score is then compared to an associated age range derived from Meindl and Lovejoy's original sample populations. Nawrocki (1998) provides regression equations for cranial and palate suture closure by using the many of the same landmarks as Meindl and Lovejoy, including some endocranial sites. Several regression equations are provided, along with the **R-squared** values that indicate how well the variables in the equation account for age. Nawrocki's (1998) equations also offer a method for estimating age in partial crania, because the equations consist of combinations of a number of variables, ranging from as few as two landmarks to as many as 27.

Early on, researchers questioned the accuracy of age estimation from cranial sutures, stating that the estimates are unreliable (Brooks 1955). Practitioners continue to debate the relationship between age and the physiological process of suture ossification today. Generally, it is considered an unreliable or, at best, an imprecise predictor of age at death, only used if no other methods are available. The age ranges associated with the composite scores are so broad (sometimes >30 year intervals) that they are not particularly useful in a forensic context. Nonetheless, some forensic anthropology cases consist of only a skull without dentition, and in these instances, suture closure may be the only option for aging. However, the practitioner should be familiar with pathological conditions that cause premature suture closure before blindly applying this method. Current research focuses on the microanalysis and histological investigation of the sutures in order to better understand the varying timetables of suture union.

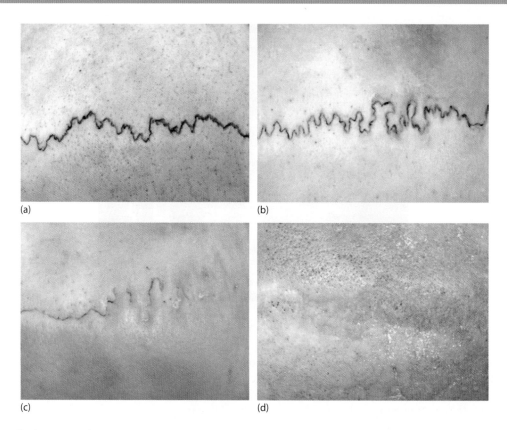

Figure 10.5 Fusion states of cranial sutures. (a) open, (b) beginning fusion (some bridging beginning to occur), (c) significant fusion (>50% bridging), (d) obliteration.

One site of suture closures in the cranium that is strongly correlated with skeletal age is the spheno-occipital synchondrosis, or basilar suture (Figure 10.6). This growth center is located at the base of the cranium between the sphenoid and the basilar portion of the occipital bone, just anterior to the foramen magnum. Although outdated literature proposes that this suture fuses in the mid-1920s (between 17- and 25 years of age), recent research using a modern forensic case sample from the Unites States demonstrates that it fuses around puberty (Shirley and Jantz 2011).

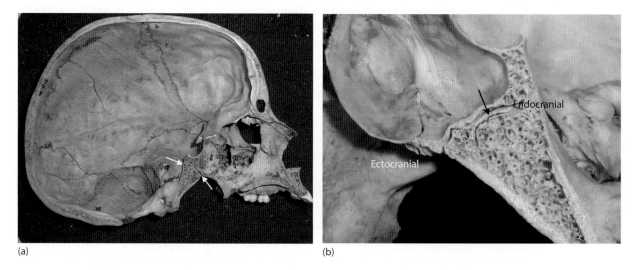

Figure 10.6 Fusion of the spheno-occipital synchondrosis. (a) Mid-sagittal section of the skull. The space between the black and white arrows illustrates an open spheno-occipital synchondrosis and (b) a fusing synchondrosis (epiphyseal plate indicated by the arrow) showing the mechanism of fusion, which proceeds from the endocranial surface to the ectocranial surface.

These gross observations of the suture are confirmed by computed tomography (CT) studies of living individuals in Australian (Lottering et al. 2015) and Japanese (Okamoto et al. 1996) populations, to name a couple. The spheno-occipital synchondrosis has received much attention in the research literature, and standards are available for a number of populations worldwide, using a variety of assessment techniques (dry bone, radiographs, and CT scans).

Gross morphology of the auricular surface

The sacroiliac joint is the joint between sacrum and posterior ilium. This joint is a key load-bearing joint that transfers weight between the upper and lower extremities. A number of researchers have documented age-related degenerative changes at the joint surface and surrounding areas of ligamentous attachment on the ilium. Three key regions of the posterior ilium are illustrative of age-related changes: the morphological appearance of the iliac auricular surface, bony changes at the apex of the joint surface, and the bony activity in the retroauricular area (Figure 10.7). The first characteristic identified is the morphology of the auricular surface, specifically the transverse organization of that surface. As a subadult, bony billows of low relief are organized in a horizontal, or transverse, orientation on the joint surface. With time, the billows become replaced by tiny striae (thin lines), traces of which may remain throughout life. Eventually, the surface loses transverse organization and changes texture. The surface texture transitions from fine-grained to coarse-grained sand paper and eventually densifies with advancing age. The surface displays signs of microporosity early on, and later, larger holes (macroporosity) become evident (Lovejoy et al. 1985). In the latter phases of auricular surface aging techniques, the bone beneath the articular surface begins to erode beneath, and the resulting subchondral destruction may be visible (Lovejoy et al. 1985; Osborne et al. 2004).

The next region on the posterior ilium used for age estimation is the apex of the auricular surface. In young individuals, the apex is often a sloping angle with a smooth margin. Over time, the margins of the auricular surface become irregular, resulting in bony buildup and lipping at the apex. This process can make the apex appear more prominent and sharp (Lovejoy et al. 1985). Apical changes in advanced age groups can be mild to moderate and can have considerable variation throughout the phases of age estimation (Lovejoy et al. 1985; Osborne et al. 2004).

Similar to the apex, the bone of the retroauricular area shows distinct signs of morphological change over time. In subadults and younger adults, the area is generally smooth. With age and activity, it transitions to a coarse texture and develops an uneven and irregular surface. Later in life, prolific **enthesophytes** can occur at either a high or low relief. These growths represent ossified ligamentous and tendinous muscle attachments. At this stage, the entire posterior portion of the ilium can appear disorganized and irregular, owing to both erosion of the bone and extra bony growth (Lovejoy et al. 1985). See Figure 10.8a–f for illustrations of the degenerative changes described here.

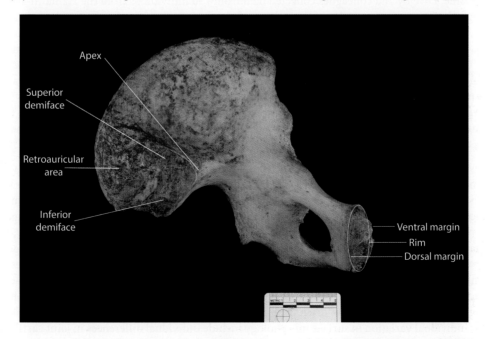

Figure 10.7 Key areas of morphological change on the auricular surface (left side of image) and pubic symphysis (right side).

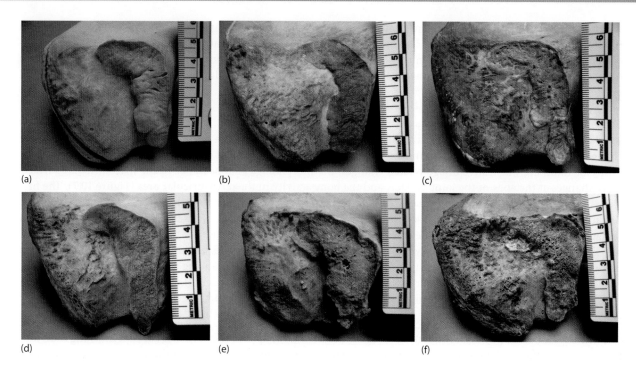

Figure 10.8 (a) Fine-grained, billowed surface with no apical or retroauricular changes; (b) Fine-grained surface with billows and striae; (c) striae, slight apical changes and retroauricular activity; (d) course grained, striae around apex and on inferior demiface, microporosity, moderate retroauricular activity; (e) loss of transverse organization, some densification, course granularity, microporosity, moderate apical and retroauricular activity; (f) microporosity, macroporosity, severe apical and retroauricular activity, course granularity, no transverse organization. (Courtesy of University of Tennessee Forensic Anthropology Center, Knoxville, TN.)

As an element for age estimation, the auricular surface has proved to be problematic for researchers. Lovejoy et al. (1985) first noticed, despite little movement throughout life, that the iliac portion of the sacroiliac joint showed signs of change over time. Lovejoy and colleagues collected data from 750 individuals from the Hamann-Todd Collection and archaeological specimens. A majority of the individuals in the sample were of unknown age, sex, and ancestry, and as a result, the age was estimated using other methods (Lovejoy et al. 1985; Komar and Buikstra 2008). Using the estimated ages and the sacroiliac morphology, the researchers developed an auricular surface age estimation method that is still in use today. The eight phases of the method are composed of descriptions of the subtle changes to the particular auricular surface components described earlier. Each phase is representative of a five-year age range, and none of the ranges overlap. In addition, no mean ages from the sample were published with the phases (Lovejoy et al. 1985). The narrow age ranges of each phase, as well as the morphological changes that are challenging to detect, particularly for less experienced practitioners, have created considerable debate within the field about the applicability and reliability of this method.

Subsequent testing of the Lovejoy et al. method proved that it was applicable to both sexes and across various ancestral backgrounds. A study by Murray and Murray (1991) confirmed that the method was applicable for use with both sexes, while also demonstrating that it is useful in the case of both Blacks and Whites. However, Osborne et al. (2004) found a low accuracy rate associated with the Lovejoy et al. method (only 33% of a test sample was aged accurately with the five-year age ranges). Osborne et al. (2004) created an alternate six-phase method for aging, based on the auricular surface, by using Terry and Bass Collection data. Each phase is associated with a predicted age range larger than the narrow five-year ranges of Lovejoy et al. method. The Osborne et al. (2004) method ranges significantly overlap one another, suggesting that there is more variation in the aging process of the auricular surface than previously thought. This method is applicable to both sexes and across ancestral groups. Osborne et al. also found that age accounts for only about 34% of the variation in the auricular surface (R-squared = 0.34). They surmise that factors responsible for individual variation in surface morphology include individual differences in joint cartilage thickness, occupational and activity-related stresses, life history variables, and the size and shape of the joint surface itself.

Buckberry and Chamberlain (2002) developed a revised method for aging the auricular surface by using a sample from several archaeological collections in England. The method utilizes all but one of the morphological features originally identified by Lovejoy and colleagues. This method assesses each feature as a separate component; component scores are added together to arrive at a composite score that is associated with an age range. Validation studies by Falys et al. (2006) and Mulhern and Jones (2005) suggest that Buckberry and Chamberlain's composite scoring method is similarly accurate to the Lovejoy et al. phase method.

Gross morphology of the pubic symphysis

Age-at-death estimation methods utilizing the pubic symphysis are based on morphological changes that occur throughout life. These include the appearance of the pubic face and changes in the bony rim and texture of the face (see Figure 10.7). Like the auricular surface, the morphology and texture of the most prominent feature, that is, the pubic symphyseal face, changes significantly in appearance over time. In subadults, the face exhibits deep ridges and furrows that run transversely across the entire face. The ridges and furrows fill in and become worn down with progressing age, ultimately leading to a flattened joint surface. The smooth texture of the bone of the face becomes coarsely granular and, in the final phases, displays porosity, pitting, and depression in the center. This later process is generally accompanied by overall erosion of the surface of the symphyseal face (Todd 1920).

The second feature paramount to age assessment of the pubic symphysis is the formation and deterioration of the rim. The dorsal rim of the pubic face forms as the ridge and furrow system on the symphyseal face flattens, and the dorsal margin is filled in with bone. The dorsal margin is the first to become delineated, as the ridge and furrow system flattens first dorsally and then proceeds ventrally. The ventral margin is the last portion of the rim to form; however, in some cases, it may never reach completion. The formation of the ventral margin is epiphyseal in nature, as new bone is laid down and fuses to the existing surface. A ventral bevel forms as epiphyseal bone is laid down onto the ventral margin. This process begins as multiple nodules of bone along the ventral rim; eventually, these nodules grow and unite to form a single epiphysis along the border. At the same time, an ossific nodule of bone on the superior portion of the face unites with the ventral epiphysis to form a single ventral bevel. The ossific nodule eventually becomes incorporated into the superior face. In some individuals, a hiatus remains at the superior portion of the ventral rim. As individuals age, the distinct rim that has formed begins to lip, deteriorate, and erode. Erosion of the rim can distort the symphyseal face, and it often loses its oval shape. In later phases, the rim disintegrates completely, leaving an irregularly shaped face with bony outgrowths on the anterolateral portions of the pubic bone (Todd 1920).

A final key change in the pubic symphysis is the change in texture of the pubic face. As with the auricular surface, the texture of the pubic symphysis changes from finely granular to coarsely granular. In addition, porosity begins to appear on the symphyseal face with advancing age, contributing to the change in texture and to the pitting and irregularity of the joint surface. Suchey and Katz (1998) also provide a detailed description and illustrations of the age-related morphological changes of the pubic symphysis discussed here. Figures 10.9 and 10.10 depict the morphological changes of the male and female pubic symphysis, respectively.

The method for analyzing the surface of the pubic symphysis has been revisited many times since Todd developed the original method in 1920. Todd (1920) describes 10 stages of grossly observable changes to the pubic symphysis related to age. Although the Todd method utilizes a limited sample, his pioneering research proved the importance of the pubic symphysis as a useful morphological element for age estimation. Meindl et al. (1985) found that the morphological changes identified by Todd are strongly correlated with age and provide accurate age estimates. In 1957, McKern and Stewart developed a component method for age estimation from the pubic symphysis. McKern and Stewart derived their data from a sample of American soldiers between the ages of 18 years and 35 years who died in the Korean War. This sample was limited to young males, and as a result, Gilbert and McKern (1973) developed a composite method for female age estimation, using the pubic symphysis.

The most popular method used today was developed by Brooks and Suchey (1990) and is based on Todd's original system (Garvin and Passalaqua 2012). Suchey and Brooks used data from a large sample of 1225 known individuals from the Los Angeles Medical Examiner's Office. The sample incorporates individuals of both sexes and a larger spectrum of ages and ancestries than previous methods. In contrast to the 10-phase Todd method, the Suchey–Brooks approach uses only six morphological phases, with significantly expanded age ranges associated with each phase and considerable age overlap between phases. Separate standards are provided for males and females. One of the main

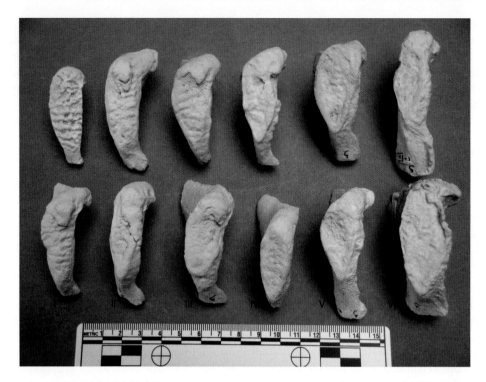

Figure 10.9 Photograph of casts of the Suchey–Brooks pubic symphyseal changes in males (casts by France Casting). Phases 1–6 are represented. The top row depicts early stages of a phase, and the bottom row depicts later stages, such that each "column" of casts represents the variation within a given phase.

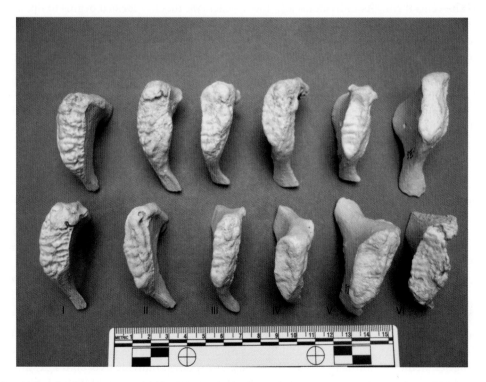

Figure 10.10 Photograph of casts of the Suchey–Brooks pubic symphyseal changes in females (made by France Casting). Phases 1–6 are represented. The top row depicts early stages of a phase, and the bottom row depicts later stages, such that each "column" of casts represents the variation within a given phase.

critiques of the Suchey–Brooks method is the extremely wide age ranges associated with the phases. This problem was most recently addressed by Hartnett (2010b) by using a modern and diverse sample from the Forensic Science Center in Arizona to identify a concise set of characteristics that defines each stage of typical symphyseal aging. A number of these characteristics are similar to those originally detailed by Todd and highlighted in the Suchey–Brooks method (face shape, rim formation, and marginal lipping). In addition to these characteristics, the Hartnett method also emphasizes tactile characteristics of the bone, such as the weight and feel of the bone, especially in the later phases. The method is composed of seven phases, with a description of the morphological characteristics of each phase. Although there are notes within a number of these phase descriptions identifying characteristics specific to either males or females, the phase descriptions are not differentiated by sex. However, sex-specific age ranges corresponding to each phase are provided. These age ranges are narrower than the age ranges given for each of the Suchey–Brooks phases and overlap one another to a lesser degree. Both the Suchey–Brooks and Hartnett methods were derived from more recent reference populations than the Todd method. Because populations change over time, it is best practice to use a technique that was derived from a contemporaneous sample when estimating age in modern forensic cases.

The methods available for age estimation from the pubic symphysis are based on the same suite of morphological features. The primary differences lie in the reference sample used to develop the method and the application of either a phase system or a component-based scoring system. Phase systems lump all of the features into a single description, whereas component systems score each feature individually. Shirley and Ramirez-Montes (2015) found lower interobserver error with component systems. These systems are user-friendly and less susceptible to subjective weighting of traits, which is apt to occur when attempting to capture complex morphologies within a single phase. A composite scoring system was developed recently for individuals under the age of 40 years, based on the developmental changes of the pubic symphyseal complex (Dudzik and Langley 2015). The method provides decision-tree-style flow charts to arrive at an age range. The age ranges are narrower than those offered by phase-based methods that combine developmental and degenerative changes in one scoring system, and the method proved to be highly accurate on a separate test sample (94% overall accuracy). Advanced readers should refer to Appendix B for a discussion on the methodological and statistical issues with skeletal age-at-death estimation.

Gross morphology of the fourth sternal rib end

Like other cartilaginous joints in the body, the sterno-costal joints show morphological signs indicative of age-related change. The articular surface of the sternal end of the rib displays the greatest change in two key areas—the rim and the pit (Figure 10.11). In younger individuals, the rim is smooth, uniform, and rounded. The edges show signs of distinct and uniform scalloping (Figure 10.12a), and the pit is shallow and, in very young individuals, billowed. In adulthood, the scallop pattern becomes uniformly smooth and takes on an arc shape (Figure 10.12b). With advancing age, the rim becomes increasingly irregular and begins to flare into a U shape, as the pit deepens. This irregular rim may exhibit bony projections, especially on the superior and inferior borders, giving the rib end a *crab claw-like* appearance (Figure 10.12c). Projections may also grow long enough to join another, forming *windows* created by the ossified cartilage (Figure 10.12d). In this later phase, the pit is deep and pitted and shows signs of erosion (Işcan et al. 1984a, 1984b, 1985). Hartnett (2010a) emphasizes that a loss in weight of the ribs can accompany deformation of the rim and pit as an individual ages. Porosity in the pit is the visual evidence of this loss of bone density over time.

The fourth rib has proven useful in age-at-death estimation. Işcan et al. (1984a, 1984b, 1985) identified significant morphological change throughout life in the sternal end of the right fourth rib. Işcan et al. used an American sample to develop a method with nine distinct phases of morphological change, eight of which have a narrow associated age range. The phase descriptions and ages for this method are applicable across Black and White ancestral populations but are separated by sex. Their research also suggests that rib numbers 3 and 5 show similar change over time and can be used in the event that the fourth rib is not available. Yoder et al. (2001) tested the Işcan method and ultimately suggested that a composite score should be used instead of the phase approach, arguing that scoring morphological changes across multiple ribs is a more accurate predictor of age than the traditional phase method. Hartnett (2010a) also reassessed the Işcan et al. method of rib aging by using the Arizona Forensic Science Center sample and developed a more concise method. Similar to her previous research with pubic symphyseal aging, the revised method utilizes a number of the characteristics emphasized by Işcan et al., with the additional emphasis on overall weight of the rib. The Hartnett method is composed of seven phases, with specific age ranges for each sex. These ranges are narrower than those originally published by Işcan et al. but broader than those suggested by the casts created for use with the Işcan et al. method.

Figure 10.11 Key areas of morphological change on the sternal end of the right fourth rib.

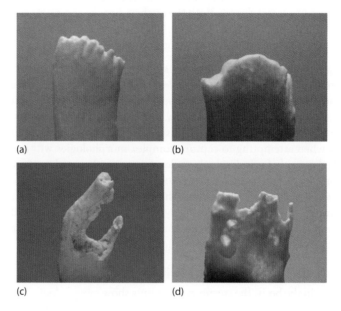

Figure 10.12 Photographs of Iscan rib end casts (casts by France Casting). (a) Scalloped rim, (b) central arc forms as scallops wear down, (c) bony projections on the superior and inferior borders, (d) windows formed by the union of ossified cartilage projections.

The application of many of these methods (pubic symphysis, clavicles, auricular surface, and ribs) is enhanced by the use of drawings and/or casts that illustrate the morphological change for each phase or component (casts available through France Casting, Fort Collins, Colorado). Several features of the dentition may also be used to estimate age at death in adults. Dental age estimation methods are discussed in Appendix A.

Other age estimation methods

In addition to conducting extensive validation tests on existing methods, during the past decade, researchers have begun to investigate new skeletal age indicators. DiGangi et al. (2009) developed a method for age estimation from the first rib, based on an earlier paper by Kunos et al. (1999). They evaluated 11 traits and found that two of the traits were most appropriate: geometric shape of the costal face and surface texture of the tubercle facet. As the sample consisted of positively identified individuals who were victims of human rights violations in Kosovo, the analysis was done solely on males. However, the authors are optimistic that the method can also be used to estimate the age of female remains, but they suggest extensive testing before drawing these conclusions. Their analysis of the data was statistically sophisticated and robust, and this aging method shows promise both as a stand-alone method and for use in conjunction with other aging methods. For more accurate estimates, the authors advocate a multifactorial approach to age-at-death estimation. In addition, the publication contains excellent descriptions and photographs for ease of scoring (DiGangi et al. 2009).

Morphological age-related changes have also been documented in the sacral auricular surface. Passalacqua (2009) developed a method for age estimation from the human sacrum. This method considers changes in the sacral auricular surface as well as fusion of the sacral vertebral bodies and fusion of the superior vertebral ring of the first sacral body. Passalacqua's method encompasses changes in the sacrum that occur across an individuals' lifespan; in other words, he considered both developmental and degenerative changes in the sacrum. Using Todd and Bass Collection sacra, he scored six traits as present or absent (S1–S2 fusion, S2–S3 fusion, microporosity, macroporosity, surface relief changes of the auricular surface, and apical changes) and one trait as *incomplete fusion*, *fused*, or *absorbed* (S1 ring fusion). The scores for these traits are recorded and compared to a table that is used to assign a phase that corresponds to an age range. The system is highly accurate and can be applied equally to males and females. Additional discussion of the method can be found in Passalacqua (2009).

Another area of the pelvis that has received recent attention as a reliable age indicator is the acetabulum (Rougé-Maillart et al. 2004; Rissech et al. 2006a, 2006b; Calce and Rogers 2011). Rissech et al. (2006a) identified seven morphological traits that correlate closely with age (acetabular groove, rim shape, rim porosity, apex activity, activity on outer edge of acetabular fossa, activity of fossa, and porosity of fossa), using a sample of 242 males from the Coimbra Collection in Portugal. This study was an expansion of earlier work by Rougé-Maillart et al. (2004) on a small Spanish and French sample. The seven variables and their morphological states are described thoroughly, and accompanying photographs are included in the article by Rissech et al. (2006a). They reported 89% accuracy, using 10-year intervals, and 67% accuracy, using 5-year intervals. Furthermore, unlike many aging methods that perform better in young age groups than in older age groups, the method does equally well across age categories. Rissech et al. (2006b) conducted a test of their method on four documented Western European collections (the Lisbon and Coimbra Collections in Portugal, University Autonòma Barcelona Collection, and St. Bride Collection in London). They achieved accuracy rates between 83% and 100%, with the lowest accuracy rates coming from the most geographically distant reference collection. A similar validation study was conducted on a Canadian collection, the Grant Collection at the University of Toronto (Calce and Rogers 2011). The authors suggested revising some of the trait descriptions and reducing the number of variable states per trait, as they found that some descriptions needed clarification and reported difficulties in deciding between variable states on account of morphological overlap. Like the previous validation study (Rissech et al. 2006b), Calce and Rogers (2011) found that the geographic distance of the reference population affects the method's accuracy and maintained that the chosen reference population must be geographically close to the test specimen, if the method is to perform reliably. However, they did find the method effective in older age groups and recommended additional testing on other geographic populations. Calce (2012) recently published a revision of the Rissech et al. (2006a) method. The revised method uses only three variables and collapses the ages into three broad ranges, corresponding to young, middle, and old adults. She also extended the method's application to include females. Calce (2012) reported 81% accuracy, with substantial observer agreement in scoring.

These new methods highlight some of the recent advances in age estimation techniques. What emerges from our experience with forensic casework is that practitioners should use multiple indicators of skeletal age at death, when possible. It is important that researchers continue to investigate new indicators, because forensic anthropology casework frequently involves fragmentary, burned, or commingled remains that may be damaged by trauma or taphonomic forces. It is also imperative that studies are accompanied by observer and method-error information and that separate and independent validation studies are conducted to test new techniques.

Other age indicators

Most experienced forensic anthropologists use a combination of the methods described in this chapter. They also gather more subjective supplemental information from the skeletons, which ultimately informs their age estimate. This type of information gathering is a natural product of years of experience in handling skeletal material and making age assessments from the skeleton. Degenerative changes that experienced osteologists look at include vertebral osteophytes, bony lipping around joint peripheries, **eburnation** on joint surfaces, **dental attrition**, bone density changes, and costal cartilage ossification. Several studies have developed scoring systems with variable states and associated age ranges for some of these morphological changes.

Dental attrition has been studied and quantified by a number of researchers (far more than can be mentioned in the scope of this chapter). Perhaps, one of the earliest dental attrition scoring systems was devised by Gustafson (1950), and since then, many researchers have conducted similar analyses and validation studies (Murphy 1959; Miles 1962; Brothwell 1963; Scott 1979; Smith 1984; Li and Ji 1995; Kim et al. 1999; Ajmal et al. 2001; Yun et al. 2007; Prince et al. 2008). Prince et al. (2008) used a Bayesian approach based on Smith's (1984) system, which was adapted from Murphy's (1959) method. They concluded that dental wear may be more appropriate for grouping individuals into large cohorts than for accurately estimating age at death. The reason for this is that substantial variation in dental wear patterns exists among individuals and populations because of differences in diet, enamel thickness, dental care and hygiene, chewing, and habitual activities (i.e., tooth grinding). Nonetheless, dental wear can supplement age estimates derived from other skeletal indicators. Furthermore, because of the durable nature and survivability of teeth, these methods can be useful, particularly for sorting commingled remains in mass disasters.

Costal and thyroid cartilage ossifications can also provide supplementary information about age at death, and several studies have developed scoring systems and accompanying age ranges for evaluating these cartilage changes (Michelson 1934; Cerny 1983; McCormick and Stewart 1988; De la Grandmaison 2003; Dang-Tran et al. 2010; Garamendi et al. 2011). Generally, the **laryngeal cartilages** are cartilaginous in younger individuals and begin to ossify with advancing age. However, completely ossified laryngeal cartilages can occur in individuals under 50 years of age, so the process is highly variable.

Degenerative changes associated with the vertebral column can also be useful for estimating age. In general, vertebral osteophytes do not appear before the age of 35 years in normal, healthy individuals. Experienced osteologists examine the condition of the vertebrae and frequently incorporate this impression into their overall age estimate. Watanabe and Terazawa (2006) developed a scoring system for osteophyte formation and devised regression equations for estimating age from osteophyte formation in males and females.

While anthropologists vary in terms of favored aging methods, practitioners should be familiar with all methods and be able to apply them with equal skill, as the condition and number of elements present in a particular forensic case will dictate the most appropriate method(s). Likewise, forensic anthropologists should understand the limitations of each method and know the sample(s) from which it was developed, as these factors may have consequences for the final age estimate. For a review of current practices in age estimation by forensic anthropologists, see Garvin and Passalaqua (2012). Also, Appendix B discusses, in detail, some of the limitations and problems associated with age-at-death estimation from skeletal remains.

SUMMARY

This chapter provides an overview of the most commonly used skeletal features and techniques for age estimation in adults and subadults. Selecting the appropriate method depends on the age category of the skeleton (adult or subadult) and on the elements available for examination. Once the methods are selected, the practitioner assesses the skeletal, or physiological, age of the remains and correlates this to chronological age. If possible, multiple methods should be used to arrive at an age estimate. Other factors that should be taken into account when selecting an appropriate method include the appropriateness of the reference sample population (sex, ancestry, age distribution, and birth cohort), method error, and the usefulness of the age ranges provided by the method.

Techniques that provide extremely large age ranges may be accurate, but they are not helpful in a forensic setting (i.e., an age range of 30–40 years does not help law enforcement in narrowing down a search for missing persons). On the other hand, overly precise age ranges are not useful, because they may omit too many possibilities (i.e., an age range of five years is precise but not likely to be as accurate as a 10-year range). Teeth are considered the gold standard for estimating age in subadult remains. Chapter 5 and Appendix A discuss dental development and age estimation from the dentition in detail. If the dentition is not available, fusion timing of skeletal epiphyses can be used to determine age. Since the timing of epiphyseal union is more susceptible to environmental influences and occurs at different times in males and females, population-specific reference samples will yield the most accurate age estimates.

Age estimation is a complicated task that requires knowledge of skeletal anatomy, variation, and familiarity with skeletal pathology. Please answer the review questions to aid you in synthesizing the concepts presented in this chapter.

Review questions

1. Explain the different methods available to determine skeletal age at death in subadults. Discuss the premise behind each method.
2. Discuss best practice in subadult age estimation. In your discussion, elaborate on age estimation from the dentition and from epiphyseal union.

3. Assuming that you are dealing with complete skeletal remains, what method would you use to determine the age of (a) a full-term fetus, (b) a child who is likely between ages three years and seven years, and (c) a pre-teen?
4. List the methods available to estimate age in young adults and late teenagers.
5. You receive a skull with dentition from law enforcement. Discuss the methods available to determine age. What is the method(s) that you would use preferentially? What if the skull did not have dentition?
6. You receive a pelvis from law enforcement (right and left os coxae and sacrum). What features on these elements are useful for age estimation?
7. Describe the nature of developmental and degenerative changes of the pubic symphysis. (What is the sequence of morphological changes that researchers have observed in this joint?)
8. Can the fourth rib cage estimation method be applied if the fourth rib is unavailable for analysis? Explain your answer.
9. After a long day of exhuming remains from a clandestine grave, you arrive in your laboratory with a complete skeleton. You ascertain that the individual is likely older than 18 years. What methods would you use to determine the skeletal age at death? What would you do differently if you ascertain that the individual was likely older than 25 years?
10. Name some of the skeletal features that experienced forensic anthropologists use to derive supplementary age information from the skeleton? In what way these might be useful for sorting remains from a plane crash?

Glossary

Articulation: the joint between two bones.

Dental attrition: the wear and tear caused by tooth-to-tooth contact. This occurs naturally with age and results in the loss of enamel on the incisive/occlusal surfaces of the teeth.

Eburnation: a polishing of the articular surface of a bone that occurs from the bone-on-bone rubbing associated with osteoarthritis; the subchondral bone becomes a smooth ivory-like mass that sometimes contains ridges or striae.

Enthesophyte: bony projections that form at the attachment of a ligament or tendon.

Hyaline cartilage: one of the three types of cartilage in the body; hyaline cartilage contains a significant portion of collagen and is usually found lining the bone surfaces at joint articulations. The nose, ears, larynx, and trachea also contain hyaline cartilage.

Laryngeal cartilages: the nine cartilages that comprise the larynx (or voice box), including the unpaired thyroid and cricoid cartilages. The thyroid cartilage has a midline prominence that is recognizable in the neck of some individuals as the Adam's apple. The thyroid and cricoid cartilages may ossify in older individuals.

Osteoarthritis: also called "degenerative joint disease," osteoarthritis is the most chronic condition of the joints. The protective hyaline cartilage on the ends of the joints wears down, causing painful bone-on-bone rubbing at the articulation. Osteophytes develop around the periphery of the joint, resulting in further pain and stiffening.

R-squared: also known as the coefficient of determination; in a linear regression, a statistical measure of how close the data are fitted to the regression line. R-squared expresses the percentage of the response variable that is explained by the linear model.

Subchondral bone: the layer of bone located immediately beneath (or deep to) the articular cartilage.

References

Ajmal M, Mody B, and Kumar G. 2001. Age estimation using three established methods. A study on Indian population. *Forensic Sci Int* 3:150–154.

AlQahtani SJ, Hector MP, and Liversidge HM. 2010. Brief communication: The London atlas of tooth development and eruption. *Am J Phys Anthropol* 142:481–490.

AlQahtani SJ, Hector MP, and Liversidge HM. 2014. Accuracy of dental age estimation charts: Schour and Massler, Ubelaker, and the London Atlas. *Am J Phys Anthropol* 154:70–78.

Baker RK. 1984. The relationship of cranial suture closure and age analyzed in a modern multi-racial sample of males and females. A Master's Thesis. California State University, Fullerton, CA.

Black S and Scheuer L. 1996. Age changes in the clavicle from the early neonatal period to skeletal maturity. *Int J Osteoarch* 6:425–434.

Brooks ST. 1955. Skeletal age at death: The reliability of cranial and pubic age indicators. *Am J Phys Anthropol* 13:567–597.

Brooks S and Suchey JM. 1990. Skeletal age determination based on the os pubis: A comparison of the Acsádi-Nemeskéri and Suchey-Brooks Methods. *Hum Evol* 3(5):227–238.

Brothwell D. 1963. *Digging up Bones*. London, U.K.: British Museum of Natural History.

Buckberry JL and Chamberlain AT. 2002. Age estimation from the auricular surface of the ilium: A revised method. *Am J Phys Anthropol* 119: 231–239.

Buikstra JE and Ubelaker DH. 1994. Standards for the data collection from human skeletal remains. *Proceedings of a Seminar at the Field Museum of Natural History*. Arkansas Archaeological Survey Research Series No. 44, Fayetteville, AR.

Calce SE. 2012. A new method to estimate adult age at death using the acetabulum. *Am J Phys Anthropol* 148(1):11–23.

Calce CE and Rogers TL. 2011. Evaluation of age estimation technique: Testing traits of the acetabulum to estimate age at death in adult males. *J Forensic Sci* 56(2):302–311.

Cerny M. 1983. Our experience with estimation of an individual's age from skeletal remains of the degree of thyroid cartilage ossification. *Acta Univ Palacki Olomuc Fac Med* 3:121–144.

Dang-Tran K, Dedouit F, Joffre F, Rougé D, Rousseau H, and Telmon N. 2010. Thyroid cartilage ossification and multislice computed tomography examination: A useful tool for age assessment? *J Forensic Sci* 55(3):677–683.

De la Grandmaison GL, Bansar A, and Durigon M. 2003. Age estimation using radiographic analysis of laryngeal cartilage. *Am J Forensic Med Pathol* 24(1):96–99.

DiGangi EA, Bethard JD, Kimmerle EH, and Konigsberg LW. 2009. A new method for estimating age at death from the first rib. *Am J Phys Anthropol* 138:165–176.

Dudzik B and Langley NR. 2015. Estimating age from the pubic symphysis: A new component-based system. *Forensic Sci Int* 257:98–105.

Falys CG, Schutkowski H, and Weston D. 2006. Auricular surface aging: Worse than expected? A test of the revised method on a documented historic skeletal assemblage. *Am J Phys Anthropol* 130:508–513.

Fazekas IG and Kosa F. 1978. *Forensic Fetal Osteology*. Budapest, Hungary: Aksdémiai Kiadó.

Garamendi PM, Landa MI, Botella MC, and Alemán I. 2011. Forensic age estimation on digital x-ray images: Medial epiphyses of the clavicle and first rib ossification in relation to chronological age. *J Forensic Sci* 56(S1):S3–S12.

Garvin HM and Passalacqua NV. 2012. Current practices by forensic anthropologists in adult skeletal age estimation. *J Forensic Sci* 57(2):427–433.

Gilbert BM and McKern TW. 1973. A method of aging the female os pubis. *Am J Phys Anthropol* 38:31–38.

Gustafson G.1950. Age determination on teeth. *J Am Dent Assoc* 41:45–54.

Hartnett KM. 2010a. Analysis of age at death estimation using data from a new, modern autopsy sample—part II: Sternal end of fourth rib. *J Forensic Sci* 55:1152–1156.

Hartnett KM. 2010b. Analysis of age at death estimation using data from a new, modern autopsy sample—part I: Pubic bone. *J Forensic Sci* 55:1145–1151.

Işcan MY, Loth SR, and Wright RK. 1984a. Metamorphosis at the sternal rib end: A new method to estimate age at death in White Males. *Am J Phys Anthropol* 65:147–156.

Işcan MY, Loth SR, and Wright RK. 1984b. Age determination from sternal ends of ribs. *J Forensic Sci* 29:1094–1104.

Işcan MY, Loth SR, and Wright RK. 1985. Age estimation from the rib by phase analysis: White females. *J Forensic Sci* 30(3):853–863.

Kim YK, Kho HS, and Lee KH. 1999. Age estimation by occlusal tooth wear. *J Forensic Sci* 45(2):303–309.

Komar DA and Buikstra JE. 2008. *Forensic Anthropology: Contemporary Theory and Practice*. New York: Oxford University Press.

Kreitner K, Schweden F, Riepert T, Nafe B, and Thelen M. 1998. Bone age determination based on the study of the medial extremity of the clavicle. *Eur Radiol* 8:1116–1122.

Kunos C, Simpson S, Russell K, and Hershkovitz I. 1999. First rib metamorphosis: Its possible utility for human age at death estimation. *Am J Phys Anthropol* 110:303–323.

Langley-Shirley N and Jantz RL. 2010. A Bayesian approach to age estimation in modern Americans from the clavicle. *J Forensic Sci* 55(3):571–583.

Li C and Ji G. 1995. Age estimation from the permanent molar in northeast China by the method of average stage of attrition. *Forensic Sci Int* 3:189–196.

Lottering N, MacGregor DM, Alston CL, and Gregor LS. 2015. Ontogeny of the spheno-occipital synchondrosis in a modern Queensland, Australian population using computed tomography. *Am J Phys Anthropol* 157(1):42–57.

Lovejoy CO, Meindl RS, Pryzbeck TR, and Mensforth RP. 1985. Chronological metamorphosis of the auricular surface of the ilium: A new method for the determination of adult skeletal age at death. *Am J Phys Anthropol* 68:15–28.

Mann RW, Symes SA, and Bass WM. 1987. Maxillary suture obliteration: Aging the human skeleton based on intact or fragmentary maxilla. *J Forensic Sci* 32:148–157.

McCormick WF and Stewart JH. 1988. Age related changes in the human plastron: A roentgenographic and morphologic study. *J Forensic Sci* 33(1):100–120.

McKern T and Stewart T. 1957. Skeletal age changes in young American males. Analysed from the standpoint of age identification. Natick, MA: Quartermaster Research and Development Center, Environmental Protection Research Division.

Meindl RS and Lovejoy CO. 1985. Ectocranial suture closure: A revised method for the determination of skeletal age at death based on the lateral-anterior sutures. *Am J Phys Anthropol* 68:57–66.

Meindl RS, Lovejoy CO, Mensforth RP, and Walker RA. 1985. A revised method of age determination using the os pubis, with a review and tests of accuracy of other current methods of pubic symphyseal aging. *Am J Phys Anthropol* 68:29–45.

Michelson N. 1934. The calcification of the first costal cartilage among Whites and Negroes. *Hum Biol* 6:543–557.

Miles AEW. 1962. Assessment of the ages of a population of Anglo-Saxons from their dentitions. *Proc R Soc Med* 55:881–886.

Moorees CFA, Fanning EA, and Hunt EE. 1963a. Formation and resorption of three deciduous teeth in children. *Am J Phys Anthropol* 21:205–213.

Moorees CFA, Fanning EA, and Hunt EE. 1963b. Age formation by stages for ten permanent teeth. *J Dental Res* 42:1490–1502.

Mulhern DM and Jones EB. 2005. Test of revised method of age estimation from the auricular surface of the ilium. *Am J Phys Anthropol* 126:61–65.

Murphy T. 1959. Gradients of dental exposure in human molar tooth attrition. *Am J Phys Anthropol* 17:179–186.

Murray KA and Murray T. 1991. A test of the auricular surface aging technique. *J Forensic Sci* 6(4):1162–1169.

Nawrocki SP. 1998. Regression formulae for the estimation of age form cranial suture closure. In K Reichs (Ed.), *Forensic Osteology: Advances in the Identification of Human Remains,* 2nd edn. Springfield, IL: Charles C. Thomas, pp. 276–292.

Okamoto K, Ito J, Tokiguchi S, and Furusawa T. 1996. High-resolution CT findings in the development of the spheno-occipital synchondrosis. *Am J Neuroradiol* 17(1):177–120.

Osborne D, Simmons T, and Nawrocki SP. 2004. Reconsidering the auricular surface as an indicator of age at death. *J Forensic Sci* 49(5):1–7.

Passalacqua NV. 2009. Forensic age at death estimation from the human sacrum. *J Forensic Sci* 54(2):252–262.

Prince DA, Kimmerle EH, and Konigsberg LW. 2008. A Bayesian approach to estimate skeletal age at death utilizing dental wear. *J Forensic Sci* 53(3):588–593.

Rissech C, Estabrook GF, Cunha E, and Malgosa A. 2006a. Using the acetabulum to estimate age at death of adult males. *J Forensic Sci* 51(2):214–229.

Rissech C, Estabrook GF, Cunha E, and Malgosa A. 2006b. Estimation of age-at death for adult males using the acetabulum, applied to four Western European populations. *J Forensic Sci* 52(4):774–778.

Rougé-Maillart CL, Telmon N, Rissech C, Malgosa A, and Rouge D. 2004. The determination of male adult age by central and posterior coxal analysis. A preliminary study. *J Forensic Sci* 49:1–7.

Schaefer M, Black S, and Scheuer L. 2009. *Juvenile Osteology: A Laboratory and Field Manual*. London, U.K.: Elsevier.

Scheuer L and Black S. 2000. *Developmental Juvenile Osteology*. London, U.K.: Academic Press.

Shirley NR and Jantz RL. 2011. Spheno-occipital synchondrosis fusion in modern Americans. *J Forensic Sci* 56(3):580–585.

Shirley NR and Ramirez-Montes PA. 2015. Age estimation in forensic anthropology: quantification of observer error in phase versus component-based methods. *J Forensic Sci* 60(1):107–11.

Schour I and Massler M. 1941. The development of human dentition. *J Am Dent Assoc* 20:379–427.

Scott EC. 1979. Dental wear scoring technique. *Am J Phys Anthropol* 51:213–218.

Smith BH. 1984. Patterns of molar wear in hunter-gatherers and agriculturalists. *Am J Phys Anthropol* 63(1):39–56.

Suchey JM and Katz D. 1998. Applications of pubic age determination in a forensic setting. In KJ Reichs (Ed.), *Forensic Osteology: Advances in the Identification of Human Remains*, 2nd edn, Springfield, IL: CC Thomas, pp. 204–236.

SWGANTH (Scientific Working Group for Forensic Anthropology). 2010. Age assessment. Issue date January 22, 2013. Electronic document available at http://swganth.startlogic.com/Age%20Rev1.pdf.

Todd TW. 1920. Age changes in the pubic bone: I, The male White pubis. *Am J Phys Anthropol* 3:286–334.

Todd T and D'Errico J. 1928. The clavicular epiphyses. *Am J Anat* 41:25–50.

Todd TW and Lyon DW Jr. 1924. Endocranial suture closure, its progress and age relationship. Part I. Adult males of White stock. *Am J Phys Anthropol* 7:325–384.

Ubelaker DH. 1989. *Human Skeletal Remains: Excavation, Analysis, Interpretation*, 2nd edn. Washington, DC: Taraxacum.

Warren MW. 1999. Radiographic determination of development age in fetuses and stillborns. *J Forensic Sci* 44(4):708–712.

Watanabe S and Terazawa K. 2006. Age estimation from the degree of osteophyte formation of vertebral columns in Japanese. *Leg Med* (*Tokyo*) 8(3):156–160.

Webb P and Suchey J. 1985. Epiphyseal union of the anterior iliac crest and medial clavicle in a modern multiracial sample of American males and females. *Am J Phys Anthropol* 68(4):457–466.

Yoder C, Ubelaker DH, and Powell JF. 2001. Examination of variation in sternal rib end morphology relevant to age assessment. *J Forensic Sci* 46:223–227.

Yun J, Lee JY, Chung J, Kho H, and Kim Y. 2007. Age estimation of Korean adults by occlusal tooth wear. *J Forensic Sci* 52(3):678–683.

Chapter 11

Stature Estimation

Natalie R. Langley

CONTENTS

History	195
Stature estimation methods	196
Additional considerations	198
Summary	199
Review questions	200
Glossary	200
References	201

> **LEARNING OBJECTIVES**
> 1. Explain how regression theory is used to predict stature.
> 2. Compare and contrast mathematical versus anatomical methods for estimating stature, and know when it is appropriate to use one method over the other.
> 3. Name some of the variables that can affect stature estimates derived from regression formulae.
> 4. Explain the importance of using an appropriate reference sample when predicting stature from a regression equation.
> 5. Explain the difference between a prediction interval and a confidence interval.
> 6. List other applications of stature estimation, aside from forensic contexts.
> 7. Describe the potential error introduced by antemortem stature records, and suggest ways in which forensic anthropologists can account for this when deriving a stature estimate.

HISTORY

Stature estimation involves estimating living height from skeletal dimensions. The theoretical underpinnings of estimating living stature from skeletal measurements date back over two centuries. Jean Joseph Sue (1710–1792) was a French anatomist who published data on maximum length of long bones and several body measurements, including stature, from 14 cadavers ranging from six weeks *in utero* to 25 years' old (Sue 1755). Sue published this data to provide artists with information about body proportions (Stewart 1979). Sue's measurements were introduced to the medico-legal community by Matthieu Joseph Bonaventure Orfila (1787–1853), a professor of legal medicine who published two textbooks on the subject (Orfila 1821–1823; Orfila and Lesueur 1831). Orfila also added measurements from 51 cadavers and 20 skeletons to Sue's original data set. To estimate stature by using the Sue–Orfila skeletal measurements, the user looked up the bone measurement and corresponding cadaver length

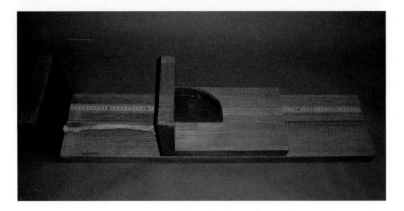

Figure 11.1 A standard osteometric board used to measure long bones.

in a table. Cadaver length can vary from living stature by as much as two inches, so a certain amount of error was incorporated into the Sue–Orfila stature estimates.

In the late nineteenth century, Paul Broca (1824–1880), a medical anthropologist, better known for his work in neuroanatomy, introduced the **osteometric board** for measuring long bones (Figure 11.1). Paul Topinard (1830–1911), one of Broca's students, devised a series of ratios of long bone lengths to stature. Topinard's ratios provided more accurate stature estimates than Orfila's look-up tables (Topinard 1885). Étienne Rollet (1862–1937) followed Topinard's work closely and published a doctoral thesis relating the average lengths of long bones from cadavers of the same length. Rollet's data were snatched up quickly and reorganized into tables by Topinard's successor, Léonce Manouvrier (1850–1927). Manouvrier's tables provide average cadaver length (as a proxy for stature), based on individuals with the same long bone lengths. He also included a correction factor for dry bone measurements, since Rollet's data were derived from measurements taken on fresh bone (Manouvrier 1893).

Shortly after Manouvrier's publication, the English biometric school released a statistical technique that would change the face of stature estimation. In 1899, Karl Pearson (1857–1936) published a monograph detailing his **regression theory**, which used the correlation between long bone length and stature to derive a linear equation that would predict stature from long bone length (Pearson 1899). Although Pearson's equations were intended for evolutionary studies, his work had a profound influence on forensic anthropology. Paul Stevenson (1890–1971) was the first to test Pearson's equations, but the equations did not perform well, because Stevenson tested them on a Chinese population, and Pearson's equations were based on Rollet's European data (Stevenson 1929). Likewise, the Chinese regression equations that Stevenson developed were not successful for estimating European stature. This initial test called attention to human variation in long bone length and body proportions and pointed to the need for population-specific standards. During the 1950s, Trotter and Gleser (1952, 1958) published regression equations for modern Americans, based on a living stature data from a large sample of males and females from various ethnic backgrounds. These equations remained the most appropriate for forensic use in the United States until the release of Fordisc 2.0 (Jantz and Ousley 1996).

Several years before the publication of Pearson's regression manuscript, Thomas Dwight (1843–1911), an American anatomist from Harvard, on whom T. Dale Stewart suitably conferred the title "Father of American Forensic Anthropology," distinguished between two methods of stature estimation: mathematical and anatomical (Dwight 1894). Dwight explained that mathematical methods rely on the relationship between bone length and body proportions to height, whereas **anatomical methods** require assembling the skeleton in anatomical position, applying a correction factor for soft tissue, and measuring height. Since the mathematical method is more susceptible to error—resulting from population differences in body proportions and pathological conditions affecting stature, including age-related decreases in stature—Dwight preferred the anatomical method.

STATURE ESTIMATION METHODS

What emerges from this brief historical synopsis is that stature estimation can be accomplished using more than one method; the nature of the forensic case usually dictates which method is most appropriate. The Scientific Working Group for Forensic Anthropology (SWGANTH) recommends incorporating information about the individual's population, sex, and temporal cohort. Ideally, the method with the smallest error should be selected over other methods, and that method should be applied reliably and objectively. The SWGANTH guidelines for stature estimation clarify the difference between anatomical and regression techniques and give recommendations for best practice (http://swganth.org).

The anatomical method of stature estimation is suitable only when the skeleton is largely complete and is the most appropriate method when sex and ancestry are not known or when **supernumerary vertebrae** are present (SWGANTH). Dwight (1894) was the first to suggest an anatomical method for estimating stature. Fully (1956) and Fully and Pineau (1960) modified Dwight's technique; Fully's technique was later revised by Raxter et al. (2006, 2007). Anatomical methods require that all of the following are present: the cranium, all vertebrae (except the atlas), the sacrum, femur, tibia, talus, and calcaneus. All elements are measured; the measurements are summed; and a correction factor for soft tissues is included in the final estimate. Occasionally, supernumerary vertebrae are present and require an additional correction factor (Lundy 1988; Raxter and Ruff 2010).

One advantage of anatomical methods of stature estimation is that they are robust to population and/or individual variation in body proportions and age-related changes in stature (Maijanen 2009), though formulae that incorporate age-related losses in stature are available (Niskanen et al. 2013; Raxter et al. 2007). Anatomical methods can be employed regardless of the sex or ancestry of the individual and provide more accurate stature estimates than regression approaches, which use only one or two skeletal elements. Limited formulae are available for calculating stature if elements are missing (Auerbach 2011). Missing element estimation is limited to the vertebral column (only when lumbar vertebrae are present) and to talocalcaneal height (only when the femur and tibiae are present); the cranium, entire vertebral column, and lower limb bones cannot be estimated (Auerbach 2011). In general, anatomical methods are not feasible if required elements are missing, as is frequently the situation with forensic anthropology casework. In this instance, the forensic anthropologist must resort to another technique.

Because many forensic anthropology cases involve incomplete skeletal remains, the most frequently used technique for stature estimation in a forensic setting is the mathematical or regression approach. This approach requires measuring bone length(s) and selecting the most appropriate formula according to sex, ancestry, and temporal cohort. Measurements from the lower limb are favored over those of the upper limb, because these dimensions correlate more strongly with stature. When present the maximum length of lower limb bones should be used preferentially, as these measurements outperform those of from the upper limb (Wilson et al. 2010). In the absence of lower limb elements, upper limb bones give adequate stature estimates. After the bone(s) are measured, the measurement(s) are plugged into the appropriate regression formula. The linear regression equation for stature estimation from a single skeletal measurement is

$$\text{Stature} = a + bx + SE$$

where:
    a is the y intercept of the line
    b is the slope
    x is the bone measurement
    SE is the standard error

Multiple regression formulae are also available for use with multiple measurements. Calculations may be done manually or automatically using a computer program such as Fordisc (Jantz and Ousley 2005).[*] The Fordisc computer program gives the user the option of using the Trotter and Gleser (1952, 1958) equations or *forensic* equations, based on the Forensic Data Bank, a collection of skeletal measurements, and biological data from modern individuals. However, when using Trotter and Gleser's formulae with tibial length, the user should note that this measurement was reported incorrectly in the original publications (1952, 1958). Jantz et al. (1995) discovered that Trotter's measurement excluded the medial malleolus, although she reported that she had included it. *Consequently, one should omit the medial malleolus from tibial length when using Trotter's formulae or use another bone.* Significant increases in adult height have been documented in many populations during the course of the last century, so the forensic equations are more appropriate than the Trotter and Gleser equations for the majority of forensic casework (Jantz and Jantz 1999; Meadows and Jantz 1995). The Trotter and Gleser equations were developed using the Smithsonian Institution's Terry Collection, consisting primarily of individuals born between 1850 and 1900. During this time period,

---

[*] Refer to Chapter 14 for a detailed discussion on the Fordisc program.

the American population was significantly shorter than it is presently, so these equations may underestimate stature in modern individuals (Meadows and Jantz 1995).

Fordisc gives the option of using a 90% or 95% prediction interval for the stature estimate. When using multiple measurements, the program returns several stature estimates, along with associated error terms and R^2 values. The user should select the estimate with the lowest standard error, highest **R^2 value**, and smallest **prediction interval**. Note that stature is reported as an interval and not as a range. A range is the difference between the maximum and minimum values in a data set of observations. Instead, stature is reported in terms of a confidence interval or a prediction interval. A **confidence interval** is a way of expressing the probability that a given regression line represents the true best-fit linear regression line. A prediction interval demarcates the area in which a specified percentage of the data surrounding the fitted regression line is expected to fall, given a certain probability. For example, with a 95% confidence interval, we can be 95% confident that the upper and lower confidence limits contain the true best-fit regression line for estimating stature. With a 95% prediction interval, we expect 95% of all data points surrounding the regression line used to estimate stature to fall within those bounds. A prediction interval is wider than a confidence interval and is the preferred and more useful parameter for reporting a stature estimate. *The formula with the smallest prediction interval should be the most accurate and precise and, therefore, should be used to estimate stature (SWGANTH).*

Another option for estimating stature uses a ratio between stature and bone length, such as the femur/stature ratio (Feldesman et al. 1990). This method is not as mathematically complex as the regression approach, and the authors report that sex and ancestry differences are negligible with the ratio approach. Criticisms of the ratio method are that it overestimates stature (particularly in taller individuals) and yields larger prediction intervals than the regression method. Ratio methods are considered less accurate than the regression approach, and the latter performs better overall (Jantz et al. 1995; Konigsberg et al. 1998).

If no complete limb bones are present to estimate stature, options are available using fragmentary limb bones and nonlimb bones. One option is to use the fragmentary element to estimate total bone length and then to apply this measurement to a regression formula (Steele and McKern 1969). However, the SWGANTH points out that this technique compounds error in the final stature estimate. An alternative is to employ a technique that estimates stature directly from the fragment; however, these methods are limited in availability for modern populations (Holland 1992; Simmons et al. 1990). Another option is to use a bone other than a limb bone, such as the cranium, vertebrae, sacrum, sternum, scapula, tarsals, metatarsals, metacarpals, and phalanges (e.g., Holland 1995; Meadows and Jantz 1992). All of these solutions will result in wider prediction intervals than using limb bones, but approximating stature from a nonlimb bone is preferable to estimating a measurement from a bone fragment, because the latter method introduces additional error.

ADDITIONAL CONSIDERATIONS

Although this chapter focuses on stature estimation in a forensic context, it should be noted that biological anthropologists use stature estimates to approach a variety of questions. Paleoanthropologists approximate the stature of hominid fossils to compare modern human morphology with the physical form of our distant relatives. Bioarchaeologists use stature as an indicator of overall health in past populations. Height has a substantial genetic component, but this aspect of a person's physical makeup is affected greatly by the environment, especially the conditions surrounding growth and development (Ulijaszek 2006). Stature increases have been documented for entire populations, as nutrition, sanitary conditions, and medical care have improved in many countries around the world (Alberman et al. 1991; Buretic-Tomljanovic et al. 2004; Fredriks et al. 2000; Hoppa and Garlie 1998; Leung et al. 1996; Meredith 1976; Ulijaszek 1993; Vercauteren and Susanne 1985). **Secular change** in long bone lengths due to stature increases has been documented in modern Americans (Jantz and Jantz 1999; Meadows and Jantz 1995). For this reason, it is imperative to use regression formulae from the appropriate temporal cohort to estimate stature accurately.

Another variable that can affect the accuracy of stature estimates is age-related stature decrease in middle-aged and older adults. Studies indicate that stature begins to decrease by the age of 45 years (Galloway 1988) and perhaps as young as age 30 (Trotter and Gleser 1951). Trotter and Gleser (1951) found that stature decreases at a linear rate by about 0.6 mm a year, but Galloway (1988) reported a larger decrease of 1.6 mm per year, and Giles (1991) found that the rate of decrease was variable and not linear. Several of these studies also investigated sex and ancestry differences in age-related stature decreases but reached variable conclusions, and, consequently, this topic warrants further investigation. Whether or not the stature estimate warrants adjustment for age depends largely on the antemortem records available for comparison.

The reliability of antemortem stature records is often questionable, particularly with self-reported stature. Willey and Falsetti (1991) found that college students' driver's licenses tend to overreport stature by nearly 1 cm. Males overreport stature to a greater degree than females; taller individuals report stature more accurately than shorter individuals; and older individuals tend to overreport stature; however, perhaps, this phenomenon is confounded by age-related stature decreases (Giles and Hutchinson 1991). Official stature measurements have also been found to vary considerably. Stature records for a single individual who was measured 19 times during a 20-year period by various medical and legal personnel ranged between 62 inches and 67 inches (Snow and Williams 1971). This variation could have been due to age, time of day, instrument calibration, or observer error, but the example illustrates the considerable variation in antemortem stature records.

Antemortem stature records can come from a variety of sources, including medical or military records (i.e., **measured stature,** or **MSTAT**) and self-reported statures, such as those on driver's licenses (i.e., **forensic stature,** or **FSTAT**). Regression equations vary as to the antemortem stature source used to derive the equations (Ousley 1995; Wilson et al. 2010), and some equations were derived using cadaver length instead of antemortem records. Discrepancies in antemortem stature estimates or measures attribute additional error to postmortem stature estimates, based on long bone length(s). Consequently, Willey (2009) recommends carefully considering the measuring context and the antemortem record source when making comparisons between skeletal-based stature estimates and antemortem records. Ousley (1995) recommends using a wide estimation interval (i.e., a 90% prediction interval) in forensic casework to compensate for discrepancies in reported stature and for the inherent error in stature estimates. Other authors recommend using an **informative prior** to approximate stature, whenever possible, which incorporates numerical information about the reference population into the calculation, but this technique is computationally intense and not always feasible (Ross and Konigsberg 2002; Wilson et al. 2010).

The SWGANTH guidelines maintain that the scarcity of research on subadult stature estimation makes these estimates uncertain. The uniqueness of subadult body proportions and **allometric changes** throughout growth and development dictate that adult regression formulae should not be used to estimate subadult stature. Formulae based on measurements of diaphyseal length from radiographs are available (Smith 2007), and the femur/stature ratio has been proposed for this purpose, as well (Feldesman 1992). Subadult stature estimation is one area where the forensic community would benefit greatly from further research.

SUMMARY

- Stature estimation involves approximating living height from skeletal dimensions.
- Forensic anthropologists use two techniques to estimate stature: mathematical/regression methods and anatomical methods.
- Anatomical methods are preferred when sex and ancestry are not known or when supernumerary vertebrae are present. These methods are less sensitive to population differences in body proportions. The anatomical method is not feasible if any of the required elements are missing (cranium, vertebrae, sacrum, femur, tibia, talus, and calcaneus).
- Regression methods are required when skeletal elements necessary to apply anatomical methods are missing or fragmentary. The regression method is susceptible to error, owing to body proportion differences between populations and pathological factors affecting height, including age-related decreases in stature.
- The regression method is the most commonly used stature estimation method in forensic anthropology, largely because of the fragmentary/incomplete nature of many skeletal cases. Regression equations are easy to apply and are available for a number of populations. The Fordisc software uses regression analysis to estimate stature.
- When using regression methods, the SWGANTH recommends incorporating information about the individual's population, sex, and temporal cohort and selecting the formula with the smallest prediction interval.
- If limb bones are not available, stature can be estimated using nonlimb bones (cranium, talus, calcaneus, metacarpals, metatarsals, and sacrum), by using regression equations based on fragments, or by estimating limb bone

length from the fragments and then using an appropriate regression equation. Approximating stature from a nonlimb bone is preferable to estimating a measurement from a bone fragment, because the latter method introduces additional error.
- Factors that can affect the accuracy of stature estimates include pathological conditions, age-related decreases in stature, population-specific variation in limb proportions, and secular change in stature and bone length.
- Antemortem stature records vary in terms of accuracy and reliability. Regression equations vary as to the antemortem stature source used to derive the equations. Practitioners should use a wide prediction interval to compensate for discrepancies in reported stature and for the inherent error in stature estimates.

Review questions

1. What are the two techniques that forensic anthropologists use to estimate stature? List the pros and cons of each method.
2. What is regression theory? Who introduced this concept? How does it apply to stature estimation?
3. List three SWGANTH recommendations for estimating stature in a forensic case.
4. Is it advisable to estimate stature from subadult remains? Why or why not? How might you do this?
5. List three variables that can affect the accuracy of a stature estimate. Which stature estimation method most effectively addresses each of these issues, and why?
6. What is the difference between measured stature (MSTAT) and forensic stature (MSTAT)? Discuss the pros and cons of using MSTAT versus FSTAT to derive stature estimation equations.
7. If you were to formulate a research project on stature estimation, what would it be? Explain the reason for your research, your primary research question, and the method(s) that you would use to address the question.

Glossary

Allometric changes: changes in body shape are not proportional to changes in body size, such that two organisms of the same size may not be the same shape.
Anatomical method: a method of estimating stature that requires assembling the skeleton in anatomical position, applying a correction factor for soft tissue, and measuring height. Elements used for the anatomical method are: cranium, vertebral column (minus the atlas), sacrum, femur, tibia, talus, and calcaneus.
Confidence interval: the interval that expresses the probability that the true parameter (i.e., stature), given the value(s) of the independent variable(s) (i.e. bone measurements), lies within the interval predicted by the regression line. A 95% confidence interval means that there is a 95% probability that the true linear regression line of the population lies within the confidence interval of the regression line calculated from the sample data (measurements).
Forensic stature (FSTAT): self-reported stature such as that on a driver's license or passport.
Informative prior: prior probability distributions are used in Bayesian statistical inference. An informative prior conveys specific information about the variable(s) in question and influences predictions/results based on this information.
Measured stature (MSTAT): stature that is measured directly (i.e. from the body and not from photos) and taken systematically following guidelines.
Osteometric board: an instrument used to measure long bones. One end of the board has an immovable upright plate that sits at a right angle to the base of the board. The measuring scale is on the base of the board. The opposite end of the board is a movable endplate. The bone is placed on the board against the immovable upright endplate, and the movable plate is adjusted to measure the length of the bone.
Prediction interval: an interval in which future observations are predicted to fall based on the regression equation, given a specific probability.
R^2 value: also known as the *coefficient of determination*. In regression analysis, this value indicates how well the regression line approximates or fits the data. The value ranges between 0 and 1 (or between 0% and 100%); the higher the R^2, the better the regression model fits the data.

Regression theory: a mathematical method first proposed by Karl Pearson in 1899. Regression analysis uses the correlation between long bone length and living stature to derive a linear equation that predicts stature from long bone length.
Secular change: short-term evolutionary change.
Supernumerary vertebrae: a common congenital anomaly of the spine in which extra vertebrae are present.

References

Alberman E, Filakti H, William S, Evans S, and Emanuel I. 1991. Early influences on the secular change in the adult height between the parents and children of the 1958 birth cohort. *Ann Hum Biol* 18:127–136.

Auerbach BM. 2011. Methods for estimating missing human skeletal element osteometric dimensions employed in the revised Fully technique for estimating stature. *Am J Phys Anthropol* 145:67–80.

Buretic-Tomljanovic A, Ristic S, and Brajenovic-Milic B. 2004. Secular change in body height and cephalic index of Croatian medical students (University of Rijeka). *Am J Phys Anthropol* 123:91–96.

Dwight T. 1894. Methods of estimating the height from parts of the skeleton. *Med Rec* 46:293–296.

Feldesman MR, Kleckner JG, and Lundy JK. 1990. Femur/stature ratio and estimates of stature in mid and late Pleistocene fossil hominids. *Am J Phys Anthropol* 83:359–372.

Feldesman MR. 1992. Femur/stature ratio and estimates of stature in children. *Am J Phys Anthropol* 87(4):447–459.

Fredriks A, Van Burren S, and Burgmeijer R. 2000. Continuing positive secular growth change in the Netherlands 1955–1997. *Pediatr Res* 47(3):316–323.

Fully G. 1956. Une nouvelle méthode de détermination de la taille. *Ann Med Legale* 35:266–273.

Fully G and Pineau H. 1960. Determination de la stature au moyen de squelette. *Ann de Med Legale* 40:145–154.

Galloway A. 1988. Estimating actual height in the older individual. *J Forensic Sci* 33:126–136.

Giles E. 1991. Corrections for age in estimating older adults' stature from long bones. *J Forensic Sci* 36:898–901.

Giles E and Hutchinson DL. 1991. Stature and age-related bias in self-reported stature. *J Forensic Sci* 36:765–780.

Holland TD. 1992. Estimation of adult stature from fragmentary tibias. *J Forensic Sci* 37(5):1223–1229.

Holland T. 1995. Brief communication: Estimation of adult stature from the calcaneus and talus. *Am J Phys Anthropol* 96(3):315–320.

Hoppa RD and Garlie TN. 1998. Secular changes in the growth of Toronto children during the last century. *Ann Hum Biol* 25(6):553–561.

Jantz RL, Hunt DR, and Meadows L. 1995. The measure and mismeasure of the tibia: Implications for stature estimation. *J Forensic Sci* 40(5):758–761.

Jantz L and Jantz R. 1999. Secular change in long bone length and proportion in the United States, 1800–1970. *Am J Phys Anthropol* 110:57–67.

Jantz RL and Ousley SD. 1996. *FORDISC 2.0: Computerized Forensic Discriminant Functions.* Knoxville, TN: University of Tennessee.

Jantz R and Ousley S. 2005. *FORDISC 3.0: Computerized Forensic Discriminant Functions.* Knoxville, TN: University of Tennessee.

Konigsberg L, Hens SM, Jantz LM, and Jungers WL. 1998. Stature estimation and calibration: Bayesian and maximum likelihood perspectives in physical anthropology. *Yearbk Phys Anthropol* 41:65–92.

Leung S, Lau J, Xu Y, and Tse L. 1996. Secular changes in standing height, sitting height and sexual maturation of Chinese—the Hong Kong growth study, 1993. *Ann Hum Biol* 23(4):297–306.

Lundy JK. 1988. Sacralization of a sixth lumbar vertebra and its effect upon the estimation of living stature. *J Forensic Sci* 33(4):1045–1049.

Maijanen H. 2009. Testing anatomical methods for stature estimation on individuals from the W. M. Bass donated skeletal collection. *J Forensic Sci* 54(4):746–752.

Manouvrier L. 1893. Le determination de la taille d'apres les grande os des membres. *Mem de la Soc d'Anthropologie de Paris* 4:347–402.

Meadows L and Jantz R. 1992. Estimation of stature from metacarpal lengths. *J Forensic Sci* 37(1):147–154.

Meadows L and Jantz R. 1995. Allometric secular change in the long bones from the 1800s to the present. *J Forensic Sci* 40(5):762–767.

Meredith H. 1976. Findings from Asia, Australia, Europe and North America on secular change in mean height of children, youths and young adults. *Am J Phys Anthropol* 44(2):315–326.

Niskanen M, Maijanen H, McCarthy D, and Junno J. 2013. Application of the anatomical method to estimate the maximum adult stature and the age-at-death stature. *Am J Phys Anthropol* 152:96–106.

Orfila MJB. 1821–1823. *Leaons de Medicine Legale*, Vols 1–2. Paris, France: Bechet Jeune.

Orfila MJB and Lesueur O. 1831. *Traite des exhumations juridiques, et considerations sur les changements physiques que les cadavres eprouvent en se pourrissant dan la terres, dan l'eau, dans les fosses d'aisance et dans le fumier*, Vols 1–2. Paris, France: Bechet Jeune.

Ousley S. 1995. Should we estimate biological or forensic stature? *J Forensic Sci* 40:768–773.

Pearson K. 1899. Mathematical contributions to the theory of evolution: On the reconstruction of the stature of prehistoric races. *Philos Trans R Soc Lond* 192:169–244.

Raxter MH, Auerbach BM, and Ruff CB. 2006. Revision of the Fully technique for estimating statures. *Am J Phys Anthropol* 130:374–384.

Raxter MH, Auerbach BM, and Ruff CB. 2007. Technical note: Revised Fully stature estimation technique. *Am J Phys Anthropol* 133(2):817–818.

Raxter MH and Ruff CB. 2010. The effect of vertebral numerical variation on anatomical stature estimates. *J Forensic Sci* 55(2):464–466.

Ross AH and Konigsberg LW. 2002. New formulae for estimating stature in the Balkans. *J Forensic Sci* 47(1):165–167.

The Scientific Working Group for Forensic Anthropology (SWGANTH). www.swganth.org.

Simmons T, Jantz RL, and Bass WM. 1990. Stature estimation from fragmentary femora: A revision of the Steele method. *J Forensic Sci* 35(3):628–636.

Smith SL. 2007. Stature estimation of 3–10 year old children from long bone lengths. *J Forensic Sci* 52(3):538–546.

Snow CC and Williams J. 1971. Variation in pre-mortem stature estimates of skeletal remains. *J Forensic Sci* 16:455–464.

Steele DG and McKern TW. 1969. A method of assessment of maximum long bone length and living stature from fragmentary long bones. *Am J Phys Anthropol* 31(2):215–227.

Stevenson PH. 1929. On racial differences in stature long bone regression formulae, with special reference to stature reconstruction formulae for the Chinese. *Biometrika* 21:303–321.

Stewart TD. 1979. *Essentials of Forensic Anthropology: Especially as Developed in the United States*. Springfield, IL: C.C. Thomas.

Sue JJ. 1755. Sur les proportions des squelette de homme, examine depuis l'age du plus tendre, jusqu' B celui de vingt cinq, soixante ans, and audel. *Acad Sci Paris Mem Mathemat Phys Present Divers Savants* 2:572–285.

Topinard P. 1885. *Elements d'anthropologie generale*. Paris, France: Delahaye and Lecrosnier.

Trotter M and Gleser GC. 1951. The effect of aging on stature. *Am J Phys Anthropol* 9:311–324.

Trotter M and Gleser GC. 1952. Estimation of stature from long bones of American Whites and Negroes. *Am J Phys Anthropol* 10(4):463–514.

Trotter M and Gleser GC. 1958. A re-evaluation of estimation of stature based on measurements of stature taken during life and of long bones after death. *Am J Phys Anthropol* 16:79–124.

Ulijaszek S. 1993. Evidence for a secular trend in heights and weights of adults in Papua New Guinea. *Ann Hum Biol* 20(4):349–355.

Ulijaszek S. 2006. The international growth standard for children and adolescents project: Environmental influences on preadolescent and adolescent growth in weight and height. *Food Nutr Bull* 27(4 Suppl Growth Standard):S279–S294.

Vercauteren M and Susanne C. 1985. The secular trend of height and menarche in Belgium: Are there any signs of a future stop. *Eur J Pediatr* 144(4):306–309.

Willey P. 2009. Stature estimation. In S Blau, and DH Ubelaker (Eds.), *Handbook of Forensic Anthropology and Archaeology*. Walnut Creek, CA: Left Coast Press Inc.

Willey P and Falsetti T. 1991. Inaccuracy of height information on driver's licenses. *J Forensic Sci* 36:813–819.

Wilson RJ, Herrmann NP, and Jantz LM. 2010. Evaluation of stature estimation from the database for forensic anthropology. *J Forensic Sci* 55(3):684–689.

CHAPTER 12

Pathological Conditions as Individuating Traits in a Forensic Context

David R. Hunt and Kerriann (Kay) Marden

CONTENTS

Skeletal responses to insult	206
Problems of attempting to diagnose pathological conditions	207
Classification of pathological conditions	210
Arthropathies	210
Trauma	212
Types of trauma	213
Fracture	213
Dislocation	214
Congenital anomalies	215
Tumors	220
Disease processes	220
Intrinsic disease processes	221
Infectious diseases	223
Bone infections	224
Summary	225
Review questions	226
Glossary	226
References	227
For more information regarding pathological conditions of the skeleton	229
For more information on the osteological paradox	229
For more information on habitual stress and occupational markers on the skeleton	230

> **LEARNING OBJECTIVES**
> 1. Discuss the process of making a positive identification by using individualistic skeletal features.
> 2. List the five atypical skeletal changes, and describe the physiological mechanisms that produce these changes.
> 3. Explain why "ribbon matching" is not a reliable approach for diagnosing skeletal conditions, and describe the best practice diagnostic approach.
> 4. Describe and provide examples of the eight broad classes of human skeletal variants.
> 5. Discuss the common forms of skeletal trauma.
> 6. Describe the process and stages of bone healing.

Forensic anthropologists are most often called in to consult when the deceased is decomposed, highly fragmented, significantly burned, or, otherwise, not identifiable through standard autopsy procedures. In these cases, the medical examiner's expectation of the anthropology consult is to provide the biological profile of the remains and to possibly recognize individualistic features that might lead to a positive identification. A number of individualistic features occur at such a low frequency that they can be considered unique, but these traits must have been documented in **antemortem** radiographs or medical records to be used for identification purposes.

A number of conditions can affect human health, but relatively few illnesses will ever affect the bone, even in patients afflicted before the advent of antibiotics and modern hygienic practices. In most of these cases, the illness would have to be a long-term, chronic struggle, rather than an acute attack. Among conditions known to leave skeletal changes, only 5%–20% of cases ever progress to skeletal expression, and these numbers are based on paleopathological data before preventive and palliative care (Ortner 2008:191). In modern cases of forensic interest, markers of disease can be expected to be even lower since, due to modern preventive and/or therapeutic treatments for many of the diseases commonly seen in historic and prehistoric archeological series (e.g., infectious diseases such as poliomyelitis and tuberculosis; deficiency diseases, including rickets and scurvy; congenital anomalies such as cleft palate and polydactyly; untreated trauma such as unreduced fractures and pseudarthroses; and sepsis or other widespread infections). Therefore, when pathological conditions are observed in skeletal remains in a forensic context, their relative rarity can make them useful for positive identification (Cunha 2006).

A frustration in identification using individualistic traits is that an antemortem medical record (i.e., radiographs, computed tomography scans, or magnetic resonance imaging scans) must exist, to which the observed pathological condition can be compared. Therefore, the features identified in the skeleton must either be something that would have caused serious-enough complications during the person's life to warrant a radiographic examination or be observable on a radiograph taken for unrelated reasons, so that it can be compared to a postmortem radiograph. When making a comparative radiographic study, the postmortem element(s) need to be in the same or very similar position to make an accurate and/or reliable comparison. Lacking this, personal accounts of the decedent's infirmity or disability may be useful, but it is important to recognize that human memory is fallible, and recollections do not automatically imply or preclude a match (Schacter et al. 2011). Furthermore, positive identification may not be possible with more commonly occurring skeletal changes, such as arthritis and periodontal disease.

More general skeletal traits, such as disease processes or patterned antemortem trauma, while not often useful for individuation, may aid in understanding the lifestyle of the decedent. Pathological changes may point to habitual drug use or dietary deficiency (see Saini et al. 2005; or Shetty et al. 2010 on dental disease related to drug abuse) or inform the cause or manner of death, such as a pattern of healed antemortem injuries in cases of habitual domestic violence (see Sauer 1998).

SKELETAL RESPONSES TO INSULT

Why does bone form and respond the way it does? Since bone is a living tissue, it is continuously formed and remodeled throughout life at relatively predictable rates, with total bone cell turnover taking place in approximately 12 years in a healthy adult (Eriksen et al. 1994). However, when bone reacts to injury or inflammation, atypical skeletal changes will occur. These changes can be classified in one of the five ways: abnormal addition, abnormal subtraction, abnormal density, abnormal size, and abnormal shape (Ortner 2003:45).

The physiological mechanisms that produce these changes are quite limited and include buildup (blastic response), breakdown (resorption, or lytic response), or a combination of these processes (Figure 12.1). Specialized cells are responsible for each of these responses—osteoblasts undertake bone deposition and osteoclasts perform bone removal. Remodeling involves the action of both of these cell types, with osteoclasts removing the existing damaged bone, while osteoblasts replacing it with new deposition. However, these actions are not equal, and under normal conditions, one osteoclast can break down the same amount of bone produced by 10 osteoblasts in the same amount of time (Johnson 1966; Snapper 1957). A notable exception to this occurs in cases of uncontrolled bone proliferation, as seen in tumors.

The initial skeletal healing response is so subtle that it is often imperceptible radiographically; nonetheless, it can be detected through direct examination of the bone surface (Marks et al. 2009). Periosteal remodeling can be as thin as

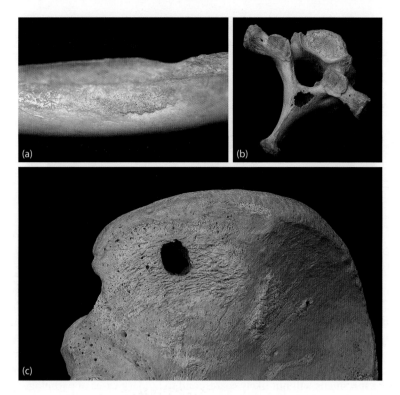

Figure 12.1 Skeletal responses to insult. (a) Osteoblastic response. New bone formation in response to infection under the periosteum (periostitis). Formation of woven bone within the pus from tuberculosis (Terry Collection). (b) Osteoclastic (lytic) response. Destruction of cortical and trabecular bone in response to an enlarging metastatic carcinoma in the dorsal spine of a thoracic vertebra. The rapid rate of expansion of the tumor does not allow for osteoblastic activity, resulting in a ragged, "moth-eaten" look of the edges of the defect (Huntington Collection). (c) Combined osteoblastic and osteoclastic response. Osteoclastic destruction of cortical and trabecular bone in response to an enlarging metastatic carcinoma from prostate cancer. Because prostate cancers are slow-growing, there is sufficient time for osteoblastic activity to occur around the lifted periosteum (Huntington Collection).

500 μm, or half a millimeter—well below the threshold of detection of standard radiographic equipment (Brogdon 1998). Although we do have diagnostic imaging technology sensitive enough to detect early healing, this is not standard equipment in most autopsy settings (Donchin et al. 1994; Dolinak and Matshes 2005). When the primary callus begins forming, both gross and radiographic methods can be used. Healthy adult skeletal response will begin within 10–14 days of the onset of injury or inflammation; however, in subadults, this response can be seen much earlier, occurring at as much as twice the rate of adult remodeling (Ortner 2003:128). Because the cells responsible for remodeling are already deployed in the dynamic matrix of actively growing subadult bone, the process begins immediately on injury (Johnstone and Foster 2001; O'Connor and Cohen 1998).

The newly deposited bone matrix is disorganized, or **woven**, whether it is in response to growth, injury, or inflammation. For this reason, it is important to observe the distribution of woven bone, especially on long bone diaphyses. Bone deposition in subadults occurs deep to the periosteum in the region of active growth, adjacent to the epiphyseal plate, or **cutback zone** (Figure 12.2). This is generally restricted to approximately 3–5 mm from the metaphysic; however, normal cutback can be as great as 10 mm in older children during rapid growth spurts. More than 10 mm, however, suggests the likelihood of a pathological condition and warrants closer scrutiny.

PROBLEMS OF ATTEMPTING TO DIAGNOSE PATHOLOGICAL CONDITIONS

Many students and novice practitioners attempt to perform diagnosis of a skeletal lesion by simple visual comparison to photographs in paleopathology atlases, or worse, using unverified online sources. This practice (referred to derisively by Don Ortner as **ribbon matching**) is an unreliable means of diagnosing skeletal conditions. Photographs are immensely useful to illustrate the textual descriptions of pathological changes, but because the possible responses of

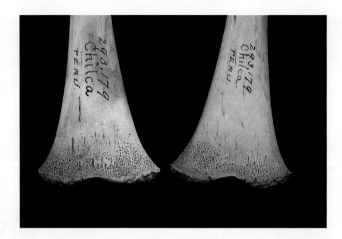

Figure 12.2 Cutback. Distal femora exhibiting normal "cutback," a region of osteoclastic activity at the metaphasis related to normal bone growth and development (archaeological, Peru).

bone are limited, many pathological changes are similar in appearance. The use of photographs should be restricted to illustration to ensure the understanding of diagnostic criteria, rather than serving as the basis for diagnosis. Identification made purely by this method will likely lead to inaccurate and/or misleading diagnoses. This is especially true, considering the fact that most of the specimens depicted in atlases are from museum collections and tend to depict the more severe expression of the conditions that they represent. Moreover, most are cases that were never treated or, at the very least, were not treated with the therapies and technologies currently available. As such, these untreated museum specimens will not likely correspond to modern cases of pathological conditions that have received medical intervention in a clinical setting. Therefore, comparing modern forensic cases with these extreme case specimens will mislead the practitioner to overlook the subtler expressions of infirmity.

Correct diagnosis of pathological conditions of the skeleton requires broad knowledge of bone biology, an understanding of pathological processes, and an appreciation of the importance of lesion distribution in determining the etiology of pathological conditions. It is imperative to have sufficient experience in observing the full range of normal skeletal variation and common anomalies, before one can begin to identify pathological changes, let alone interpret them (see Keats 1984; Keats and Anderson 2013 for radiographic illustration of the range of normal variation in the human skeleton). Until a practitioner has attained considerable experience with skeletal anomalies and pathological conditions, diagnosis should not be the goal of the investigation. Instead, the anthropologist should focus on recording a detailed description of the feature(s) by using standardized terminology, which can then be used to confer with more senior anthropological colleagues and medical personnel (see Buikstra and Ubelaker 1994 for standardized nomenclature and recording methods).

The best practice (even for the experienced practitioner) was outlined over a decade ago (Ortner 2003:49–51), and it remains the gold standard for interpreting paleopathological conditions:

- Identify and record the distribution of the lesion(s)
    - On the individual bone(s)
    - Across the skeleton as a whole
- Describe the feature, lesion, or trait by using standard but generalized terminology
    - The first step of the description should include the specifics of the individual feature(s):
        - Size
        - Shape
        - Depth or height
        - Texture
        - Density
        - Definition and shape of margins
        - Associated changes to the bony surroundings

- This descriptive information assists in classification of the lesion, using a framework of five broad categories:
  - Abnormal size
  - Abnormal shape
  - Abnormal excess of bone
  - Abnormal absence of bone
  - Abnormal bone density

Within each of these categories is a series of established subclassifications (Ortner 2003:50–51). These allow the analysts to refine their interpretation of the disease process responsible for the lesion or anomaly, which can assist in establishing a **differential diagnosis**. A differential diagnosis requires a reversal of perspective from an attempted diagnosis. Whereas a diagnosis is an exclusive process, aiming to reduce observations to a specific pathological condition, a differential diagnosis is inclusive in nature, attempting to identify all possible conditions that could produce the observed traits. It is only after an exhaustive differential diagnosis has been developed that one can begin the process of elimination to work toward a diagnosis.

As a final step to this proven analytical protocol, we would add that it is crucial to accept ambiguity. There are few skeletal pathological conditions with **pathognomonic** skeletal traits, or definitively diagnostic characteristics, that are the hallmarks of only one specific condition. Even some lesions that are generally accepted as attributable to a specific disease process (such as stellate cranial vault lesions associated with tertiary syphilis, Figure 12.3) are not pathognomonic to a single condition and can therefore be misleading without considering the totality of the skeletal features and the specific nature of their distribution. Therefore, it is essential to limit attempts at diagnosis to the evidence at hand, and this may preclude any attempt at diagnosis. Pathological analysis requires not only a tremendous amount of knowledge and experience, but also restraint.

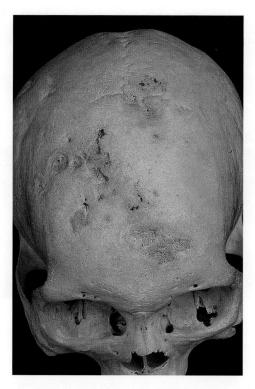

Figure 12.3 Stellate lesions. Frontal vault lesion caused by tertiary syphilis, with the infection communicating to the endocranium. Although characteristic of end-stage treponemal infection, these lesions can be confused with other infectious lesions (Huntington Collection).

CLASSIFICATION OF PATHOLOGICAL CONDITIONS

Working toward a diagnosis, some of the most common variants in the human skeleton that are utilized in forensic anthropology can be considered in eight broad classes—though not necessarily mutually exclusive. Each of these classes will be discussed in detail below:

- Arthropathies
- Trauma
- Congenital anomalies
- Tumors
- Disease processes
  - Infectious diseases
  - Intrinsic (noncommunicable) disease processes
  - Generalized infection

Each of these classes of skeletal condition has potential application in a forensic context; however, in most cases, this requires comparison with antemortem medical records—especially radiographs—or with reports and recollections regarding the decedent's ailments, pain, or limited mobility.

Arthropathies

The general term for joint disease is **arthropathy**, which simply indicates any pathological change to joints. Osteoarthritis (OA, more commonly known as simply *arthritis*) is broadly defined as joint degeneration and can be expressed either as lipping at the margins of the joint or as destruction of the subchondral bone beneath the articular surface. Articular deterioration begins with damage to the cartilage—either through biomechanical stress or by trauma—and can present in several forms (Figure 12.4). These include *porosity*, which is perforation(s) in the subchondral bone surface that

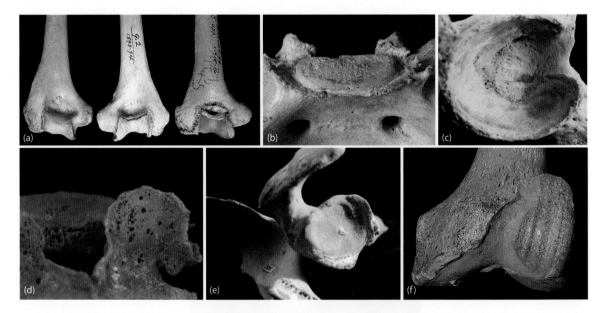

Figure 12.4 Arthritic joint destruction. (a) Lipping. Stages of arthritic lipping formation on three distal humeri: the least affected is on the left, the most severe is on the right (Huntington Collection). (b) Vertebral osteophytes. Bony arthritic projections (osteophytes) along the margin of the amphiarthrosis articulation of the first sacral element (Terry Collection). (c) Erosion. Right acetabulum exhibiting arthritic erosion of the subchondral bone, with associated porosity as well as blastic activity on the lunate surface (Huntington Collection). (d) Porosity. Articular destruction and erosion lead to porosity in the subchondral bone. Right superior articular facet on a lumbar vertebra (Huntington Collection). (e) Surface osteophytes. Right scapula with osteophyte development at the center of the glenoid fossa in response to arthritic inflammation (Huntington Collection). (f) Eburnation. Cartilage destruction leads to bone-on-bone contact, resulting in polished appearance of the affected joint surfaces (eburnation). The deep grooving and eburnation on the lateral portion of the distal femoral articulation in this case are due to patellar involvement (Huntington Collection).

can range in size from minute to two or more millimeters in width (Ortner 2003:545); **erosion**, where the cartilage is damaged enough to allow attrition of the underlying bone surface; **surface osteophytes**, which involves development of exostoses on the joint surface (See Mann and Hunt 2005); and **eburnation**, in which persistent erosion elicits a sclerotic response to the underlying bone, causing a dense, polished-looking surface, where joint surfaces meet (Hough 2007:57). Osteoarthritis sometimes occurs in younger adults and even in some children with severe metabolic or congenital diseases, but these cases are exceedingly rare. Arthritic changes can also be the result of traumatic occurrences that have damaged the cartilage and/or the bone, resulting in **necrosis** (death) of these tissues (Mankin 1982). Otherwise, OA is considered a geriatric occurrence, not seen in most normal individuals below the age of 35 years and present in most people over the age of 50 years and in 90% of persons 80 years or older (Mann and Hunt 2005:15).

Most clinical research classifies OA as either primary or secondary. Primary OA is *idiopathic*, meaning that its etiology is not understood with certainty. It is believed that primary OA may be one of the normal physiological changes of aging, wherein age-related hormonal triggers alter the amount of enzymes in the joint, resulting in inflammation and destruction of the cartilage. Secondary OA is what is most often considered the *wear and tear* of a joint through daily use (see Mann and Hunt 2005:16–17 for discussion on the *wear and tear* theory). Secondary OA is usually not of much use for the purposes of personal identification, as it is widespread and relatively uniform in expression; however, it may sometime produce unique bony outgrowths that can be compared radiographically. Primary OA is more likely to be unique and is also more likely to have received medical diagnosis and treatment, generating antemortem records and, therefore, may be useful in individuation.

Another condition that is often categorized as an **idiopathic** arthropathy is diffuse idiopathic skeletal hyperostosis (DISH). DISH is expressed as excess bone development along the anterior and posterior longitudinal ligaments of the spine, occurring across four or more vertebral bodies contiguously (Figure 12.5). This condition should not be aconfused with other diseases that affect the spine, such as tuberculosis (Figure 12.6). Its etiology remains poorly

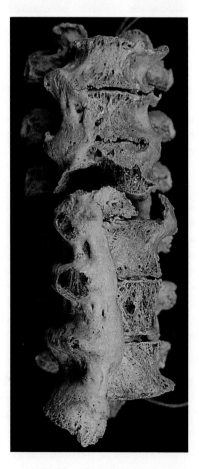

Figure 12.5 Diffuse idiopathic skeletal hyperostosis (DISH). This case involves both the lower thoracic and upper lumbar vertebrae (Terry Collection). Note that the osseous formation is only on the right side.

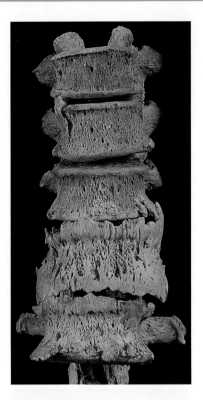

Figure 12.6 Morphological similarities between pathological conditions involving the vertebrae. Osteophytes and osteoblastic activity on the anterior bodies of the lumbar vertebrae caused by tuberculosis (Terry Collection).

understood. Although DISH primarily affects the spine and often co-occurs with vertebral osteoarthritis, it is not attributed to the same disease process and is not considered a true arthritis, as DISH does not generally affect the joint spaces (Ortner 2003:559). Despite its dramatic appearance, DISH is often subclinical and, therefore, not likely to receive antemortem diagnosis. Other than assisting in establishing or confirming the biological profile—DISH is more common in males, and its prevalence increases with age (Mann and Hunt 2005:77; Utsinger 1984)—this condition is generally not of great forensic utility, unless it has been identified incidentally in antemortem radiographs.

Anthropologists inexperienced in pathological diagnosis often confuse DISH with ankylosing spondylitis, which is a true form of spinal arthritis, or spondyloarthropathy. Like DISH, ankylosing spondylitis affects the vertebrae, is more common in males, and its cause is not well understood; however, there appears to be a genetic component (people with gene HLA-B27 have a significantly increased risk of developing ankylosing spondylitis). However, this condition differs markedly from DISH, in that it involves bony overgrowth at the joint margins, leading to fusion of the spine, creating an effect that is often referred to as *bamboo spine* (Figure 12.7). Ankylosing spondylitis involves not only the amphiarthroses (articulations of the vertebral bodies) but also the sacroiliac joints, diarthroses of the vertebrae, and costovertebral joints (Mann and Hunt 2005:88–89; Ortner 2003:571). As such, it limits mobility far more than DISH and is more likely to be clinically diagnosed.

Trauma

Trauma, by definition, implies injury to living tissue. In a forensic context, the etiology that is most often the focus is intentional injury, but accidental, therapeutic, or cosmetic alterations to bone must also be considered. Environmental influences such as nutritional deficit and endocrine conditions (i.e., hormonal disturbances) can also compromise the integrity of the bone, leading to injury.

Skeletal trauma can be classified as either antemortem or **perimortem** (occurring at or around the time of death). **Postmortem** changes are best referred to as damage rather than trauma, because traumatic injury cannot occur after death. Postmortem damage and perimortem trauma are covered elsewhere in this volume (see Chapters 4 and 13)

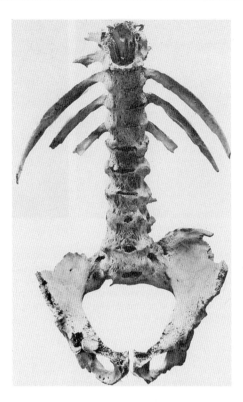

Figure 12.7 Ankylosing spondylitis. Ossification of the ligaments of the spine, creating an effect resembling bamboo. This case involves the thoracic, lumbar, and sacral elements, as well as fusion of some of the ribs (archaeological).

and in other publications (see Cunha and Pinheiro 2009; Loe 2009; Lovell 2008; Rodriguez-Martin 2006). It is imperative to keep the timing of the injury or damage in mind when assessing pathological changes. Taphonomic changes can mimic trauma (Figure 12.8) and can veil or destroy subtle evidence of pathological conditions on the skeletal elements (see Haglund and Sorg 1997, 2002; Nawrocki 2009; Sorg et al. 2012 for detailed discussion of the interpretation of taphonomic effects on bone in forensic settings).

Timing of the injury is particularly pertinent when assessing antemortem trauma that occurred at a young age. Early skeletal damage can be lost or obscured by the continual bone remodeling associated with normal growth and development. Only childhood fractures that are severe or are poorly treated are likely to deviate noticeably from normal bone size or shape. Such healed trauma can be misinterpreted as a congenital skeletal anomaly rather than being attributed to the influence of trauma on normal growth.

Types of trauma

Forces acting on bone can cause various types of injury. Some of the more common forms of skeletal trauma include *fracture*, *dislocation*, and *subluxation*.

Fracture

Fracture is classified based on the five major forces that compromise the bone: **tension** (stretching), **compression**, **torsion** (twisting), **flexion** (bending), and **shearing**. A combination of these forces causes most injuries. For example, a blow to the midshaft of a long bone may involve compression, tension, and flexion at the site of impact—and produce an incomplete, a complete, or a comminuted fracture, depending on the direction and amount of force. Similarly, torsion fractures are often a combination of compression and flexion, with the break following the normal physiological structure of the bone and/or tracing the tension imposed by tendon and muscle fibers (see Galloway 1999; Passalacqua and Rainwater 2015; Symes et al. 2012; Wedel and Galloway 2014 for more thorough discussion of fracture interpretation).

Antemortem trauma shows evidence of bone repair, whether at the microscopic level (in the very early stages of reactivity) or the macroscopic level (usually after about 10–14 days). The bone-healing process can be roughly

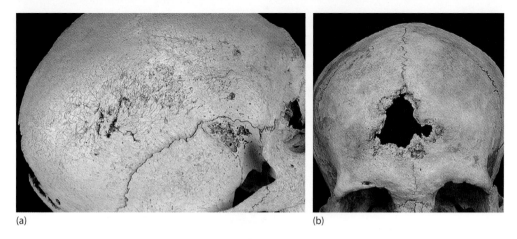

(a) (b)

Figure 12.8 Taphonomic changes. Natural processes can mimic pathological conditions and can also destroy evidence of real conditions. (a) Mimics of pathological changes (pseudopathologies). Taphonomic destruction of the cranial vault from environmental exposure of sunlight and temperature heating and cooling. Pseudopathological appearance to carcinoma lesions or scurvy (archaeological Peru). (b) Destruction of pathological changes. Taphonomic destruction of the frontal bone by environmental exposure, exacerbated by the presence of treponemal lesions that have compromised the integrity of the bone. These taphonomic effects could obscure the presence of the lesion over time (archaeological Peru).

differentiated into six stages that reflect a progression from a vascular reaction through the soft tissue response and, finally, to skeletal changes associated with fracture repair (adapted from Ortner 2003:126–128):

1. *Hematoma*: Blood flows into the injury from ruptured cortical blood vessels and the blood supply to the periosteum and endosteum. Inflammatory response stems the blood flow and initiates a clot.
2. *Formation of the fibrous callus*: The clot is infiltrated by fibrous connective tissue, which develops into a temporary, unmineralized framework.
3. *Formation of the primary bony callus*: The fibrous callus begins to ossify into loosely organized, woven bone.
4. *Transformation into the secondary bony callus*: The woven bone is replaced by dense cortical bone.
5. *Remodeling of the secondary callus*: The callus is remodeled over time, reducing the callus and returning the bone to its original shape (Figure 12.9).

A range of complications that impede the healing process or alter the outcome of healing can occur (Figure 12.10). These include necrosis; **infection**, such as osteomyelitis; deformity due to poor realignment of the fractured bone before healing; **pseudarthrosis**, resulting from a failure of the broken ends of a long bone to reunite; **myositis ossificans**, abnormal ossification of a muscle attachment caused by traumatic injury; and **ankylosis**, caused by either traumatic cartilage damage or infection of the joint space, referred to as septic arthritis (see Barnes 2012; Pearlstein et al. 2011 for detailed case presentations of septic ankylosis). Many of these complications produce skeletal responses that are easily mistaken for other conditions, such as bone tumors, tuberculosis, brucellosis, and severe arthritic changes.

Dislocation

Another form of skeletal trauma involves displacement of the bone from its normal orientation at the joint (Figure 12.11). When the articular surface of the bone is fully dislodged from the joint space, it is considered a dislocation, or a **luxation**. A partial displacement is known as a **subluxation**. Many congenital abnormalities can influence the development of a joint (such as shallow acetabulum), causing malformation that predisposes the joint to luxation or subluxation. It is important to discern such congenital anomalies from traumatically induced dislocations.

The condition of the surrounding soft tissue structures must also be considered when interpreting traumatic dislocation or subluxation. Developmental problems, muscular atrophy, nutritional deficiency, and hormonal changes can reduce the strength of the ligaments, tendons, and musculature that support and stabilize joints and lead to dislocation that could be misdiagnosed as traumatic injury. For example, rotator cuff syndrome is a luxation caused

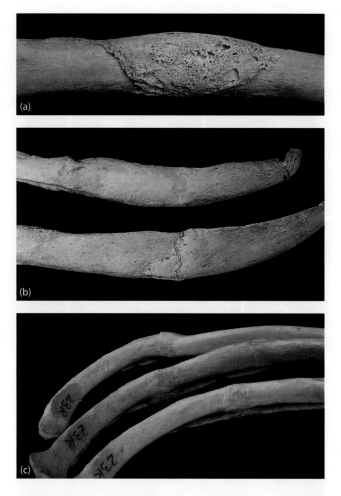

Figure 12.9 Bone repair and remodeling. (a) Formation of the primary bony callus. The fibrous callus begins to ossify into loosely organized, woven bone (Huntington Collection). (b) Transformation into the secondary bony callus. The woven bone is replaced by dense cortical bone (Terry Collection). (c) Remodeling of the secondary callus. The callus is reduced over time, returning the bone to an approximation of its original shape (Terry Collection).

by repeated stress on congenitally weakened soft tissue structure of the shoulder joint, but it is easily mistaken for traumatic dislocation of the shoulder in dry bone specimens (Figure 12.12).

Congenital anomalies

Congenital skeletal anomalies—often incorrectly referred to as *birth defects*—are embryologically derived skeletal variants caused by genetic or teratogenic (uterine environment) factors, or a combination of these (Barnes 1994a, 2008, 2012; Narotsky et al. 2004; Ortner 2003; Rogers et al. 2004). The range of these heritable skeletal defects is vast, including such varied conditions as polydactyly, cleft palate, supernumerary ribs, fused or bifid ribs, cranial suture variations such as craniosynostosis, cranial suture retention, and formation failures of the vertebrae or the sacrum (Figure 12.13). The etiology of congenital conditions stands in contrast to **acquired conditions**, which are contracted during life through infection, exposure, or injury.

While fascinating, congenital anomalies are of dubious forensic value, since most of the severe cases are not compatible with life (e.g., anencephaly, meningiocele, and spina bifida) and result in spontaneous abortion, stillbirth, or death at a very young age (Barnes 1994b, 2012). Those that are not lethal tend to be mild and asymptomatic and generally remain undetected throughout life. In order to be useful in personal identification, an anomaly must be externally visual, palpable, or cause pain or complications. Since most observable or painful skeletal anomalies today

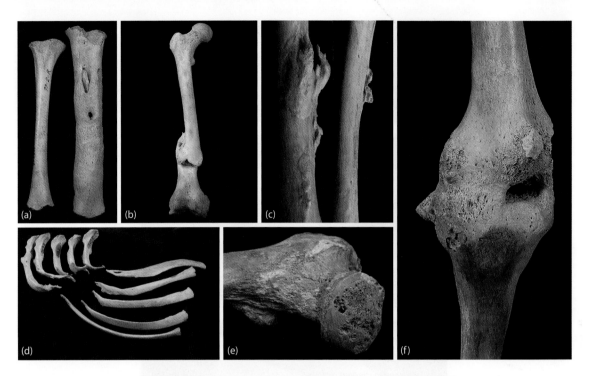

Figure 12.10 Complications of skeletal healing. (a) Infection. This case shows infection of the interior of the bone, or osteomyelitis (archaeological Peru). (b) Deformity. Unreduced fracture resulting in poor realignment of the fractured bone (Huntington Collection). (c) Myositis ossificans. Abnormal ossification of a muscle attachment caused by traumatic injury (Terry Collection). (d) Pseudarthrosis. A failure of the broken ends of a long bone to reunite, allowing movement resembling a joint (Terry Collection). (e) Necrosis. Bone death (Huntington Collection). (f) Ankylosis. Pathological joint fusion of the knee. Ankylosis, also referred to as septic arthritis, can be caused either by traumatic cartilage damage or by infection of the joint space (Huntington Collection).

Figure 12.11 Dislocation. Complete chronic hip joint dislocation (luxation) with loss of original acetabulum and remarkable arthritic activity (archaeological Egypt).

receive surgical treatment (and usually at a very early age), they may not be recognizable as congenital defects on human remains in a forensic context.

Among the most readily observed hard tissue congenital abnormalities are dental anomalies (Figure 12.14). Features of the teeth such as **winging, rotation, angulation, gemination, microdontia, agenesis, diastema, supernumerary dentition, inversion, impaction**, and **heterotopic dentition** are often evident during normal interpersonal interaction and may be seen in photographs, making them of value in identification (Figure 12.15). For a more detailed discussion of forensic odontology, refer to Chapter 6 of this volume (or see Hillson 1996; Irish and Scott 2016).

Figure 12.12 Chronic severe luxation. Rotator Cuff syndrome with complete destruction of glenoid and malformation of the proximal humeral head (Huntington Collection).

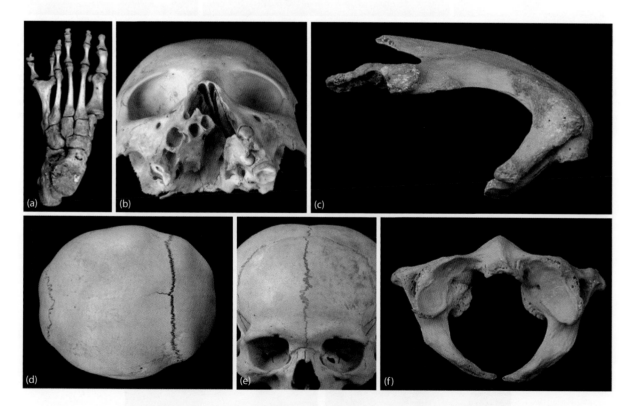

Figure 12.13 Congenital skeletal anomalies. Nonfatal congenital skeletal abnormalities can take a variety of expressions, each with a range of clinical expression. (a) Polydactyly. Extra digits of the hand or foot (NMHM Collection). (b) Cleft palate. Failure of fusion of the hard palate (archaeological Peru). (c) Fused ribs. Failure to differentiate between elements that should be distinct—in this case, first and second ribs (Huntington Collection). (d) Craniosynostosis. Premature fusion of adjacent cranial bones (archaeological Peru). (e) Metopic suture retention. Failure of fusion of the two halves of the frontal bone (archaeological Peru). (f) Neural arch defect. Failure of formation of one or more vertebral elements—in this case, the posterior arch of the atlas (Huntington Collection).

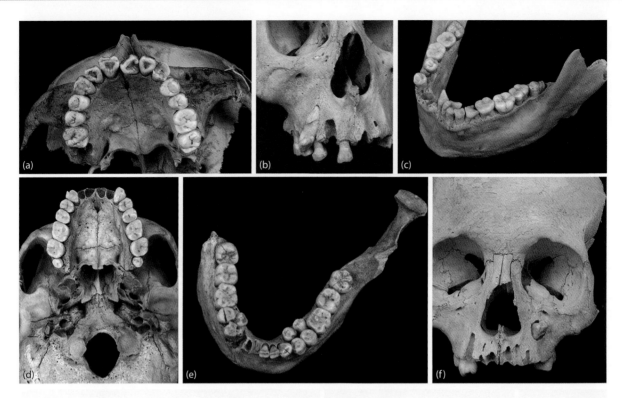

Figure 12.14 Dental anomalies. Many dental variants are likely to be observed during normal interpersonal interaction and may also be seen in photographs. (a) Winging. Winging of the maxillary central incisors (archaeological Peru). Note the shovel-shaped incisors. (b) Rotation. Right canine that is rotated 90° from the axis from normal orientation. This tooth is also involved in crowding with the premolars (archaeological Peru). (c) Retention of deciduous dentition. Retention of deciduous molars and marked anterior dental crowding (archaeological Peru). (d) Microdontia. "Peg" third maxillary molars (archaeological Peru). (e) Supernumerary dentition. Bilateral repetition of the mandibular premolars (archaeological Peru). (f) Heterotopic dentition. This right maxillary canine appears to have developed and fully erupted in the correct position in the arcade but in the wrong direction. The tooth has erupted at a 90° angle to the normal axis of growth (archaeological Peru).

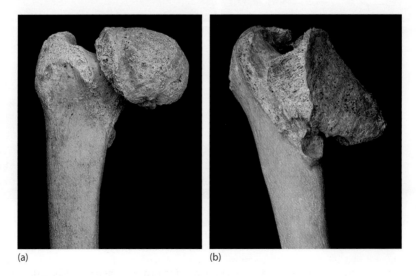

Figure 12.15 Malformation and destruction due to bone necrosis. (a). Necrotic femoral head with arthritic lipping and the beginning of head collapse (Huntington Collection). (b). A severe malformation of the femoral head due to collapse by necrosis (Huntington Collection).

More commonly occurring skeletal variants that are considered within the range of normal are scored as nonmetric traits in skeletal remains (see Finnegan 1978; Hauser and DiStefano 1989; Mann and Hunt 2013). Some examples include scapular notch, sternal foramen, olecranon foramina, parietal foramina, Wormian bones, os acromiale, patellar notch, nonsegmentation, os trigonum, and vascular etching (Figure 12.16). Some of these features are thought to reflect genetic relationships, indicating ancestry (Gill 1998; Rhine 1990) or biological relatedness. Because these are

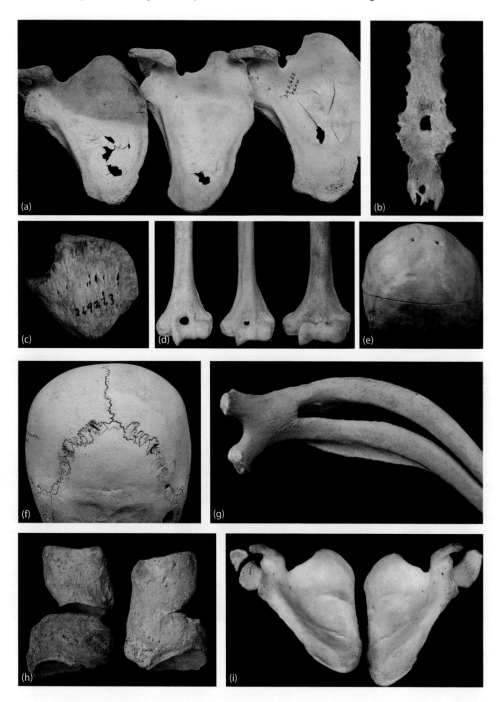

Figure 12.16 Normal skeletal variants. There is tremendous range in the normal expression of certain heritable skeletal traits, some of which may assist in determination of ancestry. (a) Scapular notch (archaeological Peru). (b) Sternal foramen (Terry Collection). (c) Patellar notch (anatomical United States). (d) Olecranon foramina (archaeological Egypt). (e) Parietal foramina (Terry Collection). Note also bilateral parietal thinning. (f) Wormian bones (archaeological Peru). (g) Bifid rib (Huntington Collection). (h) Osseous tarsal coalition (naviculocuneiform) (archaeological Mexico). (i) Os acromiale of the rt acromion (Terry Collection).

common variants, they are often of limited value for personal identification, unless they are expressed as part of a suite of anomalous features (which raises the statistical possibilities of one individual exhibiting a number of traits). Most nonmetric traits of the skeleton cause no complications and are discovered only incidentally during radiographs for unrelated medical reasons, so most are unlikely to have antemortem documentation.

Tumors

Defined simply as new growth, tumors are also referred to as neoplasms (Waldron 2009:168). Tumors may arise in bone or cartilage as primary tumors, or they can develop in other tissues and metastasize to skeletal tissue. The maturity of the tissues involved, the extent of proliferation, and the degree of localization can determine whether a tumor is malignant or benign and, therefore, may contribute to determination of cause of death (Ortner 2003: 503). However, in any forensic case, these features would be identified and evaluated by the pathologist and perhaps an oncologist, and the anthropologist would only comment on them in terms of skeletal involvement. Differentiation between the various types of skeletal neoplasms is complex and well beyond the scope of this chapter. Even a partial list of malignant and benign tumors affecting the skeleton is long. Neoplasms are diverse in expression (Figure 12.17) and are classified by the tissue type(s) affected, the anatomical origin of the tumor, and the rate of proliferation. Volumes have been written on cancers and tumors of the bone, but most are extremely rare, metastatic carcinoma being the most *common* form observed by the osteologists. However, some benign neoplasms are found in a low frequency and may be useful in personal identification, such as button osteomas and metaphyseal osteochondromas (Figure 12.18).

Disease processes

A wide range of specific diseases can affect the hard tissues of the body. While some of the conditions discussed throughout this chapter are commonly referred to as *diseases*, this section refers specifically to pathological conditions that result from disruptions of basic physiological mechanisms due to genetic, environmental, or idiopathic etiologies (intrinsic diseases) and those caused by infectious diseases.

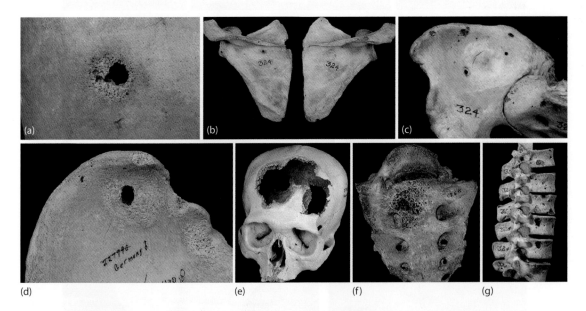

Figure 12.17 A sample of the range of expression of tumors affecting bone. Neoplasms are classified by the tissue type(s) affected, the anatomical origin of the tumor, and the rate of proliferation. Note that even tumors that share a diagnostic classification may be very dissimilar in appearance. (a) Metastatic carcinoma. Located in this case on the parietal bone (Terry Collection). Note osteoclastic destruction of the bone around the edges of the tumor location. (b) Multiple myeloma exhibited bilaterally on the scapulae (Terry Collection). (c) Multiple myeloma on the right ilium and anterior sacrum (Terry Collection). (d) Metastatic carcinoma from prostate cancer (Huntington Collection). (e) Metastatic carcinoma (archaeological Peru). (f) Metastatic carcinoma on the anterior sacrum (Terry Collection). Note the "web-like" pattern of an early stage of cortical destruction. Also, note the partial sacralization of the fifth lumbar vertebra, which is an unrelated congenital anomaly. (g) Multiple myeloma, involving several lower thoracic and upper lumbar vertebrae (Terry Collection).

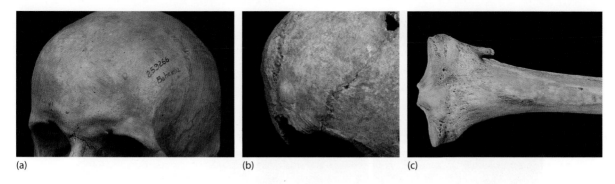

(a) (b) (c)

Figure 12.18 Benign tumors. Owing to their visibility in life, some benign tumors could potentially aid in identification. (a) Button osteoma (small) on the left frontal (historic Bohemia). (b) Button osteoma (large) on right lateral parietal (historic United States). (c) Metaphyseal osteochondroma on a right proximal tibia (archaeological Peru). Note that despite their prominence on dry bone, these particular specimens would be unlikely to have been visible in life and, therefore, are less useful for identification.

Intrinsic disease processes

Intrinsic, or noncommunicable, diseases are caused by physiological or biochemical disruptions that alter basic biological processes and leave a variety of skeletal indicators (Table 12.1; Figure 12.19). However, it is important to note the gross similarity in the skeletal responses to these various afflictions. The limited responses of skeletal tissue to a wide array of conditions must be kept in mind when attempting to develop a differential diagnosis, and this is among the primary reasons that it is crucial to accept ambiguity when interpreting pathological conditions of the skeleton.

The presence of a disease or disorder affecting the bone may assist in reconstructing elements of the decedent's lifestyle, such as malnutrition, lack of exposure to daylight, and lack of medical care. As such, these conditions may sometimes indicate cause and manner of death, as in the case of neglect or abuse of children, the elderly, or the infirm. Idiopathic disorders are intrinsic diseases for which the cause is not fully understood (Figure 12.20). These tend to be uncommon and therefore may assist in the identification of the remains. Metabolic diseases can also be useful in personal identification through comparison with family histories or by exploring possibilities of environmental exposure to certain materials, such as fluorine, mercury, and lead.

Table 12.1 Intrinsic disease categories

Category	Etiology/description	Common conditions	Illustrative figures
Circulatory Disturbances	Any condition in which the tissue receives inadequate blood supply. Several disorders involve disruption of blood flow to bone. Since bone is a living tissue that requires nutrients, reduced blood supply will cause death of the bone cells (necrosis).	Legg-Calve-Perthes disease, osteochondritis dissecans, and injuries involving large areas of the scalp.	Figure 12.10e
Hematopoietic Diseases	Wide range of conditions that affect the production or function of blood cells.	Anemias, thalassemia, sickle-cell disease, polycythemia vera, leukemia, and multiple myeloma.	Figure 12.17g, 12.19a–c
Metabolic Diseases	Any physiological disturbances that cause disruptions of the metabolism, the process by which the body converts fuel into energy.	Scurvy, rickets, osteomalacia, and osteoporosis.	Figure 12.19d–f
Endocrine Disorders	Disruptions in the production and release of the hormones that control multiple physiological processes. Some of these hormones are in direct control of the development, growth, and maintenance of skeletal tissues.	Gigantism, dwarfism, hyper- and hypothyroidism, acromegaly (Marfan's syndrome), hyper- and hypogonadism, and Cushing's syndrome.	Figure 12.19g
Idiopathic and Other Disorders	Disorders for which the etiology remains poorly understood.	Paget's disease, fibrous dysplasia, neurofibromatosis, and myositis ossificans progressive.	Figure 12.20a, b

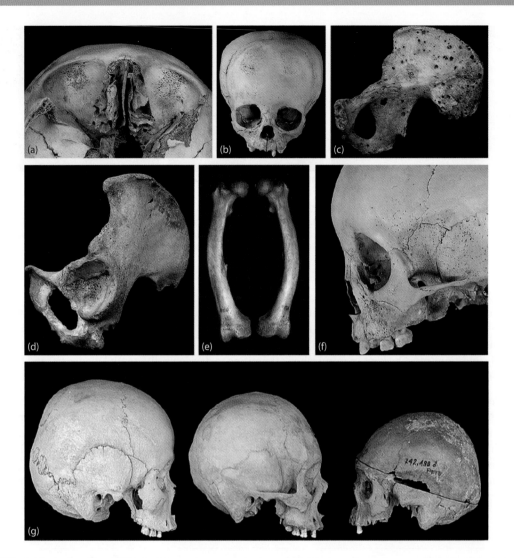

Figure 12.19 Intrinsic disorders. Intrinsic diseases have a vast range of expression, depending on which physiological processes are disrupted: hematopoietic, metabolic, or endocrine. (a) Cribra orbitalia (hematopoieic) (archaeological Peru). (b) Porotic hyperostosis (hematopoieic) (archaeological Peru). (c) Multiple myeloma (hematopoieic) (Terry Collection). (d) Osteomalacia (metabolic) (Huntington Collection). (e) Rickets (metabolic) (Terry Collection) (f) Scurvy (metabolic) (archaeological Peru). (g) Pituitary disorders (endocrine) (archaeological Peru).

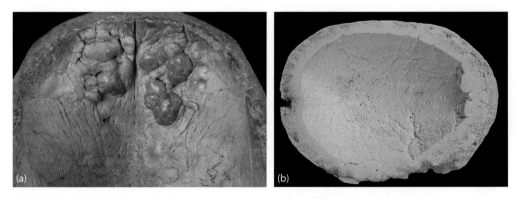

Figure 12.20 Idiopathic disorders. Conditions for which the cause is not yet well understood, but that are intrinsic in nature. (a) Hyperostosis frontalis interna (HFI) (Terry Collection). (b) Paget's disease (anatomical Switzerland).

Infectious diseases

Diseases caused by known pathogenic microorganisms (e.g., bacteria, viruses, and fungi) are considered infectious diseases. These conditions include some of the more widely recognized disease processes, such as trepanematosis, also known as syphilis (see Powell and Cook 2005); tuberculosis, or TB (see Palfi et al. 1999; Roberts and Buikstra 2003); mycosis; and parasitic infection (Table 12.2). Conversely, these better-known diseases are also among the most often misdiagnosed conditions in dry skeletal remains. Developing a differential diagnosis for each of these diseases requires experience and must include a comparison of the gross appearance of lesions, as well as their size, density, distribution, and pathophysiology (Figures 12.21 and 12.22).

Table 12.2 Infectious diseases known to affect the skeleton

Cause	Disease	Illustrative figures
Bacterial Infection		
Mycobacterium tuberculosis	Tuberculosis	Figure 12.1a, 12.6,
Mycobacterium leprae	Common name: TB	12.21c, 12.22b
	Leprosy	
Brucella abortus, Brucella canis, Brucella melitensis, Brucella suis	Brucellosis	Figure 12.22c
Treponema pallidum, Treponema pertenue, Treponema endemicum	Trepanematosis Common names: Syphilis, yaws, and bejel	Figure 12.3, 12.8b, 12.23
Viral Infection		
Poliovirus	Poliomyelitis	
Variola major, Variola osteomyelitis	Common name: Polio & smallpox	
Fungal Infection		
Coccidioides immitis, Coccidioides posadasii	Coccidioidomycosis	
Blastomyces dermatitidis	Common name: Valley fever	
	Blastomycosis	
Parasitic Infection		
Echinococcus granulosus, Echinococcus multilocularis	Echinoccocosis Common name: Tapeworm	

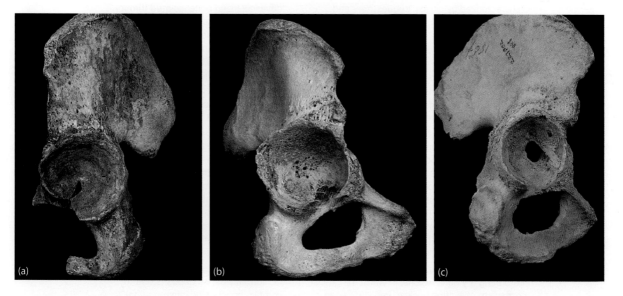

Figure 12.21 Problems with visual diagnosis: Acetabular destruction. It can be difficult to distinguish osseous changes caused by mechanical stress from those caused by congenital anomaly or infection, based on the gross appearance of the lesion alone. (a) Reactive arthritic changes due to loose movement of the hip joint, caused by congenital shallow acetabulum (archaeological Chinese). (b) Destruction and arthritic remodeling of the right acetabulum, resulting from a traumatic event (Terry Collection). (c) Destruction of the acetabulum due to tubercular infection (Huntington Collection).

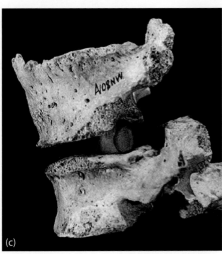

Figure 12.22 Problems with visual diagnosis: Vertebral collapse. It is important to recognize that vastly different pathological processes can produce very similar skeletal effects. (a) Vertebral body collapse due to severe osteoporosis (Terry Collection). (b) Vertebral body collapse due to destabilization of the bone structure by tubercular infection (Terry Collection). (c) Destruction of the anterior region of the vertebral body from brucellosis (archaeological Jordan). Brucellosis can also progress to vertebral body collapse; however, the effects of this infection can be discerned by destruction focused on the anterosuperior margins of the vertebral bodies.

Bone infections

Infectious diseases are of particular importance in understanding direct infection of skeletal tissues. A few generalized terms related to bone infection are important to know (Table 12.3; Figure 12.23); however, these are often

Table 12.3 Types of inflammation specific to bone

Type	Description	Characteristic traits	Illustrative figures
Osteomyelitis	Endosteal infection. Pyogenic and inflammatory. Most commonly affects the medullary cavity and is characterized by cloaca, sequestrum, and involucrum.	Cloaca (apertures for drainage), sequestrum ("island" of devascularized bone isolated through necrosis), involucrum (bony cuff of new bone that develops around the periosteum inflamed in response to accumulating pus).	Figure 12.23a
Osteitis	Inflammatory condition of the inner structures of the bone. This is a nonspecific term, meaning that several pathological conditions or traumatic injury can cause osteitis, but the suffix "-itis" indicates that the inflammation is infectious.	Primarily affects the internal composition of compact bone, including the microstructure and endosteal cortex (Ortner 2003:51).	Figure 12.23b, c
Periostitis	Inflammation of the periosteum. Indicated by subperiosteal deposition of bone, but etiology is nonspecific, as it causes conditions ranging from injury to pathological conditions. However, it can cause periosteal activity that is inflammatory but not infectious, so the use of "-itis" indicates an infectious etiology (Resnick and Niwayama, 1995).	Generalized bone deposits on outer cortical surfaces. Can be woven (recent), dense (remodeled), or sclerotic (abnormally dense deposits due to chronic inflammation).	Figure 12.23d–f
Periostosis	Inflammation of the periosteum; also, a nonspecific term for subperiosteal bone deposition. This terminology is generally preferable when interpreting dry bone lesions, in which the precise mechanism of is often unclear. Periostosis can result from any causative agent, and therefore, this term makes no assumptions about the etiology of new bone deposition, whereas the term periostitis indicates an infectious etiology (Ortner 2003:51).	Generalized bone deposits on cortical surfaces. Can be woven (recent), dense (remodeled), or sclerotic (abnormally dense deposits due to chronic, nonspecific inflammation).	

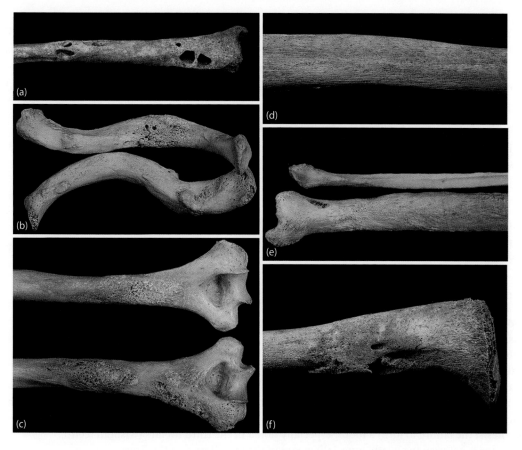

Figure 12.23 Infections specific to bone. Although there are only three major classifications of bone infection, their range of expression is tremendous. (a) Osteomyelitis in a right tibia (historic United States). Note the presence of multiple cloacae, which allow pus to drain from the internal pocket of infection, and the periostitis around the cloacae. (b) Osteitis of the right and left clavicles produced by treponemal infection (Terry Collection). (c) Osteitis of the right and left dorsal humeri as a result of smallpox (Huntington Collection). (d) Periostitis/periostosis. Thin layer of periosteal reaction on tibial diaphysis (Terry Collection). Note the striated appearance of the external cortex. This can be subtle, especially in the early stages of new bone deposition, and is easily obliterated by taphonomic processes on the external bone surface. (e) Periostitis/periostosis on the inferior half of a tibia and fibula (Terry Collection). (f) Periostitis/periostosis. Thick deposition of new bone, resulting from a blood-borne infection on the proximal shaft of an immature tibia (fetal collection). Note that the aperture in the new bone layer is not a cloaca but rather the nutrient foramen.

secondary to other pathological disorders. The challenges of discerning between pathological conditions are compounded by the nonspecific nature of many infectious disorders, so detailed descriptions of the observed lesions and their distribution is imperative.

SUMMARY

This chapter provides a cursory introduction to the enormous range of pathogenic, hereditary, environmental, and traumatic conditions that can affect the human skeleton. It is important to recognize that this encompasses only a faint scratch across the surface of the vast quantity of information comprising the study of pathological skeletal conditions. This subject has been the topic of multiple published volumes, graduate theses, and even more journal articles and has still not been exhausted. The current body of knowledge regarding abnormalities and afflictions of the skeleton could easily fill libraries, and new pathogens, vectors, disease pathways, and treatments are continually being recognized. It is our hope that the reader walks away with an appreciation of the complexity involved in the accurate diagnosis of pathological conditions and with a sense of the best approach to evaluating lesions of the skeleton. Diagnosis should not be the ultimate goal of pathological analysis, but rather, the focus should be on developing a careful description of the lesions, their location, density, and distribution, which will allow a more thorough understanding of the pathological processes and skeletal responses involved.

Review questions

1. Describe the specialized cells responsible for bone growth and repair, and explain what each cell does.
2. What are the five broad classifications of skeletal response?
3. Explain why description is a better approach to interpreting pathological conditions of the skeleton than diagnosis.
4. What is "ribbon matching," and why is it discouraged in the interpretation of skeletal lesions?
5. Explain the stages of bone healing, and describe what happens during each stage.
6. Describe the four causes of infectious disease, and for each, provide one example of a pathological condition that it causes.
7. Describe the different types of intrinsic disease, providing an example and description of each type.
8. What is the difference between a congenital condition and an acquired condition?
9. Describe the factors that must be considered in the classification of neoplasms.
10. Describe the challenges of using pathological conditions of the skeleton as individuating traits in a forensic context.

Glossary

Angulation (Dental): deviation of the tooth from normal alignment.
Ankylosis: joint fusion, usually by osseous bridging.
Antemortem: occurring before the death event.
Arthropathy: any disease or condition affecting a joint.
Compression: squeezing, compacting, or crushing force.
Cutback: area of osteoclastic activity on the subadult diaphysis near the metaphysis where the periosteum is being lifted during normal growth.
Dental agenesis: the failure to form a tooth, especially noted in the third molars (wisdom teeth).
Diastema: a space between two teeth, especially noted in the maxillary front incisors.
Differential diagnosis: the process of identifying all possible conditions that could cause the observed skeletal defect or lesion.
Dislocation: displacement of elements of a joint from their normal position. This may be temporary or permanent.
Eburnation: a polished-looking area on a synovial joint surface indicating increased density in response to persistent erosion.
Erosion: effect of OA in which the cartilage is damaged enough to allow attrition of the underlying bone surface
Etiology: the origin or cause of a disease determined by observed features.
Flexion: bending force.
Fracture: partial or complete breaking of a bone from stress or impact.
Gemination (Dental): the initial growth of a tooth bud.
Heterotopic dentition: a tooth that develops and/or erupts in the wrong position or wrong orientation.
Idiopathic: an adjective describing a condition for which the etiology is not understood with certainty.
Infection: the invasion of pathogenic agent.
Luxation: complete dislocation of a joint.
Myositis ossificans: development of bone within the connective tissue, usually of major muscles, at the site of trauma.
Necrosis: cell death due to interrupted blood supply.
Pathognomonic: specific to a certain disease of condition.
Perimortem: occurring at or around the time of death.
Porosity: perforations in the bone surface. When observed in subchondral bone, porosity results from erosion related to OA; when observed in cortical bone, porosity is caused by hypervascularization related to inflammation.
Postmortem: occurring after death.
Pseudoarthrosis: false joint.
Ribbon matching: attributing an observed lesion to a particular pathological condition on the basis of superficial visual comparison (usually to photographs of severe cases depicted in atlases).

Rotation (Dental): dislocation of a tooth in a clockwise or counterclockwise position.
Shearing: two forces working in opposite directions.
Subluxation: partial dislocation of a joint.
Supernumerary dentition: development of more teeth than expected in the normal dental arcade.
Surface osteophytes: development of exostoses on the joint surface indicative of OA.
Tension: a stretching or pulling force.
Teratogenic: interference with normal embryonic development.
Torsion: twisting.
Winging (Dental): rotation of associating teeth in opposite direction, especially noted in the maxillary and mandibular first incisors.
Woven bone: the interlaced osseous structures formed in the fibrous matrix.

References

Barnes E. 1994a. Polydactyly in the Southwest. *Kiva* 59: 419–431.

Barnes E. 1994b. *Developmental Defects of the Axial Skeleton in Paleopathology.* Niwot, CO: University of Colorado Press.

Barnes E. 2008. Congenital Anomalies. In R Pinhasi, and S Mays (Eds.), *Advances in Human Paleopathology.* Chichester, UK: Wiley. pp. 329–362.

Barnes E. 2012. *Atlas of Developmental Field Anomalies of the Human Skeleton. A Paleopathology Perspective.* Hoboken, NJ: Wiley-Blackwell.

Brogdon BG. 1998. The scope of forensic radiology. *Clin Lab Med* 18(2):203–240.

Buikstra JE, and Ubelaker DH. 1994. *Standards for Data Collection from Human Skeletal Remains.* Research Series, no. 44. Fayetteville: Arkansas Archaeological Survey.

Cunha E. 2006. Pathology as a factor of personal identity in forensic anthropology. In A Schmitt, E Cunha, and J Pinheiro (Eds.), *Forensic Anthropology and Medicine: Complementary Sciences from Recovery to Cause of Death.* Totwa, NJ: Humana Press. pp. 333–358.

Cunha E, and Pinheiro J. 2009. Antemortem trauma. In S Blau, and DH Ubelaker (Eds.), *Handbook of Forensic Anthropology and Archaeology.* Walnut Creek, CA: Left Coast Press. pp. 246–262.

Donchin YA, Rivkind AI, Bar-Ziv J, Hiss J, Almog J, and Drescher M. 1994. Utility of postmortem computed tomography in trauma victims. *J Trauma and Acute Care Sur* 37(4):552–556.

Dolinak D, and Matshes E. 2005. Child Abuse. In D Dolinak, E Matshes, and E Lew (Eds.), *Forensic Pathology: Principles and Practice.* Amsterdam: Elsevier—Academic Press.

Eriksen EF, Axelrod DW, and Melsen F. 1994. *Bone Histomorphometry.* New York: Raven Press, pp. 1–12.

Finnegan M. 1978. Nonmetric variation in the infracranial skeleton. *J Anat* 125(1):23–37.

Galloway A (Ed.). 1999. *Broken Bones: Anthropological Analysis of Blunt Force Trauma,* Springfield, IL: Charles C. Thomas.

Gill GW. 1998. Craniofacial criteria in the skeletal attribution of race. In K Reichs, (Ed.), *Forensic Osteology.* Springfield, IL: Charles C. Thomas, pp. 293–315.

Haglund WD, and Sorg MH (Eds.). 1997. *Forensic Taphonomy. The Postmortem Fate of Human Remains.* Boca Raton, FL: CRC Press.

Haglund WD, and Sorg MH (Eds.). 2002. *Advances in Forensic Taphonomy. Method, Theory, and Archaeological Perspectives.* Boca Raton, FL: CRC Press.

Hauser G, and DiStefano GF. 1989. *Epigenetic Variants of the Human Skull.* Stuttgart: Schweizerbart.

Hillson S. 1996. *Dental Anthropology.* Cambridge: Cambridge University Press.

Hough AJ. 2007. Pathology of osteoarthritis. In RW Moskowitz, RD Altman, MC Hochberg, JA Buckwalter, and VM Goldberg (Eds.), *Osteoarthritis: Diagnosis and Medical/Surgical Management.* Philadelphia: Lippincott Williams & Wilkins.

Irish JD, and Scott GR (Eds.). 2016. *A Companion to Dental Anthropology*. Hoboken: Wiley-Blackwell.

Johnson LC. 1966. *The Kinetics of Skeletal Remodeling*. National Foundation—New York: March of Dimes.

Johnstone EW, and Foster BK. 2001. The biological aspects of children's fractures. In JH Beaty, and JR Kasser (Eds.), *Rockwood and Wilkins' Fractures in Children* (5th edn). Philadelphia: Lippincott, Williams and Wilkins, pp. 1–15.

Keats TE. 1984. *An Atlas of Normal Roentgen Variants That May Simulate Disease*. Chicago: Yearbook Medical.

Keats TE, and Anderson MW. 2013. *Atlas of Normal Roentgen Variants That May Simulate Disease* (9th edn.). Philadelphia: Elsevier Saunders, pp. 485–759.

Loe L. 2009. Perimortem trauma. In S Blau, and DH Ubelaker (Eds.), *Handbook of Forensic Anthropology and Archaeology*. Walnut Creek, CA: Left Coast Press. pp. 263–283.

Lovell NC. 2008. Analysis and interpretation of skeletal trauma. In MA Katzenberg, and SR Saunders (Eds.), *Biological Anthropology of the Human Skeleton* (2nd edn). New Jersey: Wiley, pp. 341–386.

Mankin HJ. 1982. The response of articular cartilage to mechanical injury. *J Bone Joint Surg Am* 64(3):460–466.

Mann RW, and Hunt DR. 2005. *Photographic Region Atlas of Bone Disease* (2nd edn.). Springfield, IL: Charles C. Thomas.

Mann RW, and Hunt DR. 2013. *Photographic Region Atlas of Bone Disease* (3rd edn.). Springfield, IL: Charles C. Thomas.

Marks MK, Marden K, and Mileusnic D. 2009. Forensic Osteology of Child Abuse. In DW Steadman (Ed.), *Hard Evidence: Case Studies in Forensic Anthropology* (2nd edn). Upper Saddle River, NJ: Prentice-Hall. pp. 205–220.

Narotsky MG, Wery N, Hamby BT, Best DS, Pacico N, Picard JJ, Gofflot F, and Kavlock RJ. 2004. Effects of boric acid on hox gene expression and the axial skeleton in the developing rat. In EJ Massaro, and JM Rogers (Eds.), *The Skeleton: Biochemical Genetic and Molecular Interactions in Development and Homeostasis*. Totowa, NJ: Humana Press. pp. 361–372.

Nawrocki SP. 2009. Forensic Taphonomy. In S Blau, and DH Ubelaker (Eds.), *Handbook of Forensic Anthropology and Archaeology*. Walnut Creek, CA: Left Coast Press. pp. 284–294.

O'Connor JF, and Cohen J. 1998. Dating fractures. In PK Kleinman (Ed.), *Diagnostic Imaging of Child Abuse* (2nd edn.). Baltimore: Williams and Wilkins. pp. 168–177.

Ortner DJ. 2003. *Identification of Pathological Conditions in Human Skeletal Remains*. San Diego, CA: Academic Press.

Ortner DJ. 2008. Differential Diagnosis of Skeletal Lesions in Infectious Disease. In R Pinhasi, and S Mays (Eds.), *Advances in Human Paleopathology*. Chichester: Wiley. pp. 191–214.

Palfi G, Dutour O, Dedk J, and Hutas I (Eds.). 1999. *Tuberculosis: Past and Present*. Tuberculosis Foundation, Budapest: Szeged and Golden Books.

Passalacqua NV, and Rainwater CV (Eds.). 2015. *Skeletal Trauma Analysis: Case Studies in Context*. Hoboken, NJ: Wiley Blackwell.

Pearlstein K, Adia K and Hunt D. 2011. Bony ankylosis of the wrist: Four cases from the Terry and Huntington skeletal collections. Paleopathology Association annual meeting, Minneapolis, MN.

Powell ML, and Cook D. 2005. *The Myth of Syphilis: The Natural History of Treponematosis in North America*. Gainsville: University Press of Florida.

Resnick D, and Niwayama G. 1995. *Diagnosis of Bone and Joint Disorders*. Philadephia: Saunders.

Rhine S. 1990. Nonmetric skull racing. In G Gill, and S Rhine (Eds.), *Skeletal Attribution of Race*. Maxwell Museum of Anthropological Papers No. 4. Albuquerque, NM: University of New Mexico. pp. 9–20.

Roberts CA, and Buikstra JE. 2003. *Bioarcheology of Tuberculosis*. Gainsville: University Press of Florida.

Rodriguez-Martin C. 2006. Identification and differential diagnosis of traumatic lesions of the skeleton. In A Schmitt, E Cunha, and J Pinheiro (Eds.), *Forensic Anthropology and Medicine: Complementary Sciences from Recovery to Cause of Death*. Totwa, NJ, Humana Press. pp. 197–221.

Rogers JM, Setzger RW, and Chernoff N. 2004. Toxicant-induced lumbar and cervical ribs in rodents. In EJ Massaro, and JM Rogers (Eds.). *The Skeleton: Biochemical Genetic and Molecular Interactions in Development and Homeostasis*. Totowa, NJ: Humana Press. pp. 373–383.

Saini T, Edwards PC, Kimmes, NS, Shaner JW, and Dowd FJ. 2005. Etiology of xerostomia and dental caries among methamphetamine abusers. *Oral Health and Prev Dent* 3(3):189–195.

Sauer NJ. 1998. The timing of injuries and manner of death: Distinguishing among antemortem, perimortem, and postmortem trauma. In KJ Reichs (Ed.), *Forensic Osteology: Advances in the Identification of Human Remains*. Springfield, IL: Charles C. Thomas. pp. 321–332.

Schacter DL, Guerin SA, and St. Jacques PL. 2011. Memory Distortion: An adaptive perspective. *Trends Cogn Sci* 15(10):467–474.

Shetty V, Mooney LJ, Zigler CM, Belin TR, Murphy D, and Rawson R. 2010. The relationship between methamphetamine use and increased dental disease. *J Am Dent Assoc* 141(3):307–318.

Snapper I. 1957. *Bone Diseases in Medical Practice*. New York: Grune and Stratton.

Sorg MH, Haglund WM, and Wren JA. 2012. Current research in forensic taphonomy. In D Dirkmaat (Ed.), *A Companion to Forensic Anthropology*. Chichester: Wiley-Blackwell Publishing. pp. 477–498.

Symes SA, L'Abbe EN, Chapman EN, Wolff I, and Dirkmaat DC. 2012. Interpreting traumatic injury to bone in medicolegal investigations. In D Dirkmaat (Ed.), *A Companion to Forensic Anthropology*. Chichester: Wiley-Blackwell Publishing. pp. 340–389.

Utsinger PD. 1984. Diffuse idiopathic skeletal hyperostosis (DISH, ankylosing hyperostosis). In RW Moskowitz, DS Howell, VM Goldberg, and JJ Mankin (Eds.), *Osteoarthritis. Diagnosis and Management*. Philadephia: WB Saunders. pp. 225–233.

Waldron T. 2009. *Palaeopathology*. New York: Cambridge University Press.

Wedel V, and Galloway A (Eds). 2014. *Broken Bones: Anthropological Analysis of Blunt Force Trauma* (2nd edn). Springfield, IL: Charles C. Thomas.

For more information regarding pathological conditions of the skeleton

Schmitt A, Cunha E, and Pinheiro J (Eds). 2006. *Forensic Anthropology and Medicine: Complementary Sciences from Recovery to Cause of Death*. Totwa, NJ: Humana Press.

Steinbock RT. 1976. *Paleopathological Diagnosis and Interpretation*. Springfield, IL: Charles C. Thomas.

Firestein GS, Kelley WN, Budd RC, Gabriel SE, McInnes IB, and Odell JR. 2012. *Kelley's Textbook of Rheumatology*. Philadelphia, PA: Elsevier Health Sciences.

Isenberg DA and Renton P. 2003. *Imaging in Rheumatology*. Oxford, UK: Oxford University Press.

McLean FC and Urist MR. 1968. *Bone: Fundamentals of the Physiology of Skeletal Tissue*. Chicago: Chicago University Press.

Roberts CA, and Manchester K. 2005. *The Archaeology of Disease*. Gloucestershire: Sutton Publishing.

Zimmerman MR, and Kelly MA. 1982. *Atlas of Human Paleopathology*. New York: Praeger Publishing.

For more information on the osteological paradox

Cohen MN, Wood JW, and Milner GR. 1994. The osteological paradox reconsidered. *Curr Anthropol* 35(5):629–631.

Wood JW, Milner GR, Harpending HC, Weiss KM, Cohen MN, Eisenberg LE, Hutchinson DL, Jankauskas R, Cesnys G, and Katzenberg MA. 1992. The osteological paradox: Problems of inferring prehistoric health from skeletal samples. *Curr Anthropol* 33(4):343–358.

Wright LE and CJ Yoder. 2003. Recent progress in bioarchaeology: Approaches to the osteological paradox. *J Archaeol Res* 11(1):43–70.

For more information on habitual stress and occupational markers on the skeleton

Kennedy KA. 1998. Markers of occupational stress: conspectus and prognosis of research. *Int J Osteoarchaeology* 8(5):305–310.

Merbs CF. 1983. Patterns of activity-induced pathology in a Canadian Inuit population (*Musée National de l'Homme. Collection Mercure. Commission Archéologique du Canada). Publications d'Archéologie. Dossier Ottawa* 119:1–199.

Rose JC, Condon KW, and Goodman AH. 1985. Diet and dentition: Developmental disturbances. In RI Gilbert Jr. and JH Mielke (Eds.), *The Analysis of Prehistoric Diets*. Orlando: Academic Press, pp. 281–305.

Wilczak CA, Kennedy KA, and Mostly MOS. 1998. Technical aspects of identification of skeletal markers of occupational stress. In KJ Reichs (Ed.), *Forensic Osteology Advances in the Identification of Human Remains*, Springfield, IL: Charles C Thomas. pp. 461–490.

CHAPTER 13

Analysis of Skeletal Trauma

Natalie R. Langley

CONTENTS

Introduction	232
The timing of traumatic events	232
Bone biomechanics	234
High-velocity trauma	236
Low-velocity trauma	241
Blunt-force trauma: Cranium and facial skeleton	242
Blunt-force trauma: Axial and appendicular skeleton	244
Sharp-force trauma	246
Pioneering virtual research in skeletal trauma	249
Summary	249
Review questions	250
Glossary	251
References	251

LEARNING OBJECTIVES

1. Explain how forensic anthropologists use bone biomechanics to interpret skeletal trauma.
2. Distinguish between antemortem, perimortem, and postmortem trauma, and explain why the perimortem interval in bone differs from that in the soft tissues.
3. Describe the biomechanics of gunshot wound production, and recognize the signature features of gunshot entrance and exit wounds.
4. Explain the sequence of fracture production in traumatic injuries to the cranial vault, and contrast the biomechanical causes for these fractures in high- and low-velocity injuries.
5. Describe the difference between blunt- and sharp-force trauma in terms of biomechanics and wound appearance.
6. Compare and contrast bone injuries caused by knives, saws, and axes.
7. Discuss the ways in which trauma research can be conducted, and list the pros and cons of each method.

INTRODUCTION

Over the years, the primary role of forensic anthropologists has expanded from providing a biological profile to assisting with complicated scene recoveries (e.g., mass graves, mass disasters, and fire scenes) and interpreting skeletal trauma. Today, trauma consults are common requests for anthropologists. A working knowledge of **bone biomechanics** is imperative for understanding and interpreting skeletal trauma. Trained practitioners use these principles to distinguish low-velocity from high-velocity traumas. Depending on the nature and location of the trauma, the anthropologist may also be able to determine direction, number, and sequence of impacts, as well as weapon class. Knowledge of bone biology, pathology, and taphonomy facilitates an interpretation of the timing of the trauma (antemortem, perimortem, or postmortem) and the differentiation between traumatic lesions of forensic importance from those associated with disease processes, taphonomic influences, or human variation.

The anthropologist's skeletal trauma assessment is critical in skeletal, fragmentary, badly decomposed, or severely burned cases in which soft tissue examination is not possible. Even in fully fleshed cases, a collaborative effort between forensic pathologists and anthropologists yields a more thorough documentation of the events surrounding death (Christensen et al. 2015). In some cases, the anthropologist's opinion contributes essential information about the cause and manner of death; in other cases, the anthropologist's report corroborates the pathologist's interpretation of soft tissue trauma.

This chapter explains how forensic anthropologists use the basic tenets of bone biomechanics to distinguish the timing, nature, and circumstances of skeletal trauma. High-velocity traumas include injuries caused by firearms (handguns, rifles, and shotguns) and shrapnel from blast injuries. Low-velocity traumas refer to injuries caused by anything other than a fired bullet or projectile. Hammers, baseball bats, knives, and axes are examples of weapons commonly encountered in low-velocity trauma cases. Falls from heights, vehicle crashes, and pedestrian hits, as well as fractures resulting from strangulation and fistfights, are also included in this category (Berryman et al. 2012b). Low-velocity traumas are further divided into two broad categories (blunt and sharp traumas), as each leave distinctive signatures on bone.

THE TIMING OF TRAUMATIC EVENTS

Determining the timing of skeletal trauma is essential to the anthropologist's interpretation of the events surrounding a death. "Antemortem trauma" refers to trauma that occurred before death (e.g., weeks, months, and years before death). Antemortem trauma shows signs of healing and remodeling, ranging from **woven bone** to a well-developed **bony callus** (Figure 13.1). The edges of an antemortem defect are smooth, compared with sharp or jagged edges associated with perimortem trauma and postmortem trauma, respectively (Figure 13.2). Antemortem trauma does not inform the cause or manner of death, but it provides information about previous injuries for which medical records may exist, which may assist in identifying unknown remains.

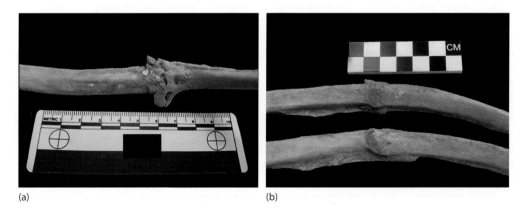

Figure 13.1 Antemortem trauma: (a) Newly developing callus on a rib fracture that likely occurred several days or a couple of weeks before the fracture and (b) well-developed callus on rib fractures that occurred months before death. (Courtesy of Natalie Langley.)

Analysis of Skeletal Trauma

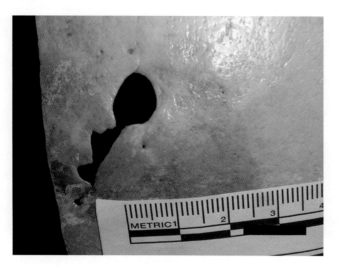

Figure 13.2 Antemortem defect. Cranial defect created by a shunt that was placed in the skull of a hydrocephalic individual. Note the smooth margins of the defect. (Courtesy of Natalie Langley, Forensic Anthropology Center, University of Tennessee, Knoxville.)

"Postmortem trauma" refers to bony injuries that occurred after death. Postmortem defects are typically caused by scavenging animals, but they may also be the result of intentional postmortem alteration of remains (e.g., dismemberment). If the postmortem damage occurs after the bone has lost most of its collagen content and therefore its elasticity, the fracture edges appear jagged and irregular and may be discolored relative to the rest of the bone. Postmortem fractures are more likely to be complete fractures, as opposed to incomplete or **greenstick fractures**. Damage caused by scavengers has diagnostic features, ranging from tooth punctures caused by carnivores to parallel striae caused by rodents (Figure 13.3). Please refer to Chapter 4 of this volume for a detailed discussion of postmortem alterations caused by carnivores and rodents.

"Perimortem trauma" refers to trauma that occurs at or around the time of death. Perimortem defects reflect the elastic nature of fresh bone, which has more collagen and a higher moisture content than dry bone (Figure 13.4). Perimortem fractures do not show signs of healing, but the edges of the fracture are not as jagged as postmortem fractures of dry bone. Occasionally, signs of hemorrhage may be detectable in the bone. The complicating factor in interpreting perimortem trauma is that bone retains its elastic properties long after death (Galloway et al. 1999; Weiberg and Wescott 2008). Depending on the environment, bone may display fracture characteristics of fresh bone for months. In other words, the perimortem interval of bone is longer than that of soft tissues, rendering anthropological analyses of perimortem trauma sometimes tenuous. A number of researchers have attempted to refine

Figure 13.3 Postmortem damage. Sternal ends of left ribs, with postmortem animal activity indicated by jagged, splintered, and crushed margins. (Courtesy of Natalie Langley.)

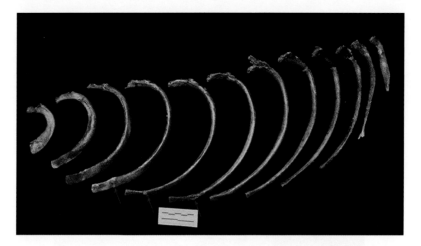

Figure 13.4 Perimortem fractures. Multiple incomplete, or greenstick, fractures of the sternal and vertebral ends of the ribs in a pedestrian who was hit by a car. The red arrows indicate buckle fractures. (Courtesy of Natalie Langley, Forensic Anthropology Center, University of Tennessee, Knoxville.)

the classification of perimortem fractures, but the unanimous conclusion is that the elastic perimortem interval in bone complicates fracture interpretation in some cases (Cappella et al. 2014; Galloway et al. 1999; Jordana et al. 2013; Ubelaker and Adams 1995; Weiberg and Wescott 2008; Wheatley 2008). If the timing of fractures that appear perimortem is questionable, the cautious approach is to report that the fractures occurred while the bone was still elastic or that they have the appearance of fresh bone fractures.

BONE BIOMECHANICS

Bone morphology, structure, and composition determine its mechanical properties, which, in turn, affect how bone responds to loading (Currey, 2002; Cowin 1989; Nordin and Frankel 1989). The old adage "form follows function" is useful when thinking about how bone responds to mechanical loads. Muscles and bones are designed to carry out specific functions (e.g., movement and support). The material properties of a bone or portion of a bone are a reflection of its functional role. For example, the femur is optimally designed to endure the daily stresses imposed by walking and standing. Owing to this functional demand, the cortical bone of the femoral shaft is distributed, so that the femur is stronger under compressive **axial loading** than under **tensile loading**. In other words, the shape, orientation, and amount of cortical and trabecular bones are designed optimally to support and carry out daily functional demands. Figure 13.5 illustrates the direction of force relative to the loading axis in several simple loading scenarios.

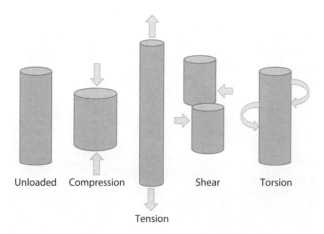

Figure 13.5 Simple loading scenarios. The cylinder represents a bone shaft, and the arrows indicate the direction of force in several simple loading scenarios. Changes in the cylinder illustrate bone's response to each type of loading. (Courtesy of Natalie Langley.)

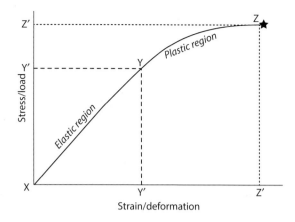

Figure 13.6 Stress-strain curve. A stress-strain (load-deformation) curve illustrating the elastic region (between x and y) and plastic region (between y and z) of a hypothetical material before failure (indicated by the star). (Courtesy of Natalie Langley.)

Different muscle groups exert opposing forces on a single bone, such that most bones undergo combined loading scenarios, even during relatively uncomplicated activities. Training in gross anatomy, especially musculoskeletal anatomy, aids greatly in understanding bone biomechanics and interpreting skeletal trauma.

Bone is a dynamic material. Bone is **anisotropic**, meaning that the direction of loading influences its strength properties. In our example of the femur in the previous paragraph, this bone is subjected to axial loading most of the time, so it is strongest in compression and weakest in tension. Tensile forces caused by excessive bending or torsion may fracture the femur, especially if these forces are applied at a high rate of speed. Depending on the velocity, rate, duration, and direction of a given load, bone may shatter like a brittle material or deform like a ductile material. Glass is an example of a brittle material, and aluminum is an example of a ductile material. Ductile materials are capable of absorbing more energy before failure than brittle materials. Because of the capacity of bone to react differently to various loading scenarios, we say that it is a **viscoelastic** material.

Figure 13.6 is a stress-strain (or load-deformation) curve that compares the behavior of a ductile material with a brittle material under similar loading conditions. A ductile material undergoes a period of elastic deformation before failure (the portion of the curve from x to y). If the load is removed during elastic deformation, the material regains its original structure and/or shape. If loading persists, the material sustains plastic deformation and will not return to its original shape, but instead, it will remain permanently deformed (Berryman et al. 2012a). The portion of the curve between y and z represents the plastic region. Continued loading causes the material to fail (the point of failure, or fracture, is denoted with a star). In comparison with ductile materials, brittle materials undergo little, if any, plastic deformation before failure. Currey (2002), McGowen (1999), and Vogel (2003) provide thorough discussions of bone biomechanics; the advanced reader may reference these texts for a more in-depth understanding of the subject.

Factors that affect the mechanical behavior of bone can be categorized as intrinsic or extrinsic variables (Berryman and Symes 2002). Intrinsic variables are related to the material and structural properties of bone, such as bone microstructure, cortical thickness, amount of trabecular bone, and bone size and shape (Berryman et al. 2012a). Extrinsic variables are associated with load properties (e.g., the magnitude, direction, rate, and duration of the load). In other words, intrinsic variables correspond to bone properties, and extrinsic variables are attributed to the object causing the injury. Understanding the relationship between intrinsic and extrinsic variables is key to interpreting skeletal trauma.

The remainder of this chapter discusses several common categories of skeletal trauma. These categories are divided by velocity (high-velocity vs. low-velocity injuries). Each category is subdivided further by weapon category (handguns and rifles, shotguns, blunt trauma, and sharp trauma) and is discussed by body region (cranium, thoracic cage, and long bones). We categorize injuries by velocity and weapon class or mechanism of injury for ease of description, but in reality, skeletal trauma is best interpreted with respect to a biomechanical continuum affected primarily by several key variables, including force, velocity, and surface area (Kroman 2010). Thinking of bone trauma in this manner enables practitioners to recognize expected patterns readily and to interpret unexpected patterns that manifest when one or more variables do not present in the usual manner.

HIGH-VELOCITY TRAUMA

Bone is a viscoelastic material and behaves like a brittle or ductile material, depending on the rate, duration, and velocity of a given load or force. High-velocity loading causes bone to react as a brittle material: the bone does not undergo plastic deformation but instead fails almost instantaneously and shatters like glass. The undeformed bone fragments produced by the fractures can be reassembled easily, much like fitting the pieces of a puzzle together (Figure 13.7) (Berryman et al. 2012a).

In terms of extrinsic variables, the wounding capacity of a bullet is derived from the kinetic energy transferred to the body tissues (Berryman et al. 2012a). Kinetic energy is equal to $(m * v^2)/2$, where "m" is the mass of the moving object and "v" is its velocity. Since velocity is squared, higher-velocity projectiles have the capacity to be exponentially more destructive than lower-velocity projectiles. If a bullet exits the body, some of the wounding power exits as well, and only a portion of its kinetic energy is transferred to the body tissues. Given this, bullet design plays a significant role in the "stopping power" and lethalness of projectile injuries (e.g., hollow point bullets designed to deform and remain in the body are more lethal than jacketed bullets that may exit).

Gunshot wounds in the cranium typically are easiest to interpret. The cranial vault is a relatively rigid and closed structure, owing to its primary role to protect the brain. Many handgun injuries have entrance and exit wounds with signature features that make them easily distinguishable from one another and permit the anthropologist to ascertain the bullet's trajectory. In addition, increased intracranial pressure caused by the entrance wound creates more extensive fracturing around this injury. These fractures relieve a significant portion of the intracranial pressure, and the fracturing associated with the corresponding exit wound is usually less severe. A "textbook" cranial gunshot wound has internal beveling associated with the **endocranial** surface of the entrance defect and external beveling on the **ectocranial** surface of the exit defect (Figure 13.8). These trademarks are most obvious on the flat bones of the cranial vault because of the material properties of these bones: two layers of cortical bone (the endo- and ectocranial surfaces), with a layer of dense spongy bone (referred to as diploë) between them.

Of course, there are exceptions to all rules, which is why it is imperative to interpret skeletal trauma in light of the interaction of intrinsic and extrinsic variables and with an understanding of what might happen if one variable presents in an atypical manner. For example, a bullet that strikes the cranial vault with a tangential trajectory creates a "keyhole" defect. Keyhole entrance wounds have external beveling on one edge of the defect and a well-defined circular outline on the other side (Figure 13.9) (Dixon 1982; Berryman and Symes 2002). However, the bullet trajectory is evident, even in these atypical wounds: the "entrance" is elucidated by the unbeveled edge, and the "exit" is marked by external beveling. Radiating fractures may also initiate from the initial bullet impact site and advance in

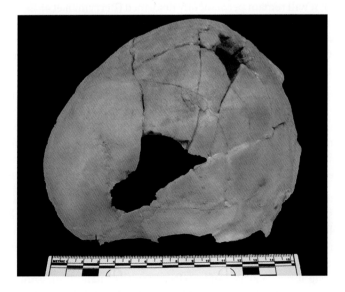

Figure 13.7 High-velocity trauma. Cranial vault reconstructed with adhesive after a gunshot wound shattered the vault bones into many fragments. (Courtesy of Natalie Langley, Forensic Anthropology Center, University of Tennessee, Knoxville.)

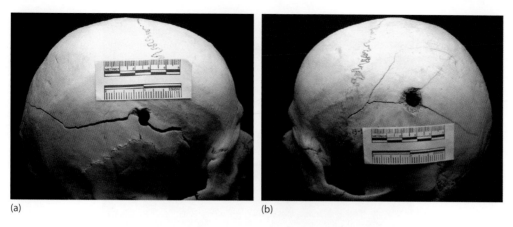

Figure 13.8 (a) Entrance wound with radiating fractures and (b) Associated exit wound with external beveling and radiating and concentric fractures. (Courtesy of Natalie Langley, Forensic Anthropology Center, University of Tennessee, Knoxville.)

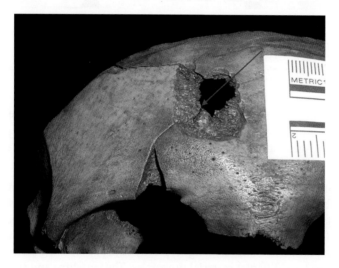

Figure 13.9 Keyhole defect. Keyhole defect on the frontal bone. The arrow indicates the approximate direction of the bullet's path: posterior to anterior, as indicated by the external beveling on the anterior margin. (Courtesy of Natalie Langley, Forensic Anthropology Center, University of Tennessee, Knoxville.)

the direction of the bullet's path (Berryman et al. 2012). Keyhole defects also have been reported in association with exiting bullets (Dixon 1984). Another exception to the typical beveling pattern observed in cranial gunshot wounds is observable in the thin bones of the facial skeleton, with their numerous sinuses and lack of diploë. These bones do not exhibit beveling and may fracture extensively and be effectively shattering (Berryman et al. 2012a).

Fracture lines may also assist in distinguishing an entrance wound from an exit wound or elucidate the sequence of multiple injuries. As mentioned earlier, the extent of fracturing is more severe for entrance wounds. In addition, fracture lines do not cross existing fractures, so fractures from an exit wound terminate into fractures from the corresponding entrance wound, because the entrance fractures traverse the vault faster than the bullet travels through the skull. Likewise, fractures from subsequent injuries terminate into previous fractures. This concept was introduced in 1903 by a forensic pathologist named Puppe, who recognized that intersecting fractures can be used to determine the sequence of blunt trauma impacts in the skull. Madea and Staak (1988) generalized Puppe's theory to gunshot trauma (Berryman et al. 2012a).

When examining skull fractures, it is also important to understand the sequence in which fractures are generated. Radiating fractures are the first fractures that form when a bullet enters the cranium. Multiple radiating fractures may extend from the gunshot wound in a stellate pattern. As the intracranial pressure increases, the cranial

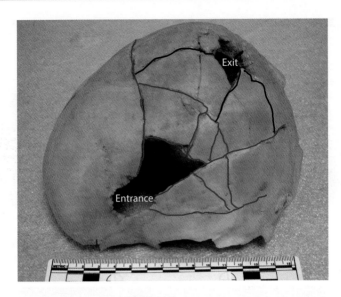

Figure 13.10 Gunshot wound fractures. The cranial vault from Figure 13.7 is used to illustrate fracture generation from gunshot wounds. The red lines depict radiating fractures associated with the entrance wound. The blue lines show second- and third-generation concentric heaving fractures caused by the increased intracranial pressure from the entering projectile. The black lines indicate radiating fractures from the exit defect that terminate into fractures caused by the entrance. (Courtesy of Natalie Langley, Forensic Anthropology Center, University of Tennessee, Knoxville.)

fragments between the radiating fractures are heaved outward, placing tensile strain on the inner table of the skull at a distance from the entrance site, which causes first-generation concentric fractures to propagate perpendicular to the radiating fractures. Second-generation concentric fractures may form, given sufficient energy and intracranial pressure (Figure 13.10).

Gunshot wounds in other areas of the body do not exhibit the same features as the cranial vault. The thoracic cage is composed of a more mobile set of skeletal elements connected by a variety of joints (intervertebral, sternocostal, and costovertebral joints), muscles (intercostal muscles and deep and superficial back muscles), and ligaments. Bullets may travel through intercostal spaces without wounding bones. Furthermore, intrathoracic pressure is not a significant factor in fracture production, and the lungs typically collapse once the pleural cavity is penetrated. The bones of the thoracic cage also have different material properties. The vertebrae and sternum are largely trabecular bones. Like the skull, the ribs are flat bones, but they are more elongated and tubular in shape. They also lack a layer of diploë between the dorsal and ventral layers of cortical bone and instead have a less dense layer of trabecular bone (Langley 2007). The rib cage is bordered posteriorly by the thin cortical bone of the scapular bodies and superiorly by the clavicles, and these bones may also be injured in gunshot wounds of the thorax. The unique intrinsic properties of the bony elements of the thoracic cage produce gunshot wounds with distinctive features. See Box 13.1 for a case example of thoracic gunshot wound.

BOX 13.1 THE EVIDENCE PREVAILS

Perpetrators may attempt to cover up their crimes in a number of ways, including burial, disposal in bodies of water, dismemberment, and burning. The belief that fire can consume a body completely and destroy evidence of a crime is optimistic but is rarely the case. Fire leaves distinctive signatures on bones that can be distinguished from other types of trauma. (See Chapter 18 for a discussion of thermal alteration of remains.)

In this case example, two individuals were killed on a boat, and the boat was set on fire. Police extinguished the flaming vessel and recovered the burned and commingled remains of a male and female, believed to be the victims of murder. The remains were sent to an anthropologist for sorting and trauma analysis.

(Continued)

BOX 13.1 (*Continued*) THE EVIDENCE PREVAILS

Approximately 30% of the male's remains and 10% of the females' remains were available for analysis; 5% of the remains could not be sorted. The cranium and extremities were mostly absent, but a number of thoracic and lumbar vertebrae were complete and hardly burned, as these bones are protected by large muscle masses.

Once the vertebrae were cleaned and assembled, a defect was observed between the bodies of the sixth and seventh thoracic vertebrae (Figure 13.11). The defect was consistent with a gunshot wound. External beveling on the right lamina of the sixth thoracic vertebra suggested that the bullet traveled from anterior to posterior. In addition, a defect was found in the head of the right seventh rib, which articulates with the sixth and seventh thoracic vertebral bodies. The defect was also consistent with a gunshot wound traveling from anterior to posterior, probably caused by the same bullet that injured the vertebrae. The bullet caused a gutter fracture of the right side of the vertebral bodies, penetrated the head of the right seventh rib, and exited the lamina of the right sixth vertebra.

The signature characteristics of the gunshot trauma persisted despite the perpetrator's efforts to cover up his crime with fire. This information was provided to the medical examiner and police and allowed them to pursue the suspect and the weapon.

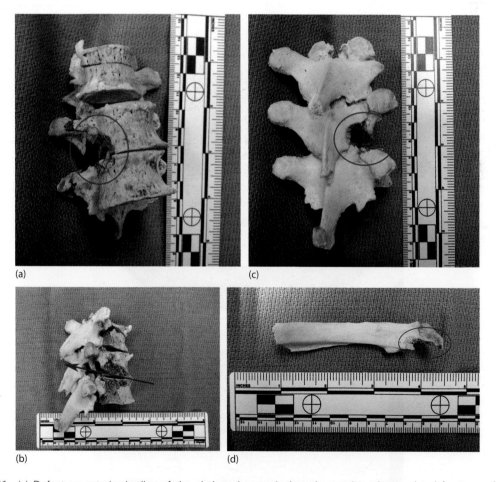

Figure 13.11 (a) Defect on anterior bodies of the sixth and seventh thoracic vertebrae (note related fractures of transverse processes). (b) Gutter fracture, with arrow showing direction of the bullet's path. (c) Exit defect on lamina of sixth thoracic vertebra (note radiating fractures and external beveling on the left margin of the defect). (d) Gunshot wound on the head of right seventh rib (direction of fire indicated by fragments displaced in direction of bullet's path).

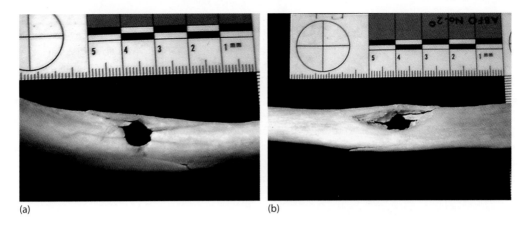

Figure 13.12 Gunshot wound in a rib. (a) Entrance defect on external surface, with radiating fractures. (b) Exit defect, with beveling on pleural surface of the same rib. (Courtesy of Natalie Langley.)

It is not possible to determine the sequence of gunshot wounds in the thoracic cavity by examining the skeletal elements, but experts can deduce direction of fire. A case study on the exhumed remains of Dr. Carl Austin Weiss (accused assassin of Louisiana Governor Huey P. Long, who was shot at least 20 times in the chest) revealed several distinctive features associated with gunshot wounds in ribs that indicate direction of fire: radiating fractures, displaced bone fragments, depressed fractures, beveling, and overall fracture patterns (Ubelaker 1996) (Figure 13.12). Langley (2007) corroborated these observations in an analysis of 87 gunshot wounds to the chest. Depressed bone fragments around the entrance site and displaced bone splinters at the exit site occur frequently in the ribs and indicate direction of the bullet's path through the rib (Langley 2007) (Figure 13.13). Owing to more flexible nature of the chest cavity compared with the cranial cavity, significant intracavity pressure does not build up, and therefore, concentric heaving fractures are absent in gunshot wounds in ribs.

Gunshot wounds in long bones have received little attention in the research literature, but reports indicate that these injuries exhibit beveling, which can be used to interpret directionality. Impacts to bone shafts typically cause massive comminution. Because the fractures associated with the entrance defect travel around the bone shaft faster than the bullet, a distinct exit defect is absent (Langley 2007). Wounds on long bone epiphyses may exhibit more typical beveling, depressed fractures, or massive fragmentation, depending on the area struck and the velocity of the projectile. Angle of impact also affects the appearance of the beveling (Smith and Wheatley 1984), and keyhole fractures have been documented in long bone shafts (Berryman and Gunther 2000).

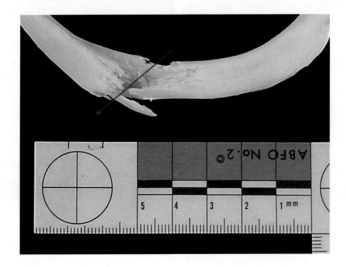

Figure 13.13 Displaced bone fragment. Bone fragment displaced in the direction of the bullet's path through the rib (direction indicated with red arrow). (Courtesy of Natalie Langley.)

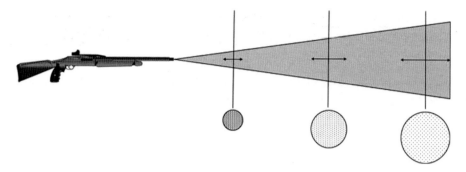

Figure 13.14 Shot dispersal pattern. At close distances, the shots are a conglomerate mass (far left). As distance between muzzle and target increases, the shots disperse (far right). (Courtesy of Natalie Langley.)

Handguns and rifle barrels have grooves that spiral the bullet, increasing the muzzle velocity and aiming precision. In contrast, shotguns are smooth bore weapons, and the barrel is not rifled. The ammunition differs greatly, as well, usually consisting of multiple pellets packed into a single shell. Of course, this classification system is simplistic, as some handguns can fire shotgun shells, and shotguns can fire single slugs. Nonetheless, conventional shotguns produce distinctive wounds, which can be distinguished readily from handgun injuries. The muzzle velocity of shotguns is comparable to handguns, but the pellets begin to disperse and decelerate quickly on exiting the barrel. The primary factor in the extent of injury produced by a shotgun blast is the distance from the muzzle to the target. At close ranges, the pellets enter the body as a conglomerate mass that imparts massive injury to surrounding tissues, causing extensive fracturing and fragmentation of osseous structures (especially the skull). At far distances, the shot disperses and enters the body at a considerably lower velocity, causing multiple separate injuries that are less destructive than handguns or rifles that fire solitary bullets (Berryman et al. 2012a) (Figure 13.14). At sufficient distances, shotgun pellets lack the energy to penetrate the body and may cause only minor injuries.

The **Scientific Working Group for Forensic Anthropology (SWGANTH)** trauma analysis document cautions against estimating caliper or gauge from wound dimensions. While it may be possible to discern large from small caliber cranial gunshot wounds, significant overlap in wound dimensions from various bullet calipers suggests that caliber estimation is not a sound practice (Ross 1996). Numerous confounding extrinsic variables can affect wound dimensions, including firing distance, angle of entry, body position at the time of impact, surface treatment and strength characteristics of the bullet, preexisting fractures, and intermediate targets that may deform the bullet (DiMaio 1993; Berryman et al. 1995). The SWGANTH also maintains that estimating muzzle to target distance is an unacceptable practice. In cases where gunshot injuries are suspected, but fracture patterns or injury features are not diagnostic, it is a good idea to X-ray the remains to check for radiodense particles (i.e., lead and other metals) that may have become embedded in the bone. Gunshot residues and primers may also be detectable on the bone (Berryman et al. 2010). In any event, the case report should avoid conjecture and state only what the bony evidence supports. Scaled photographs and diagrams with labeled fractures and defects are helpful. Descriptions of injury patterns and mechanisms are preferred to overly specific interpretations that may cause more harm than good in the outcome of a case.

LOW-VELOCITY TRAUMA

The second broad category of skeletal trauma involves objects or weapons traveling at a lower rate of speed than high-velocity projectiles. Low-velocity traumas include blunt-force trauma, sharp-force trauma, and deceleration injuries (falls from heights and automobile accidents). The major difference between these subcategories is the surface area of the weapon or injuring object. A trademark of low-velocity trauma is that bone undergoes plastic deformation, because the rate of loading is sufficiently slow for the bone to permanently deform before failure. In contrast to high-velocity trauma, where bone reacts like a brittle material and shatters, bone has the properties of a ductile material when the velocity of the load is sufficiently reduced. Plastically deformed bone fragments are difficult to reconstruct, because they do not fit together perfectly. Reconstructive efforts for the sake of analysis may require toothpicks or modeling clay to fill in the gaps between warped fragments.

Blunt-force trauma: Cranium and facial skeleton

The skull is commonly affected in blunt-force injuries inflicted by hammers, baseball bats, or other blunt objects. Head injuries also occur in motor vehicle collisions and sports injuries. Anthropological analyses of blunt impacts to the skull aim to ascertain the point of impact, number of impacts, sequence of the impacts, the amount of force, and possible weapon class (Berryman et al. 2012b). Blunt impacts to the cranial vault may produce radiating and concentric fractures, but the mechanism of fracture is different from those produced by gunshot wounds. A blunt impact to the outer cortical surface cranial vault creates tension on the inner table. Bone fails first on the inner table, producing radiating fractures that propagate to the external table and create triangular-shaped pieces of bone. As the impacting object advances, the triangular fragments are pushed into the cranial cavity, creating tension on the outer table of the skull. First-generation concentric fractures propagate perpendicular to the radiating fractures at the areas of maximum tension on the outer skull table (Figure 13.15). Second-generation concentric fractures propagate, given sufficient force (Berryman and Symes 2002; Berryman et al. 2012b). If the skull is positioned between an impacting object and another surface (e.g., a tire and the pavement), the "impact" against the surface on which the skull is resting may produce a **contrecoup fracture** (Hein and Schulz 1990). The traumas caused by this type of injury manifest as two impacts, but they are caused by a single traumatic episode (Berryman et al. 2012b) (Figure 13.16). This loading scenario usually produces a **hinge fracture**, traversing the cranial base, as well.

The forensic community has learned much from experimental studies of skull fractures. The earliest work on cranial vault impacts was done by in the mid-twentieth century (Gurdjian et al. 1945, 1947, 1949, 1950a, b). Gurdjian applied a lacquer to skulls and then used a drop tower to produce blunt impacts and visualize the mechanism of fracture. These "stress coat" experiments suggested that blunt impacts produce an area of in-bending at the impact site and areas of out-bending at other locations on the vault. Gurdjian explained that maximum tension occurs at the areas of out-bending; fractures initiate at these locations and travel back toward the impact site. A test of this research using fresh cadaver heads, a drop tower, and high-speed photography showed that the fractures initiate at the impact site and radiate *away* from this location (Kroman 2007; Kroman et al. 2011). The impact creates maximum tensile stress on the inner table of the skull, resulting in radiating fractures that propagate away from the site of impact. Kroman's research contradicted earlier studies and caused researchers to question the mechanism of skull fracture.

However, recent experimental research illustrates that fracture interpretation is not this simple, because many variables influence the production of skull fractures associated with blunt impacts. A team of researchers at Michigan

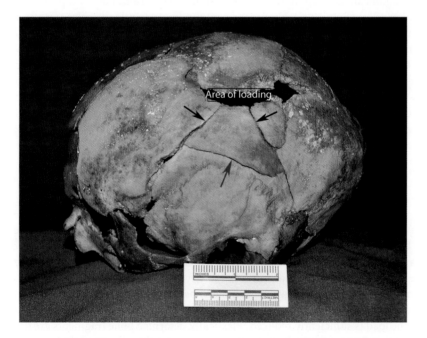

Figure 13.15 Blunt trauma. Left lateral view of skull, showing radiating (black arrows) and concentric fractures (red arrow). (Courtesy of UT Forensic Anthropology Center.)

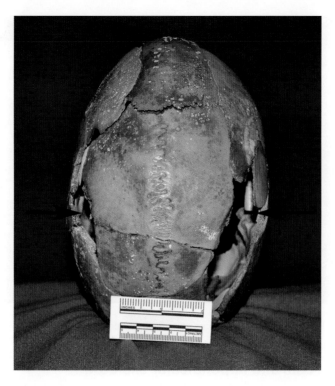

Figure 13.16 Ontrecoup fracture. Two impact areas as a result of loading from two different directions (see arrows) from a single traumatic episode. One impact was from a tire, and the other impact was from the opposing surface—the pavement. (Courtesy of Forensic Anthropology Center.)

State University used infant porcine heads to investigate pediatric fracture biomechanics (Baumer et al. 2009, Baumer et al. 2010, Powell et al. 2012, Powell et al. 2013). Using a drop tower and high-speed photography, they showed that fractures in infant porcine skulls initiate *away* from the point of impact (usually at sutures) and travel back to the impact site. The intrinsic properties of infant skulls are significantly different from adult skulls (e.g., more collagen, no diploë, and gaping sutures and fontanelles; in short, infant skulls are not as rigid as adult skulls). These differences generate a different fracture mechanism in infant skulls. Whether results from the pig skull tests can be generalized to human skulls is yet to be tested, but the experiments suggest that it is possible and remind us that we cannot use a "one size fits all" typology for fracture interpretation.

In 1901, René LeFort used human cadavers to investigate the mechanism of facial fractures. The bones of the facial skeleton are thinner than those of the vault, producing unique and complex fracture patterns. LeFort (1901a–c) devised a classification system for facial fractures that is still used today. Facial fractures exhibit one or a combination of three patterns: LeFort I, II, and III fractures (Figure 13.17). LeFort I fractures are produced by a direct blow to the anterior portion of the alveolar process of the maxilla. This force separates the alveolar process from the body of the maxilla. LeFort II fractures, or pyramidal fractures, propagate across the maxilla, lower orbital floor, and nasal bones. More centrally directed blows cause LeFort II fractures. LeFort III fractures (transverse facial or cranial disjunction fractures) traverse the upper orbits and nasal bones, separating the bones of the face from the vault. All three LeFort fractures terminate after traversing the pterygoid process of the sphenoid bone (Berryman et al. 2012b). LeFort II and III fractures frequently occur in combination (Wedel and Galloway 2014). LeFort fractures are based on areas of buttressing and lines of least resistance in the facial skeleton; consequently, gunshot wounds can produce fractures along these same lines (Berryman et al. 2012b). Other fractures that occur commonly in facial impacts are tripod fractures, or zygomaticomaxillary complex fractures. These fractures result from a direct blow to the malar eminence, or cheek bone, and exhibit three components: a fracture of the zygomatic arch, a fracture of the lateral orbital wall, and fractures of the inferior orbital rim and anterior maxilla near the zygomaticomaxillary suture.

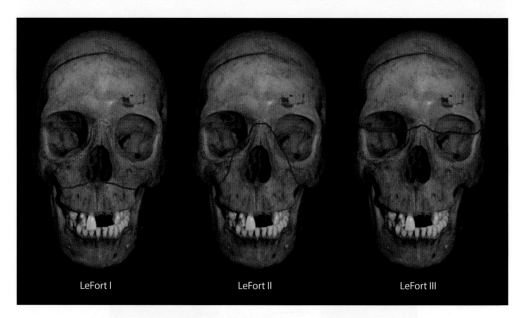

Figure 13.17 LeFort I, II, and III fractures. These fractures often occur in combination.

Blunt-force trauma: Axial and appendicular skeleton

Thorax and limb fractures occur in automobile accidents, pedestrian hits, sports injuries, and falls from heights. As with skull injuries, sites of tensile and compressive loading give clues about the mechanism of fracture. Reconstructing highly fragmented bones with adhesive facilitates fracture interpretation, but remains should be photographed in their fragmentary state before reconstruction. Long bones are easier to reassemble than skull bones, because the plastic deformation is not as severe. Once the bones are reconstructed, forensic anthropologists diagnose the type of fracture (e.g., transverse, oblique, spiral, and butterfly) and then consider loading scenarios that produce that fracture etiology (e.g., bending, torsion, and compression) (Berryman et al. 2012b). The clinical literature, particularly orthopedic literature, has an abundance of information on musculoskeletal injuries with known causes, so practitioners should consult this resource to inform their interpretations. Clinicians favor classification systems, which can be useful in narrowing down a fracture and its possible causes, but it is more important for forensic anthropologists to describe fractures and explain injury biomechanics than to categorize fractures (Wedel and Galloway 2014).

Long bones may fracture on the diaphysis or epiphysis, and fractures may be complete or incomplete. Long bones are stronger in compression than in tension or shear, and the type of loading dictates the fracture appearance. For example, a long bone loaded with simple three-point bending typically produces a butterfly fracture. In three-point bending, the two ends of the bone are fixed and a load is applied between the fixed points, producing maximum tension on the side of the shaft, opposite to the applied load. The bone fails at the site of maximum tension, and fracture lines travel back toward the compressively loaded side, creating a wedge (Berryman et al. 2012b) (Figure 13.18). Long bones subjected to excessive torsional forces exhibit spiral fractures. Avulsion fractures occur when a ligament is forcibly ripped from its attachment point on the bone, taking a piece of the bone with it as it detaches (Figure 13.19). Buckle fractures are incomplete fractures characterized by bulging or buckling of the outer cortex of a long bone or a rib (see Figure 13.4). These fractures are most common in young individuals with highly elastic bones.

Fractures of the limb girdles are affiliated with considerable force, with the exception of the clavicle. The clavicle is one of the most frequently fractured bones in the body, requiring only several pounds of pressure to fracture, if loaded along its weakest axis. Clavicle fractures are common in automobile accidents, in cycling accidents, and in sports and are usually produced by a fall onto an outstretched arm or a lateral impact to the shoulder. Scapular and pelvic fractures can be caused by direct blows or indirectly by forces transmitted from the humerus and femur, respectively (Berryman et al. 2012b). Scapular fractures are rare, as the scapula is relatively stable and protected by a number of muscles. Falls from heights, pedestrian hits, and high-speed automobile or motorcycle crashes may fracture the scapula. The same can be said of pelvic fractures, with the exception of isolated pelvic fractures in elderly individuals. Decreased bone density,

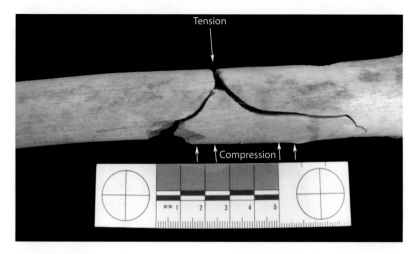

Figure 13.18 Butterfly fracture. The bone fails first in tension, and the fractures travel back toward the compression side to create a wedge. (Courtesy of Forensic Anthropology Center.)

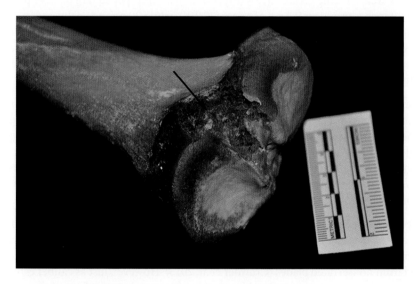

Figure 13.19 Avulsion fracture. Avulsion fracture of posterior tibial plateau at the attachment site of the posterior cruciate ligament (PCL). (Courtesy of Forensic Anthropology Center.)

particularly in postmenopausal females, makes the pelvis susceptible to fractures, even in simple falls. As with limb bone fractures, the location and nature of fractures of the limb girdles indicate the likely mechanism of trauma.

Thorax fractures are common in individuals of all ages. Owing to their propensity to puncture internal organs (e.g., lungs, liver, and spleen), rib fractures are frequently fatal and are often encountered in forensic contexts. Rib fractures in various stages of healing in infants are indicative of multiple episodes of abuse and should be investigated carefully to rule out other causes (Love et al. 2011). Resuscitation efforts may also fracture ribs, especially in the elderly. Automobile accidents, pedestrian hits, and falls from heights frequently produce thoracic trauma. While it is possible for rib fractures to occur in isolation, other elements may also be injured, including the sternum and the vertebrae.

The thorax is considered a single structural and biomechanical entity when interpreting fractures of the ribs, vertebrae, and sternum. For example, anterio-posterior compression of the thorax produces fractures on the lateral aspects of multiple ribs, whereas lateral loading produces fractures at the sternal and vertebral ends of the ribs (Berryman et al. 2012b). The location of rib fractures provides clues about the direction of impact to the thorax. Depending on the severity and direction of the force, vertebral fractures may occur in conjunction with rib fractures, especially on the spinous and transverse processes. Many muscles connect the various parts of the vertebrae to one another and to the ribs, so forcible torsion acting on these muscles can cause them to fracture the ribs and the vertebrae to which they attach. As with long bones, buckle fractures on ribs indicate areas of high elasticity; these fractures are commonly observed on the sternal ends of ribs but may also occur on the rib shaft (see Figure 13.4). Rib fractures are reported by location (e.g., rib head, angle, and shaft) and type (e.g., transverse, oblique, and buckle).

Like high-velocity trauma interpretations, analyses of blunt trauma should not exceed what the osseous evidence supports. Reports should describe fractures and propose a mechanism but steer clear of suggesting specific scenarios (e.g., "fractures on the lateral aspects of the ribs are consistent with anterior-posterior compression of the thorax" instead of "the rib fractures were likely caused by stomping on the chest"). Experts should also keep in mind that infant and juvenile bones are more elastic and have epiphyses, which may alter the appearance and mechanism of fractures relative to what we expect to see in mature bone. Also, brittle bones in elderly individuals (especially osteoporotic females) may fracture more readily and unexpectedly than younger adults. It is a good idea to consult and cite literature (clinical, forensic, and anthropological), when appropriate. Wedel and Galloway (2014) provide an overview of blunt trauma.

Sharp-force trauma

Sharp-force trauma is a low-velocity trauma produced by an edged, pointed, or beveled object. The primary difference between sharp- and blunt-force trauma is the surface area of the object. The force produced by a sharp instrument is delivered over a smaller area, creating greater stress (expressed in pounds per square inch, or psi). Thus, 20 pounds of pressure spread over 10 inches produces 0.2 psi of stress ($20/10^2$), while 20 pounds delivered to a ½-inch surface area produces 80 psi of stress ($20/0.5^2$). The increased stress delivered by sharp objects compared with blunt objects of the same velocity creates trauma with unique characteristics. Sharp-force trauma wounds have characteristics indicative of the type of sharp instrument that produced them (Berryman et al. 2012b). The instruments are divided into classes based on these distinctive characteristics.

Kimmerle and Baraybar (2008) use the categories *short, long, light,* and *heavy* to divide sharp instruments into classes. Short-light instruments are easily wielded with one hand and are used to cut or saw. Long-heavy instruments are larger and heavier and are used to hack or chop. Pocket knives and butcher knives are short-light instruments and produce wounds ranging from simple punctures to incisions with V-shaped cross-sections, depending on the area of the blade that comes in contact with the bone and the amount of tissue that the blade must penetrate in order to produce an injury. Machetes, swords, cleavers, hatchets, and axes are long-heavy instruments that produce highly destructive chop wounds (Spitz 1993).

The terminology used to discuss cut marks caused by sharp instruments requires familiarity with the various types of cutting implements and blade design. The **kerf** is the wall and floor of a cut mark and can reveal important information about instrument class, especially when viewed under a microscope (Figure 13.20). For example, serrated blades can be distinguished from nonserrated blades (Crowder et al. 2013). However, just as caliber estimation is not recommended in gunshot wound analyses, experts should not attempt to match a specific knife to a wound, because there is significant overlap in the widths of cut marks from different types of knives (Bartelink et al. 2001).

Chop marks caused by long-heavy objects are large gaping defects that may have associated crushing and fractures (Berryman et al. 2012b). Fractures may radiate from the terminal ends of a chop wound. Chopping trauma can also cause concentric fractures. The mechanism that causes concentric fractures is similar to that observed in blunt-force trauma: the chopping action causes internal displacement of the cranial vault, creating sufficient tension on the outer table to produce fractures. The lateral margins of gaping chop marks may exhibit scoop marks (Kimmerle and Baraybar 2008) or comminuted fractures, referred to as "chattering" (Kerley 1973) (Figure 13.21).

Saw marks are a unique category of sharp force trauma produced by repeated back and forth motion through the bone. Saw marks are encountered primarily in cases of dismemberment and usually occur postmortem but have been reported as a cause of death in human rights violations (Campos Varela and Morcillo 2011), suicides (Asano et al. 2008; Gloulou et al. 2009; Grellner and Wilske, 2009; Tournel et al. 2008), and accidental deaths (Byard and Gilbert 2004; Haynes et al. 1980). Saws may be manual or powered; blades can be rounded or straight and may turn/rotate or be pushed and pulled back and forth through the bone; saw teeth may be alternating, wavy, or raker. Saws leave impressions on bone as they progress vertically and horizontally (i.e., back and forth) through the cut (Figure 13.22). Saw mark experts use five class characteristics to characterize saw types: saw size, saw set, saw shape, saw power, and direction of saw motion (Symes 1992). Each of these characteristics produces diagnostic features in bone.

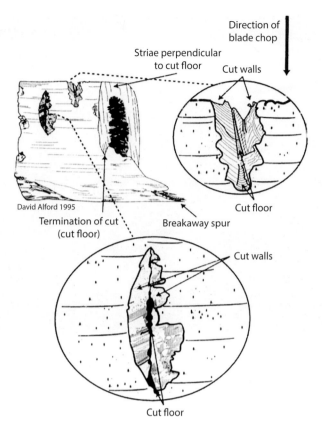

Figure 13.20 Sharp chop in a long bone, illustrating V-shaped cut floor (kerf floor) and striated cut walls (kerf walls). The walls and the floor of the cut comprise the entire kerf. (Courtesy of David Alford [image provided by Steven A. Symes].)

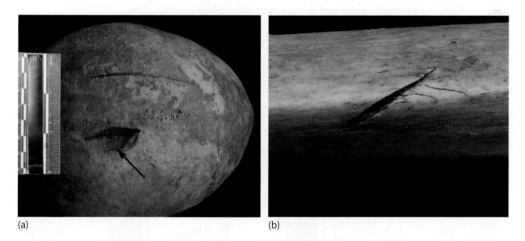

Figure 13.21 Chop marks. (a) Scoop mark (arrow) on the cranial vault, caused by an axe. (b) Chattering and linear fractures associated with an axe wound in the anterior tibia. (Courtesy of Georgia Bureau of Investigation Medical Examiner's Office.)

The terminology used to discuss saw marks is similar to that used for cut marks. Experts observe the kerf walls and kerf floor for clues about the type of saw. The kerf provides information about the five class characteristics, including the number of teeth per inch, blade thickness, tooth width, tooth orientation, blade contour (flat or curved), saw power, and blade progress through the bone (Andahl 1978; Bonte 1975; Symes 1992; Symes et al. 1998; Symes et al. 2002). Whereas the kerf floor is V-shaped in cut marks produced by knives, the kerf floor produced by saws is square and has unique features such as break-away spurs and false starts (Figure 13.23) (Symes et al. 2002). Breakaway spurs are jagged pieces of bone located at the bottom of the cut. They are fractures caused when pressure is applied

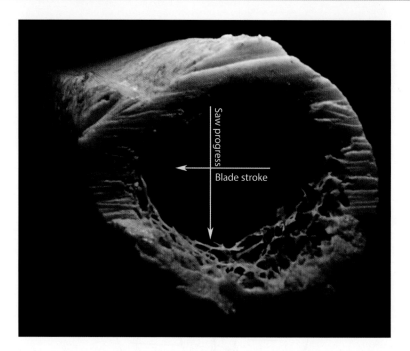

Figure 13.22 Saw impressions. This cross-sectional view of a long bone shows horizontal striations produced by the blade stroke. The vertical arrow shows the overall direction of saw progress through the bone. (Courtesy of Steven A. Symes.)

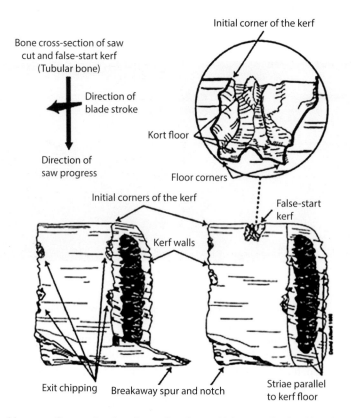

Figure 13.23 Saw mark kerf features. Saw cut in a long bone showing and false start kerf and breakaway spur. (Courtesy of David Alford [image provided by Steven A. Symes].)

to one side of the bone in order to complete a cut. False starts may also be observed near a complete saw cut. These are aborted attempts to initiate a cut. Imagine sawing a piece of wood, in which the saw gets stuck. One solution is to abandon the cut and begin another; the abandoned saw mark is referred to as a false start. False starts are more abundant with power saws than with manual saws, because the user is more likely to start a new cut than to put the saw back into the existing kerf.

Saw mark analyses yield highly specific information about the type of saw used to create cut marks. Weapon identification based solely on class characteristics is not advisable, but identifying a suspect weapon may assist efforts in locating further incriminating physical evidence (Berryman et al. 2012b). A recent study used statistical modeling to select the most accurate class characteristics for identifying saw type (Love et al. 2015). The models use information about the kerf width, floor shape, and wall shape to classify saw type accurately between 83% and 91% of the time. Empirical models with known error rates are essential for removing subjectivity from expert analyses. For more detailed information about class characteristics and saw mark analysis, readers may refer to (Symes 1992), Symes et al. 1998, 2002), and http://www.symesforensics.com/documents/1SawMarkManual_Symes-web.pdf.

PIONEERING VIRTUAL RESEARCH IN SKELETAL TRAUMA

As mentioned earlier in this chapter, experimental research is the ideal way to answer questions about skeletal trauma. Because access to the human remains to conduct this type of research has been traditionally limited, researchers have devised a number of alternative methods, including crash test dummies, ballistic gelatins, animal remains, and survey studies of existing trauma cases. The limitations of using alternative media (e.g., gelatins and dummies) are that they do not replicate the complex biomechanical properties of human tissues. During the past decade, virtual research has stepped up to provide a promising alternative to using human cadavers for experimental testing. Finite element analysis (FEA) creates a computer simulation of anatomical tissues with specific physical properties (e.g., stiffness and fatigue strength) that are subjected to various loads to observe the tissue's response. FEA models can be constructed from computed tomography (CT) scans or magnetic imaging resonance (MRI) scans. Once the models are created, they are imported into software. The various tissues are assigned biomechanical properties, and they can be used for repeated tests under any number of loading scenarios.

Finite element analysis is used frequently in automobile crash and military protective equipment testing (Doorly and Gilchrist 2006; Horgan and Gilchrist 2003, 2004; Lee and Gong 2010; Roberts et al. 2005, 2007; Shen et al. 2010). Finite element analysis has also been used to simulate gunshot injuries in pigs (Chen et al. 2010) and humans (Mota et al. 2003). Because the models can be run over and over, FEA provides statistical power to studies on the effects of caliber, trajectory (angle of impact), and velocity on gunshot wound characteristics (Mota et al. 2003; Puentes et al. 2009). As the demand grows for omitting subjectivity in forensic analyses and using extensively tested techniques with known error rates, FEA becomes increasingly popular in forensic research (Berryman et al. 2003; Mota et al. 2003; Motherway et al. 2009; Raul et al. 2006, 2008; Roth et al. 2007). Validations of FEA results with cadaveric tests or case studies will strengthen the conclusions of these studies and advance the science.

SUMMARY

This chapter divides skeletal trauma into two broad categories: high-velocity trauma and low-velocity trauma. These categories are further subdivided to discuss wounds with a unique appearance due to weapon characteristics (e.g., handguns and rifles vs. shotguns and blunt-force trauma vs. Sharp-force trauma). Although a physics-based trauma continuum produces the most meaningful skeletal trauma interpretations and accounts for any number of scenarios, forensic anthropologists must provide information about weapon category for forensic investigators, whenever possible (Berryman et al. 2012b). Without this information, investigators have little ground to pursue or eliminate possible weapons or suspects.

Forensic anthropologists contribute to a trauma analysis by providing information about the timing of skeletal trauma (antemortem, perimortem, and postmortem), the type of trauma (high- or low-velocity), and the weapon or object class. Experts should be cautious when interpreting perimortem events, because bone retains its elastic properties for weeks or months after death. High-velocity injuries are caused by projectiles traveling at a rapid rate of speed. Bone behaves like a brittle material and shatters when subjected to high-velocity loads. Projectile injuries have characteristic features that indicate the direction of fire, as well as the sequence of gunshot wounds (particularly in the cranium). Shotguns produce high-velocity wounds with characteristics that vary greatly, depending largely on the distance from the muzzle to the target and on the size of the shot in the shell.

Bone behaves like a ductile material and plastically deforms before failure when subjected to low-velocity loads. Low-velocity impacts to the cranium may exhibit some of the same features as high-velocity impacts (e.g., radiating and concentric fractures), but the mechanism creating these fractures differs in terms of the location of maximum tension and compression. Sharp-force trauma is a low-velocity trauma caused by edged, pointed, or beveled instruments such as knives, machetes, axes, and saws. The small surface area of these objects that comes in contact with the bone creates defects with unique features. Depending on the area of the body injured and on the angle at which the blade enters the body, sharp instruments may leave puncture marks or linear cut marks in bone. The appearance of the wound depends on the portion of the sharp instrument that comes in contact with the bone (if any). Cut mark and saw mark analyses examine the kerf of the cut mark to determine information about the class of knife or saw.

Forensic anthropologists use their understanding of bone biomechanics and bone's viscoelastic response to various loading scenarios to arrive at a conclusion about the mechanism of injury. When possible, anthropologists suggest a weapon class; at the least, they attempt to categorize the trauma as blunt, sharp, or gunshot trauma. They do not extend their interpretations beyond what the osseous evidence supports and avoid making overly specific statements about weapons and scenarios. When interpretation is particularly challenging or questionable, experts should describe the trauma and include helpful figures and photographs in the case report. Because remains may be returned to families, buried, or cremated before trial, careful documentation of skeletal trauma is of the utmost importance. The SWGANTH document on skeletal trauma analysis provides a list of unacceptable practices. These include providing overreaching conclusions, weapon matching, caliber estimation, estimation of the distance from muzzle to target, and use of inflammatory or suggestive terms in case reports (e.g., "victim" vs. "decedent" and "weapon" vs. "tool" or "object").

Experimental studies using human cadavers offer the optimal research design to study skeletal trauma. However, it is difficult to obtain adequate sample sizes because of limited access to human cadavers for trauma testing. Research using virtual models of human tissues may offer a viable alternative. Finite element analysis (FEA) uses models of human tissues that are assigned biomechanical properties to test the behavior of these tissues under various loading scenarios. These studies have shown promise in predicting fracture initiation, propagation, and severity. FEA models may deliver an answer to the Daubert criteria for trauma analyses by providing less subjective, extensively tested models with known error rates.

Review questions

1. Describe features that help anthropologists distinguish between antemortem, perimortem, and postmortem traumas. Explain why it may be difficult to ascertain perimortem timing of fractures.
2. Explain what the phrase "bone is viscoelastic" means. How does this affect the anthropologist's interpretation of skeletal trauma?
3. Discuss the features of a typical cranial gunshot wound. How do these features enable anthropologists to determine (a) the direction of fire and (b) the sequence of gunshot wounds (in the case of multiple gunshots)?
4. Explain how shotgun wounds differ from handgun and rifle wounds in bone. Which variables are responsible for these differences?
5. Compare and contrast the production of radiating and concentric fractures in low-velocity blunt-force trauma and high-velocity gunshot trauma.
6. Explain the differences between blunt- and sharp-force traumas, including mechanism of injury and wound features.
7. Define and describe a kerf. How do forensic anthropologists use the features of a kerf to identify sharp instrument class? What (if any) are the limitations to sharp-force trauma analysis?
8. Discuss the advantages and disadvantages of the following research models for trauma research: animal remains, human cadavers, surveys of forensic cases, synthetic materials, and finite element analysis models.

Glossary

Anisotropic: the property of being directionally dependent.
Axial loading: application of force along the long axis of a bone.
Bone biomechanics: the study of the mechanical laws relating to the material, structural, and functional properties of bone.
Bony callus: a temporary formation during the process of bone healing consisting of a mass of fibroblasts and chondroblasts.
Contrecoup fracture: a fracture occurring at a location approximately opposite to the point of impact.
Ectocranial: the outer cortical surface of the skull; the outer skull table.
Endocranial: the inner cortical surface of the skull; the inner skull table.
Greenstick fractures: incomplete fractures that indicate a high collagen content in bone; especially common in children.
Hinge fracture: a fracture across the middle cranial fossa that separates the skull base into anterior and posterior halves.
Kerf: the walls and floor of a cut mark.
Scientific Working Group for Forensic Anthropology (SWGANTH): a group of committees that provides consensus best-practice guidelines and establishes minimum standards for the forensic anthropology discipline. Formed in 2008 as a co-sponsorship of the Federal Bureau of Investigation (FBI) and the Department of Defense Central Identification Laboratory (DOD CIL), the SWGANTH committees are being incorporated into the Organization of Scientific Area Committees (OSAC).
Tensile loading: loading that induces maximum stress by stretching or pulling.
Viscoelastic: a property of materials that exhibit both viscous and elastic characteristics when undergoing deformation; the viscoelastic properties of bone are largely velocity dependent.
Woven bone: bone with haphazard organization of collagen fibers that are mechanically weak; also referred to as immature bone; most common during development.

References

Andahl RO. 1978. The examination of saw marks. *J Forensic Sci Soc* 18:31–36.

Asano M, Nushida H, Nagasaki Y, and Ueno Y. 2008. Suicide by circular saw. *Forensic Sci Int* 182(1):e7–e9.

Bartelink EJ, Wiersema JM, and Demaree RS. 2001. Quantitative analysis of sharp-force trauma: An application of scanning electron microscopy in forensic anthropology. *J Forensic Sci* 46(6):1288–1293.

Baumer TG, Passalacqua NV, Powell BJ, Newberry WN, Fenton TW, and Haut RC. 2010. Age-dependent fracture characteristics of rigid and compliant surface impacts on the infant skull: A porcine model. *J Forensic Sci* 55(4):993–7.

Baumer TG, Powell BJ, Fenton TW, and Haut RC. 2009. Age dependent mechanical properties of the infant porcine parietal bone and a correlation to the human. *J Biomech Eng* 131(11):1–6.

Berryman HE, Smith OC, and Symes SA. 1995. Diameter of cranial gunshot wounds as a function of bullet caliber. *J Forensic Sci* 40(5):751–754.

Berryman HE, and Gunther WM. 2000. Keyhole defect production in tubular bone. *J Forensic Sci* 45(2):483–487.

Berryman HE, and Symes SA. 2002. Recognizing gunshot and blunt cranial trauma. In J. Reichs Kathleen (Ed.), *Forensic Osteology: Advances in the Identification of Human Remains*, 2nd edn. pp. 333–352.

Berryman JF, Berryman HE, LeMaster RA, and Berryman CA. 2003. Numerical simulation of fracture propagation in a test of cantilevered tubular bone. Chicago, IL: 54th Annual Meeting of the American Academy of Forensic Sciences, Abstract in Proceedings of the American Academy of Forensic Sciences, Vol. IX, p. 260.

Berryman HE, Kutyla AK, and Davis JR II. 2010. Detection of gunshot primer residue on bone in an experimental setting–An unexpected finding. *J Forensic Sci* 55:488–491.

Berryman HE, Shirley NR, and Lanfear AK. 2012a. High Velocity Trauma. In MT Tersigni-Tarrant, and NR Shirley (Eds.), *Forensic Anthropology: An Introduction*. Boca Raton, FL: CRC Press.

Berryman HE, Shirley NR, and Lanfear AK. 2012b. Low Velocity Trauma. In MT Tersigni-Tarrant and NR Shirley (Eds.), *Forensic Anthropology: An Introduction*. Boca Raton, FL: CRC Press.

Bonte W. 1975. Tool marks in bones and cartilage. *J Forensic Sci* 20:315–325.

Campos Varela IY and MD Morcillo. 2011. Dismemberment: Cause of death in the Colombian armed conflict. Chicago, IL: Proceedings of the 63rd Annual Meetings of the American Academy of Forensic Sciences, Vol. 17, p. 356.

Cappella A, Amadasi A, Castoldi E, Mazzarelli D, Gaudio D, and Cattaneo C. 2014. The difficult task of assessing perimortem and postmortem fractures on the skeleton: A blind test on 210 fractures of known origin. *J Forensic Sci* 59(6):1598–1601.

Chen Y, Miao Y, Xu C, Zhang G, Lei T, and Tan L. 2010. Wound ballistics of the pig mandibular angle: A preliminary finite element analysis and experimental study. *J Biomech* 43(6):1131–1137.

Christensen AM, Passalacqua NV, Schmunk GA, Fudenberg J, Hartnett K, Mitchell RA, Love, JC, deJong J, and Petaros A. 2015. The value and availability of forensic anthropological consultation in medicolegal death investigations. *Forensic Sci Med Pathol* 11(2):438–41.

Cowin S. 1989. Mechanics of materials. In S Cowin (Ed.), *Bone Mechanics*. Boca Raton, FL: CRC Press, pp. 15–42.

Crowder C, Rainwater CW, and Fridie JS. 2013. Microscopic analysis of sharp force trauma in bone and cartilage: a validation study. *J Forensic Sci* 58(5):1119–26.

Currey JD. 2002. *Bones: Structure and Mechanics*. Princeton, NJ: Princeton University Press.

DiMaio VJ. 1993. *Gunshot Wounds: Practical Aspects of Firearms, Ballistics, and Forensic Techniques*, 2nd revised edn, New York: Elsevier Science Publishing Company.

Dixon DS. 1982. Keyhole lesions in gunshot wounds of the skull and direction of fire. *J Forensic Sci* 27:555–566.

Dixon DS. 1984. Exit keyhole lesion and direction of fire in a gunshot wound of the skull. *J Forensic Sci* 29:336–339.

Doorly MC, and Gilchrist MD. 2006. The analysis of traumatic brain injury due to head impacts arising from falls using accident reconstruction. *Comput Methods Biomech Biomed Engin* 9(6):371–377.

Galloway A, Symes SA, Haglund WD, and France DL. 1999. The Role of Forensic Anthropology in Trauma Analysis. In A Galloway (Ed.), *Broken Bones: Anthropological Analysis of Skeletal Trauma*, 1st edn. Springfield, IL: Charles C. Thomas. pp. 5–31.

Gloulou F, Allouche M, Khelil MB, Bekir O, Banasr A, Zhioua M, and Hamdoun M. 2009. Unusual suicides with band saws: Two case reports and a literature review. *Forensic Sci Int* 183(1):e7–e10.

Grellner W, and Wilske J. 2009. Unusual suicides of young women with tentative cuts and fatal neck injuries by chain saw and circular saw. *Forensic Sci Int* 190(1–3):e9–e11.

Gurdjian E, and Lissner H. 1945. Deformation of the skull in head injury: A study with the "stresscoat" technique. *Surg Gynecol Obstet* 81:679–687.

Gurdjian E, Lissner H, and Webster J. 1947. The mechanism of production of linear skull fractures. *Surg Gynecol Obstet* 85:195–210.

Gurdjian E, Webster J, and Lissner H. 1949. Studies on skull fracture with particular reference to engineering factors. *Am J Surg* 78:736–742.

Gurdjian E, Webster J, and Lissner H. 1950a. The mechanism of skull fracture. *J Neurosurg* 7:106–114.

Gurdjian E, Webster J, and Lissner H. 1950b. The mechanism of skull fracture. *Radiology* 54:313–338.

Haynes CD, Webb WA, and Fenno CR. 1980. Chain saw injuries: Review of 330 cases. *J Trauma* 20:772–776.

Hein P, and Schulz E. 1990. Contrecoup fracture of the anterior cranial fossa as a consequence of blunt force caused by a fall. *Acta Neurochir (Wein)* 105(1–2):24–29.

Horgan TJ, and Gilchrist MD. 2003. The creation of three-dimensional finite element models for simulating head impact biomechanics. *Int J Crashworthiness* 8(4):353–366.

Horgan TJ, and Gilchrist MD. 2004. Influence of FE model variability in predicting brain motion and intracranial pressure changes in head impact simulations. *Int J Crashworthiness* 9(4):401–418.

Jordana F, Colat-Parros J, and Bénézech M. 2013. Diagnosis of skull fractures according to postmortem interval: An experimental approach in a porcine model. *J Forensic Sci* 58(S1):S156–S162.

Kerley ER. 1973. Forensic anthropology. In CH Wecht (Ed.), *Legal Medicine Annual*. New York: Appleton-Century-Crofts. pp. 163–198.

Kimmerle EH and Baraybar JP. 2008. Sharp force trauma. In EH Kimmerle, and JP Baraybar (Eds.), *Skeletal Trauma: Identification of Injuries Resulting from Human Rights Abuse and Armed Conflict*. Boca Raton, FL: CRC Press. pp. 263–299.

Kroman AM. 2007. Fracture biomechanics of the human skeleton. PhD Dissertation. Knoxville, TN: University of Tennessee.

Kroman AM. 2010. *Rethinking Bone Trauma: A New Biomechanical Continuum Based Approach*. Chicago, IL: American Academy of Forensic Sciences 62nd Annual Meeting.

Kroman A, Kress T, and Porta D. 2011. Fracture propagation in the human cranium: A re-testing of popular theories. *Clin Anat* 24:309–318.

Langley NR. 2007. An anthropological analysis of gunshot wounds to the chest. *J Forensic Sci* 52(3):532–537.

Lee HP and Gong SW. 2010. Finite element analysis for the evaluation of protective functions of helmets against ballistic impact. *Comput Methods Biomech Biomed Engin* 13(5):537–550.

Le Fort R. 1901a. Fractures de la machoire superieure. *Revue de Chirurgie* 23:208–227.

Le Fort R. 1901b. Fractures de la machoire superieure. *Revue de Chirurgie* 23:360–379.

Le Fort R. 1901c. Fractures de la machoire superieure. *Revue de Chirurgie* 23:479–507.

Madea B, and Staak M. 1988. Determination of sequence of gunshot wounds of the skull. *J Forensic Sci Soc* 28(5–6):321–328.

Love JC, Derrick SM, Wiersema JM, and Peters C. 2015. Microscopic saw mark analysis: an empirical approach. *J Forensic Sci* 1(60 Suppl):S21–26.

Love JC, Derrick SM, and Wiersema JM. 2011. *Skeletal Atlas of Child Abuse*. New York: Humana Press.

Mota A, Klug S, Ortiz M, and Pandolfi A. 2003. Finite-element simulation of firearm injury to the human Cranium. *Comput Mech* 31:115–121.

Motherway J, Doorly MC, Curtis M, and Gilchrist MD. 2009. Head impact biomechanics simulations: A forensic tool for reconstructing head injury. *Legal Med* 11:S220–S222.

Nordin M, and Frankel V (Eds.). 1989. *Basic Biomechanics of the Musculoskeletal System*. Philadelphia, PA: Lea and Febiger.

Powell B, Passalacqua N, Baumer T, Fenton T, and Haut R. 2012. Fracture patterns on the infant porcine skull following severe blunt impact. *J Forensic Sci* 57(2):312–17.

Powell BJ, Passalacqua NV, Fenton TW, and Haut RC. 2013. Fracture Characteristics of Entrapped Head Impacts Versus Controlled Head Drops in Infant Porcine Specimens. *J Forensic Sci* 58(3):678–683.

Puentes K, Taveira F, Madureira AJ, Santos A, and Magalhaes T. Three-dimensional reconstitution of bullet trajectory in gunshot wounds: A case report. *J Forensic and Legal Med* 2009; 16:407–410.

Raul JS, Baumgartner D, Willinger R, and Ludes B. 2006. Finite element modeling of human head injuries caused by a fall. *Int J Legal Med* 120(4):212–218.

Raul JS, Deck C, Willinger R, and Ludes B. 2008. Finite-element models of the human head and their applications in forensic practice. *Int J Legal Med* 122(5):359–366.

Roberts JC, O'Connor JV, and Ward EE. 2005. Modeling the effect of non-penetrating ballistic impact as a means of detecting behind armor blunt trauma. *J Trauma* 58(6):1241–1251.

Roberts JC, Merkle AC, Biermann PJ, Ward EE, Carkhuff BG, Cain RP, and O'Connor JV. 2007. Computational and experimental models of the human torso for non-penetrating ballistic impact. *J Biomech* 40(4):125–136.

Ross AH. 1996. Caliber estimation from cranial entrance defect measurements. *J Forensic Sci* 41(4):629–633.

Roth S, Raul JS, Ludes B, and Willinger R. 2007. Finite element analysis of impact and shaking inflicted to a child. *Int J Legal Med* 121(3):223–228.

Shen W, Niu Y, Bykanova L, Laurence P, and Link N. 2010. Characterizing the interaction among bullet, body armor, and human surrogate targets. *J Biomech Eng* 132(12):121001.

Smith HW, and Wheatley KK. 1984. Biomechanics of femur fractures secondary to gunshot wounds. *J Trauma* 24(11):970–977.

Spitz WU. 1993. Sharp force injury. In WU Spitz (Ed.), *Spitz and Fisher's Medicolegal Investigation of Death*. Springfield, IL: Charles C. Thomas. pp. 252–310.

Symes SA. 1992. Morphology of saw marks in human bone: Identification of class characteristics. Knoxville, TN: PhD Dissertation, Department of Anthropology, University of Tennessee.

Symes SA, Berryman HE, and Smith OC. 1998. Saw marks in bone: Introduction and examination of residual kerf contour. In KJ Reichs (Ed.), *Forensic Osteology II, Advances in the Identification of Human Remains*. Springfield, IL: Charles C. Thomas. pp. 333–352.

Symes SA, Williams J, Murray E, Hoffman J, Holland T, Saul J, Saul F, and Pope E. 2002. Taphonomic context of sharp-force trauma in suspected cases of human mutilation and dismemberment. In W. Haglund, and M. Sorg (Eds.), *Advances in Forensic Taphonomy: Method, Theory, and Archaeological Perspectives*. Boca Raton, FL: CRC Press, pp. 403–434.

Tournel G, Dédouit F, Balgairies A, Houssaye C, De Angeli B, Bécart-Robert A, Pety N, Hédouin V, and Gosset D. 2008. Unusual suicide with a chainsaw. *J Forensic Sci* 53: 1174–1177.

Ubelaker DH, and Adams BJ. 1995. Differentiation of perimortem and postmortem trauma using taphonomic indicators. *J Forensic Sci* 40(3): 509–12.

Ubelaker DH. 1996. The remains of Dr. Carl Austin Weiss: An anthropological analysis. *J Forensic Sci* 41(1):60–79.

Vogel S. 2003. *Comparative Biomechanics: Life's Physical World*. Princeton, NJ: Princeton University Press.

Wedel VL and Galloway A. 2014. *Broken Bones: Anthropological Analysis of Blunt Force Trauma*, 2nd edn. Springfield, IL: Charles C. Thomas.

Weiberg DA, and Wescott DJ. 2008. Estimating the timing of long bone fractures: correlation between the postmortem interval, bone moisture content, and blunt force trauma fracture characteristics. *J Forensic Sci* 53(5):1–7.

Wheatley BP. 2008. Perimortem or postmortem bone fractures? An experimental study of fracture patterns in deer femora. *J Forensic Sci* 53:69–72.

CHAPTER 14

Introduction to Fordisc 3 and Human Variation Statistics

Richard L. Jantz and Stephen D. Ousley

CONTENTS

Historical background	256
Forensic anthropology data bank	257
Introduction to Fordisc	258
Basic Fordisc	258
Sex and ancestry from cranial measurements	258
Sex and ancestry from postcranial measurements	260
Stature	261
Advanced topics	263
Howells' data	263
Options	264
Check for measurement errors	264
LOO cross-validate	265
Typicality probs	265
Stepwise	265
Criticisms, limitations, and opportunities	265
Criticisms	265
Limitations	266
Opportunities	267
Summary	267
Supplemental materials	267
Review questions	268
Glossary	268
References	269

> **LEARNING OBJECTIVES**
> 1. Summarize the historical events that facilitated the development of Fordisc.
> 2. Briefly describe the statistical procedures that Fordisc uses to estimate sex, ancestry, and stature.
> 3. Distinguish between posterior probabilities and typicality probabilities, and know how to interpret these portions of the Fordisc output.
> 4. Explain why the Forensic Anthropology Data Bank (FDB) is an appropriate reference sample for most Fordisc analyses.
> 5. Describe the Fordisc analyses that one can perform with cranial versus postcranial metric data.
> 6. Explain why it would be desirable to use the "stepwise" selection option in Fordisc.
> 7. Summarize some of the criticisms and limitations of the Fordisc software, and explain why it is useful to forensic anthropologists and law enforcement, despite these arguments.

Fordisc 3 (FD3) is a computer software that aids in the estimation of sex, ancestry, and stature by using measurements and techniques described in Chapters 8, 9, and 11. It uses a technique known as **linear discriminant function analysis** to compare measurements from an unknown skeleton to reference samples, in order to estimate the ancestry or sex to which the person would likely have belonged in life. It also uses **linear regression** to estimate how tall the person would have been in life. Most of the statistical procedures that Fordisc uses have been around for a long time; Fordisc simply makes them available in flexible form for analysis of a single case. We present a brief review of the historical background of these statistics and how they led to the development of the software.

HISTORICAL BACKGROUND

Linear discriminant functions were developed by Fisher and apparently first used by Barnard (1935) and formally published by Fisher (1936). A related statistic, also important in Fordisc, is the **generalized distance**, developed by Mahalanobis (1936). Although the statistical methods were in place by early to mid-twentieth century, they were little used, because they were computationally burdensome with the hand calculators of the time.

The computational burden involved in computing linear discriminants and Mahalanobis distances led to some innovative attempts to reduce computation while remaining faithful to the central elements of the more complex statistics. Rao (1952) suggested some shortcuts, which replicated the **Mahalanobis distance** but still involved considerable computation. A strategy suggested by Penrose (1954) involved decomposing distance into size and shape components and incorporating an average correlation to replace the covariance matrix required by the Mahalanobis distance. This reduced the computational burden considerably, resulting in widespread use of Penrose's distances.

Linear discriminant functions were introduced to forensic anthropology in the late 1950s (Thieme 1957; Thieme and Schull 1957) and the early 1960s (Giles and Elliot 1962, 1963). These works were on the leading edge of use of electronic computers to overcome the computational burden mentioned earlier. The two works by Giles and Elliot became classics and were used extensively by forensic anthropologists until the early 1990s, when Fordisc 1.0 became available. Giles and Elliot enabled forensic anthropologists to use precomputed discriminant coefficients by multiplying coefficients with measurements to get a score, something easily accomplished with desk calculators. As computers became more widely used, many discriminant functions were devised for sex and ancestry estimation, a process that continues to the present.

Linear regression is the technique that Fordisc uses to estimate height. Like linear discriminant functions, it was developed around the turn of the century. The term "regression" was coined by Francis Galton (Smith 1980) from his observation that offspring tend to be closer to the population mean than their parents and hence regression to the mean. Karl Pearson developed the statistical basis for regression, as well as the related correlation coefficient, derived from the more descriptive term "co-relation."

FORENSIC ANTHROPOLOGY DATA BANK

Development of Fordisc depended first and foremost on the existence of an appropriate database. The idea of developing a database of modern forensic cases belongs to Clyde Snow, using the reasoning that only forensic anthropologists have the opportunity to examine skeletal remains of modern Americans. A grant from the National Institute of Justice launched the formal beginning of the Forensic Anthropology Data Bank (FDB) (Jantz and Moore-Jansen 1988). The FDB now serves as a repository of data collected from skeletons of Americans with twentieth and twenty-first century birth years. Skeletal data consist of cranial and postcranial measurements, aging criteria such as pubic symphysis phases and rib end phases, and dental and epiphyseal development stages. Since most of the skeletal remains are eventually identified, we are also able to obtain premortem and perimortem information, the most important being age, sex, race/ancestry, height, weight, and manner of death. The databank now contains skeletal information of over 4000 individuals, many of which are no longer available for study, having been returned to their families once they have been positively identified. Box 14.1 summarizes the most common way in which information gets into the FDB.

One of the observations often made about the FDB is that the sampling framework is biased, depending, as it does, on mostly nonnatural manners of death, such as homicide and suicide. Figure 14.1 demonstrates the extent of that bias; over half of the individuals in the databank have their cause of death as homicide or suicide. What at first may seem like a problem is actually an advantage, because the unknown forensic cases to be identified are drawn from the same population as the reference sample. Another advantage of the forensic sampling framework, unfortunate as it is, is that young adults and children are often victims and hence are represented in the databank, whereas they are seldom found in anatomical and donated collections. Consequently, the forensic sampling framework provides an ideal database for the basic job that Fordisc is asked to perform.

BOX 14.1 SEQUENCE OF EVENTS REQUIRED FOR DATA FROM A FORENSIC CASE TO GET INTO FDB

1. Homicide, suicide, or some other unattended death
2. Extended postmortem interval
3. Remains discovered, authorities notified, and forensic anthropologist consulted
4. Forensic anthropologist works up the case
5. Remains positively identified
6. Family claims remains
7. Forensic anthropologist provides data to FDB

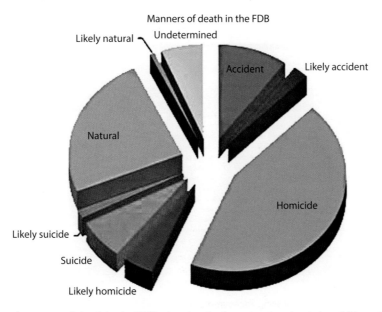

Figure 14.1 Distribution of manners of death in the FDB, showing the large number due to homicide and suicide.

INTRODUCTION TO FORDISC

In order to use Fordisc effectively, one must have a basic understanding of cranial and postcranial measurements and how to take them accurately. The measurements, techniques, and instruments required are described in Moore-Jansen et al. (1994) and Langley et al. (2016). Fordisc help files also provide measurement definitions and illustrations. One must also have a basic understanding of the statistical procedures that Fordisc uses and how to interpret the output. Sometimes, it is assumed that Fordisc absolves the user of any responsibility and simply provides the desired answer concerning sex, ancestry, or stature. Like everything about analyzing a forensic case, the ultimate interpretation is up to you. In what follows, we will attempt to guide you through what Fordisc does and how to best interpret it.

Basic Fordisc

Launching FD3 presents an opening screen, like the one shown in Figure 14.2. The basic options provided by Fordisc are given on this opening screen. These include blanks, where measurements from the skull to be analyzed can be either entered from the keyboard or loaded from a previously saved file, and a checkbox beside each measurement that allows it to be included or excluded from the analysis. Variable names in blue have missing data, which will reduce sample size, and those in red will seriously reduce sample size. Population samples of groups to which the unknown skull is to be compared have checkboxes for the user to select. Of the available reference groups, whites and blacks are predominantly forensic case; however, the samples also include donated individuals from the University of Tennessee and the University of New Mexico donated collections. The Hispanic sample consists of forensic cases, many of which were border crossers in the Southwest and none of which come from the Caribbean area. The American Indian sample is a combination of forensic cases and known historical individuals. The Asian samples are from anatomical collections in the countries from which they come. All individuals in Fordisc's database are positively identified as to sex and ancestry from antemortem information, and all have twentieth century birth years, except some of the American Indians.

Sex and ancestry from cranial measurements

We begin with an illustration of a forensic case (case UT91-42) examined by the University of Tennessee's forensic anthropologists. The authors of Fordisc have a certain emotional attachment to this case, since it is the first forensic case analyzed using Fordisc. It was recovered from a creek near Oak Ridge, Tennessee, and the original visual assessment suggested that it was a white male. Here, we compare it to white and black males and females, as indicated by the groups checked in Figure 14.2, allowing a simultaneous assessment of ancestry and sex. Once the user has selected the measurements and groups, processing is initiated by pushing the "Process" button at

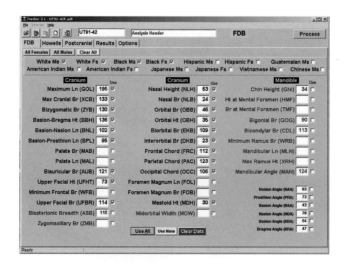

Figure 14.2 Basic Fordisc entry screen showing measurement entry blanks for the unknown case, reference samples against which it can be compared, and checkboxes to include or exclude measurements and groups.

Table 14.1 Basic FD3 output for cranial analysis

Fordisc 3.1 analysis of UT91-42

DF results using 17 variables:
AUB BBH BNL BPL DKB EKB FRC GOL NLB NLH OBB OBH OCC PAC UFHT XCB ZYB

From group	Total number	Into group				Percent correct
		BF	BM	WF	WM	
BF	38	31	3	4	0	81.6
BM	41	3	31	4	3	75.6
WF	95	9	2	74	10	77.9
WM	125	1	6	11	107	85.6

Total correct: 243 out of 299 (81.3%) ***CROSS-VALIDATED***.

Multigroup classification of UT91-42

Group	Classified into	Distance from	Posterior	Probabilities		
				Typ F	Typ Chi	Typ R
BM	**BM**	26.8	0.794	0.559	0.061	0.143 (36/42)
WF		31.0	0.097	0.118	0.020	0.031 (93/96)
WM		31.1	0.092	0.081	0.019	0.039 (121/126)
BF		34.6	0.017	0.395	0.007	0.026 (38/39)

UT91-42 is closest to BMs.

	UT91–42	Chk	Group Means			
			BF	BM	WF	WM
			38	41	95	125
AUB	121		116.1	121.1	116.7	123.6
BBH	136		130.0	138.3	134.1	141.7
BNL	102		97.9	105.5	99.5	105.5
BPL	95		98.8	103.9	92.4	96.4
DKB	23		22.9	24.4	20.0	21.3
EKB	109	++	95.5	100.9	93.9	97.7
FRC	112		107.5	113.5	109.4	114.2
GOL	195	+	178.8	189.1	178.1	187.1
NLB	24		25.0	26.1	22.5	23.7
NLH	53	+	48.4	52.4	49.3	52.7
OBB	46	+++	38.3	40.4	39.2	40.6
OBH	35	+	34.7	34.9	33.2	33.7
OCC	106	+	96.7	99.7	97.6	100.5
PAC	123	+	112.3	117.0	112.8	117.9
UFHT	74	+	67.3	72.6	67.2	71.4
XCB	133	–	133.2	136.9	135.8	141.4
ZYB	130		122.0	130.9	120.3	130.1

Note: Natural log of determinant = 38.9300.

the upper right of the screen. Basic FD3 output is shown in Table 14.1. It consists of three parts, which will be explained in turn:

1. Fordisc first gives the case identification number, the variables utilized, and a classification matrix of the reference samples selected. This classification matrix indicates how well Fordisc classified the individuals in its own reference samples. In this example, its overall correct classification rate is 81.3%, given at the bottom of the classification table. The individual rows and columns give results for specific groups. For example, of the 125 white males, 107 are correctly classified, 11 are misclassified as white females (correct by race, but incorrect by sex), 6 are misclassified as black males (incorrect by race, but correct by sex), and 1 is misclassified as black females (incorrect

for both race and sex). Note that at the bottom of the table, it is stated that the classifications were cross-validated. Cross-validation, sometimes called the leave-one-out method or jackknife method, means that each individual is removed from the sample before being classified. This prevents an individual from contributing its own information to the function. This results in an unbiased classification rate.

2. The second part of FD3 output is the most important, because it contains the classification of the unknown skull. The multigroup classification table contains three kinds of information. The "Distance From" column is the **Mahalanobis distance** (D^2) of the unknown skull from each of the reference groups, presented in order of increasing distance. A skull is normally classified into the group to which it is most similar, that is, the group with the lowest distance—in this case, black males. Additional information in the table allows us to assess the reliability of the classification. The column labeled "posterior probability" gives the probability of the skull to each group, using the assumption that it must belong to one of the groups. The posterior probability will always be highest for the group to which the skull has the lowest Mahalanobis distance. What is important in using the posterior probabilities is how they are distributed over the different groups. In this case, the posterior probability is 0.794 for black males, which we can take to mean that the individual has an almost 80% chance of being a black male, assuming that it actually belongs to one of the four groups that we have used.

Information in the remaining three columns contains various kinds of typicality probabilities. Unlike the posterior probability, typicality probabilities make no assumption that the skull must belong to one of the reference groups. The typicality probability addresses the question of whether the skull could belong to one or more of the groups. Typ F converts the Mahalanobis distance into an F ratio, and typicality is the probability associated with that F ratio. Typ F is normally the most conservative. In the present example, we see that Typ F would say that it is typical of all groups (probability > 0.05). Note that white males have the lowest typicality probability, because they have a higher Mahalanobis distance and have the largest sample size. Typ Chi is based on the chi-square distribution of the Mahalanobis distance. It ignores sample size, effectively assuming that it is very large. Typ Chi shows that the unknown skull is only typical of black males, but just barely, and is atypical of the remainder. Typ Chi will always vary directly with the Mahalanobis distance.

Typ R is arguably the most useful typicality probability, because it is empirical and makes no assumptions about distributions. It is obtained by adding the unknown skull to each group, in turn, and calculating the Mahalanobis distance of each skull to its group centroid. The example in Table 14.1 will help clarify how this probability works and what it can tell us. The unknown skull ranks 36 out of 42 ($n + 1$, where n refers to the reference sample size), meaning that 35 of the skulls that actually belong to the group are closer to the centroid than the unknown. The probability is obtained as $(42 - 36)/42 = 0.143$. This probability tells us that the unknown skull falls comfortably within the range of variation of Black males. The unknown skull would be judged atypical of the other three samples. It does fall within their range of variation but on the extreme margin.

Fordisc also lists the measurements for the unknown skull and the means for the reference samples chosen. The column labeled "chk" indicates how the individual deviates from the overall reference groups. A "+" means larger and a "−" means smaller, and the number of pluses or minuses indicates the magnitude of the deviation. In the present instance, OBB (orbital breadth) has three plus signs, meaning that it deviates two to three standard deviations from the reference samples. Large deviations may occur because some dimensions may be larger or smaller, or they could indicate measurement error. In this case, it is the former, but large deviations should be checked to rule out error.

Sex and ancestry from postcranial measurements

Fordisc offers the same analysis on postcranial measurements as that just presented using cranial measurements. Pushing the postcranial tab at the top of the screen puts up a postcranial screen like that shown in Figure 14.3. At present, the samples available for postcranial analysis are limited to blacks and whites, but Hispanics are coming soon. It often comes as a surprise, even to many experienced forensic anthropologists, that postcranial bones can discriminate ancestry as effectively as cranial bones. The reason that postcranial remains are underappreciated in this respect is apparently because it is much easier to form visual assessments of skull morphology than it is for the postcranial skeleton, not that the skull is more differentiated.

Table 14.2 shows the FD3 output for the postcranial analysis from the options chosen in Figure 14.3. The overall classification rate is only about 2% lower than that for cranial classification. Table 14.2 shows that case UT91-42 is a black male. This assessment agrees with the cranial classification, but the posterior probability is even higher, telling us that we have 94%

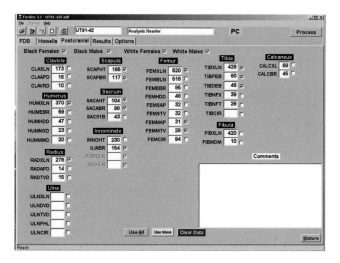

Figure 14.3 Basic Fordisc postcranial screen, which is parallel to the cranial screen, as shown in Figure 14.2.

chance of being correct, assuming that the individual belongs to one of the groups that we have selected. The typicality probabilities indicate that the skeleton falls comfortably within the range of variation of black males and white males.

Stature

Note the green button labeled "stature" at the bottom right of the postcranial screen in Figure 14.3. Pushing that button brings up the stature estimation screen, shown in Figure 14.4. The choices offered by this screen consist of different samples to be used to compute the regression and the kind of stature to be used. Groups available are black or white males or females, Hispanic males, or "any." The last should be used only if you have no idea of ancestry. Ancestry can make a substantial difference in the stature estimate because of different limb proportions.

It may be well to consider different types of stature. We often consider height to be a single number that describes an aspect of our body. It is normally included on documents such as driver's licenses, missing person's reports, descriptions of criminal suspects, and medical documents. In fact, it is a highly variable number, depending on its origin. Snow and Williams (1971) found variation as great as 5 inches in reported stature for a single individual from different sources. A person's height even varies throughout the course of a day. FD3 contains three types of stature:

1. *Nineteenth Century Cstats*: Cstats refers to cadaver statures. Choosing this option uses the cadaver statures from the Terry Anatomical Collection and included in Trotter and Gleser (1952) classic paper on estimating height. There is little agreement on how cadaver stature relates to living stature, but it is usually considered to be larger. Trotter and Gleser (1952) suggested that it exceeds living stature by 2.5 cm.

2. *Trotter Mstats:* Trotter Mstats refers to the measured statures that Trotter was able to obtain from World War II military personnel killed in the Pacific Theater. Height was measured on induction into the armed forces, and Trotter was able to measure the long bones after death. Choosing this option uses the statures and long bone lengths from which Trotter and Gleser (1952) derived their stature regression equations. We should note that the sample sizes are different than those used by Trotter and Gleser (1952), because we included some that Trotter did not use or those with one or more elements missing.

3. *Twentieth Century Fstats*: Fstats refer to forensic stature as defined by Ousley (1995). Forensic stature is a number that is written down somewhere, such as on a driver's license and other documents mentioned earlier. It has a lot in common with self-reported stature, which has been shown to be an overestimate of measured stature (Giles and Hutchinson 1991; Willey and Falsetti 1991). Forensic stature will normally result in a larger stature estimate and a larger prediction interval than the measured stature, but this is acceptable, because it is usually the forensic stature to which our skeletal estimate will be compared.

After choosing the stature data, ancestry, and sex and clicking on the "estimate" button, statistics pertaining to the stature estimate are displayed in the upper panel and a plot of height on bone length appears in the lower panel, including the **least squares regression** line and the 90% prediction interval (Figure 14.4). The prediction interval can

Table 14.2 Basic FD3 output for postcranial analysis

Fordisc 3.1 analysis of UT91-42

DF results using 13 variables:

FEMMAP	FEMMTV	FEMXLN	HUMXLN	ILIABR	RADXLN	SACABR
SACAHT	SCAPBR	SCAPHT	TIBDEB	TIBPEB	TIBXLN	

From group	Total number	Into group BF	BM	WF	WM	Percent correct
BF	25	20	0	5	0	80.0
BM	55	1	46	1	7	83.6
WF	73	14	0	56	3	76.7
WM	154	6	24	3	121	78.6

Total correct: 243 out of 307 (79.2%) *** CROSS VALIDATED ***

Multigroup classification of current case

Group	Classified into	Distance from	Posterior	Typ F	Typ Chi	Typ R
BM	**BM**	8.4	0.939	0.916	0.817	0.750 (14/56)
WM		13.9	0.061	0.477	0.383	0.355 (100/155)
BF		31.0	0.000	0.409	0.003	0.038 (26/26)
WF		40.8	0.000	0.007	0.000	0.014 (72/74)

Current case is closest to BMs.

Current	Case	Chk	Group Means BF 25	BM 55	WF 73	WM 154
FEMMAP	31		27.5	31.6	27.5	30.9
FEMMTV	28	+	23.8	27.8	24.3	27.9
FEMXLN	520	++	441.0	488.1	435.8	473.6
HUMXLN	370	++	308.8	340.7	305.1	334.6
ILIABR	154		144.3	153.4	155.2	160.9
RADXLN	278	+	235.3	267.7	227.6	253.3
SACABR	98	−	100.2	101.9	109.7	107.5
SACAHT	104		103.4	104.4	110.1	112.9
SCAPBR	117	++	94.2	109.5	94.8	108.2
SCAPHT	166	+	136.7	161.1	142.5	163.4
TIBDEB	48		45.3	51.3	46.5	52.0
TIBPEB	80	+	68.9	78.6	70.0	79.3
TIBXLN	429	+	366.4	409.2	357.8	392.9

Natural log of determinant = 47.2414.

also be set to 95% or 99% by choosing the desired figure from the box just below the stature type. For purposes of this example, we chose Fstats for black males. Note that the user can choose whether the results are expressed in inches or centimeters. We use inches in this example.

Results in the upper panel include all possible combinations of the sums of three bones, two bones, and individual bones, arranged in ascending order using R^2, or the prediction interval (the value in the left-hand column of the upper panel). Normally, one would select the estimated stature in first row as the most accurate—in this case, 72.1 in. The upper and lower prediction intervals tell us that we can be 90% confident that the stature of the individual falls between 69.5 in. and 74.8 in. Upper and lower prediction intervals are obtained by adding and subtracting the prediction interval from the estimate.

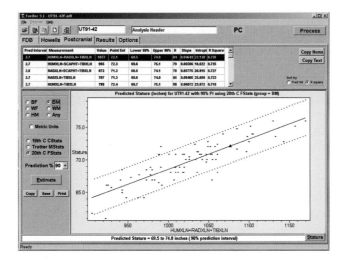

Figure 14.4 Stature screen showing choices of reference samples, kinds of statures available, and example results.

What Fordisc has done is solve the least squares regression equation. The columns labeled "slope" and "intercept" are the two constants Fordisc uses to estimate the stature of the unknown case by solving for Stature = 0.04643 * 1077 + 22.138, where 1077 is the sum of lengths of the humerus, radius, and tibia, given under the column heading "Value." Last, the column labeled R^2 gives the proportion of variation in height explained by the bone lengths. Subsequent rows present similar information for other combinations. Although the first row of the upper panel should be normally accepted as the most accurate, because it has the lowest prediction interval, it may also be instructive to examine the other estimates. The different estimates are in good agreement, the highest (72.4 in.) and the lowest (71.3 in.) differing by just over an inch. Some variation is to be expected, because different combinations of bones relate to stature in slightly different ways. In the present instance, those estimates lacking the humerus are lower than those that include it, suggesting that the humerus is relatively longer in relation to stature than the scapula, tibia, and radius.

The graph in the lower panel gives the bivariate plot of stature on bone dimensions for the reference sample, along with the fitted regression line and 90% prediction intervals. Scrolling down brings up the graph for each combination, in turn. The unknown's position is on the regression line and allows for a visual appreciation of where it falls in relation to the reference sample for bone dimensions and estimated stature. In the example given in Figure 14.4, it is apparent that the unknown is above average, but not extreme, for bone dimensions and therefore for estimated stature.

The forensic stature recorded for UT91-42 is 72 inches versus the Fordisc estimated height of 72.1 inches. In this case, the estimate is very close to the recorded forensic stature.

ADVANCED TOPICS

What has been presented thus far is the basic procedure, but Fordisc has several important options that provide some additional analytical choices. We will describe these options, when to use them, and how to use the results.

Howells' data

The main Fordisc screen is designed to deal with the majority of the situations likely to be faced in the American context. The main screen contains samples drawn from the major populations in the United States. However, it can be the case that a cranium can come from anywhere in the world, for example, trophy skulls (Sledzik and Ousley 1991), archaeological specimens, and even immigrants from different parts of the world. Alternatively, questions sometimes arise concerning museum specimens and whether they are really what the museum records say they are and therefore may be subject to repatriation (Pickering and Jantz 1995). When context is lacking, Howells' worldwide data provide a larger framework against which to compare a specimen.

Howells' data also contain many more measurements than those on the basic forensic screen, 61 in total. Howells' measurements are carefully described (Howells 1973) and have become widely adopted as a measurement protocol. One important point to note is that those measurements on the FDB screen with three-letter codes (e.g., GOL) use Howells' definitions, while those with four-letter codes (e.g., UFBR) are either not defined by Howells or are defined differently (see Langley et al. 2016; Moore-Jansen et al. 1994). Obtaining Howells' measurements requires either special instruments or a digitizer. If you have only taken the basic measurements required for the forensic screen, you can still compare your unknown skull to Howells' samples for those measurements. Using Howells' data is recommended when you suspect that the skull originates from outside the United States. In that case, Howells' data can provide some idea about the skull's geographic origin.

Options

Clicking on the options tab on the right side of the toolbar brings up a screen with several analytical options. The screen with the default options checked is shown in Figure 14.5. On the left side of this is an option called "Transformations," with choices "None" (the default), "Log," and "Shape." Using the default is normally the best course of action. The "Log" option carries out the analysis after transforming the variables into natural logarithms, and the "Shape" option carries out the calculations after transforming the variables into Darroch and Mosimann (1985) shape variables. Logarithms may, in some cases, make the data more normally distributed, particularly if the original data have a positive skew.

Darroch and Mosimann's shape variables are obtained by dividing each variable by the **geometric mean** of all variables for each individual. The geometric mean serves as a measure of **isometric size**, and dividing each variable by it yields shape variables that are free of isometric size. Choosing the shape option may be desirable if a skull or postcranial skeleton is unusually large or small and you wish to assess its relationships by using shape alone. If the question that you are asking Fordisc to answer pertains to ancestry, choosing shape may not make much difference, since ancestry variation is mostly shape-based. However, it makes a substantial difference for sex estimation, because sex dimorphism has a substantial size component. For example, using 18 raw variables to classify the white male and female reference samples yields a correct classification of 90.5%. Using the same 18 shape variables reduces the classification to 78.2%. This not only illustrates the loss of discriminatory power resulting from removing size but also illustrates that considerable sex dimorphism remains in shape alone, since 78.2% correct classification is far above the chance expectation of 50%.

Check for measurement errors

To the right of "Transformations" is a checkbox that enables the user to turn off the measurement check option. The default is on, and normally, there are no reasons to turn it off. If you enter measurements that are outside the range of normal variation, Fordisc will tell you that the measurement looks wrong and ask you if you want to change it.

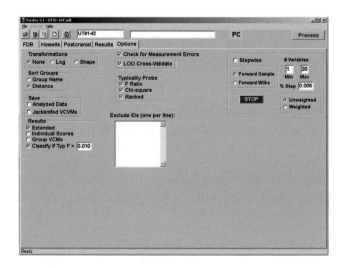

Figure 14.5 Options screen showing the various analytical options that Fordisc offers. Defaults are checked.

LOO cross-validate

"LOO" means "leave-one-out," and it is the option discussed earlier in going over the classification of the reference samples. LOO turned on is the default. If it is turned off, the classification will be done without removing the classified individual. This is known as resubstitution, and it results in a biased classification rate of the reference samples. The only advantage of turning LOO off is that processing proceeds faster, which may be important when doing a large number of groups and variables.

Typicality probs

This option presents checkboxes that enable you to turn the various typicality probabilities on or off. The default is on, and normally, there is no good reason to turn them off, since each presents some information about the skeleton being analyzed.

Stepwise

This option is an important one but also requires some care in implementation, because it is a complex topic. Stepwise solutions in discriminant analysis are a well-known way of attempting to find a subset of variables that discriminates the reference sample equally or better than the full set of variables. It can be used to prevent something known as overfitting. Overfitting means that the sample sizes are insufficient to contain all the variability in the measurements; this can result in spuriously high correct classification rates, and the functions cannot be relied on to correctly classify the unknown. The most liberal rule concerning the number of allowable variables is that the number of variables must be one less than the number of individuals in each group. The most conservative rule is that the number of individuals must be three times larger than the number of variables. When you are using the FDB screen, this will not normally be a problem, except possibly for Hispanic females. It is not advisable to use all of Howells' 61 variables, because most samples have fewer than 61 individuals.

A disadvantage of stepwise procedures is that they capitalize on sampling error. Fortunately, most samples on the FDB screen are large enough, so this may not be an important issue, but some combinations of variables (e.g., those in blue or red) and groups (e.g., Hispanic females) may create those problems and should be best avoided.

To use the stepwise option, you must check the "Stepwise" box to turn on the option. There is also a box where you can set the minimum and maximum number of variables that you want stepwise to choose. You must then choose the kind of stepwise analysis that you wish to use. The simplest and fastest is "Forward Wilks." It uses Wilks' lambda, a measure of variation among groups relative to the total. It starts with the variable with the lowest Wilks' lambda (lower values mean greater variation) and then finds the next variable that results in the greatest improvement and proceeds in this manner until the amount of improvement is trivial. This option assumes multivariate normality, a condition usually met by the FDB measurements. Choosing the "forward sample" option bases variable selection on the percentage correct in the reference samples. The "turbo" option is selected by default and speeds up the process considerably. Turbo selection uses resubstituted (and biased) classification rates to choose the best measurements but calculates LOO cross-validated classification rates at the end. You can choose the percentage improvement required for the process to continue. The default is 0.005% or 0.5%. When the minimum number of variables is set to 1, stepwise chooses the best variable and then tries all others, until adding an additional variable no longer exceeds 0.5%. If you set the minimum variables to 2, stepwise will try all possible combinations of two variables until it finds the best two and then adds others until the criterion is no longer met. The same is true of three variables, and so on. For a minimum greater than three, the computation time rises rapidly because of the large number of permutations that the program must try. If you start stepwise and it is taking longer than you want to wait, you can push the "stop" button and start over.

CRITICISMS, LIMITATIONS, AND OPPORTUNITIES

Criticisms

Fordisc was originally designed as a tool to be used by forensic anthropologists working in the United States. With the addition of Howells' data in Fordisc 2 and its expansion to all of Howells' variables in FD3, wider applications were inevitable and desirable. Also, inevitable were the tests of how well Fordisc performs in different circumstances. This is desirable when the tests are honest attempts to define limits on Fordisc's applicability. For example, Ramsthaler et al.

(2007) have asked whether Fordisc can be used to estimate sex from cranial measurements in European populations. They concluded that it presents biased results because of morphometric differences between American whites and Europeans and that in order for Fordisc to be applied to recent Europeans, reference samples from recent Europeans should be included. This is a conclusion with which we wholeheartedly agree, and we are working to make Fordisc's database more representative of Europe and other parts of the world.

Other tests of Fordisc are seemingly designed to make it fail, suggesting that the desired outcome was decided in advance of the analysis. Elliott and Collard (2009) argued that Fordisc can assess ancestry correctly only 1% of the time. They reached this absurd conclusion by imposing unrealistic expectations and misusing Fordisc's statistics to achieve results that will show either very low typicality probabilities (raising doubts about any classification) or very low posterior probabilities (raising uncertainties about possible classifications) for a new case (Siegel and Ousley 2011). One can easily imagine that regular users of Fordisc would long ago have realized such a performance deficiency and ceased using it.

Other tests of Fordisc begin with the assumption that most of the variation is within races rather than among them, following Lewontin (1972), and therefore attempts at allocating crania to their correct groups are a futile endeavor. This is the view expressed in Williams et al. (2005), who, like Elliott and Collard, imposed unrealistic expectations on Fordisc and refused to be guided by the statistics that it provides. Their conclusion is that skeletal remains cannot be sorted by geographic region except at the extremes, but forensic experience has shown that even populations in relatively circumscribed areas can be effectively allocated to the correct group, for example, North and South Japanese (Ousley et al. 2009). Numerous errors in the data of Williams et al. (2005) raise doubts about their comparisons.

Yet another direction from which criticism has come is the idea that in America, race is purely a social construct, with no biological relevance (Albanese and Saunders 2006; Goodman 1997). Much has been made of the observation that, sometimes, forensic anthropologists make incorrect ancestry assessments. Goodman (1997) argues that the Giles and Elliot (1962) functions classify at a rate no better than by chance, again a predetermined outcome seeking support. The problems with Giles and Elliot functions have little to do with the nonexistence of race but rather with the reference samples. A more realistic test of Giles and Elliot's discriminant functions revealed that they performed well, except for Native Americans (Snow et al. 1979), which is a consequence of an inappropriate reference sample.

Is it necessary for race to exist as a valid biological taxon in order to justify using Fordisc in the way forensic anthropologists do? Not in our view. What we know exist are populations, and populations exist because humans develop a variety of mechanisms that structure mating. We accept that race in America is a social category, but it also plays a major role in mating. The very terms "African-American" and "European-American" imply origins from distinctive populations, which, to a considerable degree, have remained reproductively isolated. The U.S. Census and other federal documents require self-identification of a racial and ethnic category. Asking people to place themselves into racial categories such as white, black, Native American, Asian, and Pacific Islander, or into ethnic categories such as Hispanic and non-Hispanic, is undoubtedly an oversimplification. Since the European colonization of America, these populations have evolved in various complex ways, giving rise to populations with various degrees of admixture (e.g., Parra et al. 1998). However, the U.S. Census data show that the majority of matings are within social race categories. It is a basic tenet of anthropological genetics that reproductive isolation maintains preexisting variation and allows new variation to emerge. We readily acknowledge that interracial marriages are much more common than before, American populations are becoming more complex, and ancestry estimation is becoming more difficult. To those who argue that forensic anthropologists should not make ancestry assessments because they may be wrong, we would just say that practitioners should not make dogmatic statements and should be guided by probabilities in making their assessments. Of course, we can be wrong, because the question asked pertains to social race, but the answer provided is based on biology. We would also say that it is rare when proper statistical evaluation of measurements cannot reveal useful information about the unidentified skeleton before you.

Limitations

Every method has limitations, and Fordisc is no exception. Allocating an unknown by using group means as the criterion may not yield the optimal solution. Using K-nearest neighbor, where the criterion is the individual crania most similar to the unknown, to determine the classification may be more appropriate in some cases and is being considered as a future option for Fordisc.

The greatest limitation is the lack of appropriate reference samples. American whites, blacks, and Hispanics of Mexican origin are now well represented but Caribbean Hispanics are not. We have few American-born Asians, and the exceedingly variable Native American populations are underrepresented. The most recent U.S. Census shows that the American population is changing rapidly and former barriers to intermarriage are breaking down. These new Americans will present increasingly complex challenges to forensic anthropology. It will never be possible to have adequate samples of every population in America, but the FDB and forensic anthropology offer at least some hope of keeping up with its main components.

Another challenge is incorporating more of the world's populations. Fordisc is becoming more popular abroad, and in order to be useful, it must have samples from regions where it may be applied.

Opportunities

Some of the limitations mentioned earlier can be converted into opportunities, for example, improving the sample coverage or adding new procedures. However, for students and professionals alike, Fordisc provides some interesting opportunities beyond just evaluating forensic cases. It can be used to explore human cranial and postcranial variations. For example, it is often stated that the cranium is the second best area to use for sex estimation, after the pelvis, and should be used when the pelvis is unavailable (Byers 2008). It is easy to show by using Fordisc that the cranium is inferior to most postcranial elements for metric sexing (Spradley and Jantz 2011). Alternatively, one can use the nineteenth and twentieth century Americans on the Howells' page to show that so much **secular change** has occurred in Americans over the past century that nineteenth and twentieth century American crania can be easily discriminated from one another (Jantz and Meadows Jantz 2000). Many other interesting questions can be formulated and addressed with Fordisc. Such questions will only increase as Fordisc's database and features continue to grow. We conclude with the recommendation that users of Fordisc update regularly. Updating can be accomplished using the "Internet" tab at the top of the opening screen. Updating will ensure that you are using the most current version of Fordisc with the most up-to-date database. There is also a "Submit case to the FDB" option under the Internet tab that enables you to submit a case to the FDB and participate in the effort to keep the database current.

SUMMARY

FD3 is a computer software that aids in the estimation of sex, ancestry, and stature. Fordisc uses linear discriminant function analysis to compare measurements from an unknown skeleton to reference samples, in order to estimate the ancestry or sex to which the person would likely have belonged in life, and linear regression to estimate how tall the person would have been in life. The Fordisc user does not have to be a statistician and should have a basic idea of the statistical methods used by the software and their requirements in order to properly interpret the results.

Fordisc output provides a probability that the unknown belongs to one of the reference groups selected for comparison (the posterior probability) and also tells the user how similar the unknown is to the group to which it is closest (typicality probabilities). The output gives information about how well the discriminant function performs on the reference sample and gives the user an opportunity to screen for measurement error. The software also has several advanced options that allow the user to deal with tricky cases, such as stepwise selection options and shape and log transformations. A savvy user can usually arrive at an educated conclusion by running several analyses and understanding the output.

The software uses two primary reference databases: the Forensic Anthropology Data Bank (FDB) and the Howells' worldwide craniometric data set. The FDB consists primarily of forensic cases from the United States and is therefore an ideal reference data set for forensic analyses. The Howells' data set is useful when you suspect that the unknown is not from the United States, as it contains reference samples from around the world. The software developers are working constantly to amass reference data from populations around the globe, so that Fordisc may serve as a useful tool outside of the United States.

Supplemental Materials

- Fordisc 3 Home Page: http://math.mercyhurst.edu/~sousley/Fordisc/
- Link to the latest edition of *Data Collection Procedures for Forensic Skeletal Material* (DCP 2.0, Langley et al. 2016): http://fac.utk.edu/wp-content/uploads/2016 / 03/DCP20_webversion.pdf

- Link to an instructional video accompanying the DCP 2.0 Manual:

Review questions

1. You run a craniometric data set in Fordisc and receive a 0.82 posterior probability and a 0.46 Typ R probability that the unknown is a white female. The classification matrix for the reference samples indicates that its overall correct classification rate for white, black, and Hispanic females is 85%. Comment on these probabilities, and explain how you would interpret the Fordisc results.

2. You run a craniometric data set in Fordisc and receive a 0.45 posterior probability and a 0.04 Typ R probability that the unknown is a Hispanic male. The classification matrix for the reference samples indicates that its overall correct classification rate for white, black, Asian, and Hispanic males is 62%. Comment on these probabilities, and explain how you would interpret the Fordisc results.

3. You run a craniometric data set in Fordisc and receive a 0.60 posterior probability and a 0.15 Typ R probability that the individual is a black male. The skull is relatively small, and several measurements are coming up as outliers with "--" and "---" marks for a few measurements in the "chk" column of the output. What is your interpretation of this output? What other analyses you might perform in Fordisc to double check the results from the craniometric data? Explain your reasoning.

4. A colleague from Poland has used the main Fordisc screen to classify several cold cases in his laboratory and has e-mailed you, reporting that the software returned spurious results. How do you explain his results, and what suggestions do you have for him going forward with the analyses?

5. You have provided law enforcement investigators with a stature prediction interval of 5'5"–5'10" for a missing white male. They believe that they have located a probable match for the description, but his driver's license says that he is 6'. How do you account for this discrepancy and explain to law enforcement that the driver's license stature does not rule out this person?

6. You are presenting on the utility of Fordisc at a local law enforcement training workshop, and one of the officers asks how it is possible to say that someone is from a certain ethnic group based on bone measurements. She also wants to know how we will be able to do this in the future, based on increasing numbers of interracial marriages and admixed children. How do you answer these questions?

Glossary

Generalized distance (see Mahalabonis distance): a unitless measure of the standard Euclidean distance between a point *x* and a data distribution that is based on the mean, variance, and covariance matrix of the variables.

Geometric mean: a value that indicates the central tendency or typical value of a set of numbers; the nth root of a product of *n* numbers.

Isometric: having equal dimensions or measurements; bones that are isometric increase in size at the same rate.

Least squares regression: a regression analysis that minimizes the sum of the squared deviations between the data and the model in order to estimate the unknown parameters.

Linear discriminant function analysis: a statistical analysis that uses the linear combination of one or more continuous independent variables to predict a categorical independent variable (i.e. assign it to a class).

Linear regression: a method for modeling the linear relationship between a dependent variable (x) and one or more independent variables (y) by finding the best fitting straight line through the points.

Mahalanobis distance (see generalized distance): a unitless measure of the standard Euclidean distance between a point x and a data distribution that is based on the mean, variance, and covariance matrix of the variables.

Secular change: biological changes that occur over a relatively short period of time (generations) that are thought to be the result of environmental factors (e.g., increasing stature with improvements in nutrition and healthcare).

References

Albanese J, and Saunders SR. 2006. Is it possible to escape racial typology in forensic identification? In E Schmitt, E Cunha, J Pinheiro (Eds.), *Forensic Anthropology and Medicine: Complementary Sciences from Recovery to Cause of Death*. Totowa, NJ: Humana Press, Inc. pp. 281–316.

Barnard MM. 1935. The secular variations of skull characters in four series of Egyptian skulls. *Ann Eugenics* 6:352–371.

Byers SN. 2008. *Introduction to Forensic Anthropology*. Boston, MA: Pearson/Allyn and Bacon.

Darroch JN, and Mosimann JE. 1985. Canonical and principal components of shape. *Biometrika* 72:241–252.

Elliott M, and Collard M. 2009. Fordisc and the determination of ancestry from cranial measurements. *Biol Lett* 5:849–852.

Fisher RA. 1936. The use of multiple measurements in taxonomic problems. *Ann Eugenics* 7:179–188.

Giles E, and Elliot O. 1962. Race identification from cranial measurement. *J Forensic Sci* 7:147–151.

Giles E, and Elliot O. 1963. Sex determination by discriminant function analysis of crania. *Am J Phys Anthropol* 21:53–68.

Giles E, and Hutchinson DL. 1991. Stature-related and age-related bias in self reported stature. *J Forensic Sci* 36(3):765–780.

Goodman AH. 1997. Bred in the bone. *Sci* 37:20–25.

Howells WW. 1973. *Cranial Variation in Man: A Study by Multivariate Analysis of Patterns of Difference among Recent Human Populations*. Papers of the Peabody Museum of Archaeology and Ethnology. Cambridge, MA: Harvard University.

Jantz RL, and Meadows Jantz L. 2000. Secular change in craniofacial morphology. *Am J Hum Biol* 12: 327–338.

Jantz RL, and Moore-Jansen PH. 1988. A data base for forensic anthropology. Report of Investigations No. 47, Department of Anthropology, Knoxville, TN: University of Tennessee.

Langley NR, Meadows Jantz L, Ousley SD, Jantz RL, and Milner G. 2016. *Data Collection Procedures for Forensic Skeletal Material 2.0*. Knoxville: University of Tennessee.

Lewontin RC. 1972. The apportionment of human diversity. *Evol Biol* 6:381–398.

Mahalanobis PC. 1936. On the generalized distance in statistics. *Proc Natl Inst Sci India* 2(1):49–55.

Moore-Jansen PH, Ousley SD, and Jantz RL. 1994. Data collection procedures for forensic skeletal material. Report of Investigation No. 48. Department of Anthropology, Knoxville, TN: The University of Tennessee.

Ousley SD. 1995. Should we estimate biological or forensic stature? *J Forensic Sci* 40:768–773.

Ousley SD, Jantz RL, and Freid D. 2009. Understanding race and human variation: Why forensic anthropologists are good at identifying race. *Am J Phys Anthropol* 139:68–76.

Parra EJ, Marcini A, Akey J, Martinson J, Batzer MA, Cooper R, and Forrester T et al. 1998. Estimating African American admixture proportions by use of population-specific alleles. *Am J Hum Genet* 63(6):1839–1851.

Penrose LS. 1954. Distance, size and shape. *Ann Eugenics* 18:337–343.

Pickering R, and Jantz RL. 1995. Look again before repatriating. *Fed Archaeol* 7:44.

Ramsthaler F, Kreutz K, and Verhoff MA. 2007. Accuracy of metric sex analysis of skeletal remains using Fordisc based on a recent skull collection. Int J Legal Med 121(6):477–482.

Rao CR. 1952. *Advanced Statistical Methods in Biometrical Research*. New York: John Wiley & Sons.

Siegel ND, and Ousley SD. 2011. The importance of testing and understanding statistical methods in the age of *Daubert*. Can Fordisc really classify individuals correctly only one percent of the time? *ProcAm Acad Forensic Sci* 17:364–365.

Sledzik PS, and Ousley S. 1991. Analysis of six Vietnamese trophy skulls. *J Forensic Sci* 36(2):520–530.

Smith CAB. 1980. Estimating genetic correlations. *Ann Hum Genet*. 43:265–284.

Snow CC, Hartman S, Giles E, and Young FA. 1979. Sex and race determination of crania by calipers and computer: A test of the Giles and Elliot discriminant functions in 52 forensic cases. *J Forensic Sci* 24(2):448–460.

Snow C, and Williams J. 1971. Variation in premortem statural measurements compared to statural estimates of skeletal remains. *J Forensic Sci* 16:455–464.

Spradley MK, and Jantz RL. 2011. Sex estimation in forensic anthropology: Skull versus postcranial elements. *J Forensic Sci* 56:289–296.

Thieme FP. 1957. Sex in Negro skeletons. *J Forensic Med* 4:72–81.

Thieme FP, and Schull WJ. 1957. Sex determination from the skeleton. *Hum Biol* 29:242–273.

Trotter M, and Gleser GC. 1952. Estimation of stature from long bones of American Whites and Negroes. *Am J Phys Anthropol* 10:463–514.

Willey P, and Falsetti T. 1991. Inaccuracy of height information on driver's licenses. *J Forensic Sci* 36(3):813–819.

Williams FE, Belcher RL, and Armelagos GJ. 2005. Forensic misclassification of ancient Nubian crania: Implications for assumptions about human variation. *Curr Anthropol* 46:340–346.

Section IV

Human Identification and Advanced Forensic Anthropology Applications

15 Time Since Death Estimation and Bone Weathering: The Postmortem Interval
16 Methods of Personal Identification
17 Mass Fatalities, Mass Graves, and the Forensic Investigation of International Crimes
18 Advanced Scene Topics—Fire and Commingling

Chapter 15

Time Since Death Estimation and Bone Weathering
The Postmortem Interval

Rebecca J. Wilson-Taylor and Angela M. Dautartas

CONTENTS

Decomposition	274
Variables affecting decomposition	278
Postmortem changes in the body	280
Fresh stage	281
Discoloration stage	281
Active decomposition stage (bloat)	283
Advanced decomposition (initial skeletonization) stage	284
Skeletonization stage	285
Determination of the postmortem interval	285
Forensic pathology	286
Forensic entomology	289
Entomotoxicology	290
Forensic botany	290
Diagenesis of bone	292
Chemistry	292
Morphoscopic techniques	293
Future directions	296
Decomposition research	297
Summary	301
Review questions	301
Exercises	302
Glossary	302
References	303
Suggested readings and additional resources	309

> **LEARNING OBJECTIVES**
> 1. Describe the processes of autolysis and putrefaction, and explain how they dictate the timing and extent of human decomposition.
> 2. List and describe five primary stages of decomposition.
> 3. Name the variables that affect decomposition, and explain how each of these can change the rate of decomposition.
> 4. Explain how adipocere forms.
> 5. Describe the conditions that cause mummification.
> 6. Name the methods available to estimate the postmortem interval and the scientific basis behind each method.
> 7. Describe bone diagenesis, and explain how this process enables forensic anthropologists to estimate postmortem interval.
> 8. Discuss the current state of decomposition research, and identify areas where further work is needed.

The **postmortem interval (PMI)**, or time since death, is the length of time for which an individual has been deceased until the point of discovery. An accurate and precise PMI estimation significantly contributes to forensic investigations, as it limits the possibilities of potential decedents, lending to a higher probability of identification. The PMI may also corroborate information associated with possible perpetrators and contribute to a better understanding of the circumstances surrounding a death. Understanding taphonomic processes associated with human decomposition and how environmental, cultural, or case-specific variables affect these processes is fundamental to determining a PMI. The variation inherent in human decomposition complicates precise and accurate PMI estimations and has led to extensive research and the development of multiple specialized research facilities dedicated to this effort. This chapter provides a detailed overview of the human decomposition process, research contributing to the current state of the field, and the various methods in use to estimate the PMI.

One of the first questions asked when human remains are discovered is for how long the individual has been dead. In human decomposition, the primary focus has been on determining the PMI or time since death (TSD). Many methods are available for determining the PMI for early decomposition, or in the time period shortly after death. The early stages of decomposition usually fall within the purview of forensic pathologists, because the body tissues are still present and intact. As the body deteriorates, a forensic anthropologist is typically employed in the determination of the PMI. However, the longer the PMI, the harder it becomes to estimate accurately and precisely. Several precipitating factors arise as the PMI increases, such as animal scavenging, depositional environment, and the presence of coverings. Establishing an accurate and precise PMI requires an understanding of the impact that these and other taphonomic factors may have on the sequential nature of human decomposition.

DECOMPOSITION

Human decomposition encompasses the physiochemical processes that degrade organic materials (hard and soft tissues) into their fundamental components. When the heart stops beating, cellular processes in the body do not cease immediately; these processes continue until the oxygen supply is depleted. Once oxygen is depleted, carbon dioxide levels increase and poison the cells, thereby causing cell death as pH (parts of hydrogen) levels within the cells decrease and the acidity within the body increases (Knight 2002). The increase in acidity spurs enzymes in the cell organelles, particularly lysosomes, to digest the proteins that maintain the integrity of cellular structures. The resulting cellular breakdown causes a series of events, leading to the eventual disintegration of the tissues. Certain organs (e.g., the bowel) break down faster because of their natural bacterial content, while other organs (e.g., skin) are more likely to mummify.

Two major processes dictate the timing and extent of human decomposition: autolysis and **putrefaction**. The combination of these two processes causes cells to die and tissues to break apart and leach enzymes, bacteria, and other microorganisms to feed on the now-unprotected body compartments.

Autolysis or "self digestion" involves a series of chemical reactions that break down cellular membranes and the junctions between cell membranes (Evans 1962; Gill-King 1997). The cell membrane is responsible for compartmentalizing the human body, so once these divisions are compromised, chemicals enzymes and proteins within each cell are released into the body cavities. The enzymes and proteins released as a result of the disintegration of the cell membrane cause tissue breakdown from within the cells, typically beginning in enzyme-rich organs such as the pancreas, stomach, and the liver (Gill-King 1997). An easily visible example of the breakdown of the cellular structure associated with autolysis is skin slippage. Skin slippage occurs when the epidermal and dermal layers of the skin separate, causing the more superficial epidermal layer to slough off of the body. Skin slippage is first noticeable when bullae, or blisters (Figure 15.1), form in the skin. Ultimately, skin slippage causes a "gloving" or "stocking" effect, such as the complete removal of the epidermis from the extremities (Clark et al. 1997; Figure 15.2). Head of the hair is also removed, because hair follicles, which are located in the dermis, are pulled out as the epidermis separates. The separation of the hair due to skin slippage produces a mass of hair or "hair mat" in the area that is immediately around the head, which indicates the early decomposition location of the head. This can be of great importance in contexts involving the movement of the body, because the hair mat indicates the primary decomposition area and is a great source to find missing teeth or other artifacts associated with the cranial region. Hair mats are also unlikely to be scattered by insect or animal activity, as neither group feeds on the hair.

Putrefaction follows autolysis and is highly correlated with ambient temperature and the health status of the individual. Putrefaction is the breakdown of tissues caused by microorganisms. The anaerobic environment produced by autolysis initiates putrefaction, since autolysis causes the release of microorganisms from within cells to inhabit the

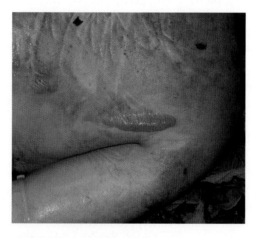

Figure 15.1 Bullae: separation of the dermal and epidermal layers of the skin, on the right posterior shoulder.

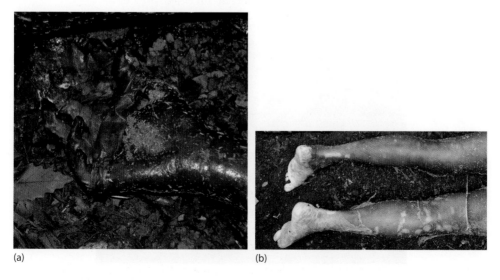

(a) (b)

Figure 15.2 (a) Gloving effect: skin slippage of the hand. (b) Stocking effect: skin slippage of the leg and foot.

circulatory system. Enteric microorganisms, specifically bacteria, released from body cells and transient microorganisms that access the body through portals of entry (e.g., respiratory and genital tracts) feed on the nutrients released by autolysis. Cells in the digestive tract break down the bacteria that assisted in extracting nutrients from food in living individuals and begin to feed on body tissues after death. Since bacteria do not need oxygen to survive, they can thrive in the anaerobic environment of a deceased body. Putrefaction is stimulated by a rapid drop in the body's pH level. In life, the body monitors and controls internal pH levels with buffering systems to avoid significant changes in pH levels. After death, the release of organic acids during the initial stages of decomposition creates a suitable environment for bacterial fermentation.

Putrefaction is responsible for the well-known sights and smells associated with human decomposition. Putrefaction is expressed initially as discoloration in the lower right quadrant of the abdomen, where the greatest amounts of enteric bacteria reside. As the bacteria feed, they produce gaseous by-products, including carbon dioxide, nitrogen, and methane, that give off distinct odors (Clark et al. 1997) and cause distention of the soft tissues in the immediate locale of bacteria, otherwise known as "bloating" (Clark et al. 1997). Bloating can occur in all locations of the body, including the extremities, but is more prominent in the abdomen and face (Figure 15.3). The abdomen has the highest accumulation of gases because of the uncontrolled enteric bacterial growth and the fact that it contains the most blood. Typically, areas with more blood are more distended because of these gases (Clark et al. 1997). For example, autopsied bodies display a higher degree of bloating in the extremities because of the lack of blood in the torso. The odorous by-products indicate that the body is in the state of putrefaction; these gases also aid in the PMI estimation (Vass et al. 2008). Furthermore, putrefaction is associated with the height of arthropod activity, specifically larval activity (Haskell et al. 1997; Hart and Whitaker 2006). This process is attractive for many animal scavengers, which can cause fragmentation and/or scattering of the remains (Haglund 1997 a, b; Klippel et al. 2005; Reeves 2009; Synstelien 2009).

The order in which soft tissues decompose differs between autolysis and putrefaction. In autolysis, organs that have the highest rate of ATP (adenosine triphosphate) production break down first. These include digestive and circulatory system organs, and their breakdown produces much of the characteristic color changes in the human body. The respiratory organs are next, followed by the kidneys and bladder of the urinary system. The brain, the skeletal muscles, and connective tissue are the last to decompose. See Tables 15.1 and 15.2 for the order of tissue decomposition due to autolysis and putrefaction, respectively.

After the anaerobic fermentation of putrefaction, remnants of decomposition in the form of **adipocere** are commonly found. Adipocere, or "grave wax," is a grayish-white substance that develops from hydrolysis or hydrogenation of lipids—a process known as **saponification** (Mant and Furbank 1957; Ubelaker and Zarenko 2011; Figure 15.4). There are two major forms of adipocere: a hard, crumbly variety produced quickly as sodium reacts with interstitial fluid and a soft paste-like variety produced slowly as potassium interacts with adipose tissue cell membranes (Vass et al. 2002). Because adipocere is developed from the body's own substances, human adipocere has a unique volatile fatty acid (VFA)

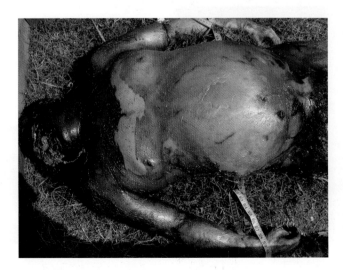

Figure 15.3 Bloating: gaseous by-products of microorganisms cause swelling of associated tissue.

Table 15.1 Order of tissue decomposition from autolysis

Order	Tissue
1	Digestive organs, blood, and circulatory system
2	Air passages and lungs
3	Kidneys and bladder
4	Brain and nervous tissue
5	Skeletal muscle
6	Connective tissue

Table 15.2 Order of tissue decomposition from putrefaction

Order	Tissue
1	Air passages and lungs
2	Digestive organs, heart, and blood
3	Brain and nervous tissue
4	Kidneys and bladder
5	Skeletal muscle
6	Connective tissue

Figure 15.4 Adipocere: a decomposition by-product often seen in association with advanced decomposition, mummified, or mostly skeletal remains.

composition as compared with other animals (Vass et al. 2002). Adipocere formation is prevalent in warm, moist, and alkaline conditions, be it natural or man-made (Forbes et al. 2005a). For example, two conditions that promote saponification are high humidity (natural) and a burial environment (typically man-made) (Mant and Furbank 1957; Mann et al. 1990; Forbes et al. 2005b, c; Ubelaker and Zarenko 2011). Also, adipocere formation can occur rapidly in submerged bodies, especially if the water is warm (O'Brien 1997). Obese individuals, females, and infants are more likely to undergo saponification because of the increased levels of adipose tissue in their bodies (Evans 1962; Gill-King 1997).

Suitable environmental conditions may also produce **mummification** or **desiccation**. Body tissues may dry out from natural processes, chemical applications, or culture practices. Most desiccation is considered partial mummification and is the drying out of the connective tissues (e.g., ligaments and tendons) and of the overlying skin. High heat, low humidity, and adequate airflow are ideal conditions for mummification. For example, a hanging context encourages mummification, because air can circulate around the entire body. Typically, mummification begins with the drying of the face, hands, and feet but can include the whole body. The drying process produces brown to golden-brown discoloration and shriveling or wrinkling of the skin. Mummification is more common during cooler months, even in the Southeastern United States, as well as in more arid climatic zones of the American Southwest. In the dry southwestern regions of the United States, the internal organs undergo rapid putrefaction, but the more superficial connective tissue mummifies. It is not uncommon to retrieve small pieces of mummified tissue with a skeleton in most temperate climates.

VARIABLES AFFECTING DECOMPOSITION

Taphonomic processes that can affect the decomposition rate of a cadaver can be classified into environmental/meteorological, individual, and cultural/case-specific factors (Table 15.3). Environmental factors can be divided further into those caused by living organisms (**biotic**) and those caused by nonliving organisms (**abiotic**). Root etchings on bone, maggot holes in skin, and carnivore gnawing are all biotic alterations to remains. Abiotic environmental factors are created by the nonliving forces in the environment such as the effects of temperature, water, and soil composition. Individual taphonomic factors are the effects that the remains bring to the decomposition process, such as health, body size, and body composition. Cultural factors are those that another person induces to a cadaver, such as embalming, burial, coverings, and trauma (Calce and Rogers 2007; Dautartas 2009; Rodriguez and Bass 1985). All these factors play a role in developing the context in which a cadaver is found and the rate at which the body decomposes in this context. There is much debate about the role that many of these factors play in the decomposition process, because even though a plethora of factors affect decomposition, the significance of each varies depending on the case (Adlam and Simmons 2007; Simmons et al. 2010). For example, a decedent's metabolic state can affect the starting temperature for the decomposition process (Spitz 2006). However, regardless of the debate, several key parameters will affect the rate of decomposition: temperature, humidity, anthropod activity, animal scavenging, and trauma (Mann et al. 1990).

The single most important criterion for decomposition, regardless of the scenario, is temperature (Mann et al. 1990; Megyesi et al. 2005; Simmons et al. 2010). Nawrocki (2013) found that temperature accounts for 80% of the variability in soft tissue decomposition. The ambient temperature around a corpse depends on the altitude, burial depth, presence of water, air movement around the body, and surrounding foliage, to name a few variables. In general, temperature dictates the speed at which chemical reactions occur. Van't Hoff's rule, "the rule of ten," states that the rate of chemical reactions increases two or more times with each 10°C increase in temperature (Gill-King 1997). When applied to decomposition rates, this suggests that the enzymes involved in decomposition work more quickly to break down tissues at higher temperatures, as opposed to lower temperatures. The amount of arthropod activity associated with a decomposing corpse is directly proportional to the temperature: the warmer the climate, the greater the insect activity (Haskell et al. 1997). The opposite can be said for cooler climates; because flies tend to be inactive below 50°F, oviposition, or egg laying, is decreased in these conditions and bacterial activity decreases (Micozzi 1991). It is important

Table 15.3 Factors affecting decomposition rates (from greatest effect to lowest effect)

Factor	Types of factor
Temperature	Environmental factors
Access by insects	Depositional factors
Burial depth	Depositional factors
Scavenging	Depositional factors
Trauma	Culturally mediated case-specific factors
Humidity level	Environmental factors
Body size	Culturally mediated case-specific factors
Embalming—chemical treatment of remains	Culturally mediated case-specific factors
Clothing	Culturally mediated case-specific factors
Deposition substrate	Environmental factors

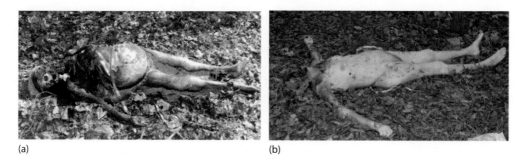

Figure 15.5 (a) Temperature is one of the most important variables affecting the rate of decomposition. A cadaver with 14-day exposure in the summer/warm months. (b) A cadaver with 14-day exposure in the winter/colder months.

to note that, even though bacteria are known to be most active in warm conditions, bacterial activity still occurs at temperatures below 4°C, but little is known as to why (Vass et al. 2002). Figure 15.5 shows winter versus summer decomposition with 14 days of exposure.

Other environmental factors that affect the decomposition process include humidity, amount of rainfall, and available water, all of which affect the decomposition rate of soft tissue decomposition as well as the subsequent **diagenesis** of the skeleton and likelihood of adipocere formation. (Behrensmeyer 1978; Ubelaker 1997). Humidity is a significant factor in that it produces a moist environment ideal for soft tissue decomposition and prolongs favorable conditions for flies, because it slows the drying of skin and connective tissues. This is the main reason why corpses decomposing in arid climates undergo mummification and have less arthropod activity (Galloway et al. 1989).

Likewise, rainfall contributes significantly to biomass reduction. An increase in larval activity is common in light rain or just after a rainfall, because rain rehydrates dried tissues. At the Anthropology Research Facility (ARF) at the University of Tennessee in Knoxville, rapid decomposition changes have been observed following a rain ending a prolonged dry period. However, little fly activity is noted in periods of heavy rain; these conditions can result in a reduction of larval activity, as the flies do not have the ability to lay eggs.

Available water near and around the body also plays an important role in decomposition. Water contributes to relative humidity; corpses contain a certain amount of water; and water can act as a locale for the deposition of a body. Water is an excellent solvent and heat regulator. Water plays a role in diluting concentrations of chemicals, which moderates the pH in soil environments and on body surfaces (Gill-King 1997). Also, water has a high specific heat, which helps stabilize temperatures. For example, if a warm body is submerged in water, decomposition will be slower than if the same body were placed on the ground surface, because the surrounding water will absorb the heat rather than aiding in dissipating that heat. Seet (2005) evaluated submerged bodies through a review of 69 cases from Eastern Tennessee, highlighting that the water temperature, depth of water, water chemistry, salinity, and aquatic life play significant roles in the decomposition of a body. Micozzi (1991) found that the partial or total submergence of a body can slow or speed decomposition, depending on the chemical composition (pH) of the water, whether the water is salty or fresh and whether a current is present. Also, the presence of water may increase saponification. More recently, Christensen and Myers (2011) suggested that pH levels in water may affect the integrity of bone; low and high pH extremes can degrade bone with lower pH levels, causing more degradation at a faster rate.

Depositional factors include the pH of the surrounding soil, which is of great importance to a burial environment. Soil pH can greatly affect the rate of decomposition and the decomposer communities found in soils (Neher et al. 2003; Tibbett et al. 2004). Typically, acidic soils (pH 3.0–5.5) are dominated by fungal communities. The decomposition process proceeds up to three times faster in acidic soil compared with alkaline soil if the rate of soft tissue loss is correlated with the measured microbial respiration in the soil (Haslam and Tibbett 2009). Neutral soils (pH 5.5–7.5) provide conditions in which bacteria have a competitive advantage. In the case of alkaline soils (pH 7.5–9.0), fungi may dominate again, especially in the conditions created by the putrefaction process (Carter and Tibbett 2003, 2006; Carter et al. 2007; Sagara et al. 2008). The pH of the surrounding soil dictates the maintenance of plant life and microbial activity in the soil, which can impact the decomposition of both soft and hard tissues.

The surrounding environment (soil) is typically more alkaline than a body, so fluids leaching from the body tend to shift the soil pH, which then causes plant death and a shift in soil **microbes**' effectiveness. This leaching of fluids from the body produces the decomposition stain, which is an indicator of the primary deposition and decomposition location. In cases of long-term exposure or burial, the soil pH also plays a role in the demineralization of bone. Acidic soils attack the inorganic mineral components of bone that help maintain bone integrity; thus, an acidic environment deteriorates bone quality.

Depositional factors also include the presence of structures limiting exposure to the elements or physical barriers that protect the body, such as clothing (Ritchie 2005; Shean et al. 1993; Srnka 2003; Ubelaker 1997). In addition, placement of the corpse can affect exposure to environmental factors, access to oxygen needed to drive chemical reactions, and accessibility to scavengers (Mann et al. 1990; Shattuck 2009). Specifically, scavenger activity can cause differential decomposition, movement, and loss of elements, regardless of the locale. Several studies have demonstrated the effects of different scavengers in a variety of deposition environments (Carson et al. 2000; Dabbs and Martin 2013; Haglund 1997; Klippel and Synstelien 2007; Reeves 2009; Sorg et al. 1997; Smith 2015). Local differences between sunlight and shade can shift temperature and suitability for arthropods (Srnka 2003). Hence, proximity to the sun or other heating elements can affect bodies in the same environment significantly. A corpse located within a climate-controlled structure, such as a house, will exhibit a more gradual systematic progression of decomposition stages because of stable temperatures and environmental conditions (Ritchie 2005). Burial depth can also play a major role in limiting arthropod activity and lowering temperatures (Rodriquez and Bass 1985) or the oxygen needed to drive the chemical reactions that break down tissues (Mann et al. 1990). When the burial depth exceeds 2 feet, temperature, arthropod activity, and oxygen levels decrease and decomposition rates are slower (Rodriquez and Bass 1985). The slower rate of decomposition is a direct result of anaerobic degradation, which is slower than degradation in an aerobic environment. Also, when a physical barrier prevents scavenger access to a corpse, biomass reduction from animal feeding is inhibited and any decrease in biomass is directly related to the actual break down of tissues. A rule of thumb in the pathological literature is that 1 week on the surface equals 2 weeks in water and 8 weeks in a burial environment (Spitz 2006).

Case-specific variables include the presence of trauma or bonding agents (e.g., rope and tape) (Haskell et al. 1997). Also, the bruising and rupturing of tissues (Calce and Roger 2007; Mann et al. 1990) caused by trauma provide additional openings for arthropod colonization, thus accelerating decomposition. Haskell et al. (1997) indicate that any artificial opening (surgery, trauma, or autopsy) in the body enhances insect activity in the region affected and speeds up the reduction of biomass in that area. This can cause **differential decomposition** in which one area of the body shows an unusual rate of decomposition. For example, raccoons have been observed creating artificial openings in the extremities of bodies and then feeding on the larvae that enter this area, which in turn causes these areas to become skeletal more quickly than would be expected for the actual PMI (Synstelien 2009). Some suggest that body size affects the decomposition process, because more material available for larval feeding increases their presence at the corpse, while others maintain that body size is not a major contributing factor to the decomposition process (Parmenter and MacMahon 2009; Simmons et al. 2010; Vass et al. 2008). However, based on personal observations at the ARF, body size does not cause significant differences in the rate at which a corpse decomposes. It plays a larger role in the amount of biomass reduced and in the amount of adipocere produced, as heavier individuals tend to have more adipose tissue. In warm, humid conditions, adipose tissue is converted to adipocere through the process of saponification, so higher amounts of adipose tissue result in more adipocere formation.

POSTMORTEM CHANGES IN THE BODY

Macroscopic postmortem changes (i.e., those visible to the investigator) follow a fairly regular pattern, regardless of the contextual environment. However, the *rate* at which these changes occur varies with the context. Postmortem changes to the body are often grouped according to when they appear. This grouping allows researchers to organize events into specific decomposition stages. Several categorization methods have been developed and are currently being tested; however, anthropologists have generally agreed on 4–5 surface decomposition stages, with a few researchers having more elaborate schemes (Bass 1997; Clark et al. 1997; Galloway et al. 1989; Marks et al. 2000, 2009; Payne 1965). All these stages focus on decomposition being a fairly linear progression of events driven by autolysis and putrefaction.

A five-stage sequence using gross anatomical changes is the most comprehensive and parsimonious (following Marks et al. 2009; Reed 1958 and outlined in Table 15.4). Other systems have been developed based on the biochemical

Time Since Death Estimation and Bone Weathering

Table 15.4 Stages of decomposition following Reed (1958) and Marks et al. (2009) and intervals following Clark et al. (1997)

Decomposition stage	Characteristics	Time interval
Fresh	Algor mortis, livor mortis, rigor mortis	1 day to approximately 1 week
Discoloration	Marbling, abdomen slight green, skin slippage	Day 1 to weeks
Bloat (active decomposition)	Distention of tissues, overall color changes, leaching of VFAs	Starts within 48 hours
Advanced decomposition (initial skeletonization)	Purging of body fluids, exposure of skeletal elements	Starts at least 1 week after death
Skeletonization	Exposure and drying of the entire skeleton	Several weeks to months after death

and entomological changes associated with a human body (Payne 1965; Tibbett and Carter 2008; Vass et al. 1992). In addition, stage sequences have been developed for buried human remains (Payne et al. 1968; Tibbett and Carter 2008). Nevertheless, surface decomposition stages include the fresh stage, the discoloration stage, the bloat or active decomposition stage, advanced decomposition or initial skeletonization, and skeletonization. All bodies generally move through these five stages (except in the case of significant animal scavenging), but the rate at which they move from one stage to the next may vary. The temperature of a specific environment largely determines the duration of a particular stage. Traditionally, forensic anthropologists have categorized the amount of decomposition observed into decomposition stages to provide a broad PMI estimate.

Fresh stage

The fresh stage is associated with **algor mortis**, **livor mortis**, and **rigor mortis** (i.e., the cooling of the body after death, the pooling of blood due to gravity, and muscle stiffening; Knight 2002). During the fresh stage, PMI techniques established by pathologists are the most reliable (e.g., body temperature). Also, initial oviposition by flies in the Calliphoridae family occurs during the fresh stage (Bass 1997). There are minimal biochemical changes to the environment surrounding the corpse, and a full standard autopsy and identification via photos are still feasible during this stage (Figure 15.6).

Discoloration stage

The discoloration stage is a highly variable stage. In warmer climates and conditions, it can be quite rapid in appearance and progression, while in cooler conditions, it is prolonged and is associated with microbial growth (mold and/or fungi)

Figure 15.6 Fresh decomposition stage characterized by general loss of color of the entire body.

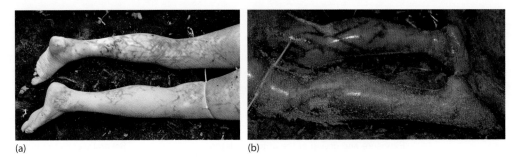

Figure 15.7 (a) Early appearance of marbling, streaking effect caused by breakdown of the circulatory system. (b) Marbling can occur and be visible throughout active decomposition.

on the skin of a corpse (Bass 1997; Tibbet and Carter 2008). Discoloration is predominantly a side effect of the decomposition of blood and associated tissues, which allows hemoglobin to interact with sulfide. This interaction causes the formation of sulfhemoglobin in the cutaneous vessels, producing a greenish color in the outer layers of skin, especially in the abdominal region (Gill-King 1997). In dry conditions, cold or warm, the face, hands, and feet exhibit a yellow-maize color. Also, the lips and eyelids recede and turn to a golden-brown color. It is common to see the yellow-maize discoloration, in conjunction with some drying of the skin, when an individual is found within a couple of days of death in a home in which the heat is turned up and the temperature is quite warm. A more expected phenomenon during the discoloration phase is **marbling**; however, this can begin as early as the fresh stage.

Marbling is a product of the breakdown of tissue and the contamination of blood within the circulatory system (Figure 15.7). Marbling is the "state of being veined like a marble" (Agnew et al. 1965: 874). It is more apparent in individuals with less subcutaneous fat, because the circulatory vessels are more visible. The various colors observed in marbling correspond to different chemical reactions occurring between hemoglobin and the now-free chemicals. For example, deoxygenated hemoglobin produces a purplish color that looks blue when viewed through the skin. When hemoglobin interacts with sulphur, a green color is produced. Most marbling is a direct result of the degradation of hemoglobin and the conversion of heme into a series of bile pigments, which account for the various colors seen in a decomposing body. The actual color expressed depends on the oxidative state of the bile (Gill-King 1997). Biliverdin will appear as a green precipitate in the skin; bilirubin will be red; and urobilin will appear brown and is associated with an increased acidic environment, so it is commonly produced later in decomposition (Gill-King 1997). The onset of marbling is determined by temperature and body composition. The appearance of marbling is brief at warmer temperatures, in which it is more common to get overall body color changes (Figure 15.8a and b). Colder temperatures, on the other hand, prolong the appearance of marbling and

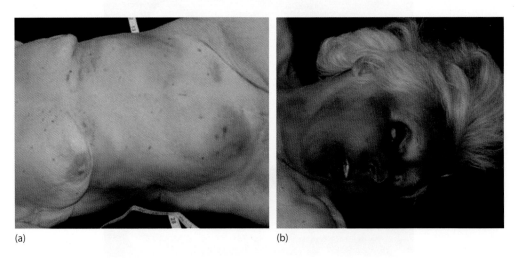

Figure 15.8 (a) Initial discoloration of the body occurs in abdomen and is usually a green-blue color. (b) Discoloration can also be seen in the hands and face in which there is an orange to golden-brown appearance due to the drying of tissue.

Figure 15.9 Autolysis, the breakdown of cells, causes the layers of the skin to separate and appears as "dishpan hands" where there is a wrinkling of the epidermal layer of the skin.

highlight its appearance. A body that has been frozen and is thawing will also demonstrate extensive brilliant marbling across the skin.

Another observation associated with the discoloration stage is the initial leaching of VFAs, or decomposition fluid, from the body, usually through the mouth and facial area. The first signs of skin slippage are also frequently observed. Pruning or wrinkling of the hands and/or feet and bullae formation on larger areas of the body are diagnostic features of autolysis, which drives the skin slippage (Figure 15.9).

Active decomposition stage (bloat)

Active decomposition is a direct result of autolysis and putrefaction and is typically associated with **bloating** (Figure 15.10a and b). This stage also marks the height of blowfly larval activity, early signs of beetle activity, and increased animal scavenging. Also associated with active decomposition is the increased release of decomposition fluid (volatile fatty acids [VFAs]) from the body, causing discoloration of the surrounding vegetation, a change in the soil pH, and an increase in nutrients available for environmental microbes (Vass et al. 1992). Decomposition fluid, primarily composed of VFAs, is the direct product of both autolysis and putrefaction, breaking down the body tissues. It is specifically formed from the decomposition of fat and muscle and is extremely important in active decomposition, until the point at which **skeletonization** is achieved. The purging and leaching of decomposition fluids from the body produce the diagnostic soil stain associated with a corpse, as the pH of the fluid interacts with

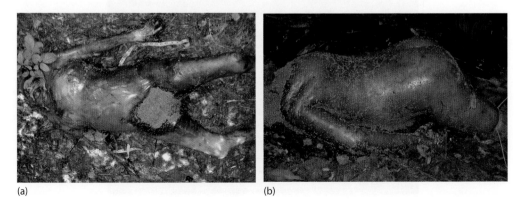

(a)   (b)

Figure 15.10 Active decomposition is associated with an overall color change that is golden to dark brown, bloating, and an abundance of larvae.

the pH of the surrounding soil. The change in pH in the soil causes a darkening, or burnt look, of the surrounding vegetation and leads to the eventual death of the vegetation, as the optimal soil pH necessary to support certain plants is compromised. The soil stain can remain for the better part of 1 year or a full cycle of seasons before renewed plant growth ensues and the stain dissipates. Skin slippage, discoloration, and bacterial growth continue throughout active decomposition, causing a major reduction in the cadaver's biomass.

Advanced decomposition (initial skeletonization) stage

Advanced decomposition (initial skeletonization) involves the purging of decomposition fluid through the natural body orifices (mouth, nose, rectum, and genital areas), causing a deflation of the body and wrinkling/crevice formation in the remaining skin, as well as vegetation death around the body (Figure 15.11). In addition, this stage causes the most significant biochemical changes in the surrounding environment because of the large amount of VFAs entering the soil (Tibbett and Carter 2008). Initial exposure of the skeleton occurs, especially in areas of the body with the least amount of soft tissue and the most amount of blowfly activity (e.g., the cranium). There is a major reduction in the amount of blowfly larvae activity, as well as migration of blowfly larvae away from the body. Advanced decomposition has been found to be less attractive to scavengers such as the raccoon (Smith 2015). The later aspects of this stage produce an almost completely exposed skeleton, with some dried tissue remaining (Figure 15.12).

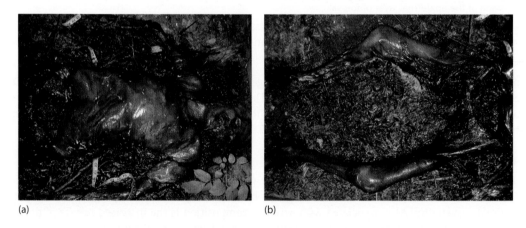

Figure 15.11 The cadaver fluids produced by the decomposition process will leach into the surrounding substrate, changing the chemical composition of the soil and causing the characteristic decomposition stain.

Figure 15.12 Advanced decomposition is marked by a deflation of the body, a darkening and drying of the outside tissue, and a purging of the cadaver fluids into the surrounding substrate.

Table 15.5 Bone weathering stages

Stage	Description
0	Bone is greasy and shows no sign of cracking or flaking; tissue may adhere to or cover the bone.
1	Bone exhibits cracking (longitudinal for long bone or mosaic for articular surfaces); residual soft tissue may be present.
2	Outer layer of cortex displays cracking and flaking; residual soft tissue may be present.
3	Outer layer of bone has roughened, fibrous patches where all cortical bone is missing; residual tissue is rarely seen.
4	Most of the outer bone surfaces is rough and fibrous, with no cortex remaining; cracks are open and have splintered.
5	Poor bone integrity, falling apart, extremely fragile, possible spongy bone exposure, ill-defined bony shape.

Source: Behrensmeyer, A.K., *Paleobiology* 4, 1978; p. 151.

Skeletonization stage

The last and longest lasting phase of decomposition is the skeletal phase. This phase is marked by a nearly complete to complete skeleton and the subsequent breakdown of the skeletal elements. Entrance into this phase is typically marked by at least 50% of the skeleton being exposed. Once skeletonized, there is nothing preventing the disarticulation, movement, or dispersal of remains across the landscape in an open environment, often resulting in the recovery of incomplete remains. Even though soft tissue decomposition ceases, diagenesis continues (e.g., the chemical, physical, and biological changes to teeth and bone). The depositional environment influences the extent of diagenesis, as bone is affected by the biochemical properties of decomposition, the depositional substrate, the local microbial activity, and plant growth (Nielson-Marsh and Hedges 2000). The most common diagenetic changes observed in bone are due to bone weathering. Bone weathering is the drying, bleaching, exfoliation/removal of the cortical surfaces, and demineralization of bone caused by sun exposure, seasonal cycles, movement, and individual composition of specific elements (Behrensmeyer 1978). The presence of these features and the extent to which elements are affected can indicate the length of time of exposure to the physical environment, especially long-term exposure. Behrensmeyer (1978) developed stages of bone weathering that have become the standard for these types of analyses (see Table 15.5). Other factors that come into play include additional scattering from animal activity and bone loss from scavengers (Haglund 1993; Klippel and Synstelien 2007; Reeves 2009; Synstelien 2009). Small rodents, specifically the gray squirrel, show a preference for dry bone that has been exposed to the environment for at least 9 months to 1 year; this rodent is one of the most significant contributors to long-term bone loss in forensic contexts (Klippel and Synstelien 2007). Bone staining is often observed once remains are skeletonized. Staining of bone can be attributed to decomposition fluids, substrate, metallic objects, and plant growth. Most staining or lack thereof is a result of contact or absence of contact with a particular substrate. The porous nature of bone allows it to absorb materials and, thus, the color from the surrounding environment. As a result, bone takes on the same or a similar color as the soil substrate. When in contact with other materials within the environment, such as certain metals (e.g., iron and copper), bone can exhibit orange-red or green staining. Also, in damp conditions, algae growth may cause green staining, much like that caused by copper. As the skeletal stage proceeds, the biochemical nutrients that were leached from the corpse in preceding stages diminish (Tibbett and Carter 2008) and plant growth is renewed after extended periods of time. Roots and other vegetation may grow within and around the bone, causing further breakdown of the skeleton (Figures 15.13 and 15.14).

DETERMINATION OF THE POSTMORTEM INTERVAL

Various methods are available to the forensic community for estimating a PMI for a deceased individual. All methods are not equally reliable, and some methods have been outdated and replaced by more recent and reliable techniques (Box 15.1). Several methods that have been used to determine a PMI include forensic entomology (Anderson 2001; Higley and Haskell 2001; Rodriguez and Bass 1983; Schoenly et al. 2007), biochemical fluid and tissue analyses (Vass et al. 1992, 2002), aroma (Vass et al. 2008), biochemical analyses of soil and decomposition substrates (Tibbett and Carter 2008; Vass et al. 1992), botanical studies (Willey and Heilman 1987), and body appearance (Bass 1997). Significant advancements have been made in PMI estimation by using biochemistry, forensic entomology, and, more recently, soil chemistry (Schoenly et al. 2007; Tibbet and Carter 2008; Vass et al. 2002). Anthropologists commonly rely on gross morphological changes to the body to determine the PMI; little advancement in statistical applications of this method has been made to date.

Figure 15.13 (a) Skeletonized remains will typically reflect the surrounding substrate by being stained in a similar color and will also show signs of wear and tear from the environment. (b) Long-term exposure of the skeleton will cause a bleaching/whitening of the bone, a cracking of cortical bone, and renewing vegetation growth in and around the elements.

Figure 15.14 The multiyear exposure of elements allows vegetation to grow through the bone, in addition to the bleaching and wearing away of the bone surfaces.

Forensic pathology

Early postmortem changes in a corpse typically represent the first several hours after death and are associated with loss of color, relaxation of sphincters and associated fecal soiling, and purging of fluids from the stomach and/or respiratory tract (Clark et al. 1997). Several methods are used to determine the PMI during this early postmortem period, where the body is still in a flaccid or limp state (Swift 2010). These methods range from measuring

> **BOX 15.1 THE CASE OF COLONEL SHY**
>
> In 1977, Dr. William M. Bass assisted in the recovery and analysis of a disturbed burial in Franklin, Tennessee. The burial was located in a family cemetery on the Shy property. Dr. Bass meticulously recovered all the skeletal elements from the burial, noting a strong odor. Decomposing remains typically have a pungent, acrid odor that is quite discernable. The body had decayed to the point that it was no longer articulated in anatomical order. The lower and upper extremities were not attached to the torso, and the skull was not present, but pink tissue was still present on the femora. Initial inspection of the remains suggested an intrusive burial, in which a secondary burial was placed above a cast iron coffin dating to the Civil War era, and eventual collapse of the coffin had apparently caused intermingling of the remains. A biological profile and a tentative PMI were assessed at the scene, indicating 6–12 months based on the odor and pink tissue. Dr. Bass based the PMI on that fact that prior burial recoveries in Tennessee from the Civil War period were poorly preserved or had a very dark soil staining. Neither of these characteristics was exhibited by the well-preserved remains from the cast iron coffin.
>
> The remains were brought to Knoxville, Tennessee, for further analysis, and Dr. Bass realized that the remains had been treated with embalming fluid. This led to further inspection of the coffin and the recovery of the fragmentary skull consisting of 17 separate bone fragments. The combination of the fragmentary skull, the State Crime Laboratory's assessment that all of the clothing was cotton, and the presence of embalmed tissue led Dr. Bass to conclude that "though my determination of age at death was entirely accurate, my estimate of length of time since death had been off by 112 years."
>
> William Shy was a colonel in the Confederate military until his death on December 16, 1864. He was killed by a Rebel attack during the battle of Compton's Hill at Nashville, Tennessee. He was killed by a Minie ball hit to the head at the age 26 years, and he was returned to his family for burial. The presence of pink tissue in the casket after so many years is a result of the unique environment that is created by cast iron coffins. These coffins create an oxygen-deprived environment that significantly retards the decomposition process. These conditions, combined with the embalming of Colonel Shy's body, preserved the remains more than Dr. Bass had initially expected.
>
> As a result of this case, Dr. Bass has sought to address the lack of understanding of human decomposition and the general dearth of literature available for forensic anthropologists to consult. It also led to the founding the Anthropology Research Facility (ARF) at the University of Tennessee in Knoxville, Tennessee (Figure 15.15).

directly observable physiochemical changes (i.e., taking rectal or core temperatures and determining lividity status; Kaliszan et al. 2009) to evaluating the chemical composition of the body fluids (such as the change in potassium in the vitreous fluid in the eye; Spitz 2006) to studying the residual reactivity of muscles to electrical stimuli (Henssge et al. 1995). A series of subsequent postmortem changes occurs within the first 48 hours after death. Traditionally, pathologists ascertain the PMI by using three soft tissue indicators: algor mortis, livor mortis, and rigor mortis (DiMaio and Dana 2006).

Algor mortis is the cooling of the body from the average body temperature (98.6°F) to the ambient/atmospheric temperature, typically a 1.5°F reduction in body temperature every hour in temperate climates, while in tropical climates, the temperature change is only 0.75°F (Spitz 2006). However, research has indicated that the decrease in body temperature is more of a sigmoid curve in which there are stepped plateaus (Henssge et al. 1995). Body temperature decreases initially by 2.0°F–2.5°F and then slows to 1.5°F. A higher temperature can remain for 3–4 hours postmortem as a result of residual metabolic activity in the body (Hutchins 1985), possibly as the result of illness. The most common method of determining body temperature is to take an inner core temperature rather than a surface temperature (Kaliszan et al. 2009). The visceral organs will have a slower, more regular decline in temperature, whereas the decline in skin temperature is quite variable. Typically, the skin cools faster, because it is one of the main pathways of heat loss for the human body. Core temperatures can be taken in the brain, liver, or the preferred rectal areas. Knight (1988) provides a thorough review of postmortem cooling, in which he acknowledges that the overall accuracy of methods used to determine the PMI from body temperature is questionable, despite the extensive research that has been conducted on the topic. Quite simply, numerous variables influence the process by which a body cools.

Livor mortis, or lividity, is the pooling or settling of liquid blood in the body to the lowest available parts of the body, owing to a combination of gravity and the cessation of blood flow. Lividity is linked with the positioning

Figure 15.15 Photograph of Colonel William Shy, who inspired Dr. Bass' development of the ARF.

of the body at death, which can provide important clues during an investigation. For example, hanging victims typically have increased lividity in forearms and lower legs compared with the rest of the body. If an area of the body is pressed against an object (such as a floor and a bed), a pale area will form, which is often referred to as **contact pallor**. Livor mortis is visible in the form of a red-purple color within 2 hours postmortem, as the quantity of deoxyhemoglobin increases in the body. Within 5 hours postmortem, there is a combination of fixed and unfixed lividity, where the body presents a mixture of areas that are permanently red-purple in color and areas that blanch if pressure is applied. Lividity is usually fixed within 8–12 hours after death, which means that the blood will not redistribute when the body is moved. In later stages of lividity, there can be splattering or splotching of the lividity, known as Tardieu spots (Spitz 2006). Once lividity is completely fixed, its usefulness in PMI estimation decreases (Figure 15.16).

Rigor mortis is the stiffening of the muscles that occurs after the initial period of flaccidity. The increase in muscle rigidity can start 1 hour postmortem in moderate temperatures and can peak after about 12 hours, or it can last for another 10–12 hours. Smaller muscles with the highest metabolic activity fix before larger muscles. This phenomenon explains why the eyelids may remain open, whereas gluteal muscles are still flaccid/soft. Rigor mortis occurs because the metabolic process of muscle contraction continues but there is no ATP (adenosine triphosphate) available to reverse the process and relax the muscles. In life, muscle contractions are initiated by the targeted release of calcium ions (Ca+) from reservoirs. Ca+ initiates muscle contraction by freeing the proteins, that is, actin and myosin, in muscle filaments to allow them to bind. Actin and myosin binding causes muscle fibers to shorten and thus contract. As the cell membrane of muscle cells breaks down after death, Ca+ is released from the cells, causing muscle contraction. Because of the lack of oxygen in a deceased individual, this process cannot be reversed, so actin and myosin remain joined. Oxygen is necessary in cellular respiration to form ATP, and ATP is necessary to break the bond between actin and myosin and to recapture the Ca+ into reservoirs. Therefore, the lack of oxygen, and thus ATP, causes the muscles to remain in the contracted state until further breakdown of the proteins occurs (Gill-King 1997). Consequently, higher ambient temperatures shorten the time that a body spends in rigor mortis.

Although a number of methods are available for determining the PMI in recently deceased, once a body moves beyond the late postmortem stage, as defined by Clark et al. (1997), other approaches should be utilized and investigated.

Figure 15.16 Lividity, caused by a settling of the blood within the body, causes a pink discoloration of the surface and is used as an early indicator in estimating the time since death.

Traditionally, these methods have been based on the sequential nature of observed gross anatomical changes (Bass 1984, 1997; Galloway et al. 1989; Payne 1965; Reed 1958). Unfortunately, observation-based techniques are highly qualitative and subjective, because they rely on the anthropologists' background and experience. There is a need in the anthropological community to devise a more scientifically rigorous approach to PMI estimation. A few mathematical models have been employed for PMI estimation that include variables, such as health, cause of death, temperature, level of ventilation and air humidity, presence of clothing, and access of the body by insects (Campobasso et al. 2001). Other authors suggest temperature-based methods (Megyesi et al. 2005; Simmons et al. 2010) or body fluid chemistry changes that provide a 95% statistical confidence interval, depending on the PMI, availability of equipment, circumstances of the death, and environment (Clark et al. 1997; Campobasso et al. 2001; Vass et al. 2002).

Forensic entomology

The forensic entomological community has developed useful ways to estimate PMI based on the developmental sequence of forensically significant arthropods commonly associated with a decomposing corpse (Goff 1993; Payne et al. 1968; Tabor Kreitlow 2010). Insects and other invertebrates feeding on and around a cadaver form a distinct faunal succession (i.e., the patterned order that distinct arthropods are found on carrion) associated with the various stages of decay. These arthropods fall into several distinct groups: **necrophagous** species, predators and parasites of the necrophagous species, omnivorous species, adventive species, and accidental species (Byrd and Castner 2010; Smith 1986).

Flies are most often the first arthropods observed on carrion and are predominantly used in PMI estimations (Haskell et al. 1997; Kulshrestha and Chandra 1987; Rodriguez and Bass 1983; Shean et al. 1993). More specifically, flies from the family Calliphoridae, otherwise known as the blowflies, have an inherent ability to detect and colonize a cadaver soon after death, making them extremely useful for estimating PMI. In warm temperatures, Diptera will be on the body within minutes of deposition and begin oviposition (i.e., egg-laying), while other flies deposit live larvae. Areas of the body that are first colonized by arthropods include the eyes, mouth, nasal cavity, anus, vagina, and areas of trauma (Sledzik 1998). Several other flies are also typically associated with decaying remains: Sarcophagidae (flesh flies), Phoridae (humpbacked fly), Sepsidae (black scavenger flies), Muscidae (house flies), Sphaeroceridae (small dung flies), and Stratiomyidae (soldier flies), depending on the time of year and region of the world (Haskell et al. 1997). In the family Piophilidae, the *Piophila casei* (cheese skipper) is found in advanced decomposition stages, when the connective tissue is drying (Goff 2010). Beetles are also extremely important to PMI estimations, especially in later stages of decomposition. The beetle orders more relevant to PMI estimation include Coleoptera, silphids (carrion beetles), dermestids (hide beetles), nitidulids (sap beetles), and clerids (checker beetles), which feed on the body, and the families of the Staphylinidae (rove beetles), Silphidae (carrion beetles), Histeridae (hister beetles), and Hydrophilidae (hydrophilid beetles), which are found on or around the body (Goff 2010). These should be distinguished from the predatory beetle species that feed on blowfly larvae earlier in the decomposition process.

There are two main approaches to the forensic entomological PMI estimations: (1) analysis of the pattern of successive colonization of different arthropods, and (2) the determination of the life stage of the specific insects, usually blowflies. The timing of specific arthropod species in relation to the decomposing cadaver provides a basis for estimating the PMI. For example, adult blowflies lay eggs and colonize a cadaver within the first 24 hours of death, but within just a few days, beetles that feed on both the body and the larvae will be present. The presence of both species on the same cadaver indicates a slightly longer PMI. More precise methods to estimate PMI typically use the lifecycle of specific insect species. Many species of insects, such as the blowfly, have well-documented developmental rates. Thus, recognition of the different immature stages of arthropod species involved, together with the knowledge of their rates of development, can give an indication of the PMI (Smith 1986). Developmental stage is most useful for early decomposition (under 15–20 days), because this is the average lifecycle of most blowflies (Haskell et al. 1997). The application of developmental stages is based on the fact that insect growth is directly correlated to ambient temperature. Thus, one would need to identify the species collected from a cadaver, note the developmental stage, obtain weather data for the deposition location, and then calculate the amount of time that would be required to reach the developmental stage observed for that species.

Most forensic entomology techniques focus on the determination of the accumulated degree days (ADD) or accumulated degree hours (ADH) needed for specific larvae to reach a certain growth stage. The ADD and ADH are heat unit values calculated from ambient air temperatures that represent the heat energy needed for larvae to reach a specific growth stage in a particular environment (Haskell et al. 1997). Although forensic entomology is arguably the most successful way to estimate PMI after the first 72 hours (Anderson and VanLaerhoven 1996), the blow fly larvae associated with a decomposing body are most valuable in the early stages of decomposition or until the time of pupation, so other methods must be employed for longer time intervals (Payne 1965). Another potential ecological-based research area is evaluating post-putrefaction fungi, which have been shown to have successive fruiting phases associated with cadaver breakdown (Sagara 1995).

Entomotoxicology

Arthropod development is impacted by multiple factors. In addition to climate considerations, various chemical substances have been shown to accelerate or delay larval growth. The subfield of entomology focusing on these effects is entomotoxicology (Introna et al. 2001). The presence of some toxins in decomposing tissue can delay the onset of insect activity and oviposition and decrease the accuracy of insect-based PMI estimations. Blowflies and flesh flies (Calliphoridae and Sarcophagidae) have been the most extensively studied (Bourel et al. 2001; Carvalho et al. 2001; Goff et al. 1997; Traqui et al. 2004). Toxins known to impact their developmental sequences include barbiturates, opium and opioids, antidepressants, benzodiazepines, cocaine, salicylates, acetaminophen, and trazodone (Levine et al. 2000; Sadler et al. 1995; Sadler et al. 1997; Wilson et al. 1993). Some toxins have a singular effect; cocaine, for example, leads to hyperdevelopment of larvae (Goff et al. 1989), while other substances have more complicated effects. Diazepam has been shown to increase the rate of early larval development and also increase the time needed for pupation and adult emergence (Carvalho et al. 2001).

It has also been suggested that larvae can be used in toxological analyses when there is no remaining soft tissue from the decedent. However, this has been shown to have limited accuracy, and some entomologists recommend focusing strictly on the impact of toxins on the behavior and growth of forensically significant species (Tracqui et al. 2004).

Forensic botany

Forensic botany is the application of botanical studies to the medico-legal context. The typical application of forensic botany to forensic anthropology contexts is the determination of body movement (i.e., whether the recovery location is the primary decomposition location). Botanical analyses can also be applied to estimate PMI; however, they are often underutilized, owing to limited botanical knowledge of investigators (Bock and Norris 1997). When utilized, these estimates rely on the recognition of flora associated with a cadaver, documentation of location, systematic collection and preservation, and determination of the growth rate for each species (Coyle et al. 2005). One of the major issues with botanical-based methods is the lack of good preservation of evidence, so estimates are often computed, despite missing information. If quality documentation and preservation techniques are employed, the flora surrounding a body can provide PMI estimations for longer time intervals at that location, so these methods are most applicable for skeletonized remains. Further, plants can be annual and/or seasonal in nature, so they can suggest time of year or particular weather conditions as to when a body was deposited. This includes aspects of a plant's lifecycle such as number of leaf falls and the presence of specific pollens (Hall 1997). Also, the amount of growth that has occurred in a plant can indicate exposure length in years, especially if plant roots are growing through naturally existing foramina of the skeleton (Box 15.2; Willey and Heilman 1987).

> **BOX 15.2 LONG-TERM DECOMPOSITION AND THE ROLE OF BOTANY**
>
> In 1995, remains were discovered in a secluded area of West Knoxville, TN. The remains were discovered by several boys playing in the area and resulted in the cranium being taken from the scene. Examination of the scene indicated well-scattered remains, with leaf litter covering much of the skeleton. The mandible was firmly secured to the ground by a root mass (Figure 15.17a). A shirt made of synthetic material was found adjacent to the mandible. The shirt had vertebrae and ribs associated with it and was also penetrated with roots. Several ribs and vertebrae were found downslope of the shirt. The innominate was secured to the ground by a root that had grown through the obturator foramen (Figure 15.17b). Bones of the lower limbs were discovered in close proximity to each other in the center of the scatter. Both femora and cranium were recovered from separate locations of the scatter.
>
> Analysis of the remains indicated that rodent and carnivore gnawings were present throughout the skeleton. The cranium had rodent gnawing on the supra-orbital margins, zygomatic bone, and maxillae. The postcrania exhibited carnivore punctures and gnawing on the ribs, vertebrae, and epiphyseal areas of the long bones. Also, the remains were stained a dark color owing to the depositional environment, soil, and leaf litter. This was particularly evident on the right side of the cranium. However, the occipital bone exhibited a green coloration from algae growth, and moss had grown on several of the vertebral elements. Desiccated tissue (dried soft tissue) adhered to the toe phalanges. No other soft tissue was present on the remains.
>
> Based on the extensive root growth, degree of skeletonization, and organic staining of the bones, it was clear that the remains had been exposed for an extended period of time. The rodent gnawing, which is consistent with gray squirrel activity, supported the assessment and suggested a PMI of greater than 1 year. A root sample, which had grown through several layers of the shirt fabric, was examined by Dr. Heilman at the University of Tennessee. He found that the root was in its fourth year of growth, thus indicating that the shirt associated with the remains had to have been at the location for no less than 4 years.

Figure 15.17 (a) Mandible that is embedded within a root mass. (b) Root that had grown through the obturator foramen of the oscoxa.

Diagenesis of bone

The breakdown of the skeleton after soft tissue decomposition and its use as a PMI estimator has long troubled forensic practitioners. Behrensmeyer's (1978) visual inspection system highlights the progressive pattern of bone weathering, suggesting its utility in estimating prolonged PMIs for skeletonized remains. However, the rate of bone weathering is highly variable among different ecosystems (e.g., temperate zones and grasslands), and other taphonomic agents often mimic the changes associated with particular stages of bone weathering (e.g., calcination due to burning). Nevertheless, some general guidelines are that bleaching and cracking caused by weathering in a cold climate are delayed, compared with arid climates, but occur more rapidly than in temperate environments (Marceau 2007). The reliance on macroscopic bone weathering observations has led many investigators to overlook the chemical-level changes associated with bone diagenesis.

Research has indicated that estimating PMI from skeletal remains is possible (Introna et al. 1999; Schwarcz et al. 2010). Schwarcz and colleagues (2010) demonstrated the utility of the citrate content of bone to determine the PMI. Using porcine and human ribs, they suggested that citrate levels in bone decrease linearly as a function of log (time) after 4 weeks postmortem. However, when tested by Kanz et al. (2014), citrate levels did not perform as well as hypothesized. Further, they showed that citrate levels varied between elements, and degradation varied between surface and buried remains. Further research in this area is needed before these methods can be applied with certainty.

Chemistry

Soil chemistry and biochemisty research show promise in the use of microorganisms and the depositional environment in the development of rapid and reliable techniques to estimate the PMI. The basis for soil analyses is to acknowledge that soil, as a complex, ever-changing combination of biological, chemical, and physical properties, is significantly changed when a cadaver comes in contact with it. Soil chemistry is of specific importance to the burial context, because it dictates how well skeletal remains are preserved. The color, consistency, and texture of soil, all play a significant role in the appearance of buried remains. However, chemical-based methods are also relevant to surface decomposition environments. When a body decomposes, it creates an influx of organic material into the surrounding soil, which creates a "cadaver decomposition island (CDI)" (Carter et al. 2007). For example, the CDI will initially have a negative effect on the surrounding vegetation, but vegetation growth is renewed as the characteristics of the CDI change.

Pioneering research by Vass et al. (1992) demonstrated that ADD correlates with the biochemical decomposition of body tissues. VFAs from the breakdown of tissues leach into the surrounding soil and change the soil chemistry. A soil sample from the decomposition location can be compared with a sterile sample from the same general locality. The ratio of the VFAs in the soil where decomposition occurred correlates with a particular ADD and is considered the maximum PMI (Vass et al. 2002). This research has even proven effective with advanced decomposition and skeletonization. Carter et al. (2008b) have indicated the utility of using ninhydrin in detecting gravesoils. Also, Spicka et al. (2008) indicate that the concentration of ninhydrin-reactive nitrogen in grave soil associated with juvenile- to adult-sized cadavers' (20–50 kg) remains at similar levels to the predepositional levels until 2 days postmortem. See Table 15.6 for an outline of the effects of decomposition on soil.

Table 15.6 Decomposition effects on soil with associated time intervals

Decomposition stage	Soil changes	Time
Fresh	Little to none	<48 hours
Bloat	Increase in pH, initial influx of nutrient sources into soil, such as ammonia, calcium, magnesium, ninhydrin-reactive nitrogen, potassium, sodium, sulfate, and VFAs	24 hours to 1 week
Active decomposition	Increased levels of all of the above	48 hours to 2 weeks
Advanced decomposition	Peak concentrations of VFAs, chemicals, and microorganisms	1 week to several months
Dry remains	Decrease in concentrations over time to eventually reestablishing "virgin" soil levels	2 weeks to 2 years

Sources: *J Forensic Sci.*, A laboratory incubation method for determining the rate of microbiological degradation of skeletal muscle tissue in soil, 2004, 560–565, Tibbett, M., Carter, D.O., Haslam, T., Major, R., and Haslam, R.; Payne, J.A., *Ecology*, 46, 592, 1965; Vass, A.A. et al., *J Forensic Sci*, 37, 1236, 1992; Spicka, A. et al., Cadaver mass and decomposition: How long does it take for a cadaver to increase the concentration of ninhydrin-reactive nitrogen in soil?, in *Proceedings of the American Academy of Forensic Sciences 61st Annual Scientific Meeting*, vol. 14, p. 178, 2008.

Microbial-based human decomposition research has grown considerably over the last few years and has shown great potential in determining PMI (Carter et al. 2008a; Finley et al. 2014). Several researchers have suggested a postmortem microbial clock that can be used to evaluate the microbial activity on a decomposing body, as well as those activities found within the soils surrounding a body (Hauther et al. 2015).

These methods eliminate the subjectivity inherent in observation-based data and have the potential to control for the increasing time error that accompanies extended decomposition stages (Tibbet and Carter 2008). Consequently, soil chemistry is a growing area of research for statistically valid PMI estimations. However, because many of these biochemical techniques require special laboratories to conduct the analysis, ongoing research is focusing on ways to convert observation-based methods and the accompanying categorical data into statistically significant PMI estimations.

Morphoscopic techniques

Megyesi et al. (2005) proposed one of the first methods to correlate gross observations of the cadaver with ADD. This research developed a scoring system for decomposition on different body areas/regions, in which the regional scores are summed to produce a total body score (TBS). Tables 15.7 through 15.9 provide the stages and associated scores for each body region. The TBS is independent of the depositional environment, because it focuses strictly on the condition of the body (Box 15.3). The TBS is input into an equation that computes the estimated ADD required

Table 15.7 Total body scoring system for the head and neck

Descriptions	Score
Fresh, no discoloration	1
Pink-white appearance with skin slippage and some hair loss	2
Gray to green discoloration; some flesh is still relatively fresh	3
Discoloration and/or brownish shades, particularly at edges; drying of nose, ears, and lips	4
Purging of decomposition fluids out of eyes, ears, nose, and mouth; some bloating of neck and face may be present	5
Brown to black discoloration of flesh	6
Caving in of the flesh and tissues of eyes and throat	7
Moist decomposition with bone exposure less than one half that of the area being scored	8
Mummification with bone exposure less than one half that of the area being scored	9
Bone exposure of more than half of the area being scored with greasy substances and decomposed tissue	10
Bone exposure of more than half of the area being scored with desiccated or mummified tissue	11
Bones largely dry but retain some grease	12
Dry bone	13

Source: Megyesi et al., *J Forensic Sci* 50, 2005; pp. 618–626.

Table 15.8 Total body scoring system for the trunk

Descriptions	Score
Fresh, no discoloration	1
Pink-white appearance with skin slippage and marbling present	2
Gray to green discoloration; some flesh is still relatively fresh	3
Bloating with green discoloration and purging of decomposition fluids	4
Post-bloating following release of the abdominal gases, with discoloration changing from green to black	5
Decomposition of tissue producing sagging of flesh; caving in of the abdominal cavit	6
Moist decomposition with bone exposure less than one half that of the area being scored	7
Mummification with bone exposure less than one half that of the area being scored	8
Bones with decomposed tissue, sometimes with body fluids and grease still present	9
Bones with desiccated or mummified tissue covering less than one half of the area being scored	10
Bones largely dry but retain some grease	11
Dry bone	12

Source: Megyesi et al., *J Forensic Sci* 50, 2005; pp. 618–626.

Table 15.9 Total body scoring system for the limbs

Descriptions	Score
Fresh, no discoloration	1
Pink-white appearance with skin slippage on hands and/or feet	2
Gray to green discoloration; marbling; some flesh is still relatively fresh	3
Discoloration and/or brownish shades, particularly at edges; drying of fingers, toes, and other extremities	4
Brown to black discoloration; skin having a leathery appearance	5
Moist decomposition with bone exposure less than one half that of the area being scored	6
Mummification with bone exposure less than one half that of the area being scored	7
Bones exposure over one half of the area being scored; some decomposed tissue and body fluids are remaining	8
Bones largely dry but retain some grease	9
Dry bone	10

Source: Adapted from Megyesi et al., *J Forensic Sci* 50, 2005; pp. 618–626.

BOX 15.3 CASE-SPECIFIC VARIABLES: HANGING IN THE OUTDOOR CONTEXT

Most suicidal hangings occur indoors, but outdoor hangings often raise more questions than answers after the investigation. A common issue with outdoor hangings is the long time delay before remains are found, often resulting in the recovery of decayed or skeletonized remains. Skeletonized remains alone make it difficult to recreate the deposition scenario and determine the PMI. There is very little published research on how hanging affects the rate of decomposition, despite detailed literature documenting the stages of decomposition related to surface scenarios. Understanding factors that could affect the decomposition process (e.g., distance from the ground, positioning of the body in relation to the asphyxiation device, and the type of noose used) may allow for more accurate evaluations of outdoor hanging scenarios.

In 2009, the Anthropological Research Facility in Knoxville, Tennessee, initiated a long-term project to address outdoor hanging scenarios. A 10-foot-high wooden device was built to bear a maximum load of over 600 pounds with a pulley/crank system to ease the force required to lift individuals into the hanging position (20 cm above the ground). The number of variables was limited, so that seasonal differences could be appreciated (clothing was eliminated; knot type was standardized; and the same type of rope was used in all trials). A cadaver was also placed on the ground near the hanging individuals to serve as the control for each trial. Observations were recorded twice daily by using video and still photography, written annotations, and data loggers. A time-lapse camera was also utilized to capture hourly photos throughout the decomposition process. Samples were collected for entomological examination, including eggs, larvae, pupae, and adults present on and around the body.

Similar decomposition events are evident for the control and hanging (e.g., skin slippage, marbling, and bloating), but a few key differences exist. These differences include entomological colonization (both species present and oviposition pattern), amount of biomass reduction before mummification, and overall rate of decay. This study confirms earlier work of Goff (1991), using a carrion model for hanging, suggesting that a hanging cadaver will have a reduced, if not complete, elimination of ground-dwelling taxa feeding on the soft tissue. In addition, few fly larvae were found in the later instar stages of development on the body surfaces of the hanging cadaver. The larvae are not able to mass, which is vital for them to maintain their body temperatures. Once the maggots started to clump into a sizeable mass, they fell off of the body and onto the ground below the body, creating a "drip zone" area that encircled the base of the body. Maggots massed in the drip zone and fed on the falling soft tissue. The ideal environment for feeding by blowfly larvae was also diminished due to rapid mummification. Hide beetles and cheese skippers were prevalent early on (within a few weeks postmortem) and remained well into the second year of the study.

Initial observations indicated similar oviposition patterns between the control and hanging cadaver; the most significant difference was temperature. The hanging cadaver maintained ambient temperature because of the 360° air circulation around the body and thus exhibited putrefaction and advanced decomposition much later in the study. This corresponds to previous surveys, indicating that decomposition rates are significantly lower in hangings at higher heights as opposed to individuals partially on the ground (Komar et al. 1999).

(Continued)

> **BOX 15.3 (Continued)   CASE-SPECIFIC VARIABLES: HANGING IN THE OUTDOOR CONTEXT**
>
> A clear difference was noted in the summer versus the winter trials. The cadavers placed in the summer had rapid decomposition events, associated with the higher temperatures, but did not exhibit the skin removal that is typically seen in summer decomposition. The outer tissues of the body mummified, but internal decomposition still occurred, resulting in the settling of the organs in the lower abdomen and the eventual disintegration of the perineum, causing the internal organs to fall into the "drip zone" below the body. This accelerated the mummification process by providing additional air circulation around the body surfaces. The loss of internal organs significantly reduced the biomass of the cadaver, thus reducing the body weight and prolonging the ability of the cadaver to remain hanging. Winter trials displayed similar early decomposition events, but much of the putrefaction needed for internal organ decomposition was slowed. This caused the retention of the internal organs and perineum, even though the organs settled into the lower abdomen. The greater biomass retention in the winter produced a heavier cadaver that reached the ground and fell after the first year of the study.
>
> This preliminary research demonstrates the continued need for observation-based studies that are actualistic in nature. **Actualistic studies** look at the process as it occurs, in order to understand better what is happening. Observing the hanging cadavers makes us rethink about blindly applying surface decomposition stages to estimate the PMI in a unique context. The hanging scenario significantly affects the rate of decomposition and encourages preservation of tissue. When a cadaver is found hanging and demonstrates significant mummification and mold growth, a prolonged time interval should be considered (Figure 15.18).

(a)   (b)

Figure 15.18 (a) Experimental: hanging cadaver after 20 months of exposure. (b) Control: surface cadaver after 20 months of exposure.

to reach the stage of decomposition represented by the TBS. The quantification of the decomposition process and the relative ease of use of the TBS have led to its adoption by researchers over the last several years. The quantification of qualitative traits inherent in the Megyesi et al. (2005) method allows for it to be statistically validated. However, several caveats must be considered when applying the Megyesi et al. (2005) technique: the technique was developed without an exact known PMI (estimates were obtained using forensic entomological techniques);

the body was evaluated and scored using pictures; and the scoring criteria produced were based on a combination of reviewing the literature for known stages of decomposition and adjusting the criteria based on photographs of actual forensic cases. In addition, cases involving animal scavenging or other atypical factors such as clothing and coverings were excluded from the developmental study. Researchers have shown mixed results in the application of the method (Suckling et al. 2015) and have proposed updated versions of the equation that address many of the inherent problems in the development of Megyesi et al.'s (2005) equation (Moffatt et al. 2015). These methods demonstrate potential for combining multiple lines of evidence and highlight the importance of temperature as a major variable affecting the decomposition rate.

Simmons et al. (2010) have added further evidence to support the validity of ADD as a standard in the estimation of the PMI. The review of research conducted at the University of Central Lancashire, combined with previously unreported research, indicates that ADD can be used as a means to compare decomposition experiments conducted under vastly different conditions. However, rabbits are the primary carrion used in studies conducted by the University of Central Lancashire. Payne et al. (1968) and, subsequently, Schoenly et al. (2007) have demonstrated the usefulness of pigs as a proxy for human decomposition, but they noted that pigs should not be considered a replacement for human-based experiments. Vass et al. (2008) detected significant biochemical differences between the bones of humans and those of other mammalian species. Further, Parmenter and MacMahon (2009) and Simmons et al. (2010) found that the size of carcass plays a significant role in the decomposition process in that smaller carcasses decompose faster than larger carcasses. Thus, some nonhuman animals may not be an appropriate proxy for humans.

Vass (2011) has demonstrated the effectiveness of ADD by combining ADD data with the percentage of body decomposition and incorporating the role of humidity in the decomposition process. This research is the culmination of observations from numerous cadavers at the Anthropology Research Facility (ARF) in Knoxville, TN, over the last 20 years. The research integrates earlier work, where Vass et al. (2002) determined the time at which decomposition ceases in ADD units, and provides estimation equations for both surface and burial contexts (see Box 15.4). This method provides a "rule of thumb" approach for the estimation of the PMI for decomposition events greater than 2 days. His formulae rely on the fact that temperature, moisture, and partial pressure of oxygen have the greatest influence on decomposition and that soft tissue decomposition ceased at 1285 ± 110 ADDs. However, much of his work is based on personal observations and has not been validated by other researchers as of now. Like Megyesi et al. (2005), Vass (2011) demonstrated advances in decomposition research and the need for statistically derived PMI estimations. Also, more research on the reliability of the percentage scoring system versus a TBS system needs to be completed, so that definitions can be clearly written for practitioners.

Future directions

The accumulated degree day (ADD) approach for estimating PMI has shown great success in entomology and has a strong foundation for use from biology and engineering perspectives (Haskell et al. 1997; Vass et al. 1992). However, this approach assumes regularity to the decomposition process when applied to soft tissue changes. Most PMI estimation methods focus on decomposition as a series of sequential events, in which PMI is the "calculation of a measurable date along a time-dependent curve back to the start point" (Henssge and Madea 2007: 182). This model simplifies the relationship between the various factors that influence decomposition to fit into a linear, sequential model. Hypothesis testing by using a linear regression approach is complicated by the fact that the variables affecting the decomposition process are continually changing and interact with one another. Therefore, PMI estimation may be more statistically rigorous if evaluated using models that view decomposition as a process that goes through several transitions; the PMI estimate could then be based on the probability of reaching a particular transition at a given time.

Further, the development of PMI equations needs to move beyond observational studies of human and nonhuman models. The chemical changes within the soil and the microbial community within and around decomposing remains are promising avenues for the development of accurate and precise PMI estimation techniques. Specifically, recent research has shown a unique microbial community in humans and a successional pattern to the microorganisms. More research using human models is needed to test the utility for PMI estimation.

> **BOX 15.4 "RULE OF THUMB" CALCULATION OF THE PMI**
>
> Dr. Arpad Vass, former researcher at the Oak Ridge National Laboratory in Tennessee, has conducted biochemical research on human decomposition at the Anthropology Research Facility in Knoxville, Tennessee, since the early 1990s. He has published several landmark papers relating to chemicals given off by a decomposing cadaver. In 2011, he published a culminating paper representing his many years of research and observations (*The Elusive Universal Postmortem Interval Formula*) to establish methods of making PMI estimations for empirical versus more subjective approaches.
>
> According to Vass (2011), PMI estimations can be calculated for surface decomposition and burial decomposition contexts. The following formula outlines the calculation for surface contexts in which there is access to oxygen. This would not be relevant for confined spaces above ground; rather, the burial formula should be implemented. This formula is designed for scenarios in which soft tissue is still present (i.e., preskeletonized stages), there is little to no adipocere development, and limited scavenging has occurred, as scavenging is known to affect rate of decay. However, it is still important to note the macroscopic changes of the body, in addition to the overall percentage of decay, because of the many context-specific variables that cannot be accounted for in the formula:
>
> $$\text{PMI (aerobic)} = [1285 \times (\text{decomposition}/100)]/[0.0103 \times \text{temperature} \times \text{humidity}]$$
>
> Where
>
> - 1285 = A constant that represents the ADD (accumulated degree days) value for which VFAs cease to be released from soft tissue. It is considered the end of soft tissue decomposition and is typically associated with skeletonization of a cadaver (Vass et al. 1992).
> - *decomposition* = A value from 1 to 100 that represents the percentage of total soft tissue decomposition. This requires an understanding of the decomposition process and of the appearance of specific features during the process. Unless one is confident in his or her estimations, a range of percentages should be used in the calculation.
> - 0.0103 = A constant that represents the effect of moisture on the decomposition rate. Prior unpublished work by Vass indicates that the humidity and moisture content released from a decomposing corpse follow a generalized linear curve. Further research is needed in other climatic zones to determine if this trend holds.
> - *temperature* = Average ambient temperature (°C) for the depositional location (can be for the day of discovery or for a period of time). It is better to use a range of days and not the average for the day of discovery, since temperatures can fluctuate considerably within and between days.
> - *humidity* = A value from 1% to 100% represents the average humidity at the depositional location (can be for the day of discovery or for a period of time).
>
> If the goal is to produce objective, quantitative PMI estimation methods, why are forensic anthropologists continuing to develop methods based on observational data? Methods that are easily administered without the need for specialized laboratory equipment are highly desired because of their practicality and cost-effectiveness. These PMI estimation methods are not going to be as objective as desired but can be statistically more rigorous.

DECOMPOSITION RESEARCH

Longitudinal studies of human decomposition in the literature have been limited by access to the appropriate number of human models to make studies statistically relevant (Mant 1953; Mann et al. 1990; Rodriguez and Bass 1983, 1985), locations to conduct this research, and graduate students willing to dedicate years to developing and conducting these experiments. The University of Tennessee, Knoxville, has been essential in human decomposition research by opening the first research facility for **actualistic research** using human cadavers. The advent of other outdoor research facilities, such Forensic Anthropology Center at Texas State at Texas State University in San Marcos, TX; the Southeast Texas Applied Forensic Science Facility at Sam Houston University in Huntsville, TX; the Forensic Osteology Research Station at Western Carolina University in Cullowhee, NC; the Forensic Investigation Research Station at Colorado Mesa State University in Grand Junction, CO; and the

Complex for Forensic Anthropology Research (CFAR) at Southern Illinois University in Carbondale, IL, combined with the over 30-year research tradition of the ARF, are needed to continue to broaden the knowledge base necessary to fully understand human decomposition and develop and validate PMI estimation methods.

Bass (1997) established the progression of decomposition events in warm, moist climates found in the southeastern United States based on actualistic studies and personal experience. He outlined the basic features of human decomposition in the greater East Tennessee area, a climate in which it is possible to get a complete skeleton in as little as 2 weeks in the summer months (Bass 1997). This speed requires the ideal complement of temperature, humidity, and arthropod activity (Mann et al. 1990). Regardless of the rate, a decomposing corpse undergoes a specific series of documented stages: fresh, fresh-bloat, bloat-decay, dry, and bone breakdown (Figure 15.19a and b; Table 15.10).

Arid environments cause rapid desiccation of soft tissues, which often results in mummification. Galloway et al. (1989) and Rhine and Dawson (1998) have established criteria and trait lists for decomposition in hot, arid conditions. Galloway et al. (1989) examined 189 cases found in southern Arizona and found that early decomposition and bloating occurred rapidly but the outer tissues became dehydrated quickly and formed a hard, mummified shell. Larval activity quickly declined, because dried tissue was not a viable environment for oviposition or feeding. Rhine and Dawson (1998) surveyed 270 decomposition-based cases from the New Mexico Office of the Medical Examiner to sequence the changes leading to skeletonization and indicated similar processes as suggested by Galloway et al. (1989). However, they highlighted the fact that the rate, and not the sequence, of events is the primary difference in decomposition in various climatic zones (Table 15.11).

Little research has been conducted on decomposition in cold weather climates, and much of the existing research utilizes carrion models. Komar (1998) provides a review of forensic cases from the Edmonton area of Canada, and Micozzi (1986, 1991) focuses on actualistic studies using carrion models to evaluate cold weather decomposition. Both studies suggest that temperature is the key factor in altering the rate of decomposition, which is further highlighted by Stokes et al. (2009). An examination of 20 cases from the medical examiner's office in Edmonton,

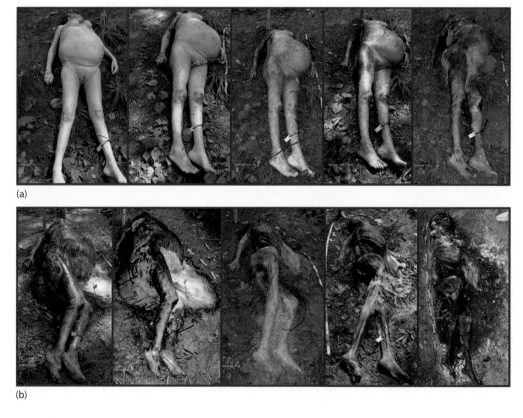

Figure 15.19 (a) Fresh to active decomposition stages. (b) Active decomposition to skeletonization stages. (Photos courtesy of Dr. Andrew Hart.)

Table 15.10 Decay rates for warm/moist climates

Time	Expectations
Day 1	Oviposition; egg masses; slight marbling; release of body fluid from mouth and genital area (from sphincter relaxation)
Week 1	Larval activity (especially in facial area) and a presence of beetles, beginning of skin slippage and hair mass production, pronounced marbling, distinct odors associated with decomposition, bloating of abdomen and extremities, mold growth, initial decomposition stain production from VFAs release and associated vegetation death, animal scavenging
Month 1 (2nd–4th week)	Reduction in larval activity and increased beetle activity; body appears to be deflating (skin is wrinkling and fluid is being purged); skin will have a leathery appearance and will be drying, variety of molds will cover remaining skin; dark, distinctive stain will be present around the body; adipocere may form (especially if moisture content is high); initial skeletonization (head, neck, and genital areas initially)
Year 1 (after first month)	Complete exposure of the skeleton; exposed elements will begin drying and the cortex will become bleached; algae or moss growth, if shaded and in a moist area; rodent gnawing in the later months; rodent and wasp nest production
Decade 1	Deterioration of the skeleton itself; continued bleaching with exfoliation of cortex (this is enhanced by periods of high moisture, followed by drying periods); cracking of the cortical bone (typically seen in the long bones, scapulae, and os coxae initially); root growth in and out of bones; staining of skeletal elements from substrate (especially from leaf falls and darker soil types); extensive rodent gnawing, causing severe bone destruction

Sources: Based on Bass, W.M., Time interval since death: A difficult decision, in Rathbun, T.A. and Buikstra, J.E. (eds.), *Human Identification: Case Studies in Forensic Anthropology*, CC Thomas, Springfield, IL, pp. 136–147, 1984; Bass, W.M., Outdoor decomposition rates in Tennessee, in Haglund W.D. and Sorg M.H. (eds.), *Forensic Taphonomy: The Postmortem Fate of Human Remains*, CRC Press, Boca Raton, FL, pp. 181–186, 1997.

Table 15.11 Decay rates for dry climates

Time	Stage of decomposition	Expectations
Day 1–7	Fresh	Lividity, egg deposition, little to no animal scavenging.
Day 1–5 to day 3–8: Larvae present by day 2; bloat present by day 2 but typically lost by day 7	Early decomposition	Skin slippage. Gray to green discoloration of overall body. Distal extremities have a brownish discoloration, bloating, and post-bloating. Purple to black discoloration of appendages.
Day 4–8 to 2 months: Mummification 2 months to 6–9 months	Advanced decomposition	Deflated, concave abdomen, bone exposure with moist decomposition. Mummification with or without internal organ retention. Adipocere development.
2 months to 1 year	Skeletonization	Greasy bones, with some body fluids or desiccated tissue but less than half the skeleton. Mostly dry bone.
>6 months: Exfoliation 12–18 months	Extreme decomposition	Bleaching and exfoliation of cortex of bone. Bone degradation, with metaphyseal loss and cancellous bone exposure.

Sources: Galloway, A. The process of decomposition. A model from the Arizona-Sonoran Desert, in Haglund, W.D. and Sorg, M.H. (eds.), *Forensic Taphonomy: The Postmortem Fate of Human Remains*, CRC Press, Boca Raton, FL, pp. 139–149, 1997; Galloway, A. et al., *J Forensic Sci* 34, 607, 1989.

Alberta, Canada, suggests that skeletonization can occur within 4 months, even in freezing temperatures (Komar 1998). Remains that experience decomposition in regions where winters are cold and harsh can undergo a putrefactive stasis during subfreezing temperatures. The stasis causes inaccurate assessment of decomposition and results in an imprecise estimation of the PMI (Dudzik et al. 2011). Micozzi (1991) found that frozen environments alter the environmental context greatly, and a thawed body may, in fact, decompose more quickly than a fresh body. The rate at which a body freezes (delayed vs. rapid) during the freeze–thaw cycle of a Montana winter and spring alters the expected decomposition pattern (Dudzik et al. 2011). Thawing causes decomposition from the "outside-in," as opposed to the decomposition from the "inside-out" typically seen in exposed surface depositions without freezing (Micozzi 1991; Parsons 2009). Micozzi (1986) concluded that decomposition in remains that have been frozen proceeds from the "outside-in" by aerobic decay, disarticulation, and foreign organisms. Decomposition of remains that have not been frozen proceeds from the "inside-out," owing to putrefaction and autolysis. Importantly, Micozzi (1991) did not observe putrefaction in temperatures colder than 4°C, which is why refrigeration slows the

Table 15.12 Decomposition rates in cold climates

Time	Stage	Expectations
Day 1–day 30	Fresh	Some marbling, general color change, and possible insect activity
Day 1–3 months	Moderate	Partial exposure of bone, loss of some bone elements from scavenging, and slight adipocere formation
3–32 months	Advanced	Mummification with or without internal organs, extensive adipocere formation, moderate bony exposure
4–42 months	Skeletal	Little wet soft tissue or desiccated tissue, with most of skeleton exposed
2 months–8 years	Completely skeletal	Fully exposed, dry skeleton with no associated tissue

Source: Komar, D.A., *J Forensic Sci*, 43, 57, 1998; p. 58, Table 18.2., and Meyer, J., Anderson, B. and Carter, D.O., *J Forensic Sci*, 58, 5 p. 1176.

decomposition process and "bloating" is much less pronounced in the winter months. However, others have indicated that decomposition occurs at 0°C due to the elevated salt content in the human body (Vass et al. 2002) and that heavy snowfall may act as an insulator, thus creating a situation in which the temperature near the carrion is higher than the ambient temperature (Parsons 2009). Also, the process of freezing and thawing of tissues results in rapid disarticulation of remains, as it weakens skin, connective tissues, and joints (Micozzi 1986, 1991, 1997). Carnivore activity is also an important factor in the decomposition and disarticulation of remains in cold climates (McKeown et al. 2011). Regardless of the reason for a cold climate (i.e., confined area, altitude, and latitude), the lower temperatures slow decomposition rates, resulting in a longer time interval to skeletonization (Bass 1997; Galloway et al. 1989; Komar 1998, see Table 15.12).

The pattern of decomposition in water is different from decomposition observed in terrestrial environments (Boyle et al. 1997; Haglund 1993). Research on soft tissue decomposition in aqueous environments indicates that water slows the decomposition process overall, but temperature, bacterial content, and water movement, all affect the rate of decomposition (Haglund 1993; MacDonnell and Anderson 1997; Seet 2005; Simpson and Knight 1985). Early work in this area focused primarily on arthropod activity, noting that the succession of specific arthropods that colonize a submerged body can be used to estimate PMI (Haskell et al. 1989; Payne and King 1972). A few key points are notable: water encourages adipocere development; autolysis and putrefaction continue with submerged remains, but the rate of these processes depends directly on the water temperature; and saltwater slows the rate of putrefaction because salinity retards bacterial growth (Cotton et al. 1987; Evans 1962). Flies are attracted to exposed areas of the body, and fish and other aquatic animals often feed on remains below the water line. Also, remains will sink initially if not prevented by clothing or other devices, but as remains become bloated, the body resurfaces. Disarticulation usually follows a sequence of hands first, followed by feet, head and lower jaw, lower legs, forearms, and, finally, the pectoral girdle, leaving the torso and pelvis (Haglund 1993). As the remains disarticulate, gases created during putrefaction escape, and the remains sink again. However, other factors are involved when considering saltwater, such as salinity, tides, currents, depth in the water column, substrate type, season, and the species and number of colonizing animals (Anderson 2010; Hobischak and Anderson 2002; Sorg et al. 1997). Sorg et al. (1997) described four roles of carcasses in an ocean environment: (1) the remains may become food for vertebrate and invertebrate scavengers; (2) they may provide shelter for other species; (3) they may attract secondary scavengers; and (4) the bones may eventually provide a source of minerals and act as a substrate for bacterial grazers. Anderson (2010) summarizes research conducted over the last decade, comparing shallow and deep saltwater environments. In both contexts, the amount of biomass lost was not solely the result of decomposition. Opportunistic scavenging and body movement expedited skeletonization, and more rapid skeletonization occurred in the deeper waters because of scavenging. The difficulty in conducting actualistic studies in aqueous environments has led to a heavy reliance on anecdotal evidence for PMI estimations. However, the creation of the Victoria Experimental Network Under the Sea and other underwater research stations may prove essential to our understanding of aqueous decomposition (Anderson 2010).

Another important body of research focuses on the burial context (Carter and Tibbett 2008; Payne 1968; Rodriguez and Bass 1985; Sagara et al. 2008; Turner and Wiltshire 1999; Vass et al. 2008; Wilson et al. 2007). The PMI is more difficult to predict with buried bodies because of the many variables at play. Factors that affect the rate of decomposition include the nature of the soil, water content, pH, and the nature of the body itself (cause of death, clothing, wrapping/coverings, and depth of the burial; Janaway 1996). Decomposition rate is strongly correlated with burial depth; the deeper the burial, the slower the rate of decomposition. The slower rate is caused by reduced oxygen available for biochemical reactions and the decreased presence of arthropods. However, the naturally occurring microbial community within the soil environment also plays an important role in the decomposition of a buried body (Parkinson et al. 2009). The soil type (e.g., sand vs. clay) directly impacts the microbes, which play a vital role in maintaining ecological homeostasis within a particular location. Long-term preservation of the buried skeleton and associated artifacts is related directly to soil chemistry (Janaway 2002, 2008). More acidic soils (low pH) are less suitable for preservation than more basic soils (high pH) (Carter and Tibbett 2008; Haslam and Tibbett 2009; Janaway 1996; Vass et al. 2008).

SUMMARY

- Decomposition of the human body follows a series of events that can be used to determine the postmortem interval (PMI) of a deceased individual.
- The physical breakdown of soft tissues relies on the processes of autolysis and putrefaction, which are responsible for many of the gross anatomical changes observed in the human body during decomposition.
- Many categorization methods exist, outlining the anatomical changes in a body. One of the most commonly applied systems divides decomposition into five stages: fresh, discoloration, active decomposition (bloat), advanced decomposition, and skeletonization.
- Adipocere formation and mummification are by-products of the decomposition process.
- Abiotic and biotic taphonomic variables dictate the state of decomposition or the state of preservation of remains.
- The major taphonomic variables affecting decomposition include temperature, humidity, arthropod activity, and scavenging.
- Diagenesis of skeletal remains significantly affects the quantity and quality of human remains recovered from a depositional environment.
- Several methods have been developed to estimate the PMI: entomology, botany, associated materials, biochemistry of tissues, soil chemistry, and observational changes.
- Forensic entomology techniques are based on the lifecycle and succession of arthropods attracted to the human body. These methods are most reliable in the early stages of decomposition.
- Biochemical changes within the body tissues and the leaching of these chemicals into the soil have demonstrated their potential to enhance PMI estimation accuracy beyond the early stages of decomposition.
- Observational changes in body appearance have been used traditionally by forensic anthropologists for PMI estimation, owing to the rapid and cost-effective nature of these assessments.
- The "Total Body Scoring" method is the predominant system used to describe macroscopic changes in the body and permits standardization of observations between research projects.
- The use of accumulated degree days (ADD) has provided an objective approach to correlate macroscopic changes in the body with temperature data with estimates the PMI.
- Recent research has statistically linked the observation of a body's appearance with environmental variables to improve the accuracy and reliability of PMI estimates.

Review questions

1. What are the primary stages of decomposition, and how does temperature affect them?
2. Compare and contrast autolysis and putrefaction.
3. Describe the major features associated with soft tissue changes during decomposition.
4. What are the major environmental factors affecting soft tissue decomposition? Explain how each could change the rate of decomposition.

5. What are the predominant factors influencing diagensis of bone and how can these be used to estimate PMI?
6. What types of cultural factors could potentially alter the decomposition rate of a body?
7. Explain the basis of forensic entomological PMI estimates.
8. Why is it important to develop statistical techniques for determining the PMI?
9. What PMI estimation methods are available for long-term decomposition events?
10. How do surface, aqueous, and burial contexts differ in relation to decomposition rates and the variables affecting these rates?
11. What is adipocere? How is it formed?
12. Compare and contrast the methods proposed by Megeysi et al. (2005) and Vass (2011). What are the merits of each, and how have these methods impacted current decomposition research?

Exercises

1. The police provide you with a photo of a decomposing body with little bony exposure in a lightly wooded area in Montana, with moderate animal scavenging present on the remains. Can you provide a PMI estimate? Why or why not? If yes, what is the stage of decomposition and the time range associated with this estimate?
2. Human remains were found in southern Virginia in March. The remains were mostly dry, had desiccated soft tissue adhering to the skeleton, and had no associated odor. The police believe that these are the remains of an individual that went missing 6 months ago. Can the remains be of the missing person, based on the condition of the remains? Why or why not?
3. A skull was found within the high grass on the sunny side of a hill in East Tennessee. There was a violet growing through the superior orbital fissure and a small root around the zygomatic arch, which has two rings. Did these remains originate on the sunny hillside? What lines of evidence can be used to estimate the PMI? What would be your PMI estimate?
4. A body was found in a shallow grave in Chattanooga, Tennessee. There was heavy plant growth and root infiltration around the remains. The bones showed evidence of wear, with epiphyseal damage, softening of the ribs, and dark staining consistent with the grave soil. The associated clothing was all but destroyed, except for nylon shorts. How long could the body have been there?
5. Remains were found in New Mexico, with most of the soft tissue present but mummified. The police believe that the remains are those of a person that went missing 3 years ago. Is the state of preservation consistent with the time interval? Why or why not?

Glossary

Abiotic: nonbiological in nature.
Actualistic studies: field experimental studies directed at understanding site formation processes, as borrowed from archaeology. Research that strives to provide an analogy for forensic investigations conducted in the field rather than a laboratory environment and without the relative strict control of variables.
Adipocere: a waxy, soap-like substance formed by the breakdown of fatty acids under moist, alkaline, and anaerobic conditions.
Algor mortis: the cooling of the body following death.
Autolysis: self-digestion, the breakdown of the cells as a result of their own enzymes.
Biotic: biological in nature.
Bloating: the distention of soft tissue resulting from the gaseous by-products of bacterial activity during decomposition.
Contact pallor: a pale area of the body resulting from compression against or contact with an object that has pushed all of the blood out from within the area and prevents pooling of blood within the area, associated with livor mortis.
Desiccation (also see mummification): the process by which tissue preserves through drying.

Diagenesis: the breakdown or conversion of the constituent parts of a material, as borrowed from geology. The physiochemical changes to bone after soft tissue decomposition.

Differential decomposition: atypical decomposition associated with the premature, irregular, or disproportional decomposition between body regions.

Livor mortis: the hypostatis or pooling of blood following death, resulting in purplish-red discoloration of the skin (lividity).

Longitudinal studies: studies that observe research subjects over a period of time to study how specific variables change.

Marbling: patterned discoloration of the skin visible during decomposition resulting from micro-organisms released within the circulatory vessels.

Microbe: a micro-organism (e.g., bacterium) that contributes to the decomposition process; associated with microbial.

Mummification (also see desiccation): the process by which tissue preserves through drying.

Necrophagous: feeding on decomposing remains.

Postmortem interval (PMI): also referred to as the time since death, the amount of time between death and discovery of deceased carrion.

Putrefaction: decomposition of organic material caused by microbial activity often producing foul-smelling byproducts.

Rigor mortis: the temporary stiffening of the human body following death that occurs soon after death, resulting from the lack of ATP to undo the protein binding associated with muscle contraction.

Saponification: anaerobic chemical reaction (hydrolysis or hydrogenation) of fats in the presence of an alkaline during decomposition that produces adipocere.

Skeletonization: complete soft tissue decomposition of a carrion in which only the hard tissues (e.g., teeth and bone) remain.

References

Adlam RE, and Simmons T. 2007. The effect of repeated physical disturbance on soft tissue decomposition—Are taphonomic studies an accurate reflection of decomposition? *J Forensic Sci* 52(5):1007–1014.

Agnew LCR, Aviado DM, and Brody JI. 1965. *Dorland's Medical Dictionary*, 24th edn. Philadelphia, PA: WB Saunders Co.

Anderson GS. 2001. Insect succession on carrion and its relationship to determining time since death. In JH Bryd, and JL Castner (Eds.), *Forensic Entomology: The Utility of Arthropods in Legal Investigations*. Boca Raton, FL: CRC Press. pp. 143–175.

Anderson GS. 2010. Decomposition and invertebrate colonization of cadavers in coastal marine environments. In J Amendt, ML Goff, CP Campobasso, and M Grassberger (Eds.), *Current Concepts in Forensic Entomology*. London, UK: Springer Press. pp. 223–272.

Anderson GS, and VanLaerhoven SL. 1996. Initial studies on insect succession on carrion in southwestern British Columbia. *J Forensic Sci* 41(4):617–625.

Bass WM. 1984. Time interval since death: A difficult decision. In TA Rathbun, and JE Buikstra (Eds.), *Human Identification: Case Studies in Forensic Anthropology*. Springfield, IL: CC Thomas. pp. 136–147.

Bass WM. 1997. Outdoor decomposition rates in Tennessee. In WD Haglund, and MH Sorg (Eds.), *Forensic Taphonomy: The Postmortem Fate of Human Remains*, Boca Raton, FL: CRC Press. pp. 181–186.

Bock JH, and Norris DO. 1997. Forensic Botany: An under-utilized resource. *J Forensic Sci* 42(3):364–367.

Bourel B, Fleurisse L, Hedouin V, Cailliez J, Creusy C, Gosset D, and Goff ML. 2001. Immunohistochemical contribution to the study of morphine metabolism in Calliphoridea larvae and implications in forensic entomotoxicology. *J Forensic Sci* 46(3):596–599.

Boyle S, Galloway A, and Mason RT. 1997. Human aquatic taphonomy in the Monterey Bay area. In WD Haglund, and MH Sorg (Eds.), *Forensic Taphonomy: The Postmortem Fate of Human Remains*, Boca Raton, FL: CRC Press. pp. 605–614.

Behrensmeyer AK. 1978. Taphonomic and ecologic information from bone weathering. *Paleobiology* 4(2):150–162.

Calce SE, and Rogers TL. 2007. Taphonomic changes to blunt force trauma: A preliminary study. *J Forensic Sci* 53(3):519–527.

Campobasso C, Di Vella G, and Introna F. 2001. Factors affecting decomposition and diptera colonization. *Forensic Sci Int* 120:18–27.

Carson EA, Stefan VH, and Powell JF. 2000. Skeletal manifestations of bear scavenging *J Forensic Sci* 45(3):515–526

Carter DO, and Tibbett M. 2003. Taphonomic mycota: Fungi with forensic potential. *J Forensic Sci* 48:1–4.

Carter DO, and Tibbett M. 2006. Microbial decomposition of skeletal muscle tissue (*Ovis aries*) in a sandy loam soil at different temperatures. *Soil Biol Biochem* 38:1139–1145.

Carter DO, and Tibbett M. 2008. Cadaver decomposition and soil: Processes. In M Tibbett, and DO Carter (Eds.), *Soil Analysis in Forensic Taphonomy: Chemical and Biological Effects of Buried Human Remains*. Boca Raton, FL: CRC Press. pp. 29–52.

Carter DO, Yellowlees D, and Tibbett M. 2007. Cadaver decomposition in terrestrial ecosystems. *Naturwissenschaften* 94:12–24.

Carter DO, Yellowlees D, and Tibbett M. 2008a. Temperature affects microbial decomposition of cadavers (*Rattus rattus*) in contrasting soils. *Appl Soil Ecol* 40:129–137.

Carter DO, Yellowlees D, and Tibbett M. 2008b. Using ninhydrin to detect grave soil. *J Forensic Sci* 53(2):397–400.

Carvalho LML, Linhares AX, and Trigo JR. 2001. Determination of drug lvels and the effect of diazepam on the growth of necrophagus flies of forensic importance in southeastern Brazil. *For Sci Int* 120(1):140–144.

Christensen AM, and Myers SW. 2011. Macroscopic observations of the effects of varying fresh water pH on bone. *J Forensic Sci* 56(2):475–479.

Clark MA, Worrell MB, and Pless JE. 1997. Post-mortem changes in soft tissue. In WD Haglund, and MH Sorg (Eds.), *Forensic Taphonomy: The Postmortem Fate of Human Remains*, Boca Raton, FL: CRC Press. pp. 151–164.

Cotton GE, Aufderheide AC, and Goldschmidt VG. 1987. Preservation of human tissue immersed for five years in fresh water of known temperature. *J Forensic Sci* 32(4):1125–1130.

Coyle HM, Lee C, Lin W, Lee HC, and Palmbach TM. 2005. Forensic botany: Using plant evidence to aid in forensic death investigation. *Croat Med J* 46(4):606–612.

Dabbs GR, and Martin DC. 2013. Geographic variation in taphonomic effect of vulture scavenging: The case of Southern Illinois. *J Forensic Sci* 58:S20–S25.

Dautartas AM. 2009. The effect of various coverings on the rate of human decomposition. M.A. Thesis, University of Tennessee, Knoxville, TN.

DiMaio VJM, and Dana SE. 2006. *Handbook of Forensic Pathology*. Boca Raton, FL: CRC Press.

Dudzik B, Parsons HR, and McKeown AH. 2011. Using the freeze-thaw cycle to determine the postmortem interval: An assessment of pig decomposition in west central Montana. In: *Proceedings from the American Academy of Forensic Sciences Meeting*, Chicago, IL, p. 375.

Evans WE. 1962 Adipocere formation in a relatively dry environment. *J Med Sci Law* 3:145–153.

Finley SJ, Benbow ME, and Javan GT. 2014. Microbial communities associated with human decomposition and their potential use as postmortem clocks. *Int J Leg Med* 129(3):623–632.

Forbes SL, Dent BB, and Stuart BH. 2005a. The effect of soil type on adipocere formation. *Forensic Sci Int* 154:35–43.

Forbes SL, Stuart BH, and Dent BB. 2005b. The effect of burial environment of adipocere formation. *Forensic Sci Int* 154:24–34.

Forbes SL, Stuart BH, and Dent BB. 2005c. The effect of the burial method on adipocere formation. *Forensic Sci Int* 154:44–52.

Galloway A. 1997. The process of decomposition. A model from the Arizona-Sonoran Desert. In WD Haglund, and MH Sorg (Eds.), *Forensic Taphonomy: The Postmortem Fate of Human Remains*. Boca Raton, FL: CRC Press. pp. 139–149.

Galloway A, Birkby WH, Jones AM, Henry TE, and Parks BO. 1989. Decay rates of human remains in an arid environment. *J Forensic Sci* 34(3):607–616.

Gill-King H. 1997. Chemical and ultrastructural aspects of decomposition. In WD Haglund, and MH Sorg (Eds.), *Forensic Taphonomy: The Postmortem Fate of Human Remains*, Boca Raton, FL: CRC Press. pp. 93–108.

Goff ML, Omori AI, and Goodbrod JR. 1989. Effect of cocaine in tissues on the development rate of Boettcherisca peregrina (Diptera: Sarcophagidae). *J Med Entomol* 26(1):91–93.

Goff ML. 1993. Estimation of postmortem interval using arthropod development and successional patterns. *Forensic Sci Rev* 5:81–94.

Goff ML. 2010. Early postmortem changes and stages of decomposition. In J Amendt, ML Goff, CP Campobasso, and Grassberger M (Eds.), *Current Concepts in Forensic Entomology*. London, UK: Springer Press. pp. 1–24.

Goff ML, Miller ML, Paulson JD, Lord WD, Richards E, and Omori AI. 1997. Effects of 3,4-methylenedioxymethamphetamine in decomposing tissues on the development of Parasarcophaga ruficornis (Diptera: Sarcophagidae) and detection of the drug in postmortem blood, liver tissue, larvae, and puparia. *J Forensic Sci* 42(2):276–280.

Haglund WD. 1993. Disappearance of soft tissue and disarticulation of human remains of aqueous environments. *J Forensic Sci* 38(4):806–815.

Haglund WD. 1997a. Dogs and coyotes: Postmortem involvement with human remains. In WD Haglund, and MH Sorg (Eds.), *Forensic Taphonomy: The Postmortem Fate of Human Remains*, Boca Raton, FL: CRC Press. pp. 367–382.

Haglung WD. 1997b. Rodents and human remains. In WD Haglund, and MH Sorg (Eds.), *Forensic Taphonomy: The Postmortem Fate of Human Remains*, Boca Raton, FL: CRC Press. pp. 405–414.

Hall DW. 1997. Forensic botany. In WD Haglund, and MH Sorg (Eds.), *Forensic Taphonomy: The Postmortem Fate of Human Remains*, Boca Raton, FL: CRC Press. pp. 353–363.

Hart AJ, and Whittaker AP. 2006. Forensic entomology: Insect activity and its roles in the decomposition of human cadavers. *Antenna* 30(4):159–164.

Haskell NH, Hall RD, Cervenka VJ, and Clark MA. 1997. On the body: Insects' life stage presence, their postmortem artifacts. In WD Haglund, and MH Sorg (Eds.), *Forensic Taphonomy: The Postmortem Fate of Human Remains*, Boca Raton, FL: CRC Press. pp. 353–363.

Haskell NH, McShaffrey DG, Hawley DA, Williams RE, and Pless JE. 1989. Use of aquatic insects in determining submersion interval. *J Forensic Sci* 34(3):622–632.

Haslam TCF, and Tibbett M. 2009. Soils of contrasting pH affect the decomposition of buried mammalian (*Ovis aries*) skeletal muscle tissue. *J Forensic Sci* 54(4):900–904.

Hauther KA, Cobaugh KL, Jantz LM, Sparer TE, and Debruyn JM. 2015. Estimating time since death from postmortem human gut microbial communities. *J Forensic Sci* 60(5):1234–1240.

Henssge C, Knight B, Krompecher T, Madea B, and Nokes L. 1995. *The Estimation of Time since Death in Early Postmortem Period*. Boston, MA: Arnold.

Henssge C, and Madea B. 2007. Estimation of the time since death. *Forensic Sci Int* 165(2–3):182–184.

Higley LG, and Haskell NH. 2001. Insect development and forensic entomology. In JJ Bryd, and JL Castner (Eds.), *Forensic Entomology: The Utility of Arthropods in Legal Investigations*. Boca Raton, FL: CRC Press. pp. 287–302.

Hobischak NR, and Anderson GS. 2002. Time of submergence using aquatic invertebrate succession and decompositional changes. *J Forensic Sci* 47(1):142–151.

Hutchins GM. 1985. Body temperature is elevated in the early postmortem period. *Hum Pathol* 16(6):560–561.

Introna F, Di Vella G, and Campobasso CP. 1999. Determination of postmoertm interval from old skeletal remains by image analysis of luminol test results. *J Forensic Sci* 44(3):535–538.

Introna F, Campobasso CP, and Goff ML. 2001. Entomotoxicology. *Forensic Sci Int* 120(1):42–47.

Janaway RC. 1996. The decay of buried human remains and their associated materials. In J Hunter, C Roberts, and A Martin (Eds.), *Studies in Crime: An Introduction to Forensic Archaeology*. London, UK: Batesford. pp. 58–85.

Janaway RC. 2002. Degradation of clothing and other dress materials associated with buried bodies of both archaeological and forensic interest. In WD Haglund, and MH Sorg (Eds.), *Advances in Forensic Taphonomy: Method, Theory, and Archaeological Perspectives*. Boca Raton, FL: CRC Press. pp. 279–402.

Janaway RC. 2008. Decomposition of materials associated with buried cadavers. In M Tibbett, and DO Carter (Eds.), *Soil Analysis in Forensic Taphonomy: Chemical and Biological Effects of Buried Human Remains*. Boca Raton, FL: CRC Press. pp. 153–202.

Kanz F, Reiter C, and Risser DU. 2014. Citrate content of bone for time since death estimation: Results from burials with different physical characteristics and known PMI. *J Forensic Sci* 59(3):613–620.

Kaliszan M, Hauser R, and Kernbach-Wighton G. 2009. Estimation of the time of death based on the assessment of postmortem processes with emphasis on body cooling. *Leg Med (Tokyo)* 11(3):111–117.

Klippel WE, Hamilton MD, and Synstelien JA. 2005. Raccoon (*Procyon lotor*) foraging as a taphonomic agent of soft tissue modification and scene alteration. In: *Proceedings of the American Academy of Forensic Sciences*, New Orleans, LA, vol. 11, p. 333.

Klippel WE, and Synstelien JA. 2007. Rodents as taphonomic agents: Bone gnawing by brown rats and gray squirrels. *J Forensic Sci* 52(4):765–773.

Knight, B. 1988. The evolution of methods for estimating the time of death from body temperature. *Forensic Sci Int* 36:47.

Knight B. 2002. *The Estimation of the Time Since Death in the Early Postmortem Period*, London: Edward Arnold.

Komar DA. 1998. Decay rates in a cold climate region: A review of cases involving advanced decomposition from the medical examiner's office in Edmonton, Alberta. *J Forensic Sci* 43:57–61.

Kulshrestha P, and Chandra H. 1987. Time since death: An entomological study on corpses. *Am J Forensic Med Pathol* 8(3):233–238.

Levine B, Golle M, and Smialek JE. 2000. An unusual drug death involving maggots. *Am J Forensic Med Path* 21(1):59–61.

MacDonnel N, and Anderson G. 1997. Aquatic forensics: Determination of the time since submergence using aquatic invertebrates. Technical Report: Canadian Police Research Center, Ottawa, Ontario, Canada.

Mann RW, Bass WM, and Meadows L. 1990. Time since death and decomposition of the human body: Variables and observations in cause and experimental field studies. *J Forensic Sci* 35(1):103–111.

Mant AK. 1953. Recent work on post-mortem changes and timing death. In K Simpson (Ed.), *Modern Trends in Forensic Medicine*. London, UK: Butterworth and Co. pp. 147–162.

Mant KA, and Furbank R. 1957. Adipocere—A review. *J Forensic Med* 4:18–35.

Marceau CM. 2007. Bone weathering in a cold climate: Forensic applications of a field experiment using animal models. M.A. Thesis. University of Alberta, Alberta, Canada.

Marks MK, Love JC, and Dadour IR. 2009. Taphonomy and time: Estimating the postmortem interval. In DW Steadman (Ed.), *Hard Evidence: Case Studies in Forensic Anthropology*, 2nd edn. Upper Saddle River, NJ: Prentice Hall. pp. 165–178.

Marks MK, Love JC, and Elkins SK. 2000. Time since death: A practical guide to physical postmortem events. In: *Proceedings of the American Academy of Forensic Sciences*, Reno, NV. pp. 181–182.

McKeown AH, Kemp WL, Dudzik B, and Parsons HR. 2011. Scavenging and its relationship to decomposition in the Northern Rockies. In: *Proceedings from the American Academy of Forensic Sciences*, Chicago, IL. p. 387.

Megyesi MS, Nawrocki SP, and Haskell NH. 2005. Using accumulated degree days to estimate the postmortem interval from decomposed human remains. *J Forensic Sci* 50(3):618–626.

Moffatt C, Simmons T, and Lynch-Aird J. 2015. An improved equation for TBS and ADD: Establishing a reliable postmortem interval framework for casework and experimental studies. *J Forensic Sci* doi: 10.1111/1556–4029.12931.

Micozzi MS. 1986. Experimental study of postmortem change under field conditions: Effects of freezing, thawing, and mechanical injury. *J Forensic Sci* 31(3):953–961.

Micozzi MS. 1991. *Postmortem Change in Human and Animal Remains: A Systematic Approach*. Springfield, IL: Charles C. Thomas.

Micozzi MS. 1997. Frozen environments and soft tissue preservation. In WD Haglund, and MH Sorg (Eds.), *Forensic Taphonomy: The Postmortem Fate of Human Remains*, Boca Raton, FL: CRC Press. pp. 171–180.

Nawrocki SP. 2013. Modeling core and peripheral processes in human decomposition: A conceptual framework. Proceeding of the 65th Annual meeting of the American Academy of Forensic Sciences, Washington, DC.

Neher DA, Barbercheck ME, El-Allaf SM, and Anas O. 2003. Effects of disturbance and ecosystem on decomposition. *Appl Soil Ecol* 23:165–179.

Nielson-Marsh CM, and Hedges REM. 2000. Patterns of diagenesis in bone I: The effects of site environments. *J Archaeol Sci* 27:1139–1150.

O'Brien TG. 1997. Movement of bodies in Lake Ontario. In WD Haglund, and MH Sorg (Eds.), *Forensic Taphonomy: The Postmortem Fate of Human Remains*. Boca Raton, FL: CRC Press. pp. 559–566.

Parkinson RA, Dias KA, Horswell J, Greenwood P, Banning N, Tibbett M, and Vass AA. 2009. Microbial Community Analysis of Human Decomposition in Soil. In K Ritz, L Dawson, and D Miller (Eds.), *Criminal and Environmental Soil Forensics*. New York: Springer. pp. 379–394.

Parmenter RR, and MacMahon JA. 2009. Carrion decomposition and nutrient cycling in a semiarid shrub–steppe ecosystem. *Ecol Monogr* 79(4):637–661.

Parsons HR. 2009. The postmortem interval: A systematic study of pig decomposition in west central Montana. MA Thesis. The University of Montana, Missoula, MT.

Payne JA. 1965. A summer carrion study of the body pig *Sus scrofa* Linnaeus. *Ecology* 46(5):592–602.

Payne JA, and King EW. 1972. Insect succession and decomposition of pig carcasses in water. *J Georgia Entomol Soc* 7(3):153–162.

Payne JA, King EW, and Beinhart G. 1968. Arthropod succession and decomposition of buried pigs. *Nature* 219:1180–1181.

Reed HB. 1958. A study of dog carcass communities in Tennessee with special reference to the insects. *Am Midl Nat* 34(3):213–245.

Reeves N. 2009. Taphonomic effects of vulture scavenging. *J Forensic Sci* 54(3):523–528.

Rhine S, and Dawson JE. 1998. Estimation of time since death in the southwestern United States. In KJ Reichs (Ed.), *Forensic Osteology: Advances in the Identification of Human Remains*, 2nd edn. Springfield, IL: Charles C. Thomas. pp. 145–159.

Ritchie GT. 2005. A comparison of human decomposition in an indoor and an outdoor environment. MA Thesis. The University of Tennessee, Knoxville, TN.

Rodriguez WC, and Bass WM. 1983. Insect activity and its relationship to decay rates of human cadavers in east Tennessee. *J Forensic Sci* 28(2):423–432.

Rodriguez WC, and Bass WM. 1985. Decomposition of buried bodies and methods that may aid in their location. *J Forensic Sci* 30(3):836–852.

Sadler DW, Fuke C, Court F, and Pounder DJ. 1995. Drug accumulation and elimination in Calliphora vicina larvae. *Forensic Sci Int* 71(3):191–197.

Sadler DW, Robertson L, Brown G, Fuke C, and Pounder DJ. 1997. Barbituates and analgesics in Calliphora vicina larvae. *J Forensic Sci* 42:481–485.

Sagara N. 1995. Association of ectomycorrhizal fungi with decomposed animal wastes in forest habitats: A cleaning symbiosis? *Can J Bot* 73(Suppl. 1):S1423–S1433.

Sagara N, Yamanaka T, and Tibbett M. 2008. Soil fungi associated with graves and latrines: Towards a forensic mycology. In M Tibbett, and DO Carter (Eds.), *Soil Analysis in Forensic Taphonomy: Chemical and Biological Effects of Buried Human Remains*. Boca Raton, FL: CRC Press. pp. 67–108.

Schoenly KG, Haskell NH, Hall RD, and Gbur JR. 2007. Comparative performance and complementarity of four sampling methods and arthropod preference tests from human and porcine remains at the Forensic Anthropology Center in Knoxville, Tennessee. *J Med Entomol* 44(5):881–894.

Seet BL. 2005. Estimating the postmortem interval in freshwater environments. MA Thesis. The University of Tennessee, Knoxville, TN.

Shattuck JK. 2009. An analysis of decomposition rates on outdoor surface variations in central Texas. MA Thesis. Texas State University, San Marcos, TX.

Shean BS, Messinger L, and Papworth M. 1993. Observations of differential decomposition on sun exposed v. shaded pig carrion in coastal Washington State. *J Forensic Sci* 38(4):938–949.

Simmons T, Adlam RE, and Moffatt C. 2010. Debugging decomposition data-comparative taphonomy studies and the influence of insects and carcass size on decomposition rate. *J Forensic Sci* 55(1):8–13.

Simpson K, and Knight B. 1985. *Forensic Medicine*, 9th edn. London, UK: Edward Arnold.

Sledzik PS. 1998. Forensic Taphonomy: Postmortem decomposition and decay. In KJ Reichs (Ed.), *Forensic Osteology: Advances in the Identification of Human Remains*, 2nded, Springfield, IL: Charles C. Thomas. pp. 109–119.

Smith JK. 2015. Raccoon scavenging and the taphonomic effects on human decomposition and PMI estimation. M.A. Thesis, University of Tennessee, Knoxville, TN.

Smith KGV. 1986. *A manual of forensic entomology*. London: British Museum.

Sorg MH, Dearborn JH, Monahan EI, Ryan HF, Sweeney KG, and David E. 1997. Forensic taphonomy in marine contexts. In WD Haglund, and MH Sorg (Eds.), *Forensic Taphonomy: The Postmortem Fate of Human Remains*, Boca Raton, FL: CRC Press. pp. 567–604.

Spicka A, Bushing J, Johnson R, Higley LG, and Carter DO. 2008. Cadaver mass and decomposition: How long does it take for a cadaver to increase the concentration of ninhydrin-reactive nitrogen in soil? In: *Proceedings of the American Academy of Forensic Sciences 61st Annual Scientific Meeting*, vol. 14. p. 178.

Spitz WU. 2006. *Spitz and Fisher's Medicolegal Investigation of Death: Guidelines for the Application of Pathology to Crime Investigation*, 4th edn. Springfield, IL: Charles C. Thomas Press.

Srnka CF. 2003. The effects of sun and shade on the early stages of decomposition. MA Thesis. The University of Tennessee, Knoxville, TN.

Stokes KL, Forbes SL, and Tibbett M. 2009. Does freezing skeletal muscle tissue affect its decomposition in soil? *Forensic Sci Int* 183:6–13.

Schwarcz HP, Agur K, and Jantz LM. 2010. A new method for determination of postmortem Interval: Citrate context of bone. *J Forensic Sci* 55(6):1516–1522.

Swift B. 2010. Methods of time since death estimation within the early post-mortem interval. *J Homicide Major Incident Invest*, National Policing Improvement Agen 6(1):97–112.

Synstelien JA. 2009. Raccoon (*Procyon lotor*) soft tissue modification of human remains. In: *Proceedings of the American Academy of Forensic Sciences*, vol. 15, p. 359.

Suckling JK, Spradley MK, and Godde K. 2015. A longitudinal study of human outsoor decomposition in Central Texas. *J Forensic Sci* doi:10.1111/1556–4029.12892.

Tabor Kreitlow KL. 2010. Insect succession in a natural environment. In JH Bryd, and JL Castner (Eds.), *Forensic Entomology: The Utility of Arthropods in Legal Investigations*, 2nd edn. Boca Raton, FL: CRC Press. pp. 251–269.

Tibbett M, and Carter DO. 2008. *Soil Analysis in Forensic Taphonomy: Chemical and Biological Effects of Buried Human Remains*. Boca Raton, FL: CRC Press.

Tibbett M, Carter DO, Haslam T, Major R, and Haslam R. 2004. A laboratory incubation method for determining the rate of microbiological degradation of skeletal muscle tissue in soil. *J Forensic Sci* 49:560–565.

Tracqui A, Keyser-Tracqui C, Kintz P, and Ludes B. 2004. Entomotoxicology for the forensic toxicologist: Much aso about nothing? *Int J Legal Med* 118(4):194–196.

Turner B, and Wiltshire P. 1999. Experimental validation of forensic evidence: A study of the decomposition of buried pigs in a heavy clay soil. *Forensic Sci Int* 101:113–122.

Ubelaker D. 1997. Taphonomic applications in forensic anthropology. In WD Haglund, and MH Sorg (Eds.), *Forensic Taphonomy: The Postmortem Fate of Human Remains*, Boca Raton, FL: CRC Press. pp. 77–92.

Vass AA. 2011. The elusive universal postmortem interval formulae. *Forensic Sci Int* 204(1):34–40.

Vass A, Barshick SA, Sega G, Caton J, Skeen JT, and Love JC. 2002. Decomposition chemistry of human remains: A new methodology for determining the postmortem interval. *J Forensic Sci* 47(3):542–553.

Vass AA, Bass WM, Wolt JD, Foss JE, and Ammons JT. 1992. Time since death determinations of human cadavers using soil solution. *J Forensic Sci* 37:1236–1253.

Vass AA, Smith RR, Thompson CV, Burnett MN, Dulgerian N, and Eckenrode BA. 2008. Odor analysis of decomposing buried human remains. *J Forensic Sci* 53(2):384–391.

Willey P, and Heilman A. 1987. Estimating the time since death using plant roots and stems. *J Forensic Sci* 32:1264–1270.

Wilson S, Hubbard S, and Pounder DJ. 1993. Drug analysis in fly larvae. *Am J Forensic Med Path* 14(2):118–120.

Wilson AS, Janaway RC, Holland AD, Dodson ID, Baran E, and Pollard AM et al. 2007. Modeling the buried human body environment in upland climes using three contrasting field sites. *Forensic Sci Int* 169:6–18.

Ubelaker DH, and Zarenko KM. 2011. Adipocere: What is known after two centuries of research. *Forensic Sci Int*. 208:167–172.

Suggested readings and additional resources

Warm, humid environments

Bass WM. 1984. Time interval since death: A difficult decision. In TA Rathbun, and JE Buikstra (Eds.), *Human Identification: Case Studies in Forensic Anthropology*. Springfield, IL: CC Thomas. pp. 136–147.

Bass WM. 1997. Outdoor decomposition rates in Tennessee. In WD Haglund, and MH Sorg (Eds.), *Forensic Taphonomy: The Postmortem Fate of Human Remains*. Boca Raton, FL: CRC Press. pp. 181–186.

Love JC. 2001. *Evaluation of decay odor as a time since death indicator. Thesis presented for the Doctorate of Philosophy*. Knoxville, TN: The University of Tennessee.

Love JC, and Marks MK. 2003. Taphonomy and time: Estimating the postmortem interval. In DW Steadman (Ed.), *Hard Evidence: Case Studies in Forensic Anthropology*. Upper Saddle River, NJL: Pearson Education, Inc. pp. 160–175.

Marks MK, Love JC, and Dadour IR. 2009. Taphonomy and time: Estimating the postmortem interval. In DW Steadman (Ed.), *Hard Evidence: Case Studies in Forensic Anthropology*, 2nd edn. Upper Saddle River, NJ: Prentice Hall. pp. 165–178.

Reed HB. 1958. A study of dog carcass communities in Tennessee, with special references to the insects. *Am Midl Nat* 59:213–245.

Sauerwein K. 2012. *Estimating the postmortem interval from the pattern of staining on skeletal remains*. In: *Proceedings of the American Academy of Forensic Sciences*, Atlanta, GA. p. 384.

Srnka CF. 2003. *The effects of sun and shade on the early stages of human decomposition Thesis Presented for the Master of Arts Degree*. Knoxville, TN: University of Tennessee.

Cold weather

Bunch AW. 2009. The impact of cold climate on the decomposition process. *J Forensic Ident* 59(1):26–44.

Decker P. 1978. Postmortem bacteriology. *Bull Ayer Clin Lab New Ser* 4(2):2–5.

Dillon LC. 1997. *Insect succession on carrion in three biogeoclimatic zones of British Columbia. Master's Thesis*. Department of Biological Sciences, Simon Fraser University, Burnaby, British Columbia, Canada

Komar KA. 1998. Decay rates in a cold climate region: A review of cases involving advanced decomposition from the medical examiner's office in Edmonton, Alberta. *J Forensic Sci* 43:57–61.

Micozzi MS. 1986. Experimental study of postmortem change under field conditions: Effects of freezing, thawing and mechanical injury. *J Forensic Sci* 31:953–961.

Micozzi MS. 1997. Frozen environments and soft tissue preservation. In WD Haglund, and MH Sorg (Eds.), *Forensic Taphonomy: The Postmortem Fate of Human Remains*. Boca Raton, FL: CRC Press. pp. 171–180.

Terneny TT. 1997. Estimation of time since death in humans using mature pigs. Masters' Thesis. Missoula, MT: Department of Anthropology, University of Montana.

Aqueous environments

Anderson GS, and Hobischak NR. 2004. Decomposition of carrion in the marine environment in British Columbia, Canada. *Int J Legal Med* 118(4):206–209.

Boyle S, Galloway A, and Mason RT. 1997. Human aquatic taphonomy in the Monterey Bay area. In WD Haglund, and MH Sorg (Eds.), *Forensic Taphonomy: The Postmortem Fate of Human Remains*. Boca Raton, FL: CRC Press. pp. 605–613.

Davis JB, and Goff ML. 2000. Decomposition patterns in terrestrial and intertidal habitats on Oahu Island and Coconut Island, Hawaii. *J Forensic Sci* 45(4):836–842.

Dix JD. 1987. Missouri's lakes and the disposal of homicide victims. *J Forensic Sci* 32(3):806–809.

Dumser TK, and Turkay M. 2008. Postmortem changes of human bodies on the Bathyal Sea floor—Two cases of aircraft accidents above the open sea. *J Forensic Sci* 53(5):1049–1052.

Ebbesmeyer CC, and Haglund WD. 1994. Drift trajectories of a floating human body simulated in a hydraulic model of Puget Sound. *J Forensic Sci* 39(1):231–240.

Haglund WD. 1993. Disappearance of soft tissue and the disarticulation of human remains from aqueous environments. *J Forensic Sci* 38:806–815.

Haglund WD, and Sorg MH. 2002. Human remains in water environments. In WD Haglund, and MH Sorg (Eds.), *Advances in Forensic Taphonomy: Method, Theory and Archeological Perspectives*. Boca Raton, FL: CRC Press. pp. 201–218.

O'Brien TG. 1997. Movement of bodies in Lake Ontario. In WD Haglund, and MH Sorg (Eds.), *Forensic Taphonomy: The Postmortem Fate of Human Remains*. Boca Raton, FL: CRC Press. pp. 559–565.

Payne JA, and King EW. 1972. Insect succession and decomposition of pig carcasses in water. *J Georgia Entomol Soc* 73:153–162.

Dry environments

Galloway A. 1997. The Process of decomposition. A model from the Arizona-Sonoran Desert. In WD Haglund, and MH Sorg (Eds.), *Forensic Taphonomy: The Postmortem Fate of Human Remains*, Boca Raton, FL: CRC Press. pp. 139–149.

Galloway A, Birkby WH, Jones AM, Henry TE, and Parks BO. 1989. Decay rates of human remains in an arid environment. *J Forensic Sci* 34(3):607–616.

Rhine S, and Dawson JE. 1998. Estimation of time since death in the southwestern United States. In KJ Reichs (Ed.), *Forensic Osteology: Advances in the Identification of Human Remains*, 2nd edn. Springfield, IL: Charles C. Thomas. pp. 145–159.

Biochemistry

Anderson B, Meyer J, and Carter DO. 2013. Dynamics of ninhydrin-reactive nitrogen and pH in gravesoil during the extended postmortem interval. *J For Sci* 58(5):1348–1352.

Benninger L, Carter D, and Forbes S (2008) The biochemical alterations of soil beneath a decomposing carcass. *Forensic Sci Int* 180:70–75

Cobaugh KL, Schaeffer SM, and DeBruyn JM. 2015. Functional and structural succession of soil microbial communities below decomposing human cadvers. *Plos one* 10(6):e0130201.

Costello EK, Lauber CL, Hamaday M, Fierer N, Gordon JI, and Knight R. 2009. Bacterial community variation in human body habitats across space and time. *Sci* 336:1694–1697.

Damann FE. 2010. Human decomposition ecology at the University of Tennessee Anthropology Research Facility [PhD Dissertation]. University of Tennessee, Knoxville, TN.

Damann FE, Tanittaisong A, and Carter DO. 2012. Potential carcass enrichment of the University of Tennessee Anthropology Research Facility: A baseline survey of edaphic features. *Forensic Sci Int* 222(1):4–10.

Damann FE, and Carter DO. 2013. Human decomposition ecology and postmortem microbiology. In J Pokines, and S Symes (Eds.), *Manual of Forensic Taphonomy*, Boca Raton, FL: CRC Press. pp. 37–51.

Hyde E et al. (2013) The living dead: Bacterial community structure of a cadaver at the onset and end of bloat stage of decomposition. *Plos One* 8:e77733.

Hopkins DW, Wiltshire PEJ, and Turner BD. 2000. Microbial characteristics of soils from graves: An investigation at the interface of soil and microbiology and forensic science. *Appl Soil Ecol* 14:283–288.

Jaggers KA, and Rogers TL. 2009. The effects of soil environment on postmortem interval. A macroscopic analysis. *J Forensic Sci* 54(6):1217–1222.

Metcalf JL, Parfrey LW, Gonzalez A, Lauber LW, Knights D, Ackerman G, Humphrey GC, Gebert MJ, Treuren WV, Berg-Lyons D, Keepers K, Guo Y, Bullard J, Fierer N, Carter DO, and Knight R. 2013. A microbial clock provides an accurate estimate of the postmortem interval in a mouse model system. *eLife* 2:e01104. DOI: 10.7554/eLife.01104.

Meyer J, Anderson B, and Carter DO. 2013. Seasonsal variation of carcass decomposition and gravesoil chemistry In a cold (Dfa) climate. *J Forensic Sci* 58(5):1175–1182.

Pechal J et al. (2013) The potential use of bacterial community succession in forensics as described by high throughput metagenomic sequencing. *Int J Legal Med* 128:193–205.

Tibbett M, and Carter DO. 2008. *Soil Analysis in Forensic Taphonomy: Chemical and Biological Effects of Buried Human Remains*. Boca Raton, FL: CRC Press.

Tibbett M, and Carter DO. 2009. Research in forensic taphonomy: A soil-based perspective. In K Ritz, L Dawson, and D Miller (Eds.), *Criminal and Environmental Soil Forensics*. New York: Springer.

Vass A, Barshick SA, Sega G, Caton J, Skeen JT, and Love JC et al. 2002. Decomposition chemistry of human remains: A new methodology for determining the postmortem interval. *J Forensic Sci* 47(3):542–553.

Vass AA, Smith RR, Thompson CV, Burnett MN, Dulgerian N, and Eckenrode BA. 2008. Odor analysis of decomposing buried human remains. *J Forensic Sci* 53(2):384–391.

Adipocere

Kahana T, Almog J, Levy J, Shmeltzer E, Spier Y, and Hiss J. 1999. Marine taphonomy: Adipocere formation in a series of bodies recovered from a single shipwreck. *J Forensic Sci* 44(5):897–901.

Lo SJ. 2007. *Factors influencing adipocere formation*. BA Hons Thesis. Simon Fraser University, Burnaby, British Columbia, Canada.

Mant AK. 1960. *Forensic Medicine: Observation and Interpretation*. Chicago, IL: The Year Book Publishers Inc.

Mellen PFM, Lowry MA, and Micozzi MS. 1993. Experimental observations on adipocere formation. *J Forensic Sci* 38:91–93.

O'Brien TG. 1994. Human soft-tissue decomposition in an aquatic environment and its transformation into adipocere. Knoxville, TN: University of Tennessee.

O'Brien TG, and Kuehner AC. 2007. Waxing grave about adipocere: Soft tissue change in an aquatic context. *J Forensic Sci* 52(2):294–301.

Taphonomic variables

Cross P, and Simmons T. 2010. The influence of penetrative trauma on the rate of decomposition. *J Forensic Sci* 55(2):295–301.

Dautartas AM. 2009. *The effect of various coverings on the rate of human decom position*. Thesis Presented for the Master of Arts Degree, University of Tennessee, Knoxville, TN.

Kelly JA, van der Linde TC, and Anderson GS. 2009. The influence of clothing and wrapping on carcass decomposition and arthropod succession during the warmer seasons In central South Africa. *J Forensic Sci* 54:1105–1112.

Mazzarelli D, Vanin S, Gibelli D, Maistrello L, Porta D, Rizzi A, and Cattaneo C. 2015. Splitting hairs: Differentiating between entomological activity, taphonomy, and sharp force trauma on hair. *Forensic Sci Med Path* 11(1):104–110.

Ritchie G. 2005. *A comparison of human decomposition in an indoor and an outdoor environment*. Thesis Presented for the Master of Arts Degree, University of Tennessee, Knoxville, TN.

Voss SC, Cook DF, and Dadour IR. 2011. Decomposition and insect succession of clother and unclothed carcasses in Western Australia. *Forensic Sci Int* 211:67–75.

Zhou C, and Byard RW. 2011. Factors and processes causing accelerated decomposition in human cadavers: An overview. *J Forensic Legal Med* 18:6–9.

Forensic entomology

Amendt J, Richards CS, Campobasso CP, Zehner R, and Hall MJR. 2011. Forensic entomology: Applications and limitations. *Forensic Sci Med Path* 7(4):379–392.

Byrd JH, and Castner JL. 2010. *Forensic Entomology: The Utility of Arthropods in Legal Investigations*, 2nd edn. Boca Raton, FL: CRC Press. pp. 251–269.

Comstock JL, Desaulniers J, LeBlanc HN, and Forbes SL. 2015. New decomposition stages to describe scenarios involving the partial and complete exclusion of insects. *Can Soc Forensic Sci J* 48 (1):1–19.

Goff ML. 1993. Estimation of postmortem interval using arthropod development and successional patterns. *Forensic Sci Rev* 5:81–94.

Michaud J-P, Schoenly KG, and Moreau G. 2012. Samplong flies or sampling flaws? Experimental design and inference strength in forensic entomology. *J Med Entomol* 49(1):1–10.

Rodriguez WC, and Bass WM. 1983. Insect activity and its relationship to decay rates of human cadavers in east Tennessee. *J Forensic Sci* 28:423–432.

Schoenly KG, Haskell NH, Hall RD, and Gbur JR. 2007. Comparative performance and complementarity of four sampling methods and arthropod preference tests from human and porcine remains at the Forensic Anthropology Center in Knoxville, Tennessee. *J Med Entomol* 44(5):881–894.

Williams RE, and Haskell NH. 2009. *Collection of Entomological Evidence at the Death Scene. Entomology and Death: A Procedural Guide*. Clemson, SC: Forensic Entomology Partners. pp. 85–101.

Forensic botany

Coyle MH. 2005. *Forensic Botany: Principles and Applications to Criminal Casework*. Boca Raton, FL: CRC Press.

Hall DW. 1997. Forensic botany. In WD Haglund, and MH Sorg (Eds.), *Forensic Taphonomy: The Postmortem Fate of Human Remains*. Boca Raton, FL: CRC Press. pp. 353–362.

Hall DW, and Byrd JH. 2012. *Forensic Botany: A Practical Guide*. Chichester, UK: John Wiley & Sons.

Morse D, Duncan J, and Stoutamire J. 1983. *Handbook of Forensic Archaeology and Anthropology*. Tallahassee, FL: Rose Printing.

Rowe WF. 1997. Biodegradation of hairs and fibers. In WD Haglund, and MH Sorg (Eds.), *Forensic Taphonomy: The Postmortem Fate of Human Remains*. Boca Raton, FL: CRC Press. pp. 337–352.

Chapter 16

Methods of Personal Identification

Angi M. Christensen and Bruce E. Anderson

CONTENTS

Introduction	314
Current techniques and practices	315
Comparative radiography	315
Nonimaged records comparison	319
Craniofacial superimposition	322
DNA and forensic anthropology	323
Additional considerations	324
Admissibility of evidence	324
Facial approximation	325
Summary	326
Questions	327
Disclaimer	327
References	328
Suggested readings	333

LEARNING OBJECTIVES

1. Describe the three categories of personal identification, and name the types of evidence typically involved in each category.
2. List the *Daubert* criteria, and explain how they affect personal identification.
3. Discuss the landmark cases that form the "trilogy" of Supreme Court decisions, and explain their impact on the admissibility of expert testimony.
4. Explain how a forensic anthropologist's expertise can assist the personal identification process.
5. Name several common methods used to compare antemortem and postmortem skeletal information for personal identification.
6. Discuss the utility of anatomical variation in personal identification from radiographic images.
7. Give examples of nonimaged records, and explain how personal identification is made with nonimaged records.
8. Explain how craniofacial approximation is used to assist the personal identification process.
9. Discuss the use of DNA in personal identification and in efforts to separate commingled remains.

INTRODUCTION

The identification of unknown individuals is an integral aspect of the medico-legal death investigation and is important in our society for both legal and humanitarian reasons. Legally, many issues depend on the ability to establish identity, including inheritance and succession to property, collection of insurance policies and pensions, detection of fraudulent deaths, accident reconstruction, remarriage, issuance of a death certificate, and other matters concerning property and business interactions. Morally, identification is often critical in closure and resolution for surviving relatives and friends, as well as being important in matters of international concern such as conflict and human rights investigations.

Confirmation of personal identification is usually the responsibility of the medico-legal authority (usually a forensic pathologist or coroner) or law enforcement officer. While other law enforcement officials, professionals, and scientists may assist in the investigative process, it is ultimately their responsibility to determine identification and issue a death certificate. Frequently, however, the actual task of identification falls upon a forensic scientist (forensic anthropologist, forensic odontologist, fingerprint examiner, and so on). Valid methods of personal identification (also called *individuation*) are based on two established and previously known facts—the identity of that individual and a record of his or her own particular uniqueness(es). Identification is then the process of verifying that the individual concerned is the same as the one that is known. A personal identification can be made on the basis of a single unique identifier and/or the presence of multiple consistent features. Forensic experts in identification often use a variety of methods and lines of evidence in order to assess identity as accurately as possible. Because all identification methods require the comparison of antemortem and postmortem data, the identification method(s) used is dictated by the postmortem condition of the body, as well as the availability, quantity, and quality of antemortem information about the deceased. The reliability of individual methods varies, but corroboration of several methods can increase the probability of a correct identification.

Personal identification is usually categorized as being tentative, circumstantial, or positive; however, currently, no consensus exists amongst forensic scientists as to the usage of these terms (Anderson 2007). A tentative identification is one in which the identity of the decedent is suspected. This may be based on associated items such as a driver's license. Circumstantial identification, also known as presumptive, possible, probable, or putative identification, is based on strong consistencies between the remains and the missing person. It meets a higher standard than tentative identification but a lower standard than positive identification (however, in some cases, a circumstantial identification can be as certain as a positive identification, if based on adequate contextual and situational information). A personal identification may also be made by a preponderance of evidence, also known as identification by multiple corresponding factors, or multiple points of similarity. This level of identification is based on a number of uncorrelated pieces of evidence, including, for example, body location, biological profile, tattoos, scars, personal effects, skeletal anomalies, and mitochondrial DNA. Identification by exclusion may be used in mass fatality situations with a *closed* population, defined by having a list of occupants, when all other individuals have been identified. Examples of such closed systems include car crashes with multiple individuals, most aircraft accidents, and other mass-fatality accidents. Positive identification is also known as personal, biological, or scientific identification and is made through the use of information that is exclusive or unique to one individual.

Identification is often a straightforward process, particularly when death was recent, remains are fresh, and many clues are readily available. In these cases, soft tissues can often provide individualizing information. Visual examination of soft tissue features is the most frequently used method of personal identification (Sopher 1972). It usually involves recent deaths with well-preserved bodies whose facial and other physical features or markings are not distorted by decomposition or injury and can be readily identified by relatives or friends, or hospital, or home deaths in which identity is not in question. This means of *identification* can be termed *positive-by-lay persons*. Although common, this method of identification is not considered scientific and is subject to considerable error. A more reliable soft tissue identification method is the comparison of friction ridges or fingerprints. Perhaps, fingerprint comparison provides the most widely used scientific means of identification in the presence of soft tissues (Sopher 1972) and is based on the belief that the probability of two individuals having identical fingerprints is extremely remote. Little research on friction ridge analysis, however, has been directed toward developing population statistics (Committee on Identifying the Needs of the Forensic Sciences Community 2009).

One challenge in identification is that of establishing the identity of bodies that are skeletal, decomposed, dismembered, or badly burned. In these cases, visual recognition is obviously out of the question, and fingerprint evidence is often unobtainable. In cases where visual or fingerprint identification is not possible, anthropological methods of identification can be of great utility, because the skeleton, including teeth, will usually survive longer than other identifiable features and hence are almost always available for examination. In these cases, forensic anthropologists are often asked to assist in the identification process, owing to their expertise in human osteology and their ability to glean biological information from the skeleton and detect subtle differences between individuals and other anomalies or features that may distinguish that person's skeleton.

A personal identification utilizing forensic anthropological evidence is based on agreement between the unidentified skeleton and facts known about the body of a putative deceased who has been selected for comparison on the basis of being missing and possibly deceased (Kerley 1977). Any distinguishing features such as prosthetics or evidence of previous medical care, fractures, congenital or traumatic deformities, unusual allometric relations, epigenetic traits, morphological peculiarities, and abnormalities may be particularly convincing evidence and can often provide the basis for a highly probable identification. The greater the number of skeletal features that match those of a sought-for individual and the more unusual the features, the greater the probability that the identification is correct (Dutra 1944). In most cases, it is even possible to quantify the strength of an osteological identification by using statistical methods (Steadman et al. 2006).

While anthropological evidence for identification can be straightforward in many cases, it is important that the skeletal information be evaluated by a qualified forensic anthropologist who is equipped with a knowledge and appreciation for how, why, and where a structure may vary and what constitutes *normal* and *non-normal* variations. Such an appreciation can be gained only through a comprehensive understanding of skeletal biology, including a working knowledge of growth and development, sources of variation, and the purpose and function of the structure(s). This knowledge may be particularly useful in cases where explainable differences occur. Differences observed between antemortem and postmortem information can be evaluated to determine whether they represent real structural differences that may be the basis for exclusion, whether there are issues with methodology, or whether they can be explained by chronological (age/time-related) factors. Explainable differences may result from trauma or surgical intervention or simply from growth and degenerative changes that accumulate slowly over time (Komar and Buikstra 2008).

Although forensic anthropologists can take measures to produce, collect, or document high-quality postmortem data, the availability, quantity or quality of antemortem data may limit the ability to make meaningful identification comparisons. It is important that anthropologists do not overemphasize the probability of correct identification based on skeletal evidence and recognize and articulate limitations of the data or identification method. The confidence placed in a correct identification is dependent not only on the quality of the data for comparison but also on the known frequency of the shared trait or characteristic (such as a fracture, pathological condition, skeletal anomaly, and surgical hardware). These frequencies are rarely known. Nor may it be known how these features vary between sex and age or by geographic location or socioeconomic status (Komar and Lathrop 2006). Furthermore, training and expertise in various identification methods may not be equivalent among practitioners. Therefore, the ethical and professional responsibility of the forensic anthropologist is to recognize the extent of their expertise and not to participate in identification procedures beyond their expertise and/or for which they lack adequate training.

Depending on the postmortem condition of the remains and the antemortem data available for comparison, several methods may be used to compare antemortem and postmortem skeletal information, including comparative radiography and photography, nonimaged records' comparison, and photographic superimposition. The following section describes these techniques in more detail. More information about current standards and practices can also be found through the Scientific Working Group for Forensic Anthropology (or SWGANTH), whose purpose is to identify and publish *best practices* for methods, techniques, and approaches used in forensic anthropology (www.swganth.org).

CURRENT TECHNIQUES AND PRACTICES

Comparative radiography

Comparative radiography in the personal identification process involves the direct visual comparison of antemortem and postmortem radiographs and matching specific visual findings or features on the antemortem and postmortem radiographs of that person (Brogdon 1998). Radiographic identification is routinely used following mass disasters and in the identification of burned, mutilated, decomposed, fragmented, skeletonized, and otherwise-unrecognizable human remains. Especially in these latter cases, radiography is sometimes the only means by which an individual's identity can be established. Fortunately, the potential for available antemortem images is great, because radiographs have become a common diagnostic tool for various medical investigations, thus increasing the availability of antemortem radiographs for comparison. Indeed, radiographic comparison has become a regular procedure in the identification of unknown remains in most forensic facilities throughout the world (DiMaio and DiMaio 1989) and is probably the most commonly used method of personal identification from skeletal remains. The method is also not limited to conventional radiography; computed tomography (CT), multislice computed tomography (MSCT), and magnetic resonance imaging (MRI) are also becoming increasingly common procedures and may be used in radiographic comparisons (Dedouit et al. 2007; Haglund and Fligner 1993; Pfaeffli et al. 2007; Sidler et al. 2007).

Radiographic investigation of human remains began soon after the discovery of X-rays in 1895, as investigators came to realize that X-rays provided a nondestructive means of examining human remains. The method was first applied to ancient, rather than forensic, specimens, with several radiographs being made of Peruvian and Egyptian mummies by the end of the nineteenth century (Rowe 1953; Petrie 1987). Other early applications of radiography in anthropology included the study of bone pathologies (Hooton 1930) and bone growth and development (Greulich and Pyle 1959). Early radiographic investigations, however, typically considered only general anthropological and pathological findings, with little or no emphasis on skeletal variability (Brothwell et al. 1968).

The earliest suggestion of the use of radiography in the identification of unknown human bodies was by Scheuller (1921), who called attention to the potential use of frontal sinus variability in this context (see Figure 16.1). Radiography soon found applications in many medico-legal and forensic anthropological investigations, including age estimation; sex assessment; ancestry estimation; stature estimation; determining whether or not remains were human; locating and recovering bullets and other foreign bodies and determining the direction, angle, and location of wounds; detecting air embolisms; detecting and aging fractures and other trauma; diagnosing tuberculosis; examining hyoid or cartilage fractures in hanging or strangulation victims; examining past medical history; illustrating dental morphologies and anomalies; separating skeletal remains from wood charcoal and other charred material; diagnosing premortem skeletal health; studying the relationship between bone and soft tissue; detecting metallic poisons such as arsenic, lead, and mercury in suspected poisoning cases; and examining burned, skeletonized, or decomposed

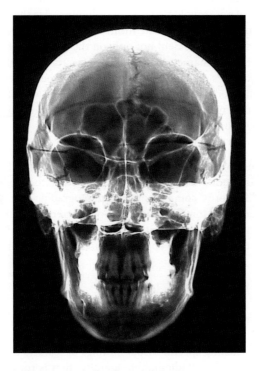

Figure 16.1 Radiograph depicting cranial features, including frontal sinuses.

individuals for the purpose of identification (Camps 1969; Eckert and Garland 1984; Fatteh and Mann 1969; Greulich and Pyle 1959; Krogman and Iscan 1986; McCormick et al. 1985; Messmer 1986; Morgan and Harris 1953; Murphy and Gantner 1982; Rowe 1953; Schmidt and Kallieris 1982; Stewart 1979). As a result, the value of radiography has become well-established in the criminal and medico-legal communities. Comparative radiography for the purpose of identification has become a commonly used technique in forensic anthropology, and it has been said to compare favorably with fingerprint identification (Murphy et al. 1980).

The typical method of comparison generally follows these steps: (1) A suitable antemortem radiograph of the putative decedent is obtained from an appropriate source. Typically, this is accomplished with the assistance of law enforcement officials who search for medical facilities possibly visited by the putative deceased and request available radiographs. (2) A radiograph of the unidentified remains is taken, with a similar scope, orientation, and magnification as the antemortem image. As discussed above, not all forensic anthropologists have commensurate training; thus, anthropologists who do not have sufficient experience taking radiographs should solicit the assistance of a radiologist. (3) The two radiographs are compared either by direct visual inspection of side-by-side films or by superimposing the images, looking for consistencies and inconsistencies in, for example, bone morphology, trabecular patterns, and orientation and placement of foreign materials such as surgical implements, enteric accretions, skull features (paranasal sinuses, sella turcica, and cranial sutures), and dental features (morphology, restorations, pathologies, and missing teeth).

Abnormal features such as anomalous or unusual development, healed fractures, deformities, degenerations, pathologies, abnormal calcifications, tumors, trauma, and prosthetic devices can be important for radiographic identification purposes (Brogdon 1998; Murphy et al. 1980), as they often contribute to a radiographic image that is unique to that individual (see Figure 16.2). Moreover, in the event of some injury or abnormality, the chance that the individual will have an antemortem radiograph for comparison is greater. This technique has been applied in various published case studies in identification involving, for example, anomalies of the feet and legs (Sudimack et al. 2002; Owsley and Mann 1989), postsurgical defects and devices (Hogge et al. 1995; Penalver et al. 1997; Sivaloganathan and Butt 1988), pelvic deformities and peculiarities (Angyal and Derczy 1998; Brogdon 1998; Rouge et al. 1993; Murphy and Gantner 1982; Varga and Takacs 1991), wrist fractures (Atkins and Potsaid 1978), and a patellar defect (Riddick et al. 1983).

Although some abnormalities may indeed be unique to an individual, the richness of normal anatomical detail revealed in radiographs may be equally, if not more, important. The widespread occurrence of nonpathological anatomical features available for comparison in most radiographs may obviate the need to use pathological or abnormal features and

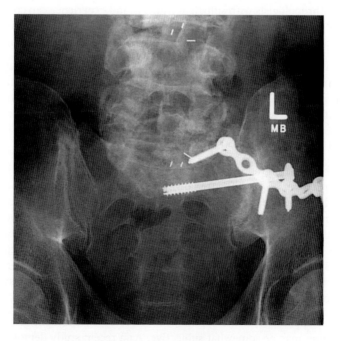

Figure 16.2 Radiograph depicting spinal and pelvic features and surgical hardware.

significantly increase the potential number of corresponding features for identification (Joblanski and Shum 1989). Numerous reported cases illustrate the use of radiography to establish individuality by using nonpathological variation of elements of the skeleton, including parts of the skull (Christensen 2005b; Culbert and Law 1927; Fatteh and Mann 1969; Joblanski and Shum 1989; Kirk et al. 2002; Murphy et al. 1980; Rhine and Sperry 1991; Sassouni 1959; Singleton 1951; Thorne and Thyberg 1953), the chest (Martel et al. 1977; Murphy et al. 1980; Singleton 1951), the sternum (Rouge et al. 1993), the abdomen (Angyal and Derczy 1998; Joblanski and Shum 1989; Murphy et al. 1980), costal cartilage (Marek et al. 1983), the spine (Brogdon 1998; Fatteh and Mann 1969; Jensen 1991; Kahana et al. 1998, 2002; Murphy et al. 1980; Owsley et al. 1993; Singleton 1951; Stevens 1966; Valenzuela 1997), the clavicle (Adams and Maves 2002; Marek et al. 1983; Sanders et al. 1972), the scapula (Ubelaker 1990), the hand and wrist (Greulich 1960; Koot 2003), the pelvis (Singleton 1951), the femur (Dutra 1944), and the ankle and foot (Kade et al. 1967; Singleton 1951).

Identity can be established in these cases by comparison of the size, location, and external cortical contours and bone surfaces, including spines, processes, exostoses, tubercles, sutures, foramina, and medullary cavities (Kerley 1977); metric analysis (Sassouni 1959; Thorne and Thyberg 1953); internal architecture of the bones such as trabecular lattice pattern (Joblanski and Shum 1989; Kahana and Hiss 1994; Kahana et al. 1998; Mann 1998); vascular grooves (Brogdon 1998), unique areas of lucency; bone density; and cranial sinus shapes and patterns. Many, if not all, parts of the skeleton are useful for identification, and while those that tend to be more variable may be more reliable, nearly every bone in the body could be used for personal identification (Hogge et al. 1993).

Comparison of antemortem and postmortem dental radiographs is a well-documented, accepted, and widely used procedure and is considered one of the most effective means of identification of unknown bodies (Sainio et al. 1990). Because teeth and dental restorations are more resistant to destruction than osseous elements, they are more likely to be available for comparison. The method's reliability is a product of the numerous combinations of restorations, prosthetics, diastemata, and carious lesions that an individual may possess that can be highly visible in radiographic images. Nondental elements, such as the morphology of the maxillary sinus inferior margin (floor) and the mandibular canal, are also frequently imaged in radiographic films made by dentists and can be utilized in the antemortem-postmortem radiograph comparison. There is some debate as to whether it is appropriate for a forensic anthropologist to compare antemortem dental radiographs with skeletal remains. While it is almost certainly the case that specific statements regarding dental therapeutic procedures and treatments are best left to dentists, teeth and surrounding structures are part of the skeleton, and thus, comparisons of bone and tooth structure and morphology are well within the expertise of the forensic anthropologist. As is true with forensic anthropology's overlap with radiology, the scientific community should acknowledge the overlap between forensic anthropology and forensic odontology and concede that forensic anthropologists with sufficient training and experience can competently perform personal identification based on the dentition. Those forensic anthropologists who practice in jurisdictions where forensic odontologists also practice are fortunate, but not all medico-legal jurisdictions utilize both disciplines.

There is no recognized number of *points of concordance* that must be demonstrated to support a personal identification, but the antemortem and postmortem radiographs should match in sufficient detail to conclude that they are from the same individual *and with no unexplainable differences*. When making statements regarding identity (or probability of correct identification) based on comparative radiography, consideration should be given to the uniqueness or population frequencies of particular skeletal features, if known. Figure 16.3 illustrates an example of a comparison of antemortem and postmortem radiographs.

An important concern regarding radiographic comparison is that the morphology of adult bone is not stable but continually remodeled and restructured in response to changes in function. The stability of bone is related to the stability of the loading regimes to which it is subjected, as well as age, which is associated with a loss of cortical bone. These factors may lead to time- and age-related changes in bone structure and therefore in radiograph images. However, osseous remodeling generally follows the existing architecture within the skeleton, and as the bone cells are replaced (approximately seven years for complete osseous turnover), the morphological intricacies are usually maintained. One study (Sauer et al. 1988) demonstrated that aspects of the postcranial skeleton generally chosen to compare for identification are quite stable, and the ability to make a positive identification from postcranial axial material may not diminish, even after two-and-a-half decades.

Another difficulty with this comparative method is that it involves a visual comparison, with the consequence that the final identification decision may be somewhat subjective. As a recent study demonstrated, however, even very

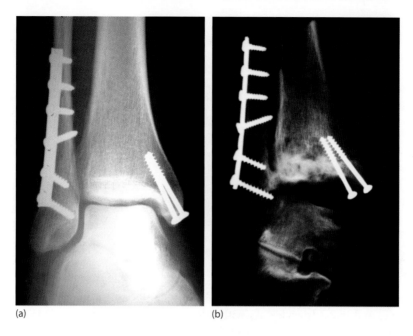

Figure 16.3 Comparison of antemortem (a) and postmortem (b) radiographs of an ankle.

small frontal sinus radiograph pairs can be matched with a high rate of success, even by those will little or no experience (Smith et al. 2010). Nonetheless, other studies as well as professional standards support the need for trained interpreters (SWGANTH, Hogge et al. 1993). However, it must again be emphasized that when performing these comparisons in the medico-legal world, experience is monumental. No errors are allowable, because mistakes can mean that families of a decedent are provided with a misidentification. While an inexperienced examiner may be able to correctly match 19 out of 20 radiographs for a *high rate of success* on a exercise, that single misidentification cannot be tolerated in the medico-legal arena and would not be expected if the comparison were performed by a more experienced examiner.

Historically, research in comparative radiography involved the publication of successful case reports (some examples of which are listed above). More recently, efforts have included the investigation of validity, individuality, reliability, and error estimation (Christensen 2004, 2005a, 2005b; Koot et al. 2005), as well as attempts to facilitate and standardize the comparison process (Cox et al. 2009; Ribeiro 2000; Yoshino et al. 1987).

Nonimaged records comparison

Nonimaged records comparison is a personal identification method involving the comparison of antemortem records such as notes, charts, and other recorded information to features or characteristics of skeletal remains. When antemortem radiographs or other medical images are not available, other information contained in medical and/or dental records may include documented features or patterns of the skeleton that may be individuating and therefore useful for identification (see Figure 16.4). Such information may include records of surgical procedures (including implanted devices), dental patterns (including supernumerary and occult teeth), injury treatments, and pathologies and anomalies. Examples of these nonimaged record comparisons include a history of a broken bone that required medical attention, a surgical procedure such as a ventriculosostomy (Anderson and Gilson 2005) or orthopedic hardware, orthodontic braces worn as a child that resulted in a recognizable pattern of caries beneath the bracket or band, the retention of an orthodontic appliance after the therapy is complete, and an unusually rare pattern of missing teeth. For each of these examples, it is preferred to have antemortem records from a dental office or medical facility for comparison with the postmortem findings, but sometimes, family recollection must suffice.

The procedure for identification by written records comparison typically involves (1) identifying of the pattern, injury, pathology, or anomaly that the practitioner believes may be useful for identification; (2) obtaining antemortem medical or dental charts and/or notes on the suspected deceased (similar to the method of locating imaged

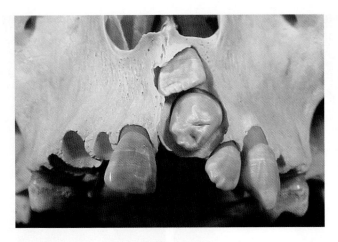

Figure 16.4 Unusual dental pattern that may be useful for personal identification.

records following a presumptive identification); and (3) comparing the record to features of the skeletal remains, looking for consistencies and inconsistencies. As with other comparative methods, some inconsistencies may be explainable, while those that are not must be the basis for exclusion.

Surgical implants (e.g., surgical implements, artifacts, and appliances) often have serial numbers and stamps of the manufacturer and may be linked with an individual or associated with a particular shipment of surgical implements that can be tracked to a specific area during a certain time period (see Figure 16.5). Once the manufacturer's symbol and lot and/or serial number from the device have been located, local or national registries or the manufacturer may be contacted for information associated with that number. It may be possible to link the device to an individual if there is a written or otherwise documented record of the device information in the patient's record (Mackinnon and Mundorff 2006). In other cases, however, it may only be possible to narrow the search, since registries may not necessarily provide adequate information to associate the surgical implement with a particular individual. It is also common to contact sales representatives of the manufacturer of the appliance (e.g., Zimmer, Synthes, and Astromedica) directly for information on the geographical regions that a particular surgical implant may have been marketed.

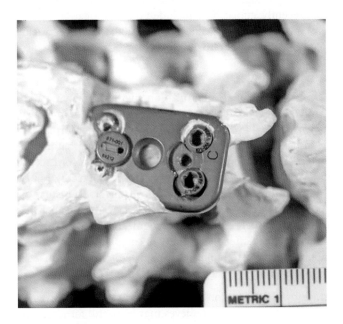

Figure 16.5 Surgical implant with visible manufacturer information.

It is our experience that someone in these offices is always eager to perform some *detective work* and thus possibly aid in locating the appropriate medical records.

Adams (2003b) reported that reference data sets show a high diversity present in dental patterns and found that even a small number of common (charted, nonradiographic) dental characteristics may produce a rare dental pattern that is useful for identification. Indeed, the diversity of dental patterns has been said to be comparable to the diversity found in mitochondrial DNA (Adams 2003a). A computer program (OdontoSearch; available at www.cilhi.army.mil) that uses reference databases and allows practitioners to calculate frequency information regarding the incidence of specific dental traits in the general population has even been developed (Adams 2003b).

Records of pathologies, anomalies, and injuries may also be very useful for identification, even in the absence of an imaged record. Identification proceeds by describing and differentially diagnosing the lesion, defect, wound, or anomaly on the skeleton and by conducting appropriate comparisons to the antemortem record. Examples of pathological conditions include chronic infectious disease, metabolic disorders, neoplastic diseases, developmental defects, degenerative joint conditions, ossified arterial plaque, cartilage or other soft tissue, and autoimmune diseases. The antemortem record may be compared to observations of gross findings in the skeleton or bone microscopy (London et al. 1994). Examples of anomalies include accessory bones, bipartite bones, sternal, septal, and other apertures, bifid and/or supernumerary ribs, vertebral shifts and other axial anomalies, prominent features (e.g., everted gonia, bilobed chin, and unusually large or small facial features), cranial asymmetry or modification, premature closure of cranial sutures, premature ossification of cartilage, supernumerary teeth, extra dental roots, fused teeth, dental agenesis, and polydactly. In the case of injuries, descriptions of the injury type, location, and severity in the medical record may be examined for consistency or inconsistency with the presence of fracture calluses or other injury-related lesions on the skeletal evidence. Once recognized, two criteria must be met in order for a pathological condition, anomaly, or injury to be potentially useful for identification purposes (www.swganth.org). The trait must have been recognized and documented *in vivo*, and the relative frequency of the condition must be known. The more uncommon the condition, the more potential it may have in contributing to identification. Unfortunately, the frequency of different pathological conditions or anomalies is typically not documented adequately in the clinical literature. Again, experience comes into play, as the more seasoned forensic scientist may have a better understanding of the relative occurrence of some of these discriminators.

Examiners should be cautious in excluding individuals, owing to the absence of a condition in the medical record. Conditions identified on the skeletal evidence may have been subclinical and therefore undocumented at the time of death or at the time the medical record was prepared. Also, if the medical records indicate a condition in the skeleton (such as an osseous fracture), and no corresponding condition can be found on the skeletal evidence, this inconsistency should not necessarily be the basis for exclusion. In some osseous fractures, the healing process completely, or nearly completely, remodels the bone to the point where it may not be possible to correctly determine that a fracture was once present. Thus, exclusions based on the lack of evidence are typically not recommended. However, if the antemortem injury was exceptionally severe, some indication of the healing process in the affected bone would be expected, and thus, the lack of any indication of a remodeled callus could be the basis for an exclusion.

Practitioners should also take caution that medical and dental charts and notes may not necessarily be accurate representations of an individual's medical or dental record. Every effort should be made to verify the medical record, especially if there is conflicting information. Problems may arise due to insufficient recording, recording mistakes, or insurance fraud. In addition, differences may exist in the recording systems used. Moreover, although there are usually written or typed summaries of medical problems and treatments, these summaries tend to be generalized and lack extensive detail (e.g., fracture to distal right fibula). Consequently, the nature of these records may mean that consistencies provide only circumstantial evidence.

In some cases, there is no basis for assumption of identification and no circumstantial clues to suggest identity, and even, a tentative identification cannot be made. In these cases, an anthropological analysis may be used to estimate biological information about the individual, including the ancestral affinity, sex, age, dental characteristics, and living stature of an individual. The estimation of a postmortem interval (PMI), or time since death, can also be a useful descriptor in evaluating whether an unidentified person might match a description of a missing person, if the length of disappearance is known. Estimating these biological and taphonomic characteristics was the topic of previous chapters in this volume and will not be described further here. This assessment, though useful in narrowing the pool of potential candidates in the search for identity, is never a positive identification; however, concordance between these antemortem and postmortem data many times directs the obtainment of the types of antemortem records than can affect such an identification. To wit, the biological information is compared with documented information or databases of missing persons (e.g., www.NamUs.gov), and individuals may be included or excluded from further consideration based on the fit with the biological profile. In the absence of other antemortem information on a presumptive deceased, a biological profile consistent with documented information about the presumed victim can provide corroborative evidence of identification.

As with radiographic comparisons, much of the early research involved case reporting. More recent research has involved the investigation of diversity and population frequencies of specific dental patterns and skeletal conditions (Adams 2003a; Komar and Lathrop 2006; Nozawa et al. 2002).

Craniofacial superimposition

Craniofacial superimposition (also known as *photographic superimposition* or *video superimposition*) is an identification method applied when an investigation has suggested that a set of remains relates to a particular missing person for whom photographs, videos, or other images (and no other antemortem data) are available (Ubelaker 2000). This method involves the superimposition of an image of a known individual (the suspected decedent) over a radiograph, photograph, negative, or drawing of the skull and/or the comparison of the photograph and skull through the use of digitized images (Ubelaker et al. 1992).

One of the earliest uses of craniofacial superimposition was a study of the remains of Johann Sebastian Bach (His 1895). The method was first applied in forensic context in the 1930s. Although the details of exactly how the comparison was conducted are not available, it involved the comparison of photographs viewed through a stereoscope with the three-dimensional head shape of the submitted skull (Glaister and Brash 1937; Ubelaker 2000). By the 1980s, the technique of craniofacial superimposition was somewhat standardized by using a skull, mirrors, metrical scales, and a camera. During this time, this identification technique was routinely utilized by the U.S. Army's Central Identification Laboratory (CILHI) in Hawaii and by the Human Identification Laboratory at the University of Arizona (Anderson 2010, pers. comm.). Also, by the 1980s, video cameras were being utilized, instead of still cameras at the Human Identification Laboratory at the University of Florida (Austin-Smith 1999). Craniofacial superimposition has been applied in numerous documented cases, using various images and comparison methodologies (Banjeree 1964; Basauri 1967; Delfino et al. 1983; Dorion 1983; Engel 1954; Koelmeyer 1982; Sen 1962; Verma 1981; Webster 1955).

The ideal image for this method is a close-up facial view, with the individual directly facing the camera and the teeth revealed. The skull is then imaged using, where possible, comparable scaling, positioning, lighting, perspective, and angle as the original portrait. The two images are then superimposed, usually by selecting some facial anthropometric landmark in the subject photograph and cranial anthropometric landmark on the skull (see Figure 16.6).

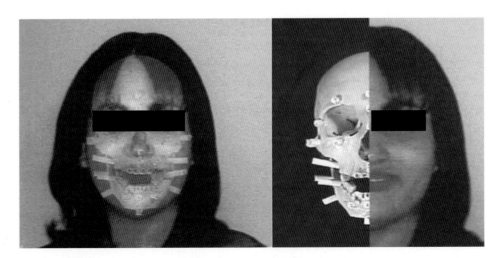

Figure 16.6 Craniofacial superimposition.

The practitioner then examines the superimposed images, looking for congruence of the lines and curves of the face and dental structures (Webster et al. 1986), or any other consistencies and inconsistencies in structures that are visible in both the antemortem and postmortem materials. An individual can be included if there is concordance between bony and/or dental landmarks, while exclusion results when there are irreconcilable disagreements between them. It will often be the case that some of the landmarks will perfectly match, some will partially match, and others will not match, and the practitioner will then make a decision about whether the skull corresponds to the photograph or not.

This approach is most often used to corroborate other identification methods or as a tool to include or exclude a possible identification (Kerley 1977), but positive identification is possible, depending on the uniqueness of the observed features (Austin-Smith and Maples 1994; Ubelaker 2000). Craniofacial superimposition should be conducted by experts with thorough knowledge of human cranial anatomy and variation (www.swganth.org) and possibly computer science. Empirical testing of the method has demonstrated that the probability of misidentification is relatively low (Austin-Smith and Maples 1994). One of the difficulties of this method is trying to find a good fit between a two-dimensional representation of the face and a three-dimensional skull. While early attempts often involved laborious and time-consuming manual superimposition methods that may incorporate significant error, researchers are now developing computer-assisted methods (Ricci et al. 2006; Ubelaker et al. 1992), as well as designing automated superimposition methods (Ibanez et al. 2009).

DNA AND FORENSIC ANTHROPOLOGY

Since the 1950s, scientists have realized that an individual's DNA (located in the cellular nuclei and cytoplasm of all living organisms) encodes information about the individual's inherited characteristics and, moreover, that the nuclear code is unique to each individual. Nuclear DNA analysis allows identification by reference to the alleles and nucleotide sequences of heritable traits and markers contained in any human nucleated cell, while the cytoplasmic mitochondrial DNA (mtDNA) allows for the determination of maternal relatives. Skeletal analyses by the forensic anthropologist may guide the molecular investigation by utilizing the biological profile to narrow potential comparisons, by determining sample selection from a single skeleton and by choosing samples from comingled cases. Utilizing the forensic anthropologist in these ways saves the families of the unidentified time and saves the particular medico-legal jurisdiction money.

Obtaining DNA for identification purposes is not generally the responsibility of a forensic anthropologist; however, a growing number of these *molecular anthropologists* are applying their expertise to medico-legal matters. The great advantage to having anthropologists perform both the skeletal and molecular analyses, even if not by the same anthropologist, is that the close proximity of skeletal and molecular analyses (especially within the same or adjacent laboratories) can save a great amount of time because of more immediate lines of communication. Typically, however, the forensic anthropologist is simply asked to collect samples or identify potentially useful skeletal material. Forensic anthropologists are frequently in the best position to decide on how DNA may be utilized to resolve the issues of identity involving skeletonized remains. Many forensic anthropologists routinely make decisions regarding which skeletal elements should be sampled when dealing with commingled remains. Selecting duplicated elements, such as sampling three right femora, will undoubtedly yield three unique genomic profiles, but by using the skills of a competent skeletal biologist, the selection of other skeletal elements can be included or excluded, based on skeletal morphology. For instance, if three adults were commingled and one was deemed morphologically to be significantly more robust than the other two, fewer samples from the robust individual might be needed because of morphological similarity, and more samples from the two more gracile individuals could be sampled to resolve the commingling between these two.

When skeletal remains of the same individual are recovered across time and/or space, DNA analysis can aid the forensic anthropologist, while at the same time being aided by the forensic anthropologist. A thorough examination of a partial skeleton by the forensic anthropologist reveals those portions that are missing. In the event that those skeletal elements are subsequently recovered, they can be tentatively associated with the principal remains on morphological bases. However, DNA testing can be utilized to confirm the tentative placement of the additional remains. This feedback loop of using DNA to confirm morphological suspicions and of using morphological suspicions to select those skeletal elements to undergo molecular testing is healthy for the field of forensic anthropology. Simply put, DNA analysis may support forensic anthropological investigations, not only in the confirmation of identity but also in the efforts to separate commingled remains or reassociate remains (Mundorf et al. 2014). Contrary to those who see DNA as a threat to the field of forensic anthropology, there are many good reasons to argue that DNA analyses and results will serve to enrich the field by confirming (or refuting) associations based on morphological grounds by identifying compromised remains that could not be identified in any other matter, thus confirming (or refuting) the various assessments associated with the biological profile. This issue of compromised remains is also important to understand which tissue types will and will not produce a full DNA profile.

While the cost of testing for nuclear DNA can be prohibitive, recent developments surrounding the President's Initiative for DNA has provided free DNA analyses for any unidentified remains. However, with the tremendous backlog of casework yet to be completed by the DNA laboratories performing this U.S. Government-sponsored molecular testing, the time necessary to complete a case has become time prohibitive for families and the medico-legal community. Suffice it to say that DNA analyses, whether genomic or mitochondrial, are more costly than other methods of identification such as fingerprints, dental, radiological, and anthropological. Blind hits in large DNA databases (e.g., CODIS) have the potential to match an unidentified sample to an antemortem sample of that person or to a family reference sample (single or multiple family members); however, the usual scenario that utilizes DNA in the identification process involves a presumptive, or tentative, identification with comparative samples from the individual or family reference samples.

ADDITIONAL CONSIDERATIONS

Admissibility of evidence

Forensic experts, including anthropologists, radiologists, and pathologists, are expected to meet structured standards regarding the validity, reliability, and application of their methods (Committee on Identifying the Needs of the Forensic Sciences Community 2009; Saks and Koehler 2005), including personal identification methods based on comparisons of skeletal evidence. The pursuits of forensic scientists and practitioners differ from those of purely academic physical anthropologists, who are largely research-driven and not concerned with medico-legal issues. In addition to performing research and acquiring knowledge as an end unto itself, forensic practitioners must also consider the applications and implications of their findings in relation to legal matters. Several landmark rulings in U.S. courts have shaped how forensic scientists conduct research, evaluate methods, and present results in a court of law (Christensen 2004; Christensen and Crowder 2009; Steadman et al. 2006).

The first such ruling was issued in *Frye v. United States* (1923). The case involved the attempted admission of systolic blood pressure deception testing (*lie detector*), which was a new technique at the time, thereby leaving the court unsure as to how to assess its validity. The court offered the following opinion on the standard for admissibility of scientific expert witness testimony: "Just when a scientific principle or discovery crosses the line between the experimental and demonstrable stages is difficult to define. Somewhere in this twilight zone the evidential forces of the principle must be recognized, and while courts will go a long way in admitting expert testimony deduced from a well-recognized scientific principle or discovery, the thing from which the deduction is made must be sufficiently established to have gained general acceptance in the particular field in which it belongs" (*Frye v. United States* 1923). The court concluded that the lie detector had not yet gained the required standing and scientific recognition among authorities in the fields of physiology and psychology to be considered scientific (McCormick 1972). This *general acceptance test* for admissibility of scientific evidence became known as the *Frye Standard* and was the dominant standard in the majority of courts, facilitated in large part by the fact that the rule was easy to apply and required little scientific sophistication on the part of the judges.

Over time and with advancements in science, many courts and legal commentators began to modify or ignore the Frye standard. One of the key concerns was that new scientific evidence, though sound, often failed the Frye test. Critics argued that *general acceptance* should not be the criterion for the admissibility of scientific evidence and that any relevant conclusions supported by a qualified expert witness should be admissible, unless there are other reasons for exclusion (McCormick 1972).

In 1975, Congress enacted the *Federal Rules of Evidence* (1975), which was the first modern and uniform set of evidentiary rules for the trial of civil and criminal cases in federal courts. *Rule 702* specifically addressed expert witness testimony, stating that: "If scientific, technical or other specialized knowledge will assist the trier of fact to understand the evidence or to determine a fact in issue, a witness qualified as an expert by knowledge, skill, experience, training or education may testify thereto in the form of an opinion or otherwise." The enactment of the *Federal Rules of Evidence*, however, did not eliminate the uncertainty in the courts but rather led to a mixed use of *Frye* and the *Federal Rules of Evidence*, or some hybrid of the two. The confusion continued until the U.S. Supreme Court decided *Daubert v. Merrell-Dow Pharmaceuticals, Inc.* (1993).

The parents of Jason Daubert and Eric Schuller filed suit against the pharmaceutical company Merrell-Dow, claiming that their children's birth defects were the result of the mother's prenatal ingestion of the prescription drug Bendectin, which is designed to alleviate morning sickness. Crucial scientific evidence showing a causal relationship between the use of the drug during pregnancy and birth defects was deemed inadmissible, because it did not meet the *Frye* standard. The Supreme Court ruled that *Federal Rules of Evidence Rule 702* superseded the *Frye* test, thus determining the standard for relevance and reliability of expert testimony. Furthermore, the court ruled that it is the trial judge who must ensure that all scientific testimony is relevant and reliable. To assist the trial judges, the court provided the following general guidelines for assessing the admissibility of scientific testimony, which have collectively become known as the *Daubert criteria* or *Daubert standard*:

1. Has the theory or technique been tested?
2. What is the known or potential rate of error?
3. Do standards exist for the control of the technique's operation?
4. Has the theory or technique been subjected to peer review and publication?
5. Has the theory or technique been generally accepted within the relevant scientific community?

Two other U.S. Supreme Court cases have supplemented and clarified the *Daubert* decision. Confronted with some of the more difficult issues regarding expert testimony, the Supreme Court's decisions in *General Electric Co v. Joiner* and *Kumho Tire Co v. Carmichael* were explanatory of *Daubert*. In *General Electric Co v. Joiner*, the Supreme Court explained that methodology and conclusions are not completely distinct from each other but are linked together and that an expert's conclusion should be excluded in the event that valid reasoning does not support it. Overall, the *Joiner* decision questioned and clarified the language in *Daubert* and solidified the burden of admissibility for the trial courts. In *Kumho Tire Co v. Carmichael*, the court reaffirmed the gatekeeper's role of the judge and held that both scientific and nonscientific expert witness testimony must be reliable and relevant to be admissible and noted that there must be a flexible approach for assessing expert testimony, because types of evidence may vary across disciplines.

These landmark cases form the *trilogy* of Supreme Court decisions set the legal standard for evaluating the admissibility of expert testimony. In 2000, *Federal Rule 702* was amended to include "... if (1) the testimony is based upon sufficient facts or data, (2) the test is the product of reliable principles and methods, and (3) the witness has applied the principles and methods reliably to the facts of the case." The implication for forensic anthropology and for evidence pertaining to personal identification is that anthropologists should employ methods of data collection, comparison, and interpretation that are considered valid and reliable. The strength of conclusions should not be overstated, and limitations of the selected technique(s) should be considered and presented appropriately.

Facial approximation

Facial approximation (also known as craniofacial reconstruction, facial reconstruction, forensic reconstruction, and facial reproduction) is an artistic reproduction of the soft tissue features of an individual (Komar and Buikstra 2008) and involves the estimation or artistic reproduction of possible facial features based on the underlying skeletal structures. The practice of facial approximation is intended to capture public attention with regard to an unidentified person when all scientific leads have been exhausted, but it is never a means of personal identification.

Facial approximation was first attempted in Europe in the late nineteenth century and was a combined effort of anatomists and a sculptor using facial tissue depth measurements taken from cadavers (His 1895; Kollman and Buchly 1898). The technique was made famous by the reconstruction of Johann Sebastian Bach (His 1895) and was first applied in a forensic case in 1913 (Wilder and Wentworth 1918). The tissue depth method, which involves the building up of soft tissues on the skull (including muscle, fatty and connective tissue, glandular, and skin) to depths at specific osseous landmarks determined by needle probe or ultrasound, was further developed by Rhine and Campbell (1980). Another method, the anatomical method, uses muscle origins and insertions to rebuild each muscle and was introduced in the 1920s by Russian anthropologist Gerasimov (1971). The tissue depth method is more widely used today in attempts to identify the deceased, perhaps because the anatomical method requires more expertise in anatomy and biomechanics and is rather time-consuming, whereas the tissue depth method is quicker and requires less anatomical training.

Knowledge of human cranial anatomy and variation, as well as artistic ability, is essential for the production of useful facial images. The production of facial approximations should therefore be a joint effort between anthropologists, anatomists, and artists. Anthropological analysis is critical in providing the demographic information such as age, sex, and ancestry. Because of their familiarity with skeletal landmarks, anthropologists are often responsible for the placement of appropriate tissue depth markers on the skull. Anthropological collaboration is also helpful in calling attention to details that may be useful to incorporate into the approximation such as size and shape of supraorbital ridges, shape characteristics of the nasal area, evidence of healed trauma, and unusual dental features (Ubelaker 2000).

The work then shifts to an artist, who marries the skull image and all other information in an appropriate manner (Ubelaker 2000). This step requires advanced artistic abilities and training and may not be practical or even possible for most forensic anthropologists to perform. Indeed, many practitioners of facial approximation, including Frank Bender (https://www.frankbender.us), are trained in fine art rather than forensics and regard the practice as more artistic than scientific. Artistic approximations may include two- or three-dimensional drawings done over frontal and lateral photographs of the skull or clay reconstructions applied directly to the skull or to a replica of the skull. Although reproductions were traditionally manual, computer-assisted approaches are becoming more prevalent (Shahrom et al. 1996; Turner et al. 2005; Ubelaker and O'Donnell 1992). The selection of which method to use may be up to the artist, but each method has advantages. Two-dimensional approximations are generally easier to change and are cheaper, while three-dimensional approximations are more time-consuming but generally appear more life-like (Caldwell 1986). The availability of images on a computer provides an advantage in facilitating the final comparison (Ubelaker and O'Donnell 1992). Although the skull itself is the basic reference for all techniques, quality photography and evidence collection are important for the accuracy of the approximation (Powers 2005).

One of the greatest limitations of the technique is that there are no credible ways to estimate many of the most distinctive and individuating structures of the face, including eye color, ears, lips, parts of the nose, and head and facial hair color/length/form, because they do not have an underlying bony structure from which to glean reliable information. These facial details are often considered crucial in the recognition of a human face by other individuals, but they must be interpreted from parts of the skull and other anthropological data available or may often be simply an artistic rendition or guess.

After an approximation is complete and publicized, a visual recognition by a relative, friend, or acquaintance may be made. This tentative identification is not a final identification and must be verified subsequently through scientific means. Although popular in the media and impressive to the public (probably because of the attention received by successful attempts while the unsuccessful ones get little attention), the method lacks scientific rigor, is associated with a high degree of error, and should be used for only investigative purposes. Suffice it to say that the results of any facial approximation are simply a lead that may be followed, and nothing more. Many anthropologists are critical of the approach (Stephan 2003) and emphasize that statements regarding the power or value of facial imagery methods should be presented carefully. Recent research in this area has been aimed at assessing the accuracy of approximations and evaluating craniofacial tissue depth measurements (Quatrehomme et al. 2007; Stephan and Simpson 2008a and b).

SUMMARY

- Personal identification is important in our society for both legal and humanitarian reasons.
- Personal identification should involve a multidisciplinary approach and strive for as many lines of evidence as possible.
- Anthropologists are often asked to assist in the personal identification process, because the skeleton usually survives longer than other identifiable features and because forensic anthropologists can glean biological information from the skeleton and detect subtle differences among individuals.
- Anthropological methods of investigating personal identification include radiographic comparison, nonimaged records comparison, and craniofacial superimposition.

- Several U.S. Supreme Court rulings have shaped how anthropologists conduct research, evaluate methods, and present results in a court of law.
- Personal identification should employ methods of data collection, comparison, and interpretation that are considered valid and reliable.
- Facial approximation is not a means of personal identification; this methodology is merely an investigative lead.
- DNA analyses can be aided by forensic anthropologists through the proper selection of samples.
- Forensic anthropology can be aided by DNA analyses by confirming associations and tentative identifications, based on morphological analyses.

Questions

1. Why is personal identification important?
2. Who is responsible for determining a personal identification?
3. What determines the method of personal identification used?
4. What is the most frequently used method of personal identification? When is it typically used, and what are some of the problems associated with it?
5. What are the different "levels" of identification, and what types of evidence do they typically involve?
6. Why is it important for forensic anthropologists to have a comprehensive understanding of skeletal biology?
7. What is the Scientific Working Group for Forensic Anthropology (SWGANTH)?
8. Do all forensic anthropologists have equivalent training in personal identification methods? Explain.
9. What are some examples of features that can be used for identification by comparative radiography?
10. Why is normal anatomical variation seen in radiographs important to the personal identification process, and what are some examples of normally varying features that are often used for personal identification?
11. Why is it not always possible to make a personal identification from a surgical implant?
12. What criteria must be met in order for a pathological condition, anomaly, or injury to be potentially useful for identification purposes?
13. What identification methods can be used when there is no antemortem medical or dental information available?
14. How do age- or time-related changes in the skeleton affect anthropological methods of personal identification?
15. When would craniofacial superimposition be used as a means of personal identification?
16. What are the *Daubert* criteria, and how do they affect personal identification?
17. What is the purpose of facial approximation, and why can it not be used as a means of personal identification?
18. What should be the strategy when selecting samples for DNA analyses when dealing with skeletal elements that may represent multiple individuals?
19. How might the results of DNA analyses improve the field of skeletal analyses by forensic anthropologists and forensic odontologists?
20. What can forensic scientists do to minimize the possibility of contamination when resecting skeletal samples to be used in DNA analyses?
21. What is NamUs?

Disclaimer

Names of commercial manufacturers are provided for only identification purposes, and inclusion does not imply endorsement of the manufacturer or its products or services by the Federal Bureau of Investigation (FBI). The views expressed are those of the author's and do not necessarily reflect the official policy or position of the FBI or the U.S. Government.

References

Adams BJ. 2003a. The diversity of adult dental patterns in the United States and the implications for personal identification. *J Forensic Sci* 48(3):497–503.

Adams BJ. 2003b. Establishing personal identification based on specific patterns of missing, filled, and unrestored teeth. *J Forensic Sci* 48(3):487–496.

Adams BJ and Maves RC. 2002. Radiographic identification using the clavicle of an individual missing from the Vietnam conflict. *J Forensic Sci* 48(2):369–373.

Anderson BE. 2007. Comment on Statistical Basis for Positive Identification in Forensic Anthropology. *Am J Phys Anthropol* 133(1):741–742.

Anderson BE and Gilson TE. 2005. Modern Day Trepanation: The Ventriculostomy utilizing an Intracranial Pressure Bolt. *Proceedings of the 57th Annual Meeting of the American Academy of Forensic Sciences*, New Orleans.

Angyal M and Derczy K. 1998. Personal identification on the basis of antemortem and postmortem radiographs. *J Forensic Sci* 43:1089–1093.

Atkins L and Potsaid MS. 1978. Roentgenographic identification of human remains. *J Am Med Assoc* 240:2307–2308.

Austin-Smith D. 1999. Video superimposition at the C.A. Pound Laboratory 1987 to 1992. *J Forensic Sci* 44(4):695–699.

Austin-Smith D and Maples WR. 1994. The reliability of skull/photograph superimposition in individual identification. *J Forensic Sci* 39(2):446–455.

Banjeree A. 1964. Camera identifies human skull. *IPJ* 2:42.

Basauri C. 1967. A body identified by forensic odontology and superimposed photographs. *Int Criminal Police Rev* 204:37.

Brogdon BG. 1998. Radiological identification of individual remains. In BG Brogdon (Ed.), *Forensic Radiology*, pp. 149–187. Boca Raton, FL: CRC Press.

Brothwell EF, Molleson T, and Metreweli C. 1968. Radiological aspects of normal variation in earlier skeletons: An exploration study. In DR Brothwell (Ed.), *The Skeletal Biology of Earlier Human Populations*, pp. 149–172. New York: Pergamon Press.

Caldwell PC. 1986. New Questions (and some answers) on the facial reproduction techniques. In KJ Reichs (Ed.), *Forensic Osteology: Advances in the Identification of Human Remains*, pp. 229–255. Springfield, MA: Charles C Thomas.

Camps FE. 1969. Radiology and its forensic application. In FE Camps (Ed.), *Recent Advances in Forensic Pathology*, pp. 149–160. London: J. & A. Churchill.

Christensen AM. 2005a. Assessing the variation in individual frontal sinus outlines. *Am J Phys Anthropol* 127(3):291–295.

Christensen AM. 2005b. Testing the reliability of frontal sinus outlines in personal identification. *J Forensic Sci* 50(1):18–22.

Christensen AM. 2004. The impact of *Daubert*: Implications for testimony and research in forensic anthropology (and the use of frontal sinuses in personal identification). *J Forensic Sci* 49(3):427–430.

Christensen AM and Crowder CM. 2009. Evidentiary standards for forensic anthropology. *J Forensic Sci* 54(6):1211–1216.

Committee on Identifying the Needs of the Forensic Sciences Community, National Research Council. 2009. *Strengthening Forensic Science in the United States: A Path Forward*. Washington, DC: National Academies Press.

Cox M, Malcolm M, and Fairgrieve SI. 2009. A new digital method for the objective comparison of frontal sinuses for identification. *J Forensic Sci* 54(4):761–772.

Culbert WL and Law FL. 1927. Identification by comparison of roentgenograms of nasal accessory sinuses and mastoid processes. *J Am Med Assoc* 88:1634–1636.

Daubert v. Merrell Dow Pharmaceuticals, Inc., 509 U.S. 579 (1993).

Delfino YP, Colonna M, Introna F, Potente M, and Vacca E. 1983. Identification by computer aided skull-face superimposition. Paper presented in the *Forensic Anthropology Symposium* (*MY Iscan, Organizer*) XIth International Congress of Anthropological and Ethnological Sciences, August 20–25, Vancouver, British Columbia.

Dedouit R, Telmon N, Costagliola R, Otal P, Florence LL, Joffre F, and Rouge D. 2007. New identification possibilities with postmortem multislice computed tomography. *Int J Leg Med* 121:507–510.

DiMaio DJ and DiMaio VJ. 1989. *Forensic Pathology*. New York: Elsevier Saina Publishing.

Dorion RBJ. 1983. Photographic superimposition. *J Forensic Sci* 28(3):724–734.

Dutra FT. 1944. Identification of person and determination of cause of death from skeletal remains. *Arch Pathol* 38:339–349.

Eckert WB and Garland N. 1984. The history of the forensic applications in radiology. *Am J Forensic Med Pathol* 5:53–56.

Engel CE. 1954. Superimposition. *Med Radiogr Photogr* 30:20–23.

Fatteh AV and Mann GT. 1969. The role of radiology in forensic pathology. *Med Sci Law* 9:27–30.

Federal Rules of Evidence. 1975, 2000.

Frye v. United States, 293 F. 1013 (C.A.D.C 1923).

General Electric Co. v. Joiner, 522 US 136 (1997).

Gerasimov MM. 1971. *The Face Finder*. Philadelphia, PA: JB Lippencot.

Glaister J and Brash JC. 1937. *Medico-Legal Aspects of the Ruxton Case*. Baltimore, MD: William Wood and Co.

Greulich WW. 1960. Skeletal features visible in the roentgengrams of hand and wrist which can be used for establishing individual identification. *Am J Roentgen Rad Ther Nucl Med* 83(4):756–764.

Greulich WW and Pyle SI. 1959. Radiographic atlas of skeletal development of the hand and wrist. Palo Alto, CA: Stanford University Press.

Haglund WD and Fligner CL. 1993. Confirmation of human identification using computerized tomography (CT). *J Forensic Sci* 38(3):708–712.

His W. 1895. Anatomische Forschungen uber Johann Sebastian Bach's Gebeine und Antlitz nebst Bemerkungen uber dessen Bilder. *Abhandlungen der mathematisch-physikalischen Klasse der Koniglichen Sachsischen Gesellschaft der Wissenschaften* 22:379–420.

Hooton EA. 1930. *Indians of the Pecos Pueblo: A Study of Their Skeletal Remains*. Andover, MA: Yale University Press.

Hogge JP, Messmer JM, and Fierro MF. 1995. Positive identification by post-surgical defects from unilateral lamdoid synostectomy: A case report. *J Forensic Sci* 40(4):688–691.

Hogge JP, Messmer JM, and Doan QN. 1993. Radiographic identification of unknown human remains and interpreter experience level. *J Forensic Sci* 39:373–377.

Ibanez O, Ballerini L, Cordon O, Damas S, and Santamaria J. 2009. An experimental study on the applicability of evolutionary algorithms to craniofacial superimposition in forensic identification. *Inf Sci* 179(23):3998–4028.

Jensen S. 1991. Identification of human remains lacking skull and teeth. A case report with some methodological considerations. *Am J Forensic Med Pathol* 12(2):93–97.

Joblanski NG and Shum BS. 1989. Identification of unknown human remains by comparison of antemortem and postmortem radiographs. *Forensic Sci Int* 42:221–230.

Kade H, Meyers H, and Wahlke JE. 1967. Identification of skeletonized remains by x-ray comparison. *J Crim Law* 58:261–264.

Kahana T and Hiss J. 1994. Positive identification by means of trabecular bone pattern comparison. *J Forensic Sci* 39:1325–1330.

Kahana T, Goldin L, and Hiss J. 2002. Personal identification based on radiographic vertebral fractures. *Am J Forensic Med Pathol* 23(1):36–41.

Kahana R, Hiss J, and Smith P. 1998. Quantitative assessment of trabecular bone pattern identification. *J Forensic Sci* 43(6):1144–1147.

Kahana T and Hiss J. 1997. Identification of human remains: Forensic radiology. *J Clin Forensic Med* 4:7–15.

Kerley ER. 1977. Forensic anthropology. In CG Tedeschi, WG Eckert, and LG Tedeschi (Eds.), *Forensic Medicine, Vol. 2*, Philadelphia, PA: W.B. Saunders Company.

Kirk NJ, Wood RE, and Goldstein M. 2002. Skeletal identification using the frontal sinus region: A retrospective study of 39 cases. *J Forensic Sci* 47(2):318–323.

Koelmeyer TD. 1982. Videocamera superimposition and facial reconstruction as an aid to identification. *Am J Forensic Med Pathol* 3(1):45–47.

Kollman J and Buchly W. 1898. Die Persistenz der Rassen und die Reconstruction der Physiognomie prahistorischer Schadel. *Archiv fur Anthropologie* 25:329–359.

Komar D and Lathrop S. 2006. Frequencies of morphological characteristics in two contemporary forensic collections: Implications for identification. *J Forensic Sci* 51(5):974–978.

Komar DA and Buikstra JE. 2008. *Forensic Anthropology: Contemporary Theory and Practice*. New York: Oxford University Press.

Koot MG. 2003. Radiographic human identification using bones of the hand: A validation study. *Proceedings of the 55th Annual Meeting of the American Academy of Forensic Sciences*, February 17–22, Chicago.

Koot MG, Sauer NJ, and Fenton TW. 2005. Radiographic human identification using bones of the hand: A validation study. *J Forensic Sci* 50(2):263–268.

Krogman WM and Iscan MY. 1986. *The Human Skeleton in Forensic Medicine*, 2nd edn. Springfield, MA: CC Thomas.

Kumho Tire Co. v. Carmichael, 526 U.S. 137 (1999).

London MR, Libbey NP, Shemin DG, and Chazan JA. 1994. Renal osteodystrophy and dialysis artifacts as indicators of identification. *Forensic Science International* 65:81–96.

MacKinnon GE and Mundorff AZ. 2006. The World Trade Center—September 11, 2001. In T Thompson and SM Black (Eds.), *Forensic Human Identification: An Introduction*, pp. 485–499. Boca Raton, FL: CRC Press.

Mann RW. 1998. Use of bone trabeculae to establish positive identification. *Forensic Sci Int* 98(1–2):91–99.

Marek Z, Kusmiderski J, and Lisowski Z. 1983. Radiograms of the paranasal sinuses as a principle of identifying catastrophe victims and unknown skeletons. *Arch Kriminol* 172(1–2):1–6.

Martel W, Wicks JD, and Hendrix RC. 1997. The accuracy of radiological identification of humans using skeletal landmarks: A contribution to forensic pathology. *Radiology* 124:681–684.

McCormick CT. 1972. *McCormick's Handbook of the Law of Evidence*. 2nd edn. Cleary EW (Ed.) St. Paul, MN: West Publishing Company.

McCormick WM, Stewart JH, and L.A. Langford. 1985. Sex determination from chest plate roentgengrams. *Am J Phys Anthropol* 68:173–195.

Messmer JM. 1986. Radiographic identification. In MF Fierro (Ed.), *CAP Handbook for Postmortem Examination of Unidentified Remains*, pp. 68–75. Skokie, IL: College of American Pathologists.

Morgan TA and Harris MC. 1953. The use of X-rays as an aid to medicolegal investigations. *J Forensic Med* 27:9–18.

Mundorf AZ, Shaler R, Bieschke ET, Mar-Cash E. 2014. Marrying anthropology and DNA: Essential for solving complex commingling problems in cases of extreme fragmentation. In BJ Adams and JE Byrd (Eds.), *Commingled Human Remains, Methods in Recovery, Analysis and Identification*. San Diego: Academic Press.

Murphy WA and Gantner GE. 1982. Radiologic examination of anatomic parts and Skeletonized remains. *J Forensic Sci* 27:9–18.

Murphy WA, Spruill FG, and Gantner GE. 1980. Radiologic identification of unknown human remains. *J Forensic Sci* 25:727–735.

Nozawa H, Watanabe-Suzuki K, Seno H, Ishii A, and Suzuki O. 2002. An autopsy case of human skeletal remains with a numerical variation in the thoraco-lumbar vertebrae. *Leg Med* 4:123–126.

Owsley DW and Mann RW. 1989. Positive identification based on radiographic examination of the leg and foot. *J Am Podiatr Med Assoc* 79(10):511–513.

Owsley DW, Mann RW, Chapman RE, Moore E, and Cox WA. 1993. Positive identification in a case of intentional extreme fragmentation. *J Forensic Sci* 38(4):985–996.

Penalver JJ, Kahana T, and Hiss J. 1997. Prosthetic devices in positive identification of human remains. *J Forensic Sci* 47:400–405.

Petrie WMF. 1987. *Deshasheh 1897. Mem 15.* London: Egyptian Exploration Fund.

Pfaeffli M, Vock P, Dirnhofer R, Braun M, Bollinger SA, and Thali MJ. 2007. Post-mortem radiological CT identification based on classical ante-mortem x-ray examinations. *Forensic Sci Int* 171:111–117.

Powers R. 2005. Remains to be seen! *J Forensic Identif* 55(6):687–696.

Quatrehomme G, Balaguer T, Staccini P, and Alunni-Perret V. 2007. Assessment of the accuracy of three-dimensional manual craniofacial reconstructions: A series of 25 controlled cases. *Int J Legal Med* 121:469–475.

Rhine S and Sperry K. 1991. Radiographic identification by mastoid sinus and arterial pattern. *J Forensic Sci* 36:272–279.

Rhine JS and Campbell HR. 1980. Thickness of facial tissues in American Blacks. *J Forensic Sci* 25:847–858.

Ribeiro FA. 2000. Standardized measurements of radiographic films of the frontal sinuses: An aid to identifying unknown persons. *Ear Nose Throat J* 79(1):26–33.

Ricci A, Marella GL, and Apostol MA. 2006. A new experimental approach to computer-aided face/skull identification in forensic anthropology. *Am J Forensic Med Pathol* 27(1):46–49.

Riddick L, Brogdon BG, Laswell-Hoff J, and Delmas B. 1983. The accuracy of radiologic identification of charred human remains through use of the dorsal defect of the patella. *J Forensic Sci* 28(1):263–267.

Rouge D, Telmon N, Arrue P, Larrouy G, and Arbus L. 1993. Radiographic identification of human remains through deformities and anomalies of post-cranial bones: A report of two cases. *J Forensic Sci* 38(4):997–1007.

Rowe JH. 1953. Technical aids in anthropology: A historical survey. In AL Krober (Ed.), *Anthropology Today: An Encyclopedic Inventory*, pp. 895–940. Chicago: University of Chicago Press.

Sainio P, Syrjanen SM, and Komakow S. 1990. Positive identification of victims by comparison of ante-mortem and post-mortem dental radiographs. *J Forensic Odontosomatol* 8:11–16.

Saks MJ and Koehler JJ. 2005. The coming paradigm shift in forensic identification science. *Science* 309:892–895.

Sanders I, Woesner ME, Ferguson RA, and Noguchi TT. 1972. A new application of forensic radiology: Identification of deceased from a single clavicle. *Am J Roengenol Radium Ther Nucl Med* 115(3):619–622.

Sassouni V. 1959. Cephalometric identification: A proposed method of identification of war dead by means of radiographic cephalometry. *J Forensic Sci* 4:1–10.

Sauer NJ, Brantely RE, and Barondess DA. 1988. The effects of aging on the comparability of antemortem and postmortem radiographs. *J Forensic Sci* 33(5):1223–1230.

Scheuller A. 1921. Das Rontgenogram der Stirnhohle: ein Hilfsmitten fur die Identitasbestimmung von Schadeln. *Monatsschrift feur Ohrenheilkunde und Laryngo-Rhinologie* 5:1617–1620.

Schmidt G and Kallieris D. 1982. Use of radiographs in the forensic autopsy. *Forensic Sci Int* 19:263–270.

Scientific Working Group for Forensic Anthropology (SWGANTH). https://www.swganth.org.

Sen NK. 1962. Identification by superimposed photographs. *Int Crim Police Rev* 162:284.

Shahrom AW, Vanezis P, Chapman RC, Gonzales A, Blenkinsop C, and Rossi ML. 1996. Techniques in facial identification: Computer-aided facial reconstruction using a laser scanner and video superimposition. *Int J Leg Med* 108(4):194–200.

Sidler M, Jackowski C, Dirnhofer R, Vock P, and Thali M. 2007. Use of multislice computed tomography in disaster victim identification—Advantages and limitations. *Forensic Sci Int* 169:188–128.

Singleton AC. 1951. The roentgenologial identification of victims of the "Noronic" disaster. *Am J Roengenol Radium Ther* 66:375–384.

Sivaloganathan S and Butt WP. 1988. A foot in the yard. *Med Sci Law* 28(2):150–156.

Smith VA, Christensen AM, and Myers SW. 2010. The reliability of visually comparing small frontal sinuses. *J Forensic Sci* 55(6):1413–1415.

Sopher IM. 1972. The dentist, the forensic pathologist, and the identification of human remains. *J Am Dent Assoc* 85(6):1324–1329.

Steadman DW, Adams BJ, and Konigsberg LW. 2006. Statistical basis for positive identification in forensic anthropology. *Am J Phys Anthropol* 131(1):15–26.

Stephan CN. 2003. Anthropological facial reconstruction—Recognizing the fallacies, "unembracing" the errors and realizing method limits. *Sci Justice* 43(4):193–200.

Stephan CN and Simpson EK. 2008a. Facial soft tissue depths in craniofacial identification (Part 1): An analytical review of published adult data. *J Forensic Sci* 53(6):1257–1272.

Stephan CN and Simpson EK. 2008b. Facial soft tissue depths in craniofacial identification (Part 2): An analytical review of published sub-adult data. *J Forensic Sci* 53(6):1273–1279.

Stevens PJ. 1966. Identification of a body by unusual means. *Med Sci Law* 6:160–161.

Stewart TD. 1979. *Essentials of Forensic Anthropology*. Springfield, MA: Thomas.

Sudimack JR, Lewis BJ, Rich J, Dean DE, and Fardal PM. 2002. Identification of decomposed human remains from radiographic comparisons of an unusual foot deformity. *J Forensic Sci* 47(1):218–220.

Thorne H and Thyberg H. 1953. Identification of children (or adults) by mass miniature radiography of the cranium. *Acta Odontol Scandinavica* 11(2):129–140.

Turner WD, Brown REB, Kelliher TP, Tu PH, Taister MA, and Miller KWP. 2005. A novel method of automated skull registration for forensic facial approximation. *Forensic Sci Int* 154:149–158.

Ubelaker DH. 1990. Positive identification of American Indian skeletal remains from radiographic comparison. *J Forensic Sci* 35(2):466–472.

Ubelaker DH. 2000. A history of Smithsonian-FBI collaboration in forensic anthropology, especially in regard to facial imagery. *Forensic Sci Commun* 2(4).

Ubelaker DH, Bubniak E, and O'Donnell GE. 1992. Computer-assisted photographic superimposition. *J Forensic Sci* 37:750–762.

Ubelaker DH and O'Donnell GE. 1992. Computer-assisted facial reproduction. *J Forensic Sci* 37:155–162.

Valenzuela A. 1997. Radiographic comparison of the lumbar spine for positive identification of human remains. A case report. *Am J Forensic Med Pathol* 18(2):215–217.

Varga M and Takacs P. 1991. Radiographic personal identification with characteristic feature in the hip joint. *Am J Forensic Med Pathol* 12:328–331.

Verma MS. 1981. Skull-photograph superimposition: A case study. *Paper presented at the 35th Annual Meeting of the American Academy of Forensic Sciences*, February 17–20, Los Angeles, CA.

Webster G. 1955. Photography as an aid to identification: The Plumbago Pit case. *Police J* 28:185.

Webster WP, Murray WK, Brinkous W, and Hudson P. 1986. Identification of human remains using photographic reconstruction. In KJ Reichs (Ed.), *Forensic Osteology – Advances in the Identification of Human Remains*, pp. 256–289. Springfield, MA: Charles C Thomas.

Wilder HH and Wentworth B. 1918. Personal identification: Methods for the identification of individuals, living or dead. Boston, MA: Gorham Press.

Yoshino M, Miyasaka S, Sato H, and Seta S. 1987. Classification system of frontal sinus patterns by radiography. Its application to identification of unknown skeletal remains. *Forensic Sci Int* 34(4):289–299.

Suggested readings

Holland TD, Mann RW, and Ubelaker DH. 1999. Advances in Personal Identification in Mass Disasters. US Army Central Identification Laboratory, Hawaii and the Smithsonian Institution.

Stewart TD. 1971. *Personal Identification in Mass Disasters*. Washington, DC: Smithsonian Institution.

Thompson T and Black SM. 2006. *Forensic Human Identification: An Introduction*. Boca Raton, FL: CRC Press.

CHAPTER 17

Mass Fatalities, Mass Graves, and the Forensic Investigation of International Crimes

Pierre Guyomarc'h and Derek Congram

CONTENTS

Introduction	335
Main frameworks for the forensic investigation of international crimes	336
Preliminary notes on human rights and mass graves	336
Humanitarian forensic action	337
Forensic anthropology as a service to prosecution	338
Mass fatalities and emergency action	339
Recent developments in forensic anthropology and investigations of international law	340
Final words on the role of the forensic anthropologist in the international law context	341
Summary	342
Review questions	343
Glossary	343
References	344

LEARNING OBJECTIVES

1. Describe the role of an anthropologist in the investigation of international crimes.
2. Explain the differences between humanitarian and prosecution-mandated missions, and give an example of each.
3. Discuss the role of "Truth and Reconciliation Commissions."
4. Describe how forensic anthropologists assess the value of their role in mass grave excavations.
5. Explain the role of forensic anthropologists in Disaster Victim Identification (DVI).
6. Discuss the need for international standards in forensic practice.
7. Discuss the ethical dilemmas that forensic anthropologists may face during the course of investigating international crimes and the importance of conducting an objective analysis.

INTRODUCTION

Forensic anthropology and archaeology are most commonly associated with domestic homicides, individual crime scenes, and the analysis of single skeletons. However, the number of scenarios that employ the expertise of forensic anthropologists is expanding. These include the rapidly increasing number of those who die as undocumented migrants (e.g., those fleeing armed conflict in Syria and those around the U.S.-Mexico border), civilian and combatant victims of on-going armed conflict (e.g., Syria, Yemen, and Ukraine), and groups that are victimized by criminal gangs and repressive governments (e.g., 43 students-teacher who disappeared after being detained in Iguala, Mexico, in 2014, *The Guardian* 2015). There are different reasons why these latter types of forensic work are less known, perhaps principally because many of the victims in these contexts are socially marginalized and do not capture the public attention like *reality* or true crime shows and fictionalized forensic work that focus on domestic crimes. Other reasons include limited incentives for practitioners in international investigations compared with practitioners who work primarily in academia and consult on domestic cases. Scientific dissemination via publication is necessary for advancing one's academic career and for research funding, but there are disincentives or even prohibitions on publication in international crimes investigations. Almost always, there is a need to exercise discretion during judicial investigations or a requirement of confidentiality applied during humanitarian work that assists governments to fulfill their legal obligations to investigate *missing persons* in post-conflict contexts. In more difficult situations, governments deny killings and prevent their investigation, either because they fear negative publicity about their failure to protect citizens or because government agents are responsible for enforced disappearances and/or killings. Many anthropologists who dedicate a majority or all their time to this type of work are faced with a constant stream of casework and lack the opportunity to write up their experience, conduct research, and explore more theoretical implications of what they do and how they do it. In sum, we hear about and read less from these colleagues than we do from those who typically consult out of academia or municipal medical legal institutions on routine cases that come into their laboratories, and that is unfortunate. This is not to denigrate in any way those who are tasked with domestic work and those in academia (for it is especially the latter who mentored all of us in the discipline), simply to say that our perception of the practice of forensic anthropology is skewed toward the domestic crime context, and the professional literature has an Anglo (North American/UK) bias.

This chapter discusses the increasing number and types of cases, where forensic archaeologists and anthropologists aid the search for and identification of persons in humanitarian contexts or of victims of grave *human rights* violations or *mass disasters*. Given the scale of death, these victims are often buried in *mass graves* or are in a state of preservation, requiring the expertise of a forensic anthropologist. The main frameworks for the forensic investigation of international crimes are developed along with the various anthropological approaches to these investigations. This relatively new and underdocumented field of applied anthropology is represented by a growing number of national and international institutions that focus on prosecution, humanitarian, or mass disasters responses. The chapter includes practical examples that emphasize the constant evolution of forensic anthropology and its innovative application beyond the context of domestic casework.

MAIN FRAMEWORKS FOR THE FORENSIC INVESTIGATION OF INTERNATIONAL CRIMES

Preliminary notes on human rights and mass graves

Although we often hear the term *human rights* used in association with forensic anthropological investigations of mass killings, human rights represent so much more than that (there are 30 *rights and freedoms* in the United Nations Universal Declaration on Human Rights, which was drafted under the direction of Eleanor Roosevelt in 1948). Rosa Parks became famous for resisting a law that infringed upon the human rights of black Americans, but forensic anthropologists played no part in that case (although biological anthropologists certainly contributed to the racialization of people, which was used to justify discriminatory laws, see Gould 1981). These rights and freedoms are diverse and cover issues of privacy, citizenship, marriage, property ownership, and expression. When we speak of human rights as they relate to forensic anthropology, we generally refer to the violations of the right to life and the right to justice, as protected by the Geneva Conventions, which place limits on actions during war, and the Convention on the Prevention and Punishment of the Crime of Genocide. Sometimes, however, forensic anthropologists are engaged with the investigation of historic deaths or killings for which there is no criminal medico-legal mandate. In these situations, we use the term *human rights* or *humanitarian* to distinguish these investigations from those that are part of a formal, *criminal* legal process (e.g., Kimmerle 2012). This is not to mean that domestic forensic work has nothing to do with humanitarianism or human rights; instead, these descriptors refer to the primary mandate of an investigation. Recognizing this somewhat-misleading separation, we have opted to accept the distinction for this chapter, but we felt that they warranted some explanation. It is also important to note that human rights are cultural constructs and their definitions evolve over time. Accordingly, the role of forensic anthropologists in these contexts varies even as the discipline itself evolves as an applied science. Ultimately, it is easy to describe what forensic anthropologists have done in certain places at distinct moments in the past, yet it is difficult to state in a definitive way what forensic anthropology *is* and what it *does* as part of investigations into violations of human rights, criminal, or humanitarian law.

Mass graves are often assumed to be the product of violations of international criminal law, but they are also sometimes produced following natural disasters by well-intentioned, though misinformed, governments that are (wrongly) concerned about public health risks posed by dead bodies. Whatever the motivation, the creation of mass burials violates certain laws and virtually all contemporary cultural traditions, and so they ought to be investigated, barring valid cultural objections by victim families. There are several book chapters and articles that discuss the technical challenges of excavating mass graves. Rather than repeating what has been said elsewhere, we recommend the following: Bernardi and Fondebrider (2007), Djurić and Starović (2015), Etxeberría et al. (2015), Skinner (2007), Skinner et al. (2002), Tidball-Binz and Hofmeister (2015), Tuller (2012), Tuller and Hofmeister (2014), Tuller and Đurić (2006). Readers of these works will notice that mass graves pose not only unique and varying methodological challenges but also political ones (see also Skinner and Sterenberg 2005; Steele 2008). Those directing mass grave excavations must understand that contextual factors may affect methodological approaches.

Humanitarian forensic action

Forensic work with a humanitarian mandate is necessary and justified through key legal texts. International Humanitarian Law (IHL), which dictates the rules to be applied in armed conflicts, covers several obligations concerning the dead that the parties (e.g., governments and armed groups) have to respect. The International Committee of the Red Cross (ICRC), as custodian of IHL, works toward the respect of these rules in countries involved in international and noninternational armed conflicts. Founded in 1863, the ICRC aims to protect populations, and it initiated the adoption of the Geneva Conventions in 1949, with additional protocols in 1977 (International Committee of the Red Cross 1995). These treaties form the core of IHL and outline the obligations to search for the dead, maintain their dignity, identify and return remains to families, note the location of gravesites, and facilitate access to victims and grave sites. Such obligations require forensic expertise, and in the case of mass graves and skeletonized remains, the role of forensic anthropologists and archaeologists is essential. The issue of missing persons (i.e., individuals with whom their families have lost contact as a result of an armed conflict or other violent situations) is complex, and for those who have died due to violent events, the time elapsed between death and discovery usually implies that remains are partially or completely skeletonized. The recovery and identification of human remains, therefore, must be performed with anthropological expertise. The anthropological process carefully documents the removal and condition of the remains, with the objective of alleviating suffering caused by uncertainty about the fate of relatives and loved ones (this primary need is the *right to know*). Once the recognizable soft tissues have decomposed, the authorities in many places rely on presumptive identification methods based on recognition of clothing or personal objects. Such presumptive identifications are often mistaken. The humanitarian application of the discipline focuses on the respect for dead bodies and assisting families of the missing with the *right to know*.

There has been a strong development of forensic anthropology in humanitarian work during the past decade, specifically following the recognized need of this expertise in the conclusions of the International Conference of Governmental and Non-Governmental Experts on missing persons in 2003 (International Committee of the Red Cross 2003). The forensic unit of the ICRC was created in 2003 and has increased operations across the world, relying on greater than 30 staff experts (mainly anthropologists, archaeologists, odontologists, and pathologists) in 2016. The ICRC forensic specialist provides support to the local authorities and promotes the application of best forensic practices not only for the management of the dead in emergencies and armed conflicts but also in postconflict contexts to clarify the fate of the missing. Rather than substituting local personnel, the ICRC uses a sustainable approach that builds and improves the local forensic capacity. This stands in contrast to international forensic operations that call on foreign scientists to fly in and conduct the work, often with little background on the historical, social, and political contexts of the conflict and typically engaging little with local experts (Figure 17.1).

A good example of a successful humanitarian operation is the Commission on Missing Persons (CMP) in Cyprus (http://www.cmp-cyprus.org/). Created in 1981, this bicommunal (Greek and Turkish Cypriot) body is supported by the United Nations, and it has recently dramatically increased forensic operations after a thorough collection of antemortem data and circumstantial information. Almost half of all persons missing from past conflicts have been exhumed as of 2015 and more than 600 have been positively identified. The success of the CMP is mainly due to its exclusive humanitarian purpose and the integration of a sustainable team of forensic anthropologists and archaeologists to ensure controlled recovery and positive identification.

Figure 17.1 Author Dr. Pierre Guyomarc'h, a forensic specialist with the International Committee of the Red Cross, and Dr. Mercedes Salado, of the Argentine Forensic Anthropology Team, provide advice to local authorities at a forensic training session. (Courtesy of ICRC/Hassan Chaaban.)

Humanitarian forensic action implies a focus on the population's essential needs and the right to know, but it excludes the notion of judicial prosecution. Information on the cause and manner of death becomes secondary at best, and the mandate of international organizations often prevents the investigation of such aspects for the sake of the basic principle of neutrality.

Forensic anthropology as a service to prosecution

Probably, the best-known use of forensic anthropologists in international crimes investigation for the purpose of prosecution is in the former Yugoslavia. From the mid-1990s until 2001, teams of archaeologists, anthropologists, pathologists, odontologists, crime scene police, and others consulted to the Office of the Prosecutor of the International Criminal Tribunal for the former Yugoslavia (ICTY). A few other anthropologists consulted to the defense counsel of those on trial at ICTY. This work is well represented in the academic and popular literature and does not need to be related in detail here (e.g., Koff 2004; Sterenberg 2009; Stover and Peress 1998). Initially, forensic anthropologists were engaged with the search for and excavation of mass graves and the skeletal analysis of bodies exhumed, with the primary purpose of collecting evidence of mass killings. This evidence has been presented in multiple cases, and expert reports and transcripts of expert testimony are available on the ICTY court records website (http://icr.icty.org/). Starting in 2002, ICTY began to rely more on information collected by regional/national forensic teams that often worked in association with archaeologists and anthropologists working with a humanitarian mandate for the International Commission on Missing Persons. In Bosnia, which suffered the greatest human losses during the different wars of separation from Yugoslavia, thousands of people are still missing, yet forensic grave excavations and DNA identifications have slowed dramatically, owing to a lack of information on undiscovered burial sites (Wagner and Kešetović 2016).

The chief prosecutor for ICTY also directed investigations for the International Criminal Tribunal for Rwanda. Despite some literature on the work of forensic anthropologists in Rwanda, only two mass graves were ever excavated there, and the quality of work has been questioned in court (Rosenblatt 2015). These two *ad hoc* tribunals have now completed investigations, and their last use of forensic anthropologists for field work was in 2003. Some will still testify at ICTY in the last remaining trials. The Rwandan tribunal completed operations in 2015, and ICTY is anticipating completion of trials in 2017.

As the *ad hoc* tribunals close their doors, another international legal organization—this time a permanent one—has been established and is increasing operations. Governed by the Rome Statute of 2002, the International Criminal Court (ICC) is independent and can investigate crimes against humanity, war crimes, and *genocide* (International Criminal Court 2011). On only a few occasions has the prosecutor's office called upon the services of a forensic anthropologist, but the incidence appears to be increasing. There are currently three forensic anthropologists, and one pathologist with forensic anthropological expertise, listed as experts at the ICC (List of experts before the ICC as of August 18, 2015). One of these experts (author DC) recently testified about archaeological and anthropological investigations in the trial of the *Prosecutor v. Bosco Ntaganda* for alleged crimes committed in the Democratic Republic of Congo (International Criminal Court 2015). At the time of writing this chapter (February 2016), the ICC was actively investigating crimes in eight countries, all in Africa. The office of the prosecutor of the ICC was also conducting preliminary investigations in eight other territories, including the Ukraine and Palestine, which ratified the Rome Statute and accepted ICC jurisdiction for alleged crimes recently committed in their territories.

The focus of both the UN *ad hoc* tribunals and the ICC is the prosecution of perpetrators, but there have been secondary effects to their work, including opening the way to national mechanisms addressing broader post-conflict needs. Sometimes, authorities create special commissions to address post-conflict issues: investigation of disappearances, documentation of torture, reparation to victims and families, and so on. *Truth and Reconciliation Commissions* (TRC) may include the goal of excavating mass graves to enable proper victim identification and reburial, but these operations suffer from political and technical pressures (see http://www.usip.org/publications/truth-commission-digital-collection for a list of commissions and links to reports). Some countries protect the perpetrators through amnesty laws. There is often a general reluctance to dig up the past and *reopen wounds* that could allegedly reignite conflicts. The validity of such deleterious effects is difficult to estimate, but in most cases, these effects serve as political excuses to avoid investigation. International organizations (e.g., UN, ICRC, and Physicians for Human Rights) are often involved in transitional justice efforts and promote TRCs, insisting on the need for integrating forensic anthropologists in mass grave excavations and victim identification.

In the case of mass graves, a partial excavation may be sufficient to provide essential elements to the prosecution for charges of crimes against humanity or genocide, where the investigation is broader than the death of only one person or a couple of people. The analysis of a sample of bodies—assessing age and sex of victims, perimortem trauma, and evidence of cause of death—might meet the prosecutorial threshold to demonstrate a crime such as genocide. Partial mass grave excavations occurred with graves of Jewish victims from World War II (e.g., Rosenblatt 2015; Wright et al. 2005) and Iraq (Stoyke in Congram and Sterenberg 2009). Despite this meeting a legal threshold, the forensic anthropologist might question the ethics of limited excavation (particularly given the expense required simply to prepare and start excavation relative to the cost of *finishing the job*, once a team is already in place). It is important to consider the value—juridical and social, and toward a grander sense of justice—of completing excavation and more comprehensive attempts at victim identification for the local population. This value is best judged by consulting with the victim community, because, in some cases, unearthing the remains represents a disrespectful or even harmful act against the victims (as some argued in the case of graves of Jewish victims from WWII). Moreover, this dichotomy is complex, since the application of the discipline in such investigations is relatively new, and both the mandate of the hiring institution and the scope of the mission may jeopardize an objective and scientific approach.[*] In the short term, limited excavation and analysis might suffice immediate purposes. However, establishing good relations with a victim community is a better strategy for longer-term investigations, where there might be multiple mass graves from slightly different events or time periods. This will help reconcile differences in the meaning of justice from both the prosecutorial perspective and the perspective of the victim community.

Mass fatalities and emergency action

The involvement of forensic anthropologists in Disaster Victim Identification (DVI) has also expanded during the past decades. Support to authorities that are overwhelmed by the number of victims after a mass disaster (e.g., tsunami, earthquake, and plane crash) may provide the management of the dead in a timely manner and with more positive identifications (and fewer misidentifications). The response to the 2004 Indian Ocean tsunami involved forensic anthropologists in the search for remains and identification attempts in Thailand. Anthropologists also worked in Haiti following the 2010 earthquake and a few worked in Ukraine following the shooting down of a

---

[*] See Chapter 3 in Rosenblatt (2015) for an excellent discussion of moral and political dilemmas of mass grave excavation in different contexts.

commercial Malaysian Airlines plane with 295 people on board (MH17). These types of events involve multiple governments, external organizations, militaries, and civilians rushing in to deal with mass fatalities that have overwhelmed local resources. It should not be surprising, then, that a significant amount of disorder might ensue, including political battles for control over resources, responsibility, space, and even, tragically, victim bodies. Such issues motivated the creation and dissemination of guidelines (e.g., INTERPOL DVI Guide, http://www.interpol.int/fr/INTERPOL-expertise/Forensics/DVI) that emphasize the need for coordination of forensic personnel toward the proper identification of human remains in DVI. Whether an operation is driven by a humanitarian or forensic scope, an integrated approach in the identification of human remains is necessary to ensure an optimized success rate.

In mass fatalities contexts, despite great potential to assist, equal opportunity exists to cause further damage. Victims of different nationalities are often treated distinctly, with some being thoroughly analyzed and others being trucked off to pits for unceremonious mass burial. This differentiation should give assistance personnel pause to critically assess how (or if) they are helping. Likewise, the shortage of expertise—due to unavailability or political decision to commit resources, which is also seen in some humanitarian-oriented investigations—can result in an anthropologist being the only person conducting a postmortem examination. This is hardly ideal, for, in most jurisdictions, only a medical doctor can legally ascertain a cause of death. The important thing to remember is that the anthropologists (or any scientist) must not attempt to do work for which they are not qualified. Routine documentation of the condition of the remains and injuries in the form of detailed notes and good-quality photographs are sufficient for a number of purposes. This documentation could form the basis for a later declaration of cause of death by someone who has the authority to do so. Although the priority response in mass fatalities incidents is usually humanitarian, some events require additional care in field operations, as a judicial dimension may be critical (e.g., the investigation of the MH17 crash and the security issues surrounding the crash site delayed the recovery of victims' remains).

RECENT DEVELOPMENTS IN FORENSIC ANTHROPOLOGY AND INVESTIGATIONS OF INTERNATIONAL LAW

A reflection of the growing importance and influence of forensic sciences in international criminal investigation is the recent creation (2015) of the American Academy of Forensic Sciences Humanitarian and Human Rights Resource Center. The Center promotes the application of forensic scientists to humanitarian and human rights projects (http://www.aafs.org/resources/humanitarian-human-rights-resource-center/). Several members of the Anthropology Section of the Academy are involved at the Center, as members of either the advisory committee or the four different subcommittees.

A recent advance in the forensic sciences in many countries is the development of standards. Several organizations offer accreditation for forensic anthropologists. Although the principal organization has long been the American Board of Forensic Anthropology, more recent, regionalized certification or accreditation schemes are available in Latin America through the Latin American Forensic Anthropology Association (*Asociacion Latinoamericana de Antropologia Forense*, http://alafforense.org/certificaciones-alaf/) or in tiered systems available through the Royal Anthropological Institute in the United Kingdom (https://www.therai.org.uk/forensic-anthropology) and the Forensic Anthropology Society of Europe (http://www.forensicanthropology.eu/index.php/activities/fase-certification-process).

This is an important development—despite differences in the testing required for certification—as these organizations work to establish the professionalization of forensic anthropology and place the responsibility of anthropological analysis with those who are properly trained.[*] Regional certification systems are necessary, at present, given the distinct academic training and professional experience of anthropologists in different parts of the world. Nevertheless, several factors are bringing the discipline toward an internationalized and more unified practice: the ICRC advises governments on forensic methods and obligations to the missing; anthropologists from different countries jointly investigate international law; and publications that address work in different regions are increasing in number. Although *international standards* in forensic practice are oft-cited, it is fair to say that these do not quite exist yet. We hope that they will soon be in place.

---

[*] See Obenson (2014), Congram (2014), and Christensen et al. (2015) in *Forensic Science, Pathology, and Medicine* for debate about this subject.

FINAL WORDS ON THE ROLE OF THE FORENSIC ANTHROPOLOGIST IN THE INTERNATIONAL LAW CONTEXT

Forensic anthropology in almost all of its forms is about human rights. Anthropologists are called upon to search for missing persons and analyze unidentified remains, so that they might be identified, and often, a cause of death might be ascertained. This works toward humanitarian *and* judicial goals and responds to violations of the rights of individuals, families, and communities. In particular contexts, however, where there are deaths on a large scale, the use of forensic anthropological expertise can change, and the context often mandates the application.

When we hear of mass graves and mass fatalities, we often think of countries enveloped in armed conflict or prone by geography and politics to suffer greater loss of life due to natural disasters. Typically, we think of the latter disasters as clustered in the Global South (e.g., Baraybar and Blackwell 2014) and/or confined to *less developed* countries. This is because wealthier countries are often better equipped, though not necessarily prepared, to prevent and respond to disasters.

The terrorist attacks of 9/11 in the United States and the earthquake, tsunami, and nuclear plant disaster in Japan in 2011, however, show that even the richest countries are not immune to mass fatality events. Mass migration from Iraq and Syria across the Mediterranean also demonstrates a lack of will and/or ability of European countries to prevent or cope with thousands of migrant deaths. Such deaths due to gross violations of human rights occur in the global north, although in wealthier countries, the crimes tend to be structural, rather than being direct and physical. In 2015, Luis Fondebrider, co-founder and director of the Argentine Forensic Anthropology Team, was giving a lecture to an audience at the University of Toronto in Canada. A professor asked him how [Canadian] students could get involved in investigations of the thousands of victims of enforced disappearance in Argentina. Fondebrider's response was poignant and, to most in the audience, surprising: "I think your indigenous people need you more than we do." At just about the same time as Fondebrider's lecture, a court-mandated Truth and Reconciliation Commission was preparing to present a report on so-called Indian Residential Schools that had operated in Canada for over 100 years, until 1996. There is good evidence that over 4000 Indigenous children died while students at these schools, although the number of deaths might be more than 6000. Most were buried around the schools in what are now largely unmarked graves (Maass 2016). One month after Fondebrider's talk, the Truth and Reconciliation Commission of Canada released a report stating that: "… the central goals of Canada's Aboriginal policy were to eliminate Aboriginal governments; ignore Aboriginal rights … and, through a process of assimilation, cause Aboriginal peoples to cease to exist as distinct legal, social, cultural, religious, and racial entities in Canada. The establishment and operation of residential schools were a central element of this policy, which can best be described as 'cultural genocide'" (Truth and Reconciliation Commission Canada 2015:1). Fondebrider saw the parallel between places like Argentina or Guatemala and Canada, but most Canadians (and others in similar states) believe that mass human rights' violations, enforced disappearance, and genocide come only by bullets or machetes, leading to mass graves. Forensic anthropology is increasingly becoming more than just the application of the scientific methods of archaeology and skeletal biology to medicolegal investigation. Some are even advocating a "more 'social' forensic anthropology … [which can] work toward remedying the unfavorable conditions that transformed people into victims in the first place" (Baraybar and Blackwell 2014:36). In places where human rights' violations and mass graves have been politically neglected and victim communities have been marginalized, forensic anthropologists are increasingly defining themselves as advocates as well as scientists (Congram et al. 2014).

In cases that appear to warrant criminal investigation, but authorities decide not to pursue one, forensic anthropologists might take on the role of advocate. While doing so might be completely appropriate, this change of role and perspective warrants consideration. We must always remember that becoming an advocate does not mean abandoning an objective approach to the case. We might believe that we can calculate reliable odds of an alleged crime being true *a priori*, as when we assume that disappeared protesters who were detained during an antigovernment rally have been killed and buried in a mass grave by the authorities of an authoritarian regime known for brutality of its citizens. Nevertheless, as a matter of course, we should not be seeking to *corroborate a story* or a witness statement (e.g., Kimmerle 2012:421, 424); rather, we should be seeking evidence to test an allegation. We see similar language throughout the forensic literature, and we can be fairly confident that, in fact, our colleagues are exercising objective discretion during analysis, but careful choice of language is important, if only because it can lead to the suggestion of bias, which might become a point of argument in court and detract from the important facts of a case. Further, the more we use language like this, the more likely we are to begin to think, rather than simply speak, this way.

For investigations of mass killings and common burials, there is a tendency to see the role of the forensic scientist as a knight in shining armor: one who unerringly seeks justice for the weak by fighting evil oppressors. Tragically, some practitioners also adopt this persona without critically assessing their motives and the nature of investigations. This can be subtle, as when subconsciously seeking to prove an allegation of human rights' abuses (whereas a good scientist tests hypotheses and is as open to refutation as confirmation). Alternatively, this can be overt, as when an anthropologist joins the forces of an invading army in a one-sided investigation of crimes committed by the vanquished, while ignoring the mass abuses and deaths caused by the military with which she or he is working.

Developing this further, we must always keep in mind the myriad ways in which people die—the manner and mode—particularly during armed conflict. Forensic anthropologists work as investigators of death and disappearances and examine the remains of those who have been extra-judicially executed, *judicially* executed, those who have died in vehicle accidents, and also those who have died due to disease (possibly the product of desperate conditions created by conflict or repressive governments). Some disappearances are self-imposed, such as those of Radovan Karadzic and Ratko Mladic, both of whom were indicted by ICTY and remained at large for years. Others are victims of natural disasters (e.g., Indian Ocean tsunami of 2004 and the 2015 earthquake in Nepal), and yet others, as mentioned previously, are well-intentioned but misguided cultural responses to mass fatalities (e.g., unmarked mass burials of the dead in Haiti). Violent deaths do not *necessarily* imply a crime (e.g., the euphemistic *collateral damage* that happens during shelling and misguided drone strikes that are happening, tragically, with increasing regularity). Ultimately, it is the job of the anthropologist to examine trauma and work toward positive identification. Culpability for crimes, if being sought in a case, is decided by judges or juries in a court, not by forensic scientists. This is not to say, however, that forensic scientists should not be exercising cautious judgment when deciding which cases to get involved in.

A growing number of forensic anthropologists dedicate the majority of their time not to case work but to training and advising governments on how to do casework, recognizing that it is the legal responsibility of states to resolve cases of missing persons. Indeed, there is a sound logic to this approach, for why would a state dedicate limited resources to the complex, lengthy, and controversial project of searching for and identifying the dead, when well-funded foreign experts will do the job for them. The temporary importation of forensic expertise with restricted objectives can remove incentive for states to develop and fund indigenous capacity, which can serve as an excuse for inaction in the instance of future deaths.

SUMMARY

In this chapter, we have discussed a range of situations where the forensic anthropologist is engaged in the search for missing persons, death investigation, and analysis of unidentified human remains. The range of criminal, legal, and humanitarian measures engaging forensic anthropologists around the world is diverse. They include an African Union-backed court to try former dictator of Chad Hissène Habré (Extraordinary African Chambers), ongoing (since 1984!) investigations of missing persons and trials in Argentina, the NGO-led search for victims of genocide in Guatemala and of political mass killings in Spain, the strictly humanitarian identification efforts in Cyprus or related to the U.S. war dead, and, significantly, the expanding work of the ICRC. Just as the range of contexts broadens, so does *what forensic anthropologists do* and *how they do it*.

This expansion of opportunities for forensic anthropologists is exciting but not without risk. As interest increases in the use of forensic anthropology for the investigation of international crimes or mass fatality response, anthropologists are placed in new environments with few well-established rules and norms. It would be ideal for all forensic anthropologists to receive a comprehensive education first in conventional archaeology and biological anthropology, followed by specialized training in forensics through domestic forensic casework mentored by an experienced, well-established forensic anthropologist. These things provide a firm foundation for working in more complex contexts of mass disaster response and international criminal investigations in diverse, fast-changing environments. Increasing professionalization of the discipline and growing accreditation processes are positive developments. Even with all of this preparatory training and experience, however, forensic anthropologists are learning that having a broader understanding of socio-political contexts and exercising critical judgment about what they do are necessary tools in the practice of a global, humane, and just forensic anthropology.

Review questions

1. Describe the role of an anthropologist in the investigation of international crimes.
2. Explain the differences between humanitarian- and prosecution-mandated missions.
3. What are the main treaties guiding the principles of humanitarian forensic action?
4. What ethical dilemmas might forensic anthropologists face in the context of international crimes that they are less likely to have to confront in a domestic investigation context?
5. How might the professional ethics of a forensic anthropologist conflict with a particular mission's mandate?
6. Should a forensic anthropologist advocate or lobby for investigations in certain cases? Would this compromise the objectivity of their analysis? Explain your answer.

Glossary

Genocide: any of the following acts committed with an intent to destroy, in whole or in part, a national, ethnical, racial or religious group, such as the following:
   (a) Killing members of the group;
   (b) Causing serious bodily or mental harm to members of the group;
   (c) Deliberately inflicting on the group conditions of life calculated to bring its physical destruction in whole or in part;
   (d) Imposing measures to prevent births within the group;
   (e) Forcibly transferring children of the group to another group.
(Article 2 of the Convention of the Prevention and Punishment of the Crime of Genocide)

Humanitarian forensic action: support in the search for recovery, analysis, identification, and management of the remains of large numbers of victims resulting from armed conflict, disasters, migration, and other situations, within a humanitarian scope.

Human rights: rights inherent to all human beings, whatever nationality, place of residence, sex, national or ethnic origin, color, religion, language, or any other status. Humans are all equally entitled to human rights without discrimination. These rights are all interrelated, interdependent, and indivisible. Universal human rights are often expressed and guaranteed by law, in the forms of treaties, customary international law, general principles, and other sources of international law. International human rights law lays down obligations of governments to act in certain ways or to refrain from certain acts, in order to promote and protect human rights and fundamental freedoms of individuals or groups (United Nations definition).

Mass disaster: threatening event, or probability of occurrence of a potentially damaging phenomenon within a given time period and area. These include natural hazards (earthquakes, landslides, tsunamis, volcanic activity, avalanches, floods, wildfires, cyclones, disease epidemics, insect/animal plagues, etc.) and technological or "man-made" hazards (conflicts, famine, displaced populations, industrial accidents, transport accidents) (IFRC definition).

Mass grave: a burial with at least three individuals, typically placed in burial in a careless or disrespectful manner (e.g., no marking of the grave, commingling of remains, without respecting cultural norms).

Missing person: a person whose whereabouts are unknown to his/her relatives and/or who, on the basis of reliable information, has been reported missing in accordance with the national legislation in connection with an international or noninternational armed conflict, a situation of internal violence or disturbances, natural catastrophes or any other situation that may require the intervention of a competent State authority (ICRC definition).

Truth and reconciliation commission: commission created in the scope of transitional justice consequently to a dictatorship, civil war, or internal unrest, aiming at a national reconciliation. Such commissions usually investigate human rights violations, and help traumatized societies, and sometimes are tasked to address to issue of missing persons (specific mandates vary among contexts).

References

Baraybar JP and Blackwell R. 2014. Where are they? Missing, forensics, and memory. *Ann Anthropol Pract* 38(1):22–42.

Bernardi P and Fondebrider L. 2007. Forensic archaeology and the scientific documentation of human rights violations: An Argentinian example from the early 1980s. In R Ferllini (Ed.), *Forensic Archaeology and Human Rights Violations*, pp. 205–232. Springfield, IL: Charles C Thomas.

Christensen AM, Passalacqua NV, Schmunk GA et al. 2015. The value and availability of forensic anthropological consultation in medicolegal death investigations. *Forensic Sci Med Pathol* 11:438–441.

Congram D. 2014. Letter regarding "In consideration of subspecialty training in forensic anthropology for pathologists." *Forensic Sci Med Pathol* 10:662.

Congram D and Sterenberg J. 2009. Grave challenges in Iraq. In *Handbook of Forensic Anthropology and Archaeology*, S Blau and DH Ubelaker (Eds.), pp. 441–453. Walnut Creek, CA: Left Coast Press.

Congram D, Flavel A, and Maeyama K. 2014. Ignorance is not bliss: Evidence of human rights violations from Civil War Spain. *Ann Anthropol Pract* 38(1):43–64.

Djurić M and Starović M. 2015. Forensic archaeology in Serbia: From exhumation to excavation. In WJ Groen, N Marquez-Grant, and RC Janaway (Eds.), *Forensic Archaeology; A Global Perspective,* pp. 149–157. Chichester, UK: Wiley-Blackwell.

Etxeberría F, Herrasti L, Serrulla F, and Márquez-Grant N. 2015. Contemporary exhumations in Spain: Recovering the missing from the Spanish Civil War. In WJ Groen, N Marquez-Grant, and RC Janaway (Eds.), *Forensic Archaeology; A global perspective,* pp. 489–497. Chichester, UK: Wiley-Blackwell.

Gould SJ. 1981. *The Mismeasure of Man.* New York: W.W. Norton & Company.

International Committee of the Red Cross. 1995. The Geneva Conventions of 12 August 1949. Geneva: ICRC. https://www.icrc.org/eng/assets/files/publications/icrc-002-0173.pdf (accessed March 6, 2016).

International Committee of the Red Cross. 2003. ICRC Report: The Missing and their families. Summary of the conclusions arising from events held prior to the international governmental and non-governmental experts (February 19–21, 2003). Geneva: ICRC.

International Criminal Court. 2011. Rome Statute of the International Criminal Court. Enschede, The Netherlands: PrintPartners Ipskamp. http://www.icc-cpi.int/NR/rdonlyres/ADD16852-AEE9-4757-ABE7-9CDC7CF02886/283503/RomeStatutEng1.pdf (accessed September 29, 2015).

International Criminal Court. 2015. Document ICC-01/04-02/0, dtd 16 April 2015.

Kimmerle E. 2012. Forensic Anthropology; a Human Rights Approach. In MT Tersigni-Tarrant and NR Shirley (Eds.), *Forensic Anthropology; An Introduction*, pp. 421–438. Boca Raton, FL: CRC Press.

Koff C. 2004. *The Bone Woman.* New York: Random House.

List of experts before the ICC as of 18 August 2015. International Criminal Court. http://www.icc-cpi.int/iccdocs/PIDS/other/180815-List-of-Experts-Eng.pdf#search=anthropolog%2A (accessed September 29, 2015).

Maass A. 2016. Perspectives on the missing: Residential schools for Aboriginal Children in Canada. In D Congram (Ed.), *Missing Persons; Multidisciplinary Perspectives on the Disappeared,* Chapter 1. Toronto: Canadian Scholars' Press.

Obenson K. 2014. In consideration of subspecialty training in forensic anthropology for pathologists. *Forensic Sci Med Pathol* 10:114–115.

Rosenblatt A. 2015. *Digging for the Disappeared*. Stanford, CA: Stanford University Press.

Skinner M. 2007. Hapless in Afghanistan: Forensic archaeology in a political Maelstrom. In R Ferllini (Ed.), *Forensic Archaeology and Human Rights Violations*, Springfield, IL: Charles C Thomas.

Skinner M and Sterenberg J. 2005. Turf wars: Authority and responsibility for the investigation of mass graves. *Forensic Sci Int* 151:221–232.

Skinner M, York HP, and Connor M. 2002. Postburial disturbance of graves in Bosnia-Herzegovina. In WD Haglund and MH Sorg (Eds.), *Advances in Forensic Taphonomy; Method, Theory, and Archaeological Perspectives,* pp. 277–308. Boca Raton, FL: CRC Press.

Steele C. 2008. Archaeology and the forensic investigation of recent mass graves: Ethical issues for a new practice of archaeology. *Archaeologies: J World Archaeol Congr* 4(3):414–428.

Sterenberg J. 2009. Dealing with the remains of conflict: An international response to crimes against humanity, forensic recovery, identification, and repatriation in the former Yugoslavia. In S Blau and DH Ubelaker (Eds.), *Handbook of Forensic Anthropology and Archaeology,* 416–425. Walnut Creek, CA: Left Coast Press.

Stover E and Peress G. 1998. *The Graves; Srebrenica and Vukovar.* Zurich: Scalo.

The Guardian. 2015, online. Experts question Mexican investigation of 43 students' disappearance. http://www.theguardian.com/world/2015/feb/08/experts-question-mexican-investigation-of-43-students-disappearance (accessed September 18, 2015).

Tidball-Binz M and Hofmeister U. 2015. Forensic archaeology in humanitarian contexts; ICRC action and recommendations. In WJ Groen, N Marquez-Grant, and RC Janaway (Eds.), *Forensic Archaeology; A Global Perspective,* 427–437. Chichester, UK: Wiley-Blackwell.

Truth and Reconciliation Commission of Canada. 2015, online. Honouring the truth, reconciling for the future. Summary of the final report of the Truth and Reconciliation Commission of Canada. http://www.trc.ca/websites/trcinstitution/File/2015/Honouring_the_Truth_Reconciling_for_the_Future_July_23_2015.pdf (accessed October 3, 2015).

Tuller HH. 2012. Mass graves and human rights: Latest developments, methods, and lessons learned. In DC Dirkmaat (Ed.), *A Companion to Forensic Anthropology,* 1st edn, pp. 157–174. Hoboken, NJ: Wiley-Blackwell.

Tuller H and Đurić M. 2006. Keeping the pieces together: Comparison of mass grave excavation methodology. *Forensic Sci Int* 156:192–200.

Tuller H and Hofmeister U. 2014. Spatial analysis of mass grave mapping data to assist in the reassociation of disarticulated and commingled human remains. In B Adams and J Byrd (Eds.), *Commingled Human Remains: Methods in Recovery, Analysis, and Identification,* San Diego, CA: Academic Press.

Wagner S and Kešetović R. 2016. Absent bodies, absent knowledge: Forensic work of identifying Srebrenica's missing and the social experiences of families. In D Congram (Ed.), *Missing Persons; Multidisciplinary Perspectives on the Disappeared,* Chapter 2. Toronto: Canadian Scholars' Press.

Wright R, Hanson I, and Sterenberg J. 2005. The archaeology of mass graves. In J Hunter and M Cox (Eds.), *Forensic Archaeology: Advances in Theory and Practice,* pp. 137–158. London: Routledge.

Chapter 18

Advanced Scene Topics—Fire and Commingling

Joanne Bennett Devlin and Nicholas P. Herrmann

CONTENTS

Fire	348
Fire dynamics	348
Degrees of burning	349
Anthropological research	350
At the fire scene	352
In the laboratory	353
Fractures	354
Commingling	357
Recent research and novel methods	359
Summary	361
Review questions	361
Glossary	361
References	362

LEARNING OBJECTIVES

1. Discuss how fire damage to remains impacts forensic anthropology laboratory and field work.
2. Briefly discuss the dynamics of fire and the impact fire has on human remains.
3. Describe the appearance of the body tissues in each of the five degrees of burning.
4. List the unique knowledge, training, and contributions that anthropologists bring to fire scenes.
5. Explain the processes of assessing the biological profile and interpreting trauma from burned remains.
6. Recognize and describe fractures specific to the burning process.
7. Discuss how commingling occurs and the appropriate methods for dealing with commingling situations.
8. Apply the various methods of estimating the number of individuals represented in a commingled skeletal assemblage.

The request to assist in the investigation of burned remains found in residences, commercial establishments, or automobiles is a solicitation that many forensic anthropologists receive. Engaging in such necessitates modifying many of our standard approaches, including the techniques that we use at scenes and also in the laboratory. Fire affects the soft and hard tissues in a manner that no other taphonomic force does, resulting in dramatically modified materials that are difficult to discover, recover, and analyze. Burned bone looks different. It can range from black to gray to white in color; it may be warped and fragmented; and it is highly susceptible to further modification as its composition is altered.

In addition, burning impacts each and every protocol codified by Snow (1982) in his guidelines. Burning makes it difficult to differentiate bone from other materials. It further affects our ability to assess whether bone is human or nonhuman and the number of individuals who may be represented in an assemblage. Clearly, dealing with burned remains presents a unique and complex situation for practicing forensic anthropologists. This chapter provides an overview of the manner in which fire affects and modifies the hard and soft tissues, alters scene processing methods in the search and recovery for burned bones, and impacts laboratory analyses. In addition, given the potential for burned remains to represent a commingled scene, we detail the process for determining the minimum number of individuals. This chapter discusses both of these situations: fire and commingled remains. While these are potentially interconnected variables in forensic case work, the focus here will be on these as two distinct situations. Following completion of this chapter, readers will appreciate guidelines for how to best discover, recover, and analyze burned remains and how to assess commingled material and methods with which to determine the minimum number of individuals represented.

FIRE

Fire can have a tremendous impact on the soft and hard tissues of the human body resulting in substantial modification to skeletal and dental tissues. Exposure to heat begins with damage to and eventual loss of soft tissues, followed by alteration of hard tissues. Fire brings about irreversible changes in the color, form, and size of skeletal and dental elements. These signatures are unique and serve as indicators of exposure to fire. The type and degree of damage result from the interplay of two important variables: (1) the heating temperature and (2) the duration of exposure to heat (Bennett-Devlin et al. 2006). That said, fire is a dynamic entity, and as such, it is not possible to unequivocally dictate the specific impact that it can have on body tissues. It is possible, however, to discuss general patterns of damage and impacts on our lab and field work.

While no specific pattern can be used to characterize the condition of all victims in each fire event, a general sequence of modification exists. Exposure of the body to fire results initially in the reduction of soft tissues as skin, muscle, and fat actually are sources of fuel that can sustain the fire; bone, too, is a fire-sustaining fuel. While fire can bring about the complete reduction of soft tissues, the properties of bone are such that it cannot be destroyed by heat or fire alone; it is not possible to completely destroy a body solely through heat. However, heat can affect bone tissue severely, making it highly fragile and therefore susceptible to further modification (and damage) related to the fire or suppression efforts, as well as discovery and recovery attempts. However, modified, burned, and cremated remains are not beyond analysis (see for example Gocha and Schutkowski 2013), and thus, a well-led investigation and efficient analyses can potentially recognize the nature of the situation and ideally establish the biological profile of fire victims.

Anthropologists can provide several specific and unique roles in investigations involving burned human remains. First, the anthropologist can aid in the discovery and recovery, which can be followed by analyses to create a biological profile. Furthermore, the anthropologist can assist in establishing identification. While the frequency of forensic cases that involve burned remains is likely lower than that of other traumas, the particulars of heat modification are so tremendous that practicing forensic anthropologists must be aware of the signatures of burning. To facilitate the development of information needed to function successfully in each of these stages of forensic work, the anthropologist must appreciate the mechanisms by which fire occurs and grows.

FIRE DYNAMICS

An understanding of the dynamics of fire is imperative to appreciate the impact fire can have on human remains. While this is not the arena for a comprehensive discussion of fire, it is important to have a basic awareness of how fire spreads and grows and the temperatures up to which it can reach. At the most basic level, fire is a chemical reaction and, as such, it is associated with the presence and interaction of several components: *heat*, *oxygen,* and *fuel.* The relationship between these entities is often represented and referred to as the fire triangle or fire tetrahedron (with the chain reaction of this relationship—fire—at the center of the tetrahedron). The latter more accurately reflects the complex reality of combustion and fire growth, and the reader is encouraged to supplement the present discussion with more detailed explanations (see DeHaan 2002; DeHaan 2015).

Heat involves raising the temperature of an object to the lowest temperature at which it will release vapors and thus sustain combustion. The amount of air or *oxygen* (identified as the oxidizing agent) must be of a level that can sustain combustion. Finally, the *fuel* refers to the combustible materials that are present and are capable of sustaining the chemical reaction (fire). Obviously, an ignition source must also be present to initiate a burning event and to elevate the fuel to its ignition temperature. To appreciate the role of each of these components, consider the extinguishment that results if one or more of these is absent or removed. For example, reflect upon how the use of water can extinguish a fire; it reduces the temperature of combustible materials below the ignition point (removal of heat). Also, consider the practice of putting the lid on a pan in the event of a grease fire; extinguishment occurs because you have eliminated the oxygen source and altered the structure of the fire triangle. Importantly, the amount of each critical component (heat, oxygen, and fuel) that is present and available will dictate the duration and the intensity of the fire. While we recognize the products of fire as heat and light, the amount of both of these that is produced depends directly on the interaction of the parts of the fire triangle. Simply, these components are responsible for the intensity (temperature) and duration of a fire.

The temperatures attained by fires are limited by the fuel load (the nature of the combustible material and how much is present). Typically, man-made products such as plastics burn at higher temperatures than wood or natural materials. However, the amount of combustible material dictates the duration of the fire. While it is not possible to isolate these influences, it is beneficial to have an awareness of the average temperatures that are reached in a variety of scenarios: residential fires, open fires, and automobile fires. The average residential fires attain temperatures of approximately 1200°F, while campfires have lower temperatures, rarely exceeding 600°F, though the type of wood used influences the amount of heat that is generated. Data from experimentally burned automobile fires (conducted during the last decade) demonstrate that not only do cars generate high temperatures, in excess of 1800°F, but they also reach maximum temperatures after a very short period of time (see Devlin et al. 2004; Devlin and Herrmann 2013).

DEGREES OF BURNING

Clinicians commonly utilize a five-tiered system, (1st–5th degrees) to describe the degree of damage caused by exposure to heat. First-degree, or superficial, burns commonly involve reddening of the epidermis without blistering, similar to the pattern found with sun burn. Continued exposure leads to 2nd-degree (partial thickness) burns that impact the **epidermis** and the upper dermal layer and are characterized by blisters or sloughing of the epidermis. Third-degree (full thickness) burns destroy the epidermis and **dermis**, further impacting vascular and nervous supplies. Continued exposure leads to 4th-degree burns characterized by charring (blackening) of tissues. It is common with 4th-degree burning to have exposure of bone in regions where soft tissue layers are thin (e.g., shins and forehead). The bone appears black in color. The development of **pugilistic posturing** (boxer's pose), which is the flexion of major body joints due to the dehydration of muscle tissue, is also expected at this level of burning (Figure 18.1). This posture is so named because the characteristic flexion of the elbows, medial rotation of the shoulders, and flexion of the hands resemble the posture of a boxer. The legs flex at the knees and the hip, causing a lateral rotation at the hip and a bent-knee pose.

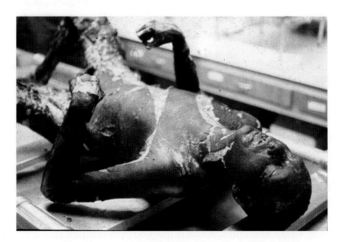

Figure 18.1 The pugilistic posture (boxer's pose).

The head is bent backward, and the circular muscles around the eyes and mouth contract such that these structures take on wide circular openings; further, the tongue swells and appears to slightly protrude. The pugilistic posture results from the presence of antagonistic muscle pairs, such that dominant muscle groups (joint flexors) contract and overpower extensors, leading to flexion at each joint. The absence of this pattern in a decedent should be of concern to an investigator. Continued exposure to heat leads to the loss of all soft tissues, or 5th-degree burning, which is more commonly referred to as *cremation*; the resultant bone tissues are called **cremains**. Bone is the predominant product, and it exhibits classic signatures of thermal modification (i.e., white in color, and highly fragmented and fragile). Little soft tissue remains, though it may be observed in the gluteal region. Similarly, teeth are impacted, and although roots and enamel may remain intact, they frequently separate from one another and are extremely fragile. For a discussion on the specific changes that occur to teeth, the reader is encouraged to review the recent works of Schmidt (2015) and Beach and coworkers (Beach et al. 2015).

ANTHROPOLOGICAL RESEARCH

Archaeologists and biological anthropologists have long been interested in examining burned bone. Beginning with macroscopic examinations of archaeologically recovered burned bone to experimental work and histological analyses, anthropologists are well versed in bone modification signatures resulting from heat. The earliest examinations were motivated by attempts to interpret skeletal collections from archaeologically derived cremation sites (Baby 1954; Binford 1963; Webb and Snow 1945; Wells 1960). Baby (1954) promoted a three-way classification of burned bone incorporating modification of element color and surface morphology by labeling bone as *completely incinerated*, *incompletely incinerated* or *nonincinerated*. The observable variation was attributed to several factors, including the duration of exposure and proximity of bone to the fire (Baby 1954). Shortly thereafter, more formal examination of cremated remains was initiated by Wells (1960), who documented the effects of cremation and suggested that the degree of distortion was, in part, associated with firing temperature. Binford (1963) extended the study of cremations, noting that particular patterns of warping and fracturing indicate the pre-incineration condition of the remains, suggesting that bone covered by soft tissue at the time of incineration will appear different from bone burned either devoid of soft tissues or even lacking the grease and marrow of *fresh bone*. Binford (1963) observed that the resultant condition and appearance of bone is attributable to not only the duration and intensity of the fire, but also the condition of the material before heat exposure. Subsequent publications focused on the condition of bone as a predictor of heating parameters. These researchers sought to establish a correlation between varied surface signatures as indicators of the degree of heat exposure through both laboratory and natural context examinations (Buikstra and Swegle 1989; Gilchrist and Mytum 1986; McCutcheon 1992; Nicholson 1993; Shipman et al. 1984; Spenneman and Colley 1989; Thurman and Wilmore 1981).

Bonucci and Graziani (1975) presented an overview of the heat exposure and its relationship to bone appearance. They noted that a progression through heating stages (from dehydration to decomposition and inversion) is associated with outwardly apparent modification, including change in the surface color of bone. With increasing exposure to heat, bone progresses through a sequence of colors, from unburned tan to shades of dark brown to black, progressing to blue and gray and finally to white (Bonucci and Graziani 1975). Bonucci and Graziani (1975:531) claim that "color itself can be used to some extent for deducing the approximate value of the temperature," correlating ochraceous to brownish colors, with exposure to 200°C–300°C (392°F–572°F); black with temperatures of 300°C–350°C (572°F–662°F); gray with temperatures of 550°C–600°C (1022°F–1112°F); and white for heating in excess of 650°C (1202°F). Black color and an increase in the fragility of bone involve combustion and progressive incineration of the organic materials of collagen and carbon. The gray and white colors (see Figure 18.2) result from continued combustion that impacts the crystalline structures.

Subsequent examinations investigated the impact of heat on bone by investigating the association between exposure temperature and surface color (e.g., Shipman et al. 1984) and the microscopic and structural impacts of heat on skeletal material (e.g., Herrmann 1977; McCutcheon 1992; Nicholson 1993; Shipman et al. 1984; Thompson 2005). Undoubtedly, best known of these works is the research led by Shipman and colleagues (1982), who examined bone and teeth burned in a muffle furnace. They standardized the description of surface colors (assessed using the Munsell soil color charts), recognizing five distinct color stages associated with specific increases in temperature. Each stage is characterized by variation in hue, value, and chroma and is marked by dominant surface colors and multiple secondary colors. Shipman associated the dominant colors of pale yellow and very pale brown with exposure to temperatures less

Figure 18.2 The generalized changes in color observed in burning. From right to left: unaltered, smoked, charred (carbonized), and calicined. The left three fragments can be classifed as calcined, but the fragment on the far left is white.

than 285°C (545°F). Temperatures from 285°C to 525°C (545°F–977°F) are associated with dominant colors of pink and black and secondary colors of very dark grayish brown and brown; "common colors are reddish brown, very dark gray-brown, neutral dark gray and reddish yellow" (Shipman et al. 1984:313). Shipman and colleagues characterized stage three, delineated by temperatures up to 645°C (1193°F), by surface colors of "neutral black, with medium blue and some reddish-yellow appearing" (1984:313) and identified Munsell colors of light gray, and secondary colors of brown and light brownish gray (Munsell Color Company 2000). Bone heated to 940°C (1724°F) exhibits neutral white and light blue-gray colors. Elements heated to temperatures in excess of 940°C exhibit neutral white and medium gray colors.

This association between heating temperature and surface color was echoed in later work by McCutcheon (1992) and Nicholson (1993). McCutcheon (1992) combined the assessment of surface color with microscopic criteria to recognize a three-level system for estimating exposure temperatures. He observed pale brown to black in specimens heated to 340°C (644°F), light brownish gray in elements heated to 600°C (1112°F), and white color dominating bone heated in excess of 650°C (1202°F) (McCutcheon 1992). Bennett (1999) demonstrated that a change in the surface color could also occur through indirect exposure to heat. Black colors were observed on bone buried beneath campfires that were maintained for 48 hours. Recorded temperatures in these subsurface deposits were less than those normally equated with direct fire exposure. The extended duration likely accounts for the appearance. Further, buried bones exhibit less distortion and more consistent surface colors than directly burned bones.

Beach et al. (2008 and 2015) investigated heat related color change of human teeth, concluding that color changes are similar to those reported for bone. Evaluating dentin and enamel colors using the Munsell charts, Beach et al. (2015:146) noted "… that temperature is the most important variable when it comes to interpreting tooth color." Of interest is the overall pattern of dental modification caused by heat exposure, such that the enamel tissue begins to flake at temperatures less than 600°C, dark grayish brown enamel color characterized teeth heated to 700°C, and separation of the crown occurred at 800°C after only 30 minutes of exposure.

While the traditional assessment of surface color, employing Munsell soil color charts has been part of the primary protocol in examining and interpreting burned bone assemblages, Devlin and Herrmann (2008) demonstrated the use of a **spectrophotometer** as a more objective means to collect bone color data. While the association between particular surface colors and exposure temperatures has become commonplace, Shipman and coworkers (1984:320) stated that the reliance on color as an indicator of exposure temperature is "an essentially imprecise criterion both because of individual differences in the ability to perceive fine color distinctions and because burnt bones may change color if they are buried." Nonetheless, they stated that surface color can be informative as to the range of temperatures to which a bone has been exposed.

In addition to the widely recognized color change, burned bone can present an interpretative challenge to forensic anthropologists, given the shrinkage and fragmentation, in addition to fracturing and warping that characterize thermally altered remains. The observable changes in surface color and element integrity are outward reflections of the modification of the chemical composition of bone in the presence of heat. Bone, as an organic material, evidences predictable changes when exposed to heat and displays a basic chronology of changes. Heating leads to the alteration of the chemical compounds and subsequent dehydration of the tissue, leading to an increase in the brittleness of the tissue. Heated bone is fragile and exhibits warping, shrinkage, fracturing, and color change.

Researchers have reported four identifiable stages in the process of cremation (Bonucci and Graziani 1975; see Correia 1997; Thompson 2005 for review). The first stage, *dehydration*, is characterized by the breaking of hydroxyl bonds, which causes water loss. As reviewed by Correia (1997) these events are associated with exposure to temperatures up to 600°C. Stage 2, *decomposition*, is characterized by the removal of the organic components and is associated with temperatures ranging from 500°C to 800°C. Exposure to temperatures ranging from 700°C to 1100°C causes the loss of carbonates that characterizes the third stage, referred to as *inversion*. The final stage, referred to as *fusion*, is associated with temperatures in excess of 1600°C and is characterized by melting of the crystals (Correia 1997). While there is disagreement as to the particulars associated with these stages (see Thompson 2004), they are more theoretical than applied, and while informative, they do not necessarily influence the actions of forensic anthropologists.

Several other approaches provide more applicable mechanisms for describing bone modification. Correia (1997) proposed the use of a four-level system of defining thermal modification: charring, partial cremation, incomplete cremation, and complete cremation. The first two levels refer to situations in which the internal organs and soft tissues, respectively, survive. Incomplete cremation refers to the presence of burned bone, and complete cremation implies bone fragments. A comparable five-level scheme of patterned destruction was presented by Glassman and Crow, termed the Crow-Glassman scale (CGS) (Glassman and Crow 1996). Although not widely employed, this scale has great potential, as it correlates levels of modification and destruction with necessary recovery efforts and analytical approaches. The first level refers to cases in which there is little skin damage, with the fatality likely associated with smoke inhalation; no specialized recovery or identification methods are required. In fact, the unique skills of the anthropologist are not required in a level 1 event. Level 2 involves deeper skin damage with charring of certain areas. Glassman and Crow (1996) noted that elements or regions, such as ears, hands, and external genitalia, may not be connected to the body and may require the assistance of specialists for recovery. Identification may rely on dental remains and will likely involve the combined efforts of the medical examiner and an odontologist. The degree of damage associated with these two levels is consistent with the modification that clinicians associate with 1st- and 2nd-degree burns. For the CGS level 3, the greater destruction is such that major elements and areas are affected with bone exposure and skeletal damage. Glassman and Crow (1996) recommend the presence of an anthropologist in the field to oversee recovery and ensure the highest level of bone and dental recovery. The reliance on the anthropologist is further required for laboratory analysis and identification. The presence of an anthropologist from recovery to analysis is also required in scenes assigned levels 4 and 5 on the Crow-Glassman scale. While levels 4 and 5, respectively, involve greater burning of bone from carbonized and blackened bone to gray and white calcined bone, the anthropologist is constant at all levels of analysis from discovery to recovery and from analysis to identification. This raises two questions: What is unique about the anthropologist, and why are anthropologists so beneficial at fire scenes?

AT THE FIRE SCENE

The role and responsibility of an anthropologist at a fire scene is identical to any other scene: maximize recovery of remains. As presented in Chapters 2 and 3 of this volume, the anthropologist has unique responsibilities and capabilities that contribute to scene work. Anthropologists are also well versed in recognizing human bones and differentiating them from animal bones. Obviously, the ability to recognize bone fragments as representing particular elements from a specific side of the body is also beneficial and facilitates an early understanding of the position of a victim (or victims) at a scene. These contributions are critical in fire scenes. Of utmost importance at fire scenes involving fatalities, the anthropologist can recognize bone and differentiate it from other materials. Burned fragmented bone can mimic burned fragmented sheetrock, a common component at residential fires. Understanding the amount of heat and duration of exposure required to alter bone is also crucial in that the anthropologist is able to predict whether the deceased is more likely to appear charred, partially cremated, incompletely cremated, or completely cremated (after Correia 1997).

The anthropologist is necessary, as Glassman and Crow (1996) promote, because anthropologists can assist in the search and recovery by using standard archaeological methods. While Chapter 3 provides details on the appropriate means of search, documentation, excavation, and recovery, it must be realized that burned bone is highly susceptible to damage, so slight modifications to the standard approach are required to work a fire scene effectively (see Dirkmaat 2002; Fairgrieve 2008). Discovering the location of a victim in a fire may require painstaking examination of the residue. Using hand tools is preferred. As with any excavation once a scene is disturbed it cannot be returned to its original state. It is highly possible to damage burned bone during the search phase, especially if heavy equipment is used. Walking through a scene can destroy cremated or calcined bone. Realizing the potential to lose all evidence is important before the search begins. If the precise location is not initially apparent, begin a systematic search from one end and slowly move through the debris by using hand tools.

Once remains are discovered, documentation is more complicated, given the expected fragmentation and distortion of partially and completely cremated bone. Taking high-quality photographs is more difficult because the post-fire color of bone appears similar to the color of the surrounding material. Cremated bone is gray-white in appearance, and at a scene with that level of destruction, the majority of the surrounding materials are often of a similar color. Following documentation of the scene, recovery and transport must take place. While the presence of some soft tissues may require the use of a body bag or other standard transport means, it is best to rely on other materials to package cremated bone to protect it from further damage. The use of paper to wrap and package burned bone is acceptable. Some practitioners make foil pouches for transporting cremated bone fragments. These pouches can be made to any size, and the foil can be labeled. Recently, it was noted that the foil will not release elements into the bone and should be considered a safe and reliable method for packing burned bone (Lewis and Christensen 2015).

IN THE LABORATORY

Earlier chapters in this book have emphasized the methods and techniques that anthropologists must use during their analyses of skeletal remains. Visual examinations can be made as to the sex of an individual (Chapter 8), and age can be estimated (Chapter 10). Ancestry can also be assessed (Chapter 9) using **macroscopic methods**. These same approaches can be employed on burned remains. Remember, compared with unaffected bone, burned bone will differ in color, form, and size; therefore, metric approaches should be avoided. While this limits some of the methods presented in this text, it does not prevent an anthropological assessment to establish the biological profile. Analyses may not be as precise as in cases where bone is not burned, but with a good representation of the skeleton, a thorough anthropological examination is possible.

Similarly, burned bone can be examined for evidence of trauma, both antemortem and more recent unhealed traumas, such as those described in chapter 13 of this volume. Bone can retain the signatures of specific trauma that occur before incineration. As Herrmann and Bennett (1999) demonstrated, bones subjected to ballistic trauma, blunt-force trauma, and sharp-force trauma before being placed in a residence scheduled to be burned retained evidence of the trauma. No suppression efforts were made, and the residence completely burned, such that much of the recovered bone was described as cremated. Cut marks from both saws and smaller scalpel blades were readily apparent on the burned fragments (see Figure 18.3). Similarly, residues and fracturing indicative of ballistic trauma were also

Figure 18.3 Sharp trauma evident on a pig (*Sus scrofa*) tibia. Note the straight margins of the cut mark (indicated by red arrow) compared with more irregular margins of the fire-induced fractures on the periphery of the bone.

recognizable following recovery. Blunt-force trauma is more difficult to identify as this is primarily characterized by fracture patterns. Given that fracturing and fragmentation also result from fire, discerning the cause (fire or trauma) is more difficult in these cases. An analyst should attempt to trace the trajectory of fracture lines. Obviously, fractures that traverse unburned portions of bone are not the result of heat and thus indicate trauma. To differentiate between heat and trauma induced damage, the anthropologist should adopt a protocol requiring radiographic examination before any analyses.

FRACTURES

Owing to the structural changes associated with the dehydrating process of burning, bone will warp and fail, resulting in several types of fractures. While some are unique to fire, not all are; some mimic fracture types associated with blunt and ballistic traumas. Symes et al. (2015) provide one of the most complete descriptive lists of the types of fractures observed in burned bone. Fractures that are only seen in burned bone include curved transverse, delamination/splintering, and patina. In addition, longitudinal, transverse, and step fractures are usually observed. Symes and colleagues (2015) also note the presence of *burned fracture lines*.

Curved transverse are also known as thumbnail fractures given the arc shape that characterizes these thermal features (Figure 18.4). Primarily identified at the extreme ends of the diaphysis, these fractures are associated with heat-induced destruction of soft tissues. As the layers of soft tissues begin shrinking and are consumed by fire, the differentially heated tissues (bone and soft tissues) cause cracks in the bone in a concentric pattern. Symes et al. (2015) stated the concave surface is the side on which the overlying tissues were located. Occasionally these processes result in a near complete circular pattern; this is more often seen on flatter regions of diaphyses. Curved transverse fractures indicate the presence of soft tissues before incineration. For example, dry and nongreasy bone that is devoid of soft tissues does not exhibit these fractures.

Another unique fire-induced fracture found in particular locations is a pattern referred to as *patina*. Named after its resemblance to aged oil paintings, this fracture is a series of intersecting irregular lines that forms a mini checkerboard (Figure 18.5). Patina is only found in areas where thin cortical bone overlays trabeculae, such as joint surfaces. While its

Figure 18.4 Curved transverse (thumbnail) fractures on a long bone shaft.

Figure 18.5 Patina (note the crackled appearance of the cortical surface).

etiology is not well appreciated, patina is primarily a superficial fracture pattern. Several more traditional linear fracture types can also develop from burning. These include longitudinal fractures that tend to follow the length of the bone with slight arcing trajectories when observed in long bones. Transverse fractures run across bones but do not completely encircle bones (Figure 18.6). An intersecting pattern of transverse and longitudinal fractures is called *step fractures*. The fracture lines tend to step between transverse or longitudinal fractures generating a margin that resembles a set of stairs. Delamination of cranial bone is common and is characterized by the spalling off of the endocranial or ectocranial

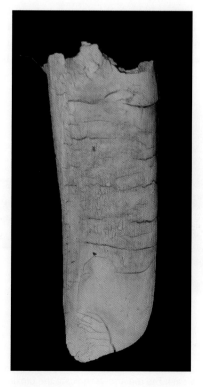

Figure 18.6 Transverse fractures caused by heat exposure.

Figure 18.7 Delamination of the skull. Arrows indicate areas of delamination, where the outer table of the skull has separated from the underlying diploe.

surfaces as the bone separates at the diploe (Figure 18.7). Similar patterns of trabecular exposure are sometimes found on epiphyseal surfaces. Symes and coworkers (2015) described *burn line fractures* as the margin between burned and unburned areas. Description of the type and location of fractures is an important aspect of analysis.

Symes and colleagues (2015) professed that understanding standard burn patterns, or deviations from them, can be of great forensic significance. They noted that it is possible to categorize burn patterns by using three diagnostic process signatures (Symes et al. 2015). Anthropologists should consider (1) body position and tissue shielding in bone, (2) color change in burned bone, and (3) burned bone fracture biomechanics (Symes et al. 2015). Toward this end, they promoted a four-term system for describing heat-altered bone. They encouraged the use of two well-known concepts: calcined (white) and charred (blackened) in describing the diagnostic color changes of burned bone. To facilitate accurate descriptions of body position and tissue shielding, they defined two new terms: heat line and burn border (Symes et al. 2015) (see Figure 18.8). Not fractures per say, these superficial features occur when the bone is not completely consumed and tissue shielding protects regions of a bone. The burn border is a white band along the burned margin caused by dehydration. The heat line is the aspect of the burn border located furthest away from the burned area. This systematic description of color change indicates the progression of fire and the subsequent

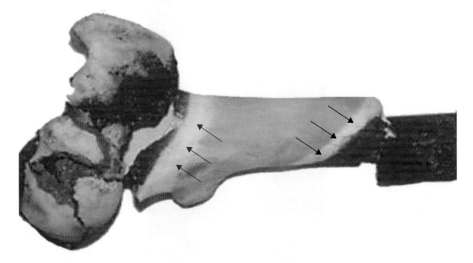

Figure 18.8 Proximal femoral shaft. Note the calcined area of the head and greater trochanter with minimal charred areas (left). The diaphysis is charred (right). The mid-section is unburned (shielded by soft tissues) and banded on both proximal and distal aspects by burn borders (narrow bright white areas). The red arrows indicate the proximal heat border. The distal heat line is indicated by the black arrows. Heat lines mark the boundary between the borders and the unburned area.

 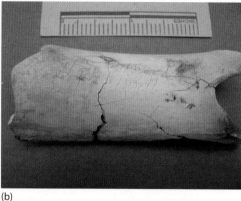

Figure 18.9 "Situational" fracture. (a) The stark contrast in color indicates differential heat exposure and signifies the break in the bone occurred after the fire. (b) The reconstructed specimen.

thermal alteration of bone (Symes et al. 2015). Additional fracturing may occur during recovery and transport of burned materials. Frequently, these post-fire fractures are recognizable, given that the exposed core is different in color than the other bone surfaces. Figure 18.9 shows a reconstructed fragment and the two portions that comprise it; notice the black (charred) core contrasted with the white (calcined) external surfaces. This distinction indicates that the fracture occurred after the fire: during overhaul of the fire scene, during search and recovery, or during transport. These *situational* fractures should be identified as such and should also be considered areas for reconstruction to be completed.

While heat-altered bone presents a unique situation for the examiner, many of the standard analyses can be completed with careful field work and precise analyses.

COMMINGLING

"Do they (the remains), represent a single individual or the commingled remains of several?" (Snow 1982:104). As the second item in his list, Snow realizes the significant impact that multiple victims can have on an investigation and the need to accurately identify the number of individuals. The presence of multiple individuals affects all facets of an investigation. How does an analyst accurately determine the number of individuals represented at a scene? If more than one individual are present, how does that affect management of the scene and subsequent methods of analysis and identification? Elements must be sorted and associated correctly in addition to determining the number of individuals represented. While it seems reasonable that sorting and counting elements should be straightforward, investigators need to be aware that such scenes are potentially complex and specific recovery methods should be used.

Often, the nature of the scene dictates whether multiple individuals are present, though in other situations it does not become apparent until analysis is underway. In large-scale mass disasters, multiple individuals are likely present and potentially commingled. The nature of the incident also impacts the condition of the individuals, resulting in relatively intact or highly fragmented remains.

Clearly the degree, or scale, of commingling in part dictates how to proceed. This is true both in the field and subsequently during laboratory analyses. Commonly, two scales are recognized: small-scale and large-scale. Small-scale commingling involves fewer elements, and the group size is smaller. Therefore, large-scale commingling can involve more fragments and more elements and represents a larger group. The labeling of a situation as small- or large-scale dictates how it is managed.

Two additional terms are used to describe the commingling. These concern the nature of what caused the commingling. Mundorff (2014) describes type I and type II commingling situations. The latter is used to describe situations in which the elements are commingled due to the nature of the event itself (e.g., crash and explosion). Type I situations arise when the collection and transport of remains of multiple individuals create the commingling. Type I commingling is preventable or at least protocols should be in place to reduce the likelihood of widespread type I commingling. Konigsberg and Adams (2014) note that, in addition to the scale of the event, the condition of the bone (e.g., the level of fragmentation and preservation) must also be considered. As has been discussed in this chapter, burned bone is fragile and more susceptible to fragmentation and thus has a greater potential for commingling.

Commingling events can result from a broad range of situations such as mass graves associated with human rights violations, large-scale fragmentation resulting from explosions, and smaller commingled cremation cases. Numerous approaches for resolving commingling issues have been proposed, with sorting of elements as a critical early step. Initial sorting must be done to remove nonosseous material, and subsequent examination should focus on separating nonhuman bone from human remains. Macroscopic, and if necessary histological, means can be used to ensure that the analytical assemblage consists of human remains. The separation and removal of nonbone and nonhuman bone must be thoroughly documented.

Macroscopic methods rely on sorting of elements with particular recognition made as to the side of the body and the size of the limb. Snow (1948) proposed a systematic methodology for managing military dead that lies at the basis of current approaches. Fragmentary elements are reconstructed and then the elements are sorted. He suggests that limb bones be "paired off on the basis of similarities of size, morphology and muscle markings" (Snow 1948:326). Sorting begins with matching elements of consistent size (e.g., a large left femur should pair with a large right femur; a large femur is more likely to be from the same individual who has a large humerus, and so forth). In effect, the analyst operates from the perspective that two bone specimens of comparable size originate from a single individual. This is considered a null hypothesis in osteometric sorting, as outlined by Byrd and Adams (2003). Snow (1948) further suggests that sorting should continue by assessing articulations at major joints. For example, the left and right innominate bones can be associated with the correct sacrum, based on features of their shared surface; subsequently, vertebrae can be associated, given attributes of individual articulations.

> Final assignment is checked by elimination and repeated trials which demonstrate the impossibility of the thigh bones belonging to any but the selected body. All possibilities are eliminated systematically. (Snow 1948:326)

Throughout this sorting process, analysts should observe unique features, pathological qualities, and other attributes that can be used to segregate remains further and separate individuals. As a follow-up, Snow and Folk (1970) presented a statistical approach for managing cases and recognizing the probability of commingling, based on recovered elements. While the computations are not necessarily utilized today, in developing this approach, Snow and Folk (1970) remind the analyst of the importance of several aspects that are relevant today. First and foremost, commingling is apparent when duplication of elements is realized. This is one of the primary tools for assessing commingled situations, especially those with little fragmentation. Two right femora must indicate that at least two individuals are represented. Assessments are difficult in cases where the individuals are similar in morphology, that is, same age and size. Commingling is further complicated when an individual, or individuals, is not represented by the entire skeleton or in cases of extreme fragmentation. In such cases, DNA analysis has been pivotal in resolving commingling and determining the number of individuals (Damann and Edson 2008; Mundorff et al. 2014).

Byrd and Adams (2003) proposed the inclusion of measurement data and statistical models to maximize the likelihood of correct sorting. While the basic premise is that proposed by Snow (1948), their method relies on statistical models to sort elements. Other straightforward methods of assessment include visual pair matching, where right and left pairs are established based on observable morphological similarities. Visual pair matching relies on an appreciation of the range of osteological variation and an understanding of acceptable or normal size variation both within and between limbs. Similarly, articulations of elements can be used to determine associations. For example, a small femoral head should articulate with an innominate possessing a small acetabulum. The analyst should also consider elimination as an informative principle when sorting commingled remains.

Physical anthropologists frequently borrow methods and quantification techniques developed by zooarchaeologists to describe and assess commingled collections, as zooarchaeologists often deal with highly fragmentary, commingled samples from various contexts. The **archaeofaunal** literature is rich with analytical approaches to assess these complex assemblages and, while several measures exist for quantifying individuals, most often, biological and forensic anthropologists seek to estimate the minimum number of individuals (MNI). An accurate determination of the minimum number of skeletal elements (MNE) in a collection of remains lies at the core of MNI assessments. Simply, the MNI is generated by sorting recovered elements to side of the body and recognizing the highest count as an indicator of the number of individuals present (Grayson 1984). This method is straightforward and generates an approximation of the minimal number of individuals represented in the assemblage, but it cannot be considered an accurate reflection of the size of the population that produced the recovered assemblage. Estimations of the original population size are possible through the application of other methods of quantification (see Adams and Konigsberg 2004; Konigsberg and Adams 2014). These include the Lincoln Index (LI) and the most likely number of individuals (MLNI) (discussed later).

In applying any of these methods, one follows the basic premise of not counting an individual more than once. The MNI is the minimum number of individuals that could represent the assemblage and is simply the maximum count when elements are sorted by element and side. In computing an MNI, the investigator must also realize that unless a proximal fragment of a bone and a distal fragment have duplicate features, they are merely counted as one. If taphonomic or age variation is apparent, then the MNI count can be affected. So, while the MNI is based on the maximum count of elements of a side of the body, it merely represents the minimum number of individuals present. A modification of MNI is the grand minimum total, where the number of left- and right-side counts of a bone is summed. The number of pairs is then subtracted from this value. It does not reflect at all on the original population size.

The Lincoln Index (LI) was developed based on animal capture and release counting techniques for estimating the population size. Promoted for skeletal analyses by Adams (1996), the LI is based on two observation counts divided by overlapping observations. The numbers of left sides and right sides are multiplied and then divided by the number of pairs. A similar approach is the *most likely number of individuals* (MLNI) computation, where one is added to the counted numbers of right and left elements, which are multiplied and then divided by the number of pairs (also + 1). From this, 1 is subtracted. The LI and MLNI can be used to calculate confidence intervals, which can be advantageous. However, these methods are not appropriate if the number of pairs cannot be tallied. The following example shows how to calculate MNI, grand minimum total, LI, and MLNI from a sample of tibiae in a commingled assemblage of remains.

You have identified the following:
104 right tibia, 97 left tibia, 69 pairs
MNI = max(L, R) = 104 individuals
Grand minimum total = (L + R) − pairs = (104 + 97) − 69 = 132 individuals
LI = (L × R)/P = (104 × 97)/69 = 146 individuals
MLNI = (L + 1) (R + 1)/(P + 1) − 1 = [(105 × 98)/70] − 1 = 146

These techniques for estimating the number of individuals represented in a commingled assemblage, predominantly those based on pair-matching, are poorly suited to applications on extremely fragmentary or cremated assemblages (Adams and Konigsberg 2004). Similarly, many osteological data collection manuals are lacking in coding systems for highly fragmentary collections. Although Buikstra and Ubelaker (1994) provide coding systems for incomplete elements, the emphasis is on large fragments. Comparable shortcomings characterize many of the database approaches such as Osteoware (Wilczak and Dudar 2012) or modifications of zooarchaeological coding systems such as that by Church and Burgett (1996) and Burgett (1990). A similar approach has been advocated by Knüsel and Outram (2004), in which elements are divided into anatomically important zones. The authors argue that the zonal method captures fragmentation patterns better than systems such as that offered by Buikstra and Ubelaker (1994).

RECENT RESEARCH AND NOVEL METHODS

More advanced and complex approaches to estimating the number of individuals from highly fragmentary assemblages are those by Marean and coworkers (Abe et al. 2002 and Marean et al. 2001) that use bony landmarks and Geographic Information Systems (GIS) software. These methods offer a corresponding fragment management system that eliminates some of quantification issues related to highly fragmentary remains. These systematic methods facilitate assessment of extremely large and highly fragmentary assemblages by allowing researchers to tally specific landmarks or bone shapes and generate element counts and subsequently MNI. However, fragments must be identified to element and placed on the bone template to be quantified. Herrmann et al. (2014a) discuss the application of both GIS-based and osteological landmark approaches in their continuing investigation of a prehistoric crematory, concluding that while the landmark approach is less time intensive, the GIS-based system has the capacity for the collection of more data. The latter approach employs the analysis of fragment shapes within a GIS. They analyzed a collection of highly fragmentary burned bone (in excess of 18 kg, with fragments less than 3 cm in diameter). Modifying the approach of Giovas (2009), they selected 17 elements for landmark analysis. Seven cranial elements (frontal, maxilla, zygomatic, mandible, temporal, parietal, and occipital), the shoulder girdle and arm (clavicle, scapula, humerus, radius, and ulna), and the pelvic girdle and legs (innominates, femur, tibia, fibula, and calcaneus) were analyzed. The number of landmarks ranged from nine on the calcaneus to 36 on the occipital, with landmarks correlated to explicitly defined points on the bone. Each identifiable fragment was assessed to determine side (if possible), and any applicable landmarks were recorded. A query provides the most frequently identified landmark and establishes the MNE for each bone; the highest MNE is then used as the landmark-based MNI. This approach is somewhat comparable to the osteometric sorting method described by Byrd and LeGarde (2014).

The GIS-based approach (Bennett Devlin et al. 2006; Herrmann et al. 2014a) requires the assignment of a fragment to its exact location on an illustration of the intact bone using shapefile templates (Marean et al. 2001). The fragments are digitized by recording the location on two-dimensional images of the complete element. The overall shape, location, and size of each fragment are recorded as the specimen is positioned on the appropriate view of the intact element. Each entry is saved as a shapefile that can be merged to demonstrate the numbers present (see Figure 18.10). In total, nearly 700 cranial fragments alone were assessed using these methods. Initial accounts of the assemblage indicated an MNI of less than 10 individuals (Sharp et al. 2003). The methods employed by Herrmann and colleagues (2014) yielded MNI values of 41, using the GIS-based approach from the left temporal bone, and a high of 40, using the landmark method on the right temporal bone. Either system provides a more accurate and likely higher count of individuals than more traditional element coding methods (Herrmann et al. 2014b) (Figure 18.10).

Quantifying fragmentary human remains is a necessary and challenging responsibility for forensic anthropologists. The level of fragmentation can vary widely. The fragmentation index can be used to describe the severity of a scene (Kontanis and Sledzik 2014). The fragmentation index (FI) is a ratio of the number of recovered remains to the number of decedents. In airline disasters, they have observed a low of FI of just over one to a high of 39. Cases with higher FI values require more analytical time and more frequently require the use of DNA for identification of the remains. Incidents with lower FI values often permit the use of more traditional identification methods (Kontanis and Sledzik 2014).

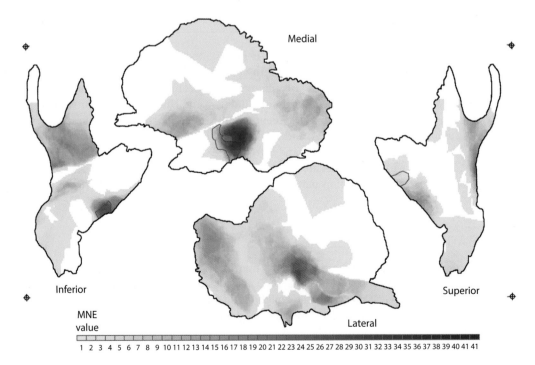

Figure 18.10 GIS-based fragment quantification by using the right temporal bone. The red outline represents a single fragment digitized into ArcGIS software. The blue raster is a summary of all fragments entered into the system. An MNE of 40 is derived from the grid.

SUMMARY

This chapter has discussed the methods that forensic anthropologists use to approach scenes with burned and/or commingled remains. Heat exposure from fire modifies the hard and soft tissues. Fire brings about irreversible changes in the color, form, and size of skeletal and dental elements. Forensic anthropologists' training in anatomy, skeletal biology, human variation, fracture biomechanics, and archaeological recovery methods allow them to interpret fire scenes and burned remains with a unique perspective. They can assist in the discovery and recovery of remains, analyze burned remains to produce a biological profile, and distinguish fractures induced by fire from those produced by other traumatic agents (e.g., gunshot wounds, sharp-force trauma, and blunt-force trauma).

Anthropologists are also specially qualified to process commingled scenes. Commingling may arise from certain events (crashes or explosions; type II commingling), or it may occur during the collection and transport of remains (type I commingling). Commingling events may also be small or large. Anthropologists try to ascertain the number of individuals represented in a commingled assemblage by using a variety of techniques, including minimum number of individuals (MNI), grand minimum total, the Lincoln Index (LI), and most likely number of individuals (MLNI). More advanced and labor-intensive methods have also been developed using bony landmarks and GIS software.

Situations that consist of either burned or commingled remains can present anthropologists with complex and taxing work. Understanding the basics of thermal damage ensures a better outcome for such cases. Similarly, the accurate determination of the number of individuals present in an assemblage is critical to assessing the nature of the events that created the collection and examining and identifying the individuals. Whether heat and high fragmentation or commingling, these cases require modification of traditional methods, but anthropologists can still assess these remains successfully.

Review questions

1. A house fire has burned all evening, and two adults are suspected to have succumbed to the fire. Why should the search and recovery team call an anthropologist to the scene? Discuss the contributions that an anthropologist can make to the recovery effort.
2. List the five degrees of burning, and explain how each affects the human body/human remains.
3. What is the relevance of the Crow-Glassman scale to scene recovery? At what stage(s) do forensic anthropologists become involved?
4. You have been called to a two-story house fire in which the second floor has collapsed onto the first floor. Explain how you would approach the search for remains. List some of the difficulties that you expect to encounter as you search. How would you approach these challenges?
5. A young woman is discovered heavily burned in her automobile outside a popular night club, and fowl play is suspected. Can the anthropologist detect skeletal trauma? How would you distinguish fractures caused by blunt trauma from those caused by the fire?
6. List and describe three types of fractures that are unique to burned bone, and explain how they can be readily distinguished from perimortem fractures.
7. Explain how the Symes et al. (2015) method of describing thermal alterations in bone elucidates burn progress and incorporates anatomy into the understanding of burning, as opposed to being purely descriptive.
8. Compare and contrast large- and small-scale comminglings.
9. Define type I and type II commingling situations. How could a fire create each of these situations?
10. You are called to a plane crash with burned and commingled remains. Discuss how you would approach this situation.
11. You have sorted through commingled remains in a mass grave and have ascertained that you have 207 left femora, 106 right femora, and 96 pairs. Calculate the MNI, LI, MLNI, and grand minimum total of individuals.

Glossary

Archaeofaunal: dealing with assemblages of animal remains recovered from archaeological sites.
Calcined bone: burned bone that is white in color and very brittle. Calcined bone has lost all water and collagen, and has been reduced to its mineral constituents.

Cremains: the cremated remains of an individual. Cremains can be created by accidental fires or funerary processes (e.g., cremation).
Dermis: the layer of skin deep to the epidermis, which is the most superficial layer. The dermis contains blood vessels, nerve endings, sweat glands, and hair follicles.
Epidermis: the outer, most superficial layer of skin composed of epithelial cells. The epidermis is nonvascular and nonsensitive (contains no blood vessels or nerves).
Macroscopic methods: methods of assessing the biological profile by visual examination of gross morphological features of the skeleton.
Pugilistic posturing: the position assumed by a burned body in which the more powerful flexor muscles contract due to the heat exposure and cause the joints to flex. This posture makes the body assume a boxer-like position, thus the pugilistic posture is sometimes called boxer's pose.
Spectrophotometer: a device that measures the intensity of light that is transmitted or emitted by particular substances.

References

Abe Y, Marean C, Nilssen P, Assefa Z, and Stone E. 2002. The Analysis of Cutmarks on Archaeofauna: A Review and Critique of Quantification Procedures, and a New Image-Analysis GIS Approach. *Am Antiquity* 67:643–663.

Adams BJ. 1996. The Use of the Lincoln/Peterson Index for Quantification and Interpretation of Commingled Human Remains. Unpublished Master's Thesis. Department of Anthropology, University of Tennessee, Knoxville.

Adams BJ, and Konigsberg LW. 2004. Estimation of the Most Likely Number of Individuals from Commingled Skeletal Remains. *Am J Phys Anthropol* 125(2):138–151.

Baby RS. 1954. Hopewell Cremation Practices. *Ohio Historical Society Papers in Archaeology* 1:1–17.

Beach JJ, Passalacqua NV, and Chapman EN. 2015. Heat-Related Changes in Tooth Color. In C Schmidt and S Symes (Eds.), *The Analysis of Burned Human Remains, second edition*. Boston, MA: Elsevier/Academic Press. pp. 139–147.

Beach JJ, Passalacqua NV, and Chapman EN. 2008. Heat-Related Changes in Tooth Color: Temperature versus Duration of Exposure. In C Schmidt and S Symes (Eds.), *The Analysis of Burned Human Remains*. Boston, MA: Elsevier/Academic Press. pp. 137–144.

Bennett J. 1999. Thermal Alteration of Buried Bone. *J Archaeological Sci* 26:1–8.

Bennett-Devlin JL, Herrmann NP, and Pollack D. 2006. GIS Analysis of Cremated and Commingled Bone. Poster presented at *71st Annual Meeting of the Society for American Archaeology*, San Juan, PR.

Bennett-Devlin JL, Kroman A, Herrmann NP, and Symes SA. 2006. Time, Temperature and Color: Heat Intensity versus Exposure Duration Part 1. Macroscopic Influence on Burned Bone. *Proc Am Acad Forensic Sci* 12:311–312.

Binford LR. 1963. An Analysis of Cremations from Three Michigan Sites. *Wisc Archaeol* 44:98–110.

Bonucci E, and Graziani G. 1975. Comparative Thermogravimetric, X-ray Diffraction and Electron Microscope Investigations of Burnt Bones from Recent, Ancient and Prehistoric Age. Atti Memorie Academia Nazionale die Lincei, Sci Fis Matem. *Nature* 59:517–534.

Buikstra JE, and Ubelaker DH (Eds.). 1994. Standards for Data Collection from Human Skeletal Remains, Proceedings of a Seminar at The Field Museum of Natural History (Arkansas Archeological Survey Research Series No. 44). Spiral ed. Arkansas Archeological Survey, Fayetteville.

Buikstra J, and Swegle M. 1989. Bone Modification Due to Burning, Experimental Evidence. In R Bonnichsen, and M Sorg (Ed.), *Bone Modification*. Orono, ME: University of Maine Press. p. 247–258.

Burgett G. 1990. The bones of the beast: Resolving questions of faunal assemblage formation processes through actualistic research. Unpublished dissertation, Department of Anthropology, The University of New Mexico, Albuquerque.

Byrd JE and Adams BJ. 2003. Osteometric Sorting of Commingled Human Remains. *J Forensic Sci* 48: 717–24.

Byrd JE, and LeGarde CB. 2014. Osteometric Sorting. In B Adams, and J Byrd (Eds.), *Commingled Human Remains: Methods in Recovery, Analysis, and Identification*. Boston, MA: Elsevier/Academic Press. pp. 167–192.

Church M and Burgett G. 1996. An Information Recording and Retrieval System for Use in Taphonomic and Forensic Investigations of Human Mortalities (abstract). *Proc Am Acad Forensic Sci* 48th Sci. Meeting 2:167–168.

Correia PM. 1997. Fire Modification of Bone: A Review of the Literature. In W Haglund, and M Sorg (Eds.), *Forensic Taphonomy: The Postmortem Fate of Human Remains*. Boca Raton, FL: CRC Press. pp. 275–293.

Damann FE, and Edson SM. 2008. Sorting and Identifying Commingled Remains of U.S. War Dead: The Collaborative Roles of JPAC and AFDIL. In B Adams, and J Byrd (Eds.) *Recovery, Analysis, and Identification of Commingled Human Remains*. Totowa, NJ: Humana Press. pp 301–315.

DeHaan JD. 2002. *Kirk's Fire Investigation*, 5th ed. Upper Saddle River, NJ: Prentice Hall.

DeHaan JD. 2015. Fire and Bodies. In C Schmidt, and S Symes (Eds.): *The Analysis of Burned Human Remains, second edition*. Boston, MA: Elsevier/Academic Press. pp. 1–15.

Devlin JB, and Herrmann NP. 2013. Taphonomy of Fire. In MTA Tersigni-Tarrant, and NR Shirley (Eds.), *Forensic Anthropology, An Introduction*. Boca Raton, Fl: CRC Press, Taylor and Francis Group. pp. 307–323.

Devlin JB, and Herrmann NP. 2008. Bone color as an Interpretive Tool of the Depositional History of Archaeological Cremains. In C Schmidt, and S Symes (Eds.), *The Analysis of Burned Human Remains: Archaeological and Forensic Approaches*. Boston, MA: Elsevier/Academic Press. pp. 109–127.

Devlin JL, Dalton MW, and Kennamer DC. 2004. Vehicle Fires: Actualistic Investigations. *Proc Am Acad Forensic Sci Annual Meetings*, Dallas, TX, 10:176.

Dirkmaat DC. 2002. Recovery and Interpretation of the Fatal Fire Victim: The Role of Forensic Anthropology. In WD Haglund, and MH Sorg (Eds.), *Advances in Forensic Taphonomy: Method, Theory, and Archaeological Perspectives*. Boca Raton, FL: CRC Press. pp. 451–472.

Fairgrieve SI. 2008. *Forensic Cremation: Recovery and Analysis*. Boca Raton, FL: CRC Press, Taylor and Francis Group.

Gilchrist M, and Mytum H. 1986. Experimental Archaeology and Burnt Animal Bone from Archaeological Sites. *Circaea* 4:29–38.

Giovas CM. 2009. The Shell Game: Analytic Problems in Archaeological Mollusc Quantification. *J Archaeol Sci* 36(7):1557–1564.

Glassman DM, and Crow RM. 1996. Standardization Model for Describing the Extent of Burn Injury to Human Remains. *J Forensic Sci* 41(1):152–154.

Grayson DK. 1984. *Quantitative Zooarchaeology*. New York: Academic Press.

Gocha TP, and Schutkowski H. 2013. Tooth Cementum Annulation for Estimation of Age-at-Death in Thermally Altered Remains. *J Forensic Sci* 58:S151–S155.

Herrmann B. 1977. On Histological Investigations of Cremated Human Remains. *J Human Evolution* 6:101–103.

Herrmann NP, and Bennett JL. 1999. A Microscopic Evaluation of Trauma on Burned Bone. *J Forensic Sci* 44:460–469.

Herrmann NP, Devlin JB, and Stanton JC. 2014a. Assessment of Commingled Human Remains Using a GIS-Based and Osteological Landmark Approach. In B Adams, and J Byrd (Eds.) *Commingled Human Remains: Methods in Recovery, Analysis, and Identification*. Boston, MA: Elsevier/Academic Press. pp. 221–238.

Herrmann NP, Devlin JB, and Stanton, JC. 2014b. Bioarchaeological Spatial Analysis of the Walker-Noe Crematory (15GD56). In AJ Osterholtz, KM Baustian, and DL Martin (Eds.) *Commingled and Disarticulated Human Remains: Working toward Improved Theory, method and Data*. New York: Springer. pp. 51–66.

Knüsel C, and Outram A. 2004. Fragmentation: The Zonation Method Applied to Fragmented Human Remains from Archaeological and Forensic Contexts. *Environ Archaeol* 9:85–97.

Konigsberg LW, and Adams BJ. 2014. Estimating the Number of Individuals Represented by Commingled Human Remains: A Critical Evaluation of Methods. In B Adams, and J Byrd (Eds.) *Commingled Human Remains: Methods in Recovery, Analysis, and Identification*. Boston, MA: Elsevier/Academic Press. pp. 193–220.

Kontanis EJ, and Sledzik PS. 2014. Resolving Commingling Issues during the Medicolegal Investigation of Mass Fatality Incidents. In B Adams, and J Byrd (Eds.), *Commingled Human Remains: Methods in Recovery, Analysis, and Identification*. Boston, MA: Elsevier/Academic Press. pp. 447–468.

Lewis L, and Christensen AM. 2015. Effects of Aluminum Foil Packaging on Elemental Analysis of Bone. *J Forensic Sci* 61(2):439–441.

Marean C, Abe Y, Nilssen P, and Stone E. 2001. Estimating the Minimum Number of Skeletal Elements (MNE) in Zooarchaeology: A Review and a New Image-Analysis GIS Approach. *Am Antiquity* 66:333–348.

McCutcheon P. 1992. Burned Archaeological Bone. In JK Stein (Ed.), *Deciphering a Shell Midden*. California: Academic Press, San Diego. pp. 347–368.

Munsell Color Company. 2000. *Munsell Soil Color Charts*. New York: Munsell Color Gretag Macbeth, New Windsor.

Mundorff AZ. 2014. Anthropologist-Directed Triage: Three Distinct Mass Fatality Events Involving Fragmentation and Commingling of Human Remains. In B Adams, and J Byrd (Eds.), *Commingled Human Remains: Methods in Recovery, Analysis, and Identification*. Boston, MA: Elsevier/Academic Press. pp. 365–388.

Mundorff AZ, Shaler R, Bieschke ET, and Mar-Cash E. 2014. Marrying Anthropology and DNA: Essential for Solving Complex Commingling Problems in Cases of Extreme Fragmentation. In B Adams, and J Byrd (Eds.), *Commingled Human Remains: Methods in Recovery, Analysis, and Identification*. Boston, MA: Elsevier/Academic Press. pp. 257–273.

Nicholson R. 1993. A Morphological Investigation of Burnt Animal Bone and an Evaluation of Its Utility in Archaeology. *J Archaeol Sci* 20:411–428.

Schmidt CW. 2015. Burned Human Teeth. In C Schmidt, and S Symes (Eds.), *The Analysis of Burned Human Remains, second edn*. Boston, MA: Elsevier/Academic Press. pp. 61–81.

Sharp W, Pollack D, Schlarb E, and Tune T. 2003. Walker-Noe: A Middle Woodland Mound in Central Kentucky. Paper presented at the 68th Annual Meeting of the Society for American Archaeology, Montreal, Canada.

Shipman P, Foster G, and Schoeninger M. 1984. Burnt Bone and Teeth: An Experimental Study of Color, Morphology, Crystal Structure and Shrinkage. *J Archaeol Sci* 11:307–325.

Snow CC. 1948. The Identification of the Unknown War Dead. *Am J Phys Anthropol* 6:323–328.

Snow CC. 1982. Forensic Anthropology. *Ann Rev Anthropol* 11:97–131.

Snow CC, and Folk ED. 1970. Statistical Assessment of Commingled Skeletal Remains. *Am J Phys Anthropol* 32:423–428.

Spenneman D, and Colley S. 1989. Fire in a Pit: The Effects of Burning on Faunal Remains. *Archaeozoologia* 3:51–64.

Symes SA, Rainwater CW, Chapman EN, Gipson DR and Piper AL. 2015. Patterned Thermal Destruction in a Forensic Setting. In C Schmidt, and S Symes (Eds.), *The Analysis of Burned Human Remains, second edition*. Boston, MA: Elsevier/Academic Press. pp. 17–47.

Thompson TJU. 2004. Recent Advances in the Study of Burned Bone and Their Implications for Forensic Anthropology. *Forensic Sci Int* 146:203–205.

Thompson TJU. 2005. Heat-Induced Dimensional Changes in Bone and Their Consequences for Forensic Anthropology. *J Forensic Sci* 50:1008–1015.

Thurman M, and Wilmore L. 1981. A Replicative Cremation Experiment. *N Am Archaeol* 2:275–283.

Webb WS, and Snow CC. 1945. The Adena People No. 2. The University of Kentucky Reports in Anthropology and Archaeology 6. Lexington, KY.

Wells C. 1960. A Study of Cremation. *Antiquity* 34:29–37.

Wilczak CA, and Dudar JC (Eds.). 2012. Osteoware Software Manual, Volume I. Washington, DC: Smithsonian Institution.

Appendix A
Application of Dentition in Forensic Anthropology

Debra Prince Zinni and Kate M. Crowley

CONTENTS

Introduction	365
Subadult age-at-death estimation from dentition	365
Adult age-at-death estimation from dentition	367
Ancestry determination from dentition	371
Sex estimation from dentition	372
Summary	373
Review questions	373
References	373

> **LEARNING OBJECTIVES**
>
> 1. Explain the biological mechanisms that facilitate age estimation from the dentition in subadults and apply these principles to estimate age from a dental radiograph.
> 2. List the age-related changes that permit adult-age estimation from the dentition.
> 3. Describe some of the features and methods that can be used to estimate ancestry from the dentition.
> 4. Explain how forensic anthropologists use the dentition to estimate sex.

INTRODUCTION

Dentition is important in the identification of unknown remains, as their postmortem longevity often surpasses that of bone, especially when remains are fragmentary or extensively damaged. Teeth can aid in developing the biological profile in commingled cases, such as those resulting from mass disasters or mass graves. Physical anthropologists use dentition to assess age, sex, and ancestry, especially in cases where the osseous remains yield little information for the biological profile.

Variation in developmental and degenerative changes of the skeletal system differ among individuals, across populations, and between the sexes (Brooks 1955; Biggerstaff 1977; Boldsen et al. 2002; Hanihara 1952; Hoppa and Vaupel 2002; Jackes 2000; Kemkes-Grottenthaler 1996, 2002; Komar 2003; Konigsberg and Frankenberg 1992; Moore-Jansen and Jantz 1986; Molleson et al. 1993; Plato et al. 1994; Prince and Ubelaker 2002; Prince 2004; Prince and Konigsberg 2008; Ross and Konigsberg 2002; Šlaus et al. 2003; Ubelaker 1989; Zhang 1982). Differences can be attributed to socio-economic status, cultural differences, genetic variation, behavioral differences, environmental factors, diet, and disease (Buckberry and Chamberlain 2002; Kemkes-Grottenthaler 2002). This variation is apparent in the skeleton and dentition, and therefore, appropriate reference samples must be employed when determining the biological profile of unknown remains.

SUBADULT AGE-AT-DEATH ESTIMATION FROM DENTITION

Dentition offers the most reliable age estimates for subadults. Both eruption into the oral cavity (Ubelaker 1989) and tooth mineralization (Demirjian et al. 1973; Moorrees et al. 1963; Mincer et al. 1993) have been utilized for the estimation of age-at-death, with the latter providing the most reliable method (Moorrees et al. 1963). However, mineralization and eruption patterns are sex and ancestry dependent; therefore, the appropriate reference sample must be utilized with any method.

Development and maturation of the human dentition consist of three periods: primary dentition period, mixed dentition period, and permanent dentition period. The primary emergence sequence of deciduous dentition is central incisors (approximately 6–12 months of age), lateral incisors (9–16 months), first molars (13–19 months), canines (16–23 months), and second molars (23–33 months; see Table A.1). The mixed dentition period extends from 6 to 12 years of age, its beginning marked by the eruption of the first permanent molar at approximately 6 years of age. Although the timing is variable, the typical eruption sequence of the permanent dentition occurs in the following three waves with resting periods between each wave: first molar, central incisor, lateral incisor at approximately 6–9 years of age; canine, first premolar, second premolar, and second molar at approximately 9–13 years of age; and the third molar between 17–22 years of age (Table A.2). The resting periods allow for the growth of the dental–facial complex and provide room for the next eruption phase.

One of the most widely used methods of subadult age estimation is that of Moorrees et al. (1963). Their method incorporates radiographs from a longitudinal study of middle-class white children from Ohio (110 females and 136 males) and Boston (51 females and 48 males). Dental formation is accessed via radiographs and is assigned a stage of development starting with an initial cusp formation to closure of the apex of the root (see Moorrees et al. 1963). The amount of mineralization is assessed and placed into the most appropriate phase of mineralization. Age is then determined by referencing tables that equate the mineralization phase with the mean age of attainment for

Table A.1 Deciduous dentition typical eruption pattern

Tooth (UNS)	Age
Central Incisors: P, O, E, F	6–12 months
Lateral Incisors: Q, N, D, G	9–16 months
First Molars: B, J, L, S	13–19 months
Canines: C, H, M, R	16–23 months
First Molars: A, J, K, T	23—33 months

Table A.2 Permanent dentition typical eruption pattern

Tooth (UNS)	Age	Wave
First Molars: #3, #14, #19, #30	6–9 years	1
Central Incisors: #8, #9, #24, #25		
Lateral Incisors: #7, #10, #23, #26		
Canines: #6, #11, #22, #27	9–13 years	2
First Premolars: #5, #12, #21, #28		
Second Premolars: #8, #9, #24, #25		
Second Molars: #2, #15, #18, #31		
Third Molars: #1, #16, #17, #32	17–22 years	3

males and females. The age range is reported as two standard deviations from the mean. Recently, AlQahtani and colleagues (2010) developed the London Atlas using bone and radiographs. Tooth development was determined according to Moorrees et al. (1963), and eruption was assessed relative to the alveolar bone. A separate validation study found that the London Atlas performs better than two other dental development charts (the Schour and Massler [1941] and Ubelaker [1989] charts) (AlQahtani et al. 2014). The London Atlas is available online in 20 languages at https://atlas.dentistry.qmul.ac.uk/?lang=english and has an accompanying free downloadable interactive software app. For older subadults, Mincer and colleagues (1993) provide a method that assesses the amount of mineralization of the third molars.

ADULT AGE-AT-DEATH ESTIMATION FROM DENTITION

Although difficult to achieve, estimation of adult skeletal age-at-death is an important identifying feature (Prince and Konigsberg 2008). Repeatability, high accuracy, and a strong correlation with age are traits of a good age indicator. Aging methods developed on indicators that are less affected by lifestyle variables increase the precision of age ranges, especially for older individuals.

Several researchers have developed techniques to determine age-at-death for adults by employing dentition and dental morphology. Adult age-estimation methods assess a number of indicators, including age-related changes in dental attrition/wear (Brothwell 1963, 1989; Lunt 1978; Lavelle 1970; Lovejoy 1985; Li and Ji 1995; Miles 1962, 1963, 1978; Molnar 1971; Ito 1972; Prince et al. 2008; Scott 1979; Smith 1984; Zuhrt 1955), secondary dentin deposits (Kvaal et al. 1994; Morse et al. 1993), cementum apposition (Charles et al. 1986; Condon et al. 1986; Großkopf 1989; Kagerer and Grupe 2001; Stott et al. 1982; Wittwer-Backofen 2000; Wittwer-Backofen and Buba 2002; Wittwer-Backofen et al. 2004), apical translucency (Bang and Ramm 1970; Prince 2004; Prince and Konigsberg 2008), periodontal recession (Borrman et al. 1995; Solheim 1992), root resorption (Borrman et al. 1995), aspartic acid racemization (Carolan et al. 1997; Helfman and Bada 1975, 1976; Masters 1986; Mörnstad et al. 1994; Ogino et al. 1985; Ohtani and Yamamoto 1991, 1992; Ohtani 1994, 1995; Ohtani et al. 1995; Ritz et al. 1990, 1993; Shimoyama and Harada 1984), color change of the root (Borrman et al. 1995; Solheim 1988; Ten Cate et al. 1977), or a combination of several of these indicators (Gustafson 1950; González-Colmenares et al. 2007; Johanson 1971; Kashyap and Koteswara Rao 1990; Kvaal et al. 1995; Lamendin and Cambray 1980; Lamendin et al. 1992; Maples 1978; Maples and Rice 1979; Matsikidis and Schultz 1982; Prince and Ubelaker 2002; Russell 1996; Solheim 1993; Ubelaker and Parra 2008). Several researchers have analyzed these features individually and multifactorially (Baccino and Zerilli 1997; Baccino et al. 1999). The most reliable methods for assessing adult age-at-death from dental remains analyze translucency of the root, cementum annuli, aspartic acid racemization on the enamel, dentin, or cementum, and secondary dentin deposits.

Gösta Gustafson, a Swedish stomatologist, was a pioneer in recognizing age-related changes to dentition and laid the foundation for utilizing dental microstructure to estimate age-at-death (Gustafson 1950). He documented age-related changes in six features of the human dentition: attrition, secondary dentin deposits, root translucency, periodontal recession, cementum annulation apposition (thickness), and root resorption. He assigned a numeric score (0–3 points) to account for change in each feature and assessed the amount of change by making longitudinal sections of the tooth. In the point system, increased score was equated with increased age. Gustafson found a strong correlation between dental microstructural features and age ($r = 0.98$) and concluded that root translucency and secondary dentin deposits were the best indicators of age. The advantage of this method is that it considers a number of dental features and uses information from several teeth. However, poor oral heath affects the age estimates by producing higher age-at-death estimates than actual age. Disadvantages of this method are that it is a destructive method, and the observer must have a thorough knowledge of dental histology to interpret the features. Although the importance of Gustafson's research was evident, many authors noted problems with the statistical methodology and offered improvements to his method (Aykroyd et al. 1997; Bang and Ramm 1970; Baccino et al. 1999; Borrman et al. 1995; Burns and Maples 1976; Dalitz 1962; Haertig et al. 1985; Johanson 1971; Kashyap and Koteswara Rao 1990; Lamendin et al. 1992; Lucy and Pollard 1995; Lucy et al. 1996; Maples 1978; Maples and Rice 1979; Marcsik et al. 1992; Metzger et al. 1980; Monzavi et al. 2003; Nkhumeleni et al. 1989; Saunders 1965; Solheim 1993; Solheim and Sundnes 1980; Ubelaker et al. 1998).

A direct relationship exists between chronological age and apical transparency: as age increases, the amount of transparency in the tooth root also increases (Gustafson 1950; Hillison 1996; Marcsik et al. 1992). This physiological feature typically does not appear before age 18 and is the "result of gradual mineralization of the peritubular dentine, which leads eventually to obliteration of the dentine tubules" (López-Nicolás et al. 1993, p. 2). Root translucency should not be confused with sclerotic dentin found in the crown as a result of pathological conditions (Mendis and Darling 1979; Pindborg 1970). Vasiliadis et al. (1983) compared apical translucency in pathological and nonpathological teeth and concluded that the development of root translucency is independent of pathological conditions.

Several authors have reported that root translucency is the best dental indicator of age and is most closely correlated to chronological age (Ajmal et al. 2001; Bang and Ramm 1970; Gustafson 1950; Johanson 1971; Kósa et al. 1983; Lorensten and Solheim 1989; López-Nicolás et al. 1990, 1993, 1996; Miles 1963; Maples 1978; Metzger 1980; Schroeder 1991; Sengupta et al. 1998, 1999; Sognnaes et al. 1985; Solheim 1989; Solheim and Sundnes 1980; Vasiliadis et al. 1983). However, translucency may be influenced by genetic, environmental, and cultural factors (López-Nicolás et al. 1996, Prince 2004). Acquisition of apical translucency may be related to a myriad of individual lifestyle variables, as well. For example, mastication and heavy-loading forces may increase the amount of translucency, which may account for the variation in translucency acquisition between individuals and populations.

Translucency of the root can be analyzed in longitudinal thin sections (see Dechaume et al. 1960; Gustafson 1950; Johanson 1971; Nalbandian et al. 1960; Sengupta et al. 1998, 1999; Solheim and Sundnes 1980, Vasiliadis et al. 1983; Whittaker and Bakri 1996) or on intact teeth (see Bang and Ramm 1970; Colonna et al. 1984; Drusini et al. 1991; Lamendin et al. 1992; Prince and Ubelaker 2002; Prince and Konigsberg 2008; Sarajlić et al. 2003; Solheim 1989). Translucency of the root can be seen macroscopically, but is enhanced with the aid of a light box, such as those used to view radiographs. Taking measurements directly from intact teeth is nondestructive, less expensive, and less time-consuming than sectioning methods, and does not require extensive knowledge of dental histology. Quantifications of apical translucency have been suggested in several different formats such as subject indices (Gustafson 1950; Johanson 1971), direct measurement from the apex toward the cement-enamel junction (CEJ) (Bang and Ramm 1970; Miles 1963; Sengupta et al. 1998, 1999), area of translucency (Lorentsen and Solheim 1989; Sengupta et al. 1998, 1999), translucency length expressed as a proportion of the total root length (Drusini et al. 1991; Lamendin and Cambray 1980; Lamendin et al. 1992; Prince and Konigsberg 2008; Prince and Ubelaker 2002; Sengupta et al. 1998, 1999; Sarajlić et al. 2003; Thomas et al. 1994), translucent area expressed as a proportion of the total root area (Drusini et al. 1991; Johnson 1968; Sengupta et al. 1998, 1999; Vasiliadis et al. 1983), computer-assisted image analysis (Drusini et al. 1991; López-Nicolás et al. 1990, 1993, 1996; Sengupta et al. 1998, 1999), and total volume (Manly and Hodge 1939; Rathod et al. 1993).

Several researchers have found significant differences in translucency between the sexes (Lorentsen and Solheim 1989; Prince and Ubelaker 2002), while others have not (Drusini et al. 1991; Lamendin et al. 1992). Lorentsen and Solheim (1989) suggested that sexual dimorphism in translucency may be attributed to differences in masticatory forces. Similarly, ancestry variation has been noted by several authors (González-Colmenares et al. 2007; Prince and Ubelaker 2002; Prince and Konigsberg 2008; Ubelaker and Parra 2008; Whittaker and Bakri 1996). As with any age indicator, taphonomic processes may affect the properties and visual assessment of apical translucency. These processes include water insults, soil conditions, temperature, and humidity, and faunal, fungal, or bacterial scavenger activity (Prince 2004; Sengupta et al. 1999).

Lamendin et al. (1992) proposed a quick and nondestructive method of adult age-at-death estimation using apical translucency and periodontal regression. Three measurements are taken from the labial surface of each tooth and recorded in millimeters: (1) root height (RH) = maximum distance from the apex of the root to the CEJ; (2) periodontal regression = maximum distance from the CEJ to the line of soft tissue attachment; and (3) translucency of the root, measured from the apex of the root to the CEJ with the aid of a light box. Lamendin et al. (1992) established the following regression equation to estimate age-at-death: $A = (0.18 * P) + (0.42 * T) + 25.53$, where A represents age in years, P represents the periodontal measurement $* 100/RH$, and T represents the translucency of the root measurement $* 100/RH$. These researchers produced a mean error of ± 10 years on their working sample and ± 8.4 years on their forensic control sample.

Although apical translucency is a reliable indicator of age, periodontal recession has serious limitations. Several intrinsic and extrinsic factors other than age influence periodontal recession, such as inflammation of the periodontium (van der Velden 1984), poor dental hygiene (Foti et al. 2001; Prince 2004; Prince and Ubelaker 2002), extrinsic irritation (physical, chemical, or mechanical; Foti et al. 2001), systemic diseases (Foti et al. 2001), and drug treatments

(Foti et al. 2001). Foti et al. (2001) found no correlation between periodontal recession and chronological age and concluded that Lamendin's technique cannot be used on teeth with periodontal disease. Supporting the conclusions made by Foti et al. (2001), Prince (2004) and, Solheim (1992) also conclude that periodontal recession is not useful as a single indicator for age-at-death estimation. Modeled after the method of Lamendin et al. (1992), Prince and Ubelaker (2002) offer a similar method for application in the United States based on a sample of 100 black males, 100 black females, 100 white males, and 100 white females. The authors provide new formulae that separate individuals by sex and ancestry, and include RH as a variable, which significantly lowers mean error.

Cementum annulation analysis has proven useful in estimating age-at-death, with some reports of mean errors as low as ±2.5 years (Wittwer-Backofen et al. 2004). Broomell (1898) was the first to note that the correlation between cementum thickness and chronological age was independent of functional and mechanical stress. The first age-at-death estimates utilizing cementum apposition measured the thickness of the band of cementum in longitudinal thin sections after Gustafson's methodology (Azaz et al. 1974; Gustafson 1950; Johanson 1971; Kashyap and Koteswara Rao 1990; Maples 1978; Nitzan et al. 1986; Solheim 1993). In mammalian aging studies, incremental cementum annuli were counted from transverse thin sections. Incremental bands of cementum are laid down in alternating light and dark bands (Figure A.1). Each pair of light and dark bands is considered to equate to 1 year of life (Kagerer and Grupe 2001; Stott et al. 1982). Age-at-death is estimated by adding the number of incremental bands to the age of eruption for the particular tooth. The counting of cementum annuli has provided reliable and accurate age estimates for seasonal animals; however, differences have been noted between human cementum annuli and other mammals in that human incremental bands are closer together and more numerous (Kvaal and Solheim 1995). Factors cited as the cause of this "annual" apposition include seasonal changes, UV-radiation dose, climatic parameters, differential food quality, and hormonal status (Kagerer and Grupe 2001).

Several researchers followed the procedure of counting cementum annuli instead of measuring the thickness of the band to estimate age-at-death in humans (Charles et al. 1986; Condon et al. 1986; Groβkopf 1989; Geuser et al. 1999; Jankauskas et al. 2001; Kagerer and Grupe 2001; Kvaal and Solheim 1995; Lipsinic et al. 1986; Miller et al. 1988; Naylor et al. 1985; Renz et al. 1997; Stott et al. 1982; Solheim 1990; Stein and Corcoran 1994; Wittwer-Backofen 2000; Wittwer-Backofen and Buba 2002; Wittwer-Backofen et al. 2004). Conflicting results on the reliability and accuracy of counting cementum annuli have been reported. Several researchers conclude that it is an unreliable technique for age estimation (Lipsinic et al. 1986; Lucas and Loh 1986; Miller et al. 1988), while others state that it is a moderately reliable technique (Charles et al. 1986; Condon et al. 1986; Stein and Corcoran 1994). Still others claim that it is a highly reliable method and that poor results are produced by incorrect procedures, rather than inherent flaws with the method itself (Groβkopf 1989; Kagerer and Grupe 2001; Stott et al. 1982; Wittwer-Backofen 2000; Wittwer-Backofen and Buba 2002; Wittwer-Backofen et al. 2004).

Conflicting results have also been reported on the effects of periodontal disease and cementum annulations. Several authors report that periodontal disease increases the error rate of this method (Condon et al. 1986; Kvaal and Solheim 1995) and that cementum annuli production is halted by periodontal disease (Kagerer and Grupe 2001). However, other

Figure A.1 Transverse thin section depicting cementum annuli.

authors report that periodontal disease has no effect on cementum annuli (Großkopf et al. 1996; Wittwer-Backofen 2000; Wittwer-Backofen and Buba 2002; Wittwer-Backofen et al. 2004). Several researchers have reported incidences of doubling cases, which refer to observing twice as many incremental lines as predicted (Condon et al. 1986; Stein and Corcoran 1994; Wittwer-Backofen 2000). Consequently, all methods available for age estimation from the dentition should be utilized to detect doubling cases (Kagerer and Grupe 2001).

Recent advances have examined the utility of amino acids for age estimation from dental materials. Enamel (Helfman and Bada 1975; Ohtani and Yamamoto 1992), dentin (Carolan et al. 1997; Helfman and Bada 1976; Masters 1986; Mörnstad et al. 1994; Ohtani and Yamamoto 1991, 1992; Ritz et al. 1990; Ritz et al. 1993; Shimoyama and Harada 1984), and cementum (Ohtani 1995; Ohtani et al. 1995) have been evaluated for their usefulness in amino acid racemization (i.e., changes in amino acid molecules over time). Aspartic acid has been found to be the most useful amino acid for this purpose. The racemization process regularly changes tissues over time and can be used to estimate age-at-death:

> The racemization of amino acids is a reversible first-order reaction, which is relatively rapid in living tissues that have a slow metabolic rate. The amino acids composing proteins are L-enantiomers. However, over the course of time, amino acids undergo racemization with an increased ratio of D-enantiomers, metamorphosing into a racemate (King and Bada 1979). Aspartic acid shows a high racemization reaction rate and is considered to provide useful information on changes occurring in living tissues over time (Ohtani 1995, p. 805).

Ohtani (1995) evaluated the correlation of the D- and L-aspartic acids in cementum with chronological age, documented the rate of the racemizing reaction, and compared the results with those obtained from similar analyses with enamel and dentin. Cementum yielded the fastest reaction, followed by dentin and then enamel. Overall, dentin had the highest correlation with age, followed by cementum and then enamel ($r = 0.992$, $r = 0.988$, and $r = 0.961$, respectively). The authors concluded that aspartic acid racemization is a precise and useful method to estimate age-at-death. Master (1986) evaluated the effects of postmortem changes to aspartic acid racemization. She concludes that aspartic acid racemization is a more accurate method of estimating age-at-death than other skeletal methods, especially in older individuals, but notes that postmortem conditions may affect the racemization rate. She suggests that further studies be conducted with a larger sample to test for such effects.

Deposition of secondary dentin was once thought to be influenced by attrition and therefore correlated with age-related changes in the dentition. The suggested mechanism is that secondary dentin is deposited in the lining of the pulp chamber to combat the loss of crown enamel. However, several authors have reported a weak correlation between attrition and secondary dentin deposits (Kvaal et al. 1995; Philippas 1961; Solheim 1993). Changes in osmotic pressure throughout the tooth have also been suggested as an influence on secondary dentin deposits (Philippas 1961). Regardless of the mechanism, secondary dentin deposition has a relatively high correlation with age (Feng 1985; Ito 1975; Kvaal et al. 1994; Lantelme et al. 1976; Moore 1970; Nitzan et al. 1986; Philippas 1961; Solheim 1992). Solheim (1992) tested three secondary dentin deposit scoring methods—Gustafson's (1947), Dalitz's (1962), and Johanson's (1971)—and found no significant difference between tooth age and chronological age. Although all three methods yielded strong correlations with chronological age, Johanson's (1971) method produced the strongest correlation between secondary dentin deposits and chronological age.

Radiographs are an excellent source for assessing age-related changes in dentition and have been used to assess Gustafson's method (Matsikidis and Schultz 1982), secondary dentin deposits (Kvaal and Solheim 1994; Morse et al. 1993), and proportions of the tooth (Kvaal et al. 1995). Techniques that utilize dental radiographs are completely nondestructive, offer simple procedures, and can be used on forensic and archaeological material as well as living individuals (Kvaal et al. 1995). In addition, taking dental radiographs is a common practice in dental clinics, so a large resource is available to researchers and practitioners.

All available skeletal and dental biological indicators should be used to derive multiple-trait age estimates. A more robust age estimate can be derived when multiple indicators corroborate an age range. In addition, interpersonal variation can be better understood when multiple indicators are analyzed. Focusing on only one or two age indicators offers a minimum understanding of the aging process and may be subject to significant error if one of the indicators is pathological. Technological, methodological, and statistical advances continually refine and improve the techniques that the physical anthropologists employ to estimate age-at-death.

ANCESTRY DETERMINATION FROM DENTITION

Although there have been numerous studies on age assessment from dentition, far fewer methods are available to assess ancestry. Most ancestry assessment methods are based on nonmetric dental traits and cusp morphology (Lasker and Lee 1957; Hinkes 1990; Hsu et al. 1997; Harris 2007). Nonmetric dental traits include shovel-shaped incisors (Figure A.2) which occur more frequently in Asian ethnic groups, Carabelli's cusp (Figure A.3) which occurs more frequently in European ancestral groups, and crenulated molars (Figure A.4) which are more common in African ancestries. Scott and Turner (1997) offer a plethora of information on dental trait frequencies across populations, and Turner et al. (1991) provide scoring procedures for nonmetric traits. The Arizona State University (ASU)

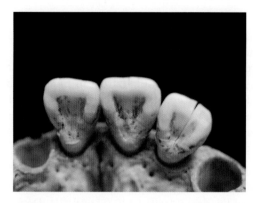

Figure A.2 Shovel-shaped incisors (teeth #8, #9, and #10).

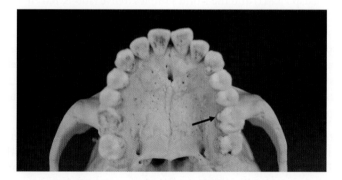

Figure A.3 Carabelli's cusp located on the mesial-lingual surface of tooth #14.

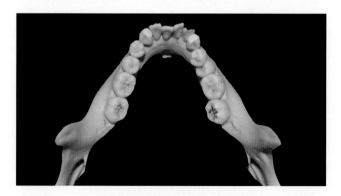

Figure A.4 Crenulated molars depicted on teeth #18, #19, #30, and #31.

Dental Anthropology System (Turner et al. 1991) provides dental casts that exhibit variation in the expression of over 20 nonmetric dental traits. When utilizing the ASU Dental Anthropology System, both the right and left dentition should be scored; the higher trait expression is utilized (Turner et al. 1991). Scoring is the same for males and females because the authors did not note significant sexual differences in dental nonmetric traits.

Edgar (2005) offers a method to discriminate nonmetric dental traits with Bayesian and logistic regression probabilities. Her method assesses eight nonmetric dental traits scored according to the ASU Dental Anthropology System (see Turner et al. 1991): maxillary canine tuberculum dentale (UCTD), mandibular anterior premolar cusp variation (LaPCV), mandibular posterior premolar cusp variation (LpPCV), first mandibular molar deflecting wrinkle (LM1DW), first mandibular molar trigonid crest (LM1TC), second mandibular molar cusp 5 (LM2C5), third mandibular molar cusp 5 (LM3C5), and first mandibular molar cusp 7 (LM1C7). Edgar's method distinguishes European American and African American ancestries.

In 2013, Edgar provided a similar method based on observations of 29 dental traits (ASU dental anthropology system) to distinguish between African, European, and Hispanic Americans (see also Edgar 2015). Logistic discriminant function equations distinguish European American and African American ancestries from Hispanic Americans and also further discriminate between African Americans and European Americans based on dental morphology. Edgar concluded that the various logistic discriminant equations yielded high classification rates and are useful in discriminating between these ancestry groups (66.7%–89.3% African or European American versus Hispanic Americans; 71.4%–100% African Americans versus European American). Due to similarities within the Hispanic group, further geographic assignment did not yield high classification rates.

Irish (2015) provides dental morphology trait frequencies for 21 ASU dental anthropology system traits for 14 world samples. Based on strong positive and negative principal component analyses and relevance to casework in the United States, the author reduced the dental traits to 10 (Shovel UI1, Double Shovel UI1, Interruption Groove UI2, Carabelli's Cusp UM1, Enamel Extension UM1, Groove Pattern LM2, Cusp Number LM1, Cusp Number LM2, Deflecting Wrinkle LM1, and Cusp 7 LM1) and the world samples to five (sub-Saharan African [SSA], Europe [EUR], China–Mongolia [CHM], North and South Native Americans [NSA], and Polynesia [POL]) in order to provide a simple method for ancestry assessment. Each of the 10 traits is scored as present or absent, and the frequency percentage is recoded for each world group. The frequency percentages are summed for each group, and the highest summed score provides the classification. The author cautions, however, that the method by itself is simplistic and that all available methods for ancestry determination should be employed in conjunction with this method (Irish 2015). In addition, Irish (2015) also states that practitioners must be well versed in the ASU dental anthropology system before attempting to employ this method.

Dental metrics have also proved useful in the assessment of ancestry (Harris and Foster 2015; Pilloud et al. 2014). Pilloud et al.'s (2014) study based on a large sample of mesiodistal and buccolingual tooth dimensions yielded high classification rates when classifying ancestry into broad geographic categories. The authors generated a 71.3 % correct classification rate, which increased to 88.1% for females and 71.9% for males when sex was known. Their results were similar to previous studies that concluded African populations possess the largest teeth, European populations possess the smallest teeth, and Asian populations exhibit intermediate sized teeth. Another study by Harris and Foster (2015) assessed mesiodistal and buccolingual crown dimensions for a sample of 52 black males, 74 black females, 94 white males, and 104 white females from the mid-Southern United States. Their results concurred with Pilloud et al. (2014) in that the black sample exhibits larger teeth than the white sample. In their sample, black males exhibit the largest crown dimensions, followed by black females, white males, and finally, white females.

SEX ESTIMATION FROM DENTITION

Sex estimation from dentition is limited primarily to sexual dimorphism in odontometrics. Most studies analyze differences between the sexes in mesiodistal and buccolingual tooth dimensions and conclude that males exhibit larger crown dimensions than females (De Vito and Saunders 1990; Garn et al. 1964, 1966, 1977; Harris and Foster 2015; Maudrich 1977; Moorrees 1957, 1959; Moorrees et al. 1957; Teschler-Nicola and Prossinger 1998); however, some classification rates are only slightly greater than chance. Another consideration is that dental dimensions are variable across populations, thus decreasing the usefulness of odontometrics in a forensic setting. However, determining the sex of an unknown individual is important for identification purposes and for correctly estimating other parameters of the biological profile. In general, females are precocious to males in epiphyseal union, dental calcification, and dental eruption.

SUMMARY

All available skeletal and dental biological indicators should be analyzed when assessing the biological profile of unknown remains. Teeth are able to withstand various postmortem environments and insults due to their highly mineralized composition. Dental traits also exhibit a high heritability; therefore forensic anthropologists can utilize dentition to develop the biological profile. Dental mineralization offers the most accurate method for age determination in subadults. Most adult dental aging methods are derived from Gustafson's (1950) features of dental attrition, secondary dentin deposits, cementum annulation apposition, and root translucency. Many authors suggest that measurement of apical translucency is the best single age indicator, although there is a concern regarding its utility with archaeological material. Two destructive analytical methods for age estimation, aspartic acid racemization and counting tooth cementum annuli, have both produced very strong correlations with age, accurate age estimates, and precise age ranges. As with apical translucency, both methods are prone to increased error with increased antiquity of the dental material and are susceptible to other taphonomic influences that affect accuracy. Taphonomic processes affect all aging methods, whether based on skeletal or dental traits. Continued research and development of techniques to counter problems pertaining to taphonomic processes are imperative to ensure best practice.

Although sexual dimorphism in crown dimension has been documented, only a limited number of studies utilize the dentition in sex determination. Ancestry assessment from dentition is also limited; however, several recent studies have provided useful methodologies for distinguishing between broad geographic groups based on both dental metrics (Harris and Foster 2015; Pilloud et al. 2014) and dental morphology (Edgar 2005, 2013, 2015; Irish 2015). These methods utilize more robust statistical analyses, thus enhancing the forensic tool kit.

Review questions

1. What two biological mechanisms related to dental development and maturation are used to develop subadult age estimation methods from the dentition?
2. What is the most reliable dental age determination method in subadults? and adults?
3. What is typically the first permanent tooth to erupt in the oral cavity? How are eruption sequences used to estimate age-at-death?
4. How would you estimate age if you had a bite-wing X-ray from a male with mixed dentition?
5. What aspects of the dentition are utilized to assess ancestry?
6. How can dentition be used to determine sex?
7. Describe the role of taphonomy in the application and development of age estimation methods using the dentition. How can researchers and practitioners address taphonomic influences in study and practice?

References

Ajmal M, Mody B, and Kumar G. 2001. Age estimation using three established methods. A study on Indian population. *Forensic Sci Int* 122:150–154.

AlQahtani SJ, Hector MP, and Liversidge HM. 2010. Brief communication: The London atlas of tooth development and eruption. *Am J Phys Anthropol* 142:481–490.

AlQahtani SJ, Hector MP, and Liversidge HM. 2014. Accuracy of dental age estimation charts: Schour and Massler, Ubelaker, and the London Atlas. *Am J Phys Anthropol* 154:70–78.

Aykroyd RG, Lucy D, Pollard AM, and Solheim T. 1997. Regression analysis in adult age estimation. *Am J Phys Anthropol* 104:259–265.

Azaz B, Ulmansky M, Moshev R, and Sela J. 1974. Correlation between age and thickness of cementum in impacted teeth. *Oral Surg* 38:691–694.

Baccino E, Ubelaker DH, Hayek LA, and Zerilli A. 1999. Evaluation of seven methods of estimating age at death from mature human skeletal remains. *J Forensic Sci* 44:931–936.

Baccino E, and Zerilli A. 1997. The two step strategy (TSS) or the right way to combine a dental (Lamendin) and an anthropological (Suchey-Brooks System) method for age determination (abstract). *ProcAm Acad Forensic Sci* 3:150.

Bang G, and Ramm E. 1970. Determination of age in humans from root dentin transparency. *Acta Odontol Scand* 28:3–35.

Biggerstaff RH. 1977. Craniofacial characteristics as determinants of age, sex and race in forensic dentistry. *Dent Clin N Am* 21:85–97.

Boldsen JL, Milner GR, Konigsberg LW, and Wood JW. 2002. Transition analysis: A new method for estimating age from skeletons, In RD Hoppa, and JW Vaupel (Eds.), *Paleodemography: Age Distributions from Skeletal Samples.* Cambridge, NY: Cambridge University Press. pp. 73–106.

Borrman H, Solheim T, Magnusson B, Kvaal SI, and Stene-Johansen W. 1995. Inter-examiner variation in the assessment of age-related factors in teeth. *Int J Legal Med* 107:183–186.

Broomell IN. 1898. The histology of cementum. *Dent Cosmos* 51:697–722.

Brooks ST. 1955. Skeletal age at death: The reliability of cranial and pubic age indicators. *Am J Phys Anthropol* 13:567–597.

Brothwell D. 1963. *Digging up Bones: The Excavation, Treatment and Study of Human Skeletal Remains.* London, U.K: British Museum (Natural History).

Brothwell D. 1989. The relationship of tooth wear to aging. In MY İşcan (Ed.), *Age Markers in the Human Skeleton*, Springfield, IL: Charles C. Thomas. pp. 303–316.

Buckberry JL, and Chamberlain AT. 2002. Age estimation from the auricular surface of the ilium: A revised method. *Am J Phys Anthropol* 119:231–239.

Burns KR, and Maples WR. 1976. Estimation of age from individual adult teeth. *J Forensic Sci* 21:343–356.

Carolan VA, Gardner ML, Lucy D, and Pollard AM. 1997. Some considerations regarding the use of amino acid racemization in human dentine as an indicator of age at death. *J Forensic Sci* 42:10–16.

Charles DK, Condon K, Cheverud JM, and Buikstra JE. 1986. Cementum annulation and age determination in Homo sapiens: I. Tooth variability and observer error. *Am J Phys Anthropol* 71:311–320.

Colonna M, Introna F, Favia G, and Pesce-Delfino V. 1984. Valutazione della trasparenza della dentina per la determinazione dell'eta: Revisione metodologica e analisi di un campione. In F deFazio, and B Vernole (Eds.), *La Laurea in Odontoiatria e Protesi dentaria. I Problemi Medico-legali in Odontostomatologia.* Rome, Italy: CIC Edizioni Internazionali. pp. 357–368.

Condon K, Charles DK, Cheverud JM, and Buikstra JE. 1986. Cementum annulation and age determination in *Homo sapiens*: II. Estimates and accuracy. *Am J Phys Anthropol* 71:321–330.

Dalitz GD. 1962. Age determination of adult human remains by teeth examination. *J Forensic Sci Soc* 3:11–21.

Dechaume M, Derobert L, and Payen J. 1960. De la valeur de la determination de l'age par examen des dentes en coupes minces. *Ann Med Leg (Paris)* 40:165–167.

Demirjian A, Goldstein H, and Tanner JM. 1973. A new system of dental age assessment. *Hum Biol* 45(2):211–227.

De Vito C, and Saunders S. 1990. A discriminant function analysis of deciduous teeth to determine sex. *J Forensic Sci* 35:845–858.

Drusini A, Callieari I, and Volpe A. 1991. Root dentine transparency: Age determination of human teeth using computerized densitometric analysis. *Am J Phys Anthropol* 85:25–30.

Edgar, HJH. 2005. Prediction of race using characteristics of dental morphology. *J Forensic Sci* 50:269–273.

Edgar, HJH. 2013. Estimation of ancestry using dental morphological characteristics. *J Forensic Sci* 58(Suppl 1):S3–S8.

Edgar, HJH. 2015. Dental morphological estimation of ancestry in forensic contexts. In GE Berg, and SC Ta'ala (Eds.), *Biological Affinity in Forensic Identification of the Human Skeletal Remains: Beyond Black and White*. Boca Raton, FL: CRC Press. pp. 191–208.

Feng J. 1985. Age determination from structure of teeth. *Acta Anthropol Sinica* 4:380–384.

Foti B, Adalain P, Signoli M, Ardagna Y, Dutour O, and Leonetti, G. 2001. Limits of the Lamendin method in age determination. *Forensic Sci Int* 122:101–106.

Garn SM, Cole PE, and Wainwright RL. 1977. Sex discriminatory effectiveness using combinations of permanent teeth. *J Dent Res* 56:697.

Garn SM, Lewis AB, and Kerewsky RS. 1964. Sex differences in tooth size. *J Dent Res* 43:306.

Garn SM, Lewis AB, and Kerewsky RS. 1966. Sexual dimorphism in the buccolingual tooth dimension. *J Dent Res* 45:1819.

Geuser G, Bondioli L, Capucci E, Cipriano A, Grupe G, Savorè C, and Macchiarelli R. 1999. Dental cementum annulations and age at death estimates. Osteodental biology of people of Portus Romae (Necropolis of Isola Sacra, 2nd–3rd cent. AD). In L Bondioli and R Macchiarelli (Eds.), *Digital Archives of Human Paleobiology*. Rome, Italy: Museo Nazionale Preistorico Etnografico "L. Pigorini," Vol. 2.

González-Colmenares G, Moreno-Rueda G, and Fernández-Cardenete JR. 2007. Age estimation by a dental method: A comparison of Lamendin's and Prince and Ubelaker's technique. *J Forensic Sci* 52(2):1156–1160.

Großkopf B. 1989. Incremental lines in prehistoric cremated teeth. A technical note. *Zeitschrift für Morphologie und Anthropologie* 77:309–311.

Großkopf B, Dender JM, and Krüger W. 1996. Untersuchengen zur Zementapposition bei Paradontitis marginalis profunda. *Deut Zahnaerztl Z*, 25:1763–1768.

Gustafson G. 1947. Åldersbestämnigar på tänder. *Odontologisk Tidskrift* 54:556–568.

Gustafson G. 1950. Age determination on teeth. *J Am Dent Assoc* 41:45–54.

Haertig A, Crainic K, and Durigon M. 1985. Medico-legal identification by the dental system. *Presse Med* 14:543–545.

Hanihara K. 1952. Age changes in the male Japanese pubic bone. *J Anthropol Soc Nippon* 62:245–260.

Harris EF. 2007. Carabelli's trait and tooth size of human maxillary first molars. *Am J Phys Anthropol* 132(2):238–246.

Harris EF, and Foster CL. 2015. Size Matters: Discrimination Between American Blacks and Whites, Males and Females, Using Tooth Crown Dimensions. In GE Berg, and SC Ta'ala (Eds.), *Biological Affinity in Forensic Identification of the Human Skeletal Remains: Beyond Black and White*. Boca Raton, FL: CRC Press. pp. 209–238.

Helfman PM, and Bada J. 1975. Aspartic acid racemization in enamel from living humans. *P Natl Acad Sci USA* 72:2891–2894.

Helfman PM, and Bada J. 1976. Aspartic acid racemization in dentine as a measure of ageing. *Nature* 262:279–281.

Hinkes MJ. 1990. Shovel-shaped incisors in human identification. In GW Gill, and JS Rhine (Eds.), *Skeletal Attribution of Race*. Albuquerque, NM: Anthropology Papers 4, Maxwell Museum of Anthropology. pp. 21–26.

Hoppa RD, and Vaupel JW. 2002. *Paleodemography: Age Distributions from Skeletal Samples*. Cambridge, NY: Cambridge University Press.

Hsu JW, Tsai P, Liu K, and Ferguson D. 1997. Logistic analysis of shovel and Carabelli's tooth traits in a Caucasoid population. *Forensic Sci Int* 89(1–2):65–74.

Irish JD. 2015. Dental nonmetric variation around the world: Using key traits in populations to estimate ancestry in individuals. In GE Berg, and SC Ta'ala (Eds.), *Biological Affinity in Forensic Identification of the Human Skeletal Remains: Beyond Black and White*. Boca Raton, FL: CRC Press. pp. 165–190.

Ito S. 1972. Research on age estimation based on teeth. *JPN J Legal Med* 26:31–41.

Ito S. 1975. Age estimation based on tooth crowns. *Int J Forensic Dent* 3:9–14.

Jackes M. 2000. Building the basis for paleodemographic analysis: Adult age Determination. In AM Katzenberg, and SR Saunders (Eds.), *Biological Anthropology of the Human Skeleton*. New York: Wiley-Liss, Inc. pp. 417–466.

Jankauskas R, Barakauskas S, and Bojarun R. 2001. Incremental lines of dental cementum in biological age estimation. *Homo* 52:59–71.

Johanson G. 1971. Age determination from human teeth. *Odontologisk Revy* 22:1–126.

Johnson CC. 1968. Transparent dentine in age estimation. *Oral Surg* 25:834–838.

Kagerer P, and Grupe G. 2001. Age-at-death diagnosis and determination of life-history parameters by incremental lines in human dental cementum as an identification aid. *Forensic Sci Int* 118:75–82.

Kashyap VK, and Koteswara Rao NR. 1990. A modified Gustafson method of age estimation from teeth. *Forensic Sci Int* 47:237–247.

Kemkes-Grottenthaler A. 1996. Critical evaluation of osteomorphognostic methods to estimate adult age at death: A test of the "complex method." *Homo* 46:280–292.

Kemkes-Grottenthaler A. 2002. Aging through the ages: Historical perspectives on age indicator methods. In RD Hoppa, and JW Vaupel (Eds.), *Paleodemography: Age Distributions from Skeletal Samples*. Cambridge, NY: Cambridge University Press. pp. 48–72.

King K, and Bada JL. 1979. Effect of in situ leaching on aminoacid racemisation rates in fossil bone. *Nature* 281(13):135–137.

Komar DK. 2003. Twenty-seven years of forensic anthropology casework in New Mexico. *J Forensic Sci* 48:521–524.

Konigsberg LW, and Frankenberg SR. 1992. Estimation of age structure in anthropological demography. *Am J Phys Anthropol* 89:235–256.

Kósa F, Szendrenyi J, and Toth V. 1983. A fogak transzparenciajanak vizsgalata az eletkor megallapitasara. *Morphologiai es Igazsagugyi Orvosi Szemle* 23:286–291.

Kvaal SI, Kolltveit KM, Thomsen IO, and Solheim T. 1995. Age estimation of adults from dental radiographs. *Forensic Sci Int* 74:175–185.

Kvaal SI, Koppang HS, and Solheim T. 1994. Relationship between age and deposit of peritubular dentine. *Gerondontology* 11:93–98.

Kvaal SI, and Solheim T. 1995. Incremental lines in human dental cementum in relation to age. *Eur J Oral Sci* 103:225–230.

Lamendin H, Baccino E, Humbert JF, Tavernier JC, Nossintchouk RM, and Zerilli A. 1992. A simple technique for age estimation in adult corpses: The two criteria dental method. *J Forensic Sci* 37:1373–1379.

Lamendin H, and Cambray JC. 1980. Etude de la translucidite et des canalicules dentinaires pour appreciation de l'age. *Soc de Med Legale* 9:489–499.

Lantelme RL, Handelman SL, and Herbison RJ. 1976. Dentin formation in periodontally diseased teeth. *J Dent Res* 55:48–51.

Lasker GW, and Lee MMC. 1957. Racial traits in human teeth. *J Forensic Sci* 2:401–419.

Lavelle CLB. 1970. Analysis of attrition in adult human molars. *J Dent Res* 49:822–828.

Li C, and Ji G. 1995. Age estimation from the permanent molar in northeast China by the method of average stage of attrition. *Forensic Sci Inte* 75:189–196.

Lipsinic FE, Paunovich E, Houston GD, and Robison SF. 1986. Correlation of age and incremental lines in the cementum of human teeth. *J Forensic Sci* 31:982–989.

López-Nicolás M, Canteras M, and Luna A. 1990. Age estimation by IBAS image analysis of teeth. *Forensic Sci Int* 45:143–150.

López-Nicolás M, Morales A, and Luna A. 1993. Morphometric study of teeth in age calculation. *J Forensic Odontostomatology* 11:1–8.

López-Nicolás M, Morales A, and Luna A. 1996. Application of dimorphism in teeth to age calculation. *J Forensic Odontostomatology* 14:9–12.

Lorensten H, and Solheim T. 1989. Age assessment based on translucent dentine. *J Forensic Odontostomatology* 7:3–9.

Lovejoy CO. 1985. Dental wear in the Libben population: Its functional pattern and role in the determination of adult skeletal age at death. *Am J Phys Anthropol* 68:47–56.

Lucas PW, and Loh HS. 1986. Are the incremental lines in human cementum laid down annually? *Ann Acad Med* 15:384–386.

Lucy D, Aykroyd RG, Pollard AM, and Solheim T. 1996. A Bayesian approach to adult human age estimation from dental observations by Johanson's age changes. *J Forensic Sci* 41:189–194.

Lucy D, and Pollard AM. 1995. Further comments on the estimation of error associated with the Gustafson dental age estimation method. *J Forensic Sci* 40:222–227.

Lunt DA. 1978. Analysis of attrition in adult human molars. In PB Butler, and J Joysey (Eds.), *Development, Function, and Evolution of Teeth*. London, U.K: Academic Press. pp. 465–482.

Manly RS, and Hodge HC. 1939. Dentistry and refractive index studies of dental hard tissues I. *J Dent Res* 18:133–141.

Maples WR. 1978. An improved technique using dental histology for estimation of adult age. *J Forensic Sci* 23:764–770.

Maples WR, and Rice PM. 1979. Some difficulties in the Gustafson dental age estimations. *J Forensic Sci* 24:168–172.

Marcsik A, Kósa F, and Kocsis G. 1992. The possibility of age determination on the basis of dental transparency in historical anthropology. In P Smith, and E Tchernov (Eds.), *Structure, Function and Evolution of Teeth*. London, U.K.: Freud Publishing House Ltd.. pp. 527–538

Masters PM. 1986. Age at death determinations for autopsied remains based on aspartic acid racemization in tooth dentin: Importance of post-mortem conditions. *Forensic Sci Int* 32:179–184.

Matsikidis G, and Schultz P. 1982. Altersbestimmung nach dem Gebiss mit Hilfe des Zahnfilms. *Zahnaerztliche Mitt* 72:2527–2528.

Mendis BR, Darling AI. 1979. Distribution with age and attrition of peritubular dentine in the crowns of human teeth. *Arch Oral Biol* 24:131–139.

Metzger Z, Buchner A, and Gorsky M. 1980. Gustafson's method for age determination from teeth—A modification for the use of dentists in identification teams. *J Forensic Sci* 25:742–749.

Miles AEW. 1962. Assessment of the ages of a population of Anglo-Saxons from their dentitions. *Proc J Roy Soc Med* 55:881–886.

Miles AEW. 1963. Dentition and the estimation of age. *J Dent Res* 42:255–263.

Miles AEW. 1978. Teeth as an indicator of age in man. In PM Butler, and KA Joysey (Eds.), *Development, Function and the Evolution of Teeth*. London, U.K.: Academic Press. pp. 455–462.

Miller CS, Dove, SB, and Cottone JA. 1988. Failure of use of cemental annulations in teeth to determine the age of humans. *J Forensic Sci* 33:137–143.

Mincer HH, Harris EF, and Berryman HE. 1993. The A.B.F.O. study of third molar development and its use as an estimator of chronological age. *J Forensic Sci* 38:379–390.

Molleson T, Cox M, Waldron AH, and Whittaker DK. 1993. The Spitalfields Project. Vol.2, The middling sort: CBA Research Report no. 86. Council for British Archaeology, York, U.K.

Molnar S. 1971. Human tooth wear, tooth function and cultural variability. *Am J Phys Anthropol* 34:175–190.

Monzavi BF, Ghodoosi A, Savabi O, and Hasanzadeh A. 2003. Model of age estimation based on dental factors of unknown cadavers among Iranians. *J Forensic Sci* 48:1–3.

Mörnstad H, Pfeiffer H, and Teivens A. 1994. Estimation of dental age using HPLC-technique to determine the degree of aspartic acid racemization. *J Forensic Sci* 39:1425–1431.

Morse DR, Esposito JV, and Schoor RS. 1993. A radiographic study of aging changes of the dental pulp and dentin in normal teeth. *Quintessence Int* 24:329–333.

Moore GE. 1970. Age changes occurring in the teeth. *J Forensic Sci Soc* 10:179–180.

Moore-Jansen PH, and Jantz RL. 1986. *A computerized skeletal data bank for forensic anthropology.* Knoxville, TN: Department of Anthropology, University of Tennessee.

Moorrees CFA. 1957. *The Aleut Dentition: A Correlative Study of Dental Characteristics in an Eskimoid People.* Cambridge, MA: Harvard University Press.

Moorrees CFA. 1959. *The Dentition in the Growing Child.* Cambridge, MA: Harvard University Press.

Moorrees CFA, Fanning EA, and Hunt EE. 1963. Age variation of formation stages for ten permanent teeth. *J Dent Res* 42:1490–1502.

Moorrees CFA, Thomsen SO, Jenson E, and Yen PKJ. 1957. Mesiodistal crown diameters of deciduous and permanent teeth. *J Dent Res* 36:39–47.

Nalbandian J, Gonzales F, and Sognnaes F. 1960. Sclerotic age changes in root dentin of human teeth as observed by optical, electron, and x-ray microscopy. *J Dent Res* 39:598–607.

Naylor JW, Miller WG, Stokes GN, and Stott GG. 1985. Cementum annulation enhancement: A technique for age determination in man. *Am J Phys Anthropol* 68:197–201.

Nitzan DW, Michaeli Y, Weinreb M, and Azaz B. 1986. The effect of aging on tooth morphology: A study on impacted teeth. *Oral Surg* 61:54–60.

Nkhumeleni FS, Raubenheimer EJ, and Monteith BD. 1989. Gustafson's method for age determination revised. *J Forensic Odonto-Stamatology* 7:13–16.

Ogino T, Ogino N, and Nagy B. 1985. Application of aspartic acid racemization to forensic odontology: Postmortem designation of age at death. *Forensic Sci Int* 29:259–267.

Ohtani S. 1994. Age estimation by aspartic acid racemization in dentin of deciduous teeth. *J Forensic Sci* 68:77–82.

Ohtani S. 1995. Studies on age estimation using racemization of aspartic acid in cementum. *J Forensic Sci* 40:805–807.

Ohtani S, Sugimoto H, Sugeno H, Yamamoto S, and Yamamoto K. 1995. Racemization of aspartic acid in human cementum with age. *Arch Oral Biol* 40:91–95.

Ohtani S, and Yamamoto K. 1991. Age estimation using the racemization of amino acid in human dentin. *J Forensic Sci* 36:792–800.

Ohtani S, and Yamamoto K. 1992. Estimation of age from a tooth by means of racemization of an amino acid, especially aspartic acid—Comparison of enamel and dentin. *J Forensic Sci* 37:1061–1067.

Philippas GG. 1961. Influence of occlusal wear and age on formation of dentin and size of pulp chamber. *J Dent Res* 40:1186–1198.

Pilloud MA, Hefner JT, Hanihara T, and Hayashi A. 2014. The use of tooth crown measurements in the assessment of ancestry. *J Forensic Sci* 59(6):1493–1501.

Pindborg JJ. 1970. *Pathology of the Dental Hard Tissues.* Philadelphia, PA: WB Saunders.

Plato CC, Fox KM, and Tobin JD. 1994. Skeletal changes in human aging. In DE Crews, and RM Garruto (Eds.), *Biological Anthropology and Human Aging. Perspectives on human Variation Over the Life Span.* Oxford, U.K: Oxford University Press. pp. 272–300.

Prince DA. 2004. Estimation of adult skeletal age-at-death from dental root translucency. PhD dissertation, The University of Tennessee, Knoxville, TN.

Prince DA, Kimmerle EH, and Konigsberg LW. 2008. A Bayesian approach to estimate skeletal age-at-death utilizing dental wear. *J Forensic Sci* 53(3):588–593.

Prince DA, and Konigsberg LW. 2008. New formulae for estimating age-at-death in the Balkans utilizing Lamendin's dental technique and Bayesian analysis. *J Forensic Sci* 53(3):578–587.

Prince DA, and Ubelaker DH. 2002. Application of Lamendin's adult dental aging technique to a diverse skeletal sample. *J Forensic Sci* 47:107–116.

Rathod H, Roberts D, Gray C, Jones SJ, and Boyde A. 1993. Autofluorescence mode confocal microscopy of dental tissues. *Proc Bone and Tooth Society* 1:39.

Renz H, Schaefer V, Duschner H, and Radlanski RJ. 1997. Incremental lines in root cementum of human teeth: An approach to their ultrastructural nature by microscopy. *Adv Dent Res* 11:472–477.

Ritz S, Schütz H-W, and Peper C. 1993. Postmortem estimation of age at death based on aspartic acid racemization in dentin: Its applicability for root dentin. *Int J Legal Med* 105:289–293.

Ritz S, Schütz H-W, and Schwatzer B. 1990. The extent of aspartic acid racemization in dentin: A possible method for a more accurate determination of age at death? *Z Rechtsmed* 103:457–462.

Ross AH, and Konigsberg LW. 2002. New formulae for estimating stature in the Balkans. *J Forensic Sci* 47:165–167.

Russell KF. 1996. Determination of age-at-death from dental remains. PhD dissertation, Kent State University, Kent, OH.

Sarajlić N, Klonowski EE, Drukier P, and Harrington R. 2003. Lamendin's and Prince's dental aging methods applied to a Bosnian population. *Proc Fifty-fourth Ann Am AcadForensic Sci* 6:239–240 (abstract).

Saunders M. 1965. Dental factors in age determination. *Med, Sci Law* 5:34–37.

Schour I, and Massler M. 1941. The development of human dentition. *J Am Dent Assoc* 20:379–427.

Schroeder HE. 1991. *Oral Structural Biology: Embryology, Structure, and Function of Normal Hard and Soft Tissues of the Oral Cavity and Temporomandibular Joints*. Stuttgart, Germany: Georg Thieme Verlag.

Scott EC, 1979. Dental wear scoring technique. *Am J Phys Anthropol* 51:213–218.

Scott GR, and Turner CG, II. 1997. *The Anthropology of Modern Human Teeth: Dental Morphology and Its Variation in Recent Human Populations*. Cambridge, NY: Cambridge University Press.

Sengupta A, Shellis RP, and Whittaker DK. 1998. Measuring root dentine translucency in human teeth of varying antiquity. *J Archaeol Sci* 25:1221–1229.

Sengupta A, Whittaker DK, and Shellis RP. 1999. Difficulties in estimating age using root dentine translucency in human teeth of varying antiquity. *Arch Oral Biol* 44:889–899.

Shimoyama A, and Harada K. 1984. An age determination of an ancient burial mound man by apparent racemization reaction of aspartic acid in tooth dentin. *Chem Lett* 1984:1661–1664.

Šlaus M, Strinović D, Škavić J, and Petrovečki V. 2003. Discriminate function sexing of fragmentary and complete femora: Standards for contemporary Croatia. *J Forensic Sci* 48:509–512.

Smith BH. 1984. Rate of molar wear: Implications for developmental timing and demography in human evolution. *Am J Phys Anthropol* 63:220.

Sognnaes RF, Gratt BM, and Papin PJ. 1985. Biomedical image processing for age measurements of intact teeth. *J Forensic Sci* 30:1082–1089.

Solheim T. 1984. Dental age estimation. An alternative technique for tooth sectioning. *Am J Foren Med Path* 5:181–184.

Solheim T. 1988. Dental attrition as an indicator of age. *Gerodontics* 4:299–304.

Solhiem T. 1989. Dental root translucency as an indicator of age. *Scand J Dent Res* 97:189–197.

Solhiem T. 1990. Dental cementum apposition as an indicator of age. *Scand J Dent Res* 98:510–519.

Solhiem T. 1992. Recession of periodontal ligament as an indicator of age. *J Forensic Odontostomatology* 19:32–42.

Solheim T. 1993. A new method for dental age estimation in adults. *Forensic Sci Int* 59:137–147.

Solheim T, and Sundnes PK. 1980. Dental age estimation of Norwegian adults—A comparison of different methods. *Forensic Sci Int* 16:7–17.

Stein TJ, and Corcoran JF. 1994. Pararadicular cementum deposition as a criterion for age estimation in human beings. *Oral Surg, Oral Med, Oral Pathol, Oral Radiol Endod* 77:266–270.

Stott GG, Sis RF, and Levy BM. 1982. Cemental annulation as an age criterion in forensic dentistry. *J Dent Res* 61:814–817.

Ten Cate AR, Thompson GW, Dickinson JB, and Hunter HA. 1977. The estimation of age of skeletal remains from colour of roots of teeth. *J Can Dent Assoc* 43:83–86.

Teschler-Nicola M, and Prossinger H. 1998. Sex determination using dental tooth dimensions. In KW Alt, FW Rösing FW, and M Teschler-Nicola (Eds.), *Dental Anthropology: Fundamentals, Limits, and Prospects*. New York, NY: SpringerWien. pp. 479–500.

Thomas GJ, Whittaker DK, and Embery G. 1994. A comparative study of translucent apical dentine in vital and non-vital human teeth. *Arch Oral Biol* 39:29–34.

Turner CG, II, Nichol CR, and Scott GR. 1991. Scoring procedures for key morphological traits of the permanent dentition: The Arizona State University Dental Anthropology System. In MA Kelley, and CS Larsen (Eds.), *Advances in Dental Anthropology*. New York: Wiley-Liss. pp. 13–31.

Ubelaker DH. 1989. *Human Skeletal Remains, Analysis, Interpretation*, revised edition. Washington, DC: Taraxacum.

Ubelaker DH, Baccino E, Zerilli A, and Oger E. 1998. Comparison of methods for assessing adult age at death on French autopsy samples (abstract). *Proc Am Acad Forensic Sci IV*, 174–175.

Ubelaker DH, and Parra RC. 2008. Application of three dental methods of adult age estimation from intact single rooted teeth to a Peruvian sample. *J Forensic Sci* 53(3):608–611.

Vasiliadis L, Darling AI, and Levers BGH. 1983. The amount and distribution of sclerotic human root dentine. *Arch Oral Biol* 28:645–649.

van der Velden U. 1984. Effects of age on the periodontium. *J Clin Periodontol* 11:281–294.

Whittaker DK, and Bakri MM. 1996. Racial variations in the extent of tooth root translucency in ageing individuals. *Arch Oral Biol* 41:15–19.

Wittwer-Backofen. 2000. Tooth cementum annulation: A validation study for individual age at death estimates. Report for the Max-Planck-Institut für Demografische Forschung, Rostock, Germany.

Wittwer-Backofen U, and Buba H. 2002. Age estimation by tooth cementum annulation: Perspectives of a new validation study. In RD Hoppa, and JW Vaupel (Eds.), *Paleodemography: Age Distributions from Skeletal Samples*. Cambridge, NY: Cambridge University Press. pp. 107–128.

Wittwer-Backofen U, Gampe J, and Vaupel JW. 2004. Tooth cementum annulation for age Estimation: Results from a large known-age validation study. *Am J Phys Anthropol* 123:119–129.

Zhang Z. 1982. A preliminary study of estimation of age by morphological changes in the symphysis pubis. *Acta Anthropol Sinica* 1:132–136.

Zuhrt R. 1955. Stomatologische Üntersuchungen an Spätmittelalterlichen Funden von Reckkahn (12–14 Jh.): I. Die Zahnkaries und ihre Folgen. *Deutsche Zahn-Mund und Kieferheilkunde* 25:1–15.

Appendix B
Age Estimation in Modern Forensic Anthropology

Bridget F. B. Algee-Hewitt

CONTENTS

Practicalities of aging: The significance of age-at-death in anthropology	382
Aging takes time: A historical overview	384
Aging today: Current trends in estimation	386
Problems with aging: Difficulties in estimating age-at-death	388
Chronological and biological age	388
Phase-aging	389
Choice of indicators	390
Choice of methods	391
Error, bias, and statistics	393
New solutions to some aging problems: Improving age estimation	396
Indicators and methods	396
Choosing an appropriate indicator	397
Choosing an appropriate method	398
Statistics	400
Prior distribution problems	401
Calibrated approaches	401
Getting better with age: Best practices in age estimation	402
Age estimation for the juvenile	403
Age estimation for the adult	405
Adding up the years: Chapter summary of age-at-death estimation	408
Review questions	408
References	409

> **LEARNING OBJECTIVES**
>
> 1. Describe the major problems with age-at-death estimation and suggest solutions to these issues.
> 2. Summarize the history of age estimation in forensic anthropology and comment on landmark researchers and methods.
> 3. Define chronological and biological age, and explain how variations in the biological process of aging complicate age estimation.
> 4. Assess some of the problems with phase-aging methods.
> 5. Discuss the advantages and disadvantages of multiple-trait methods.
> 6. Discuss how the following three phenomena bias age estimates: the calibration problem, attraction to the middle, and age mimicry.
> 7. Name the characteristics of an appropriate age indicator and method.
> 8. Compare age estimation in juveniles versus adults.

Forensic anthropology is an applied, case-driven, goal-oriented subfield within skeletal biology (Ubelaker 2008). In keeping with the complexity of our negotiations among theory, method, and application, and in response to the demands of new casework and the diversity of legal systems both home and abroad, Blau and Ubelaker (2009a) encourage a dynamic understanding of forensic anthropology. As practitioners, forensic anthropologists seek to aid the medico-legal community in making identifications for unknown, mostly skeletal, and often fragmentary human remains by first developing a biological profile (ABFA 2010). Using a combination of grossly observable and measurable morphological features on the available skeletal elements, we rely on our technical expertise to estimate demographic parameters as they relate to the population-at-large, and isolate unique traits that characterize *an* individual as *the* individual (such as sex, ancestry, age-at-death, stature, and pathology). We also look to identify those taphonomic variables that help to explain the circumstances surrounding the unknown individual's death, deposition, and decomposition (including markers of peri and postmortem trauma and approximations of time since death). We estimate these features with some degree of statistical certainty based on our application of standard criteria, casework experience, and advanced techniques supported by interdisciplinary collaboration.

PRACTICALITIES OF AGING: THE SIGNIFICANCE OF AGE-AT-DEATH IN ANTHROPOLOGY

Sex and ancestry are often the first two parameters that a forensic anthropologist must establish, given the fact that many other methods of identification are sex and population specific. Although the estimation of sex removes 50% of the possible candidates from further consideration (assuming an equal male–female ratio), ancestry may have limited evidentiary value when we consider geographical context and the population demographics (Konigsberg et al. 2009). Age-at-death, however, is arguably the most significant variable for narrowing potential *positive* identities using a missing person's list or other medico-legal documents. As an example of modern U.S. trends, consider the current missing persons cases for the state of Tennessee, obtained from the missing persons database and made available through the National Missing and Unidentified Persons System (NamUs 2010). Figures B.1 and B.2 graphically display the frequency of missing individuals in each age cohort, subdivided by sex and race, relative to the total number of cases ($n = 72$). These data represent open cases with assigned National Crime Information Center (NCIC) numbers, include juveniles and adults with reported sex and race, and have last known alive (LKA) dates spanning 30 years, from 1969 to 2009. As of May 2010, there are 72 persons reported missing in Tennessee: 36 (50%) are female; of those females, 6 (17%) are between the ages 40 and 60, and 2 (33%) are identified as nonwhite. In this example, an assignment of any remains in question to this broad *middle age* range (ages 40–60) would reduce possible female matches from 36 active case files to only six probable identities. This number has potential for even further reduction when ancestry (or reported race) is considered.

In theory, the narrower the estimated age interval, the more useful the skeletal analysis may be for medico-legal investigators in excluding cases, finding a presumptive match among open cases, and eventually establishing a positive identification. Reporting a restricted age range, however, is not always realistic and is often unwise on account of the acknowledged variability in patterns of skeletal aging (attributable to genetics, environment, health, and lifestyle). As illustrated by the TN data, the estimated age interval of 20 years (ages 40–60) offers a considerable reduction in the number of possible matches. Even such a broad interval can help law enforcement intelligently refocus their investigation into the unknown's identity; here, dropping the total 72 missing persons registered by almost 95% when only sex is *known* for the remains. A wider age range also better accounts for individual variation in age-progressive traits and is more likely to capture the unknown individual's chronological age, thus, reducing the risk of excluding the unknown's true identity from the missing persons files considered and thereby failing to make a correct match for the remains under evaluation.

Considering the importance of age-at-death in contributing a chapter to the individual's (or skeletal population's) osteobiography, it should not be surprising that age estimation is among the most critical, and yet the most problematic of endeavors when establishing a biological profile. Within the publication history of skeletal biology, few topics have generated as much heated debate as the issue of assigning age-at-death. This discussion extends beyond questioning the accuracy of the ages themselves (i.e., the point estimates or intervals) to unseating assumptions surrounding the very practice of *estimation* and its relevance. These debates take into account the imperfect correlation between skeletal maturation stages (biological age) and chronological age (calendar years since birth). The skeletal indicators, procedures, analytical techniques and statistical methods, the applicability of standard criteria to diverse groups (both modern and

Appendix B 383

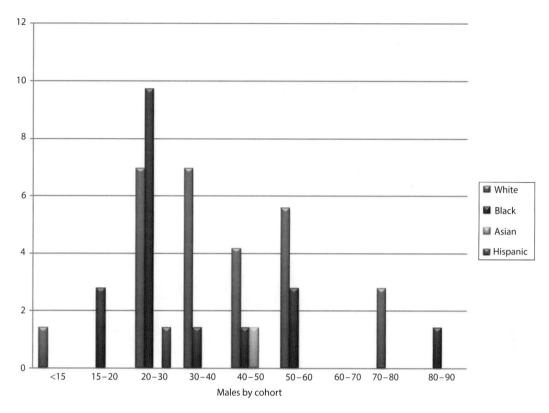

Figure B.1 Frequency of males in each age cohort (n = 36), subdivided by race, and relative to the total number of individuals reported missing in Tennessee (n = 72), after data obtained from NamUS. Cohorts are based on the age of the individual when she/he was last known to be alive. The >15 year cohort includes all juveniles, the 15–20 year cohort is a five year interval, and all other age cohorts are 10 year intervals.

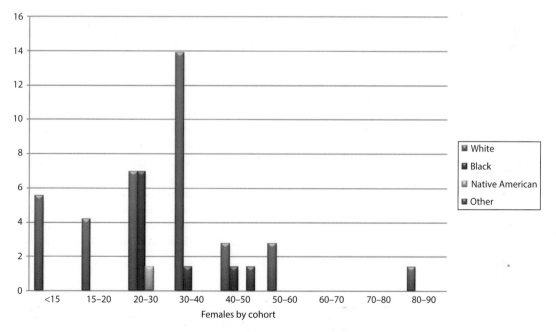

Figure B.2 Frequency of females (n = 36) in each age cohort, subdivided by race, and relative to the total number of individuals reported missing in Tennessee (n = 72), after data obtained from NamUS. Cohorts are based on the age of the individual when she/he was last known to be alive. The >15 year cohort includes all juveniles, the 15–20 year cohort is a five year interval, and all other age cohorts are 10 year intervals.

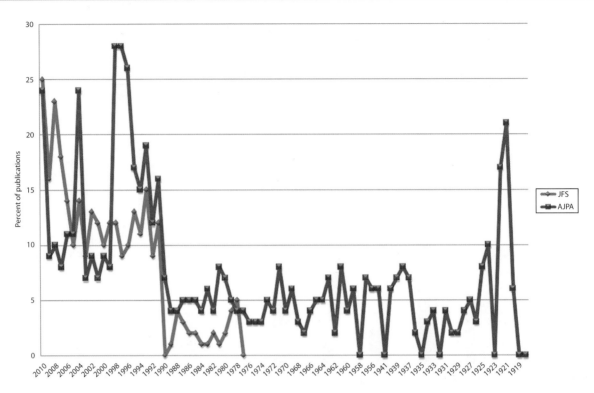

Figure B.3 Annual percent of publications addressing, in some part, age-at-death estimation for the AJPA and the JFS calculated using counts obtained from a Web of Science search. Query parameters: databases = SCI-EXPANDED, SSCI, A&HCI, CPCI-S, CPCI-SSH; topic = "age" or "aging"; time span = all years. AJPA record count = 1617 of 21,219; JFS record count = 687 of 7068.

archaeological), and even the validity of continuing this type of research are all called into question by experts heavily influenced by their philosophical views, anatomical and statistical experience, and research agendas.

A simple search for the topic *age* or *aging* in the Web of Science (2010) citation database across all years of publication for the *American Journal of Physical Anthropology* (*AJPA*) and the *Journal of Forensic Sciences* (*JFS*) shows the sustained and increasing interest in the topic of age estimation by anthropologists and our colleagues writing over the past century. The recent explosion in technologies (e.g., multicore processors and advances in computational statistics, high-resolution imaging and virtual modeling devices, and improved techniques for biochemical and molecular analyses) has forever altered our approach to all aspects of human skeletal biology, past and the present. The field of physical anthropology is experiencing a fragmentation into subspecializations, a significant shift toward interdisciplinary concerns, a burgeoning of new research directions, and the introduction of a theoretical complexity previously unattainable in the absence of our current enhanced exploratory tools. Despite this broadening of research topics, Figure B.3 shows that in the past two decades at least 10% of all *AJPA* and *JFS* publications have been concerned with age-at-death estimation and that this interest has continued to grow—exceeding 20% of all publications in the 2009–2010 year. The publication patterns illustrated in Figure B.3 are also useful as a guide for tracking the historical trends in research on age-at-death estimation, whose complex scholarly foundation rests heavily on the collective knowledge and experience of many contributors (human anatomists, [paleo]demographers, [bio]archaeologists, and forensic anthropologists alike), and whose developmental trajectory follows in similar suit with the evolution of [physical] anthropological theory and thought.

AGING TAKES TIME: A HISTORICAL OVERVIEW

Early publications concerned with skeletal morphology were firmly rooted in a descriptive, typological tradition that characterized the turn of the century's approach to decoding, organizing, and explaining our human cultural and biological history. Science was used to reinforce ideas tied to Linnaean taxonomy so that "evolutionism became cast in traditional descriptive historicism" (Armelagos et al. 1982; Armelagos and Van Gerven 2003, p. 55). Owing much to the research methods of human anatomists, the training of many founding forensic anthropologists (from Dwight and Dorsey, to Hrdlička and Hooton) as practitioners within medical schools and in pathology departments, and the continued interdisciplinary concern for classification through the identification of difference (be it in race, belief, and behavior) focused on studies documenting variation between sexes, among populations, and through time (Armelagos and Van Gerven 2003; Lovejoy et al. 1982; Stewart 1979; Ubelaker 2009).

Anatomist T. Wingate Todd is often credited for laying the foundation for the first systematic and analytical approach to age estimation using skeletal indicators (Iscan and Loth 1989). Underlying his seminal analysis of the pubic symphysis (1920) and his work with Lyon (1924) on the relationship between age and cranial suture closure is a concern for both the descriptive and the analytical. By isolating differences in pubic symphyseal morphology, linking progressive bony change to increasing age, and sorting out the observed metamorphosis into phases, Todd (1921a, 1921b, 1921c, 1923, 1930) developed a model system of age-at-death estimation, but using a single part of the pubic bone that was applicable to the classic white male, and relevant for use in both anatomical studies and legal medicine. His consideration of the pubic bone as a unit within the larger context of growth and development (i.e., by treating the symphysis as an epiphysis that undergoes buildup and breakdown throughout the lifetime), and his treatment of the cranial sutures as part of a complex within the context of functional morphology (i.e., by addressing suture closure as it relates to the shape of the braincase), and his recognition of biological variation (interpreting differences within and among tested skeletal series) cleared the way for a broadened direction in future analyses. Nonetheless, Todd's research remained constrained by his more traditional concern for documenting skeletal norms, providing static descriptions and, from these, establishing a corpus of classic criteria.

It was not until the mid-century paradigm shift toward Washburn's (1951, 1953a, 1953b) *New Anthropology*, with its interest in explaining evolutionary mechanisms and adaptation through hypothesis testing that scholarship in age-at-death estimation took up the more "dynamic approach now introduced into anthropometrical analyses" (Brooks 1955, p. 567). This new approach to understanding biological processes was problem-oriented. The concern for formulating research questions and testing hypotheses, coupled with the identification of unexpected and unexplained trends in age distributions for past populations, fostered an environment, in which researchers undertook studies on age estimation that questioned past procedures and their built-in assumptions. These new projects were directed toward testing, redeveloping, and validating methods that appreciated all degrees of variation, and served a practical purpose in the field. Brooks (1955), for example, tested Todd's methods for the pubic symphysis and cranial sutures, on both the original and a new sample produced error rates for use, and suggested modifications to the general framework for the pubic symphysis. Following this work, McKern and Stewart (1957) were the first to abandon Todd's original phase-aging system and introduced a new component scoring approach that incorporated greater within-group skeletal variation and reduced error. Although the widespread use of the McKern and Stewart method is limited by the narrow distribution of their reference sample of young, white American male soldiers who had died in the Korean War (few of whom surpassed young adulthood), their component approach has contributed considerably to questions of methodology, and forms the basis for the newly developed and very promising transition analysis (TA) coding method that is swiftly growing in popularity today (Boldsen et al. 2002; Milner and Buikstra 2006).

An interest in adopting a functional approach to understanding the skeletal *form*, emerging ideas about the adaptive process in the subfield of bioarchaeology, and the contributions of multivariate statistics to the social sciences, further refined the direction of research into age-at-death estimation (Armelagos 2003, 2008; Buikstra 1977; Enlow 1996; Moss and Young 1960; Moss 1968). With the 1970s questions, although still problem-oriented, were backed by a wealth of raw data, and analyses were being driven by greater accessibility to quantitative methods (Armelagos and Van Gerven 2003). This shift from qualitative descriptions to quantitative statistical analyses encouraged efforts in data synthesis and method validation. At this same time, the term *bioarchaeology* (after Clark 1972) was introduced to physical anthropology for population studies using human remains (Buikstra 1977). Bioarchaeology brought with it a more holistic, population-oriented, and statistically grounded approach to age estimation. The emergence of multiindicator or summary methods for age estimation reflect this emphasis on conducting broader-based anthropological analyses by redirecting the practitioner's focus from an individual skeletal marker and to the skeletal as a whole. Although new to method refinement and rigorous evaluation, the multiindicator approach had been a topic of discussion in earlier work. Todd (1920) and later Brooks (1955) had recommended the use of more than one indicator and posited improved age assessment despite initial findings to the contrary (specifically, Brooks found that the cranial sutures deviated unreliably from the age estimates using the pubic symphysis). This shift toward multiple age markers was also foreshadowed in application by McKern (1957) who pooled observations from the pubic symphysis, skeletal epiphyses, and cranial sutures. These methods grew in popularity because they promised a more complete picture of the aging process by considering all possible morphological variation as it is dispersed throughout the skeleton. For example, the multifactorial summary age (MSA) method of Lovejoy et al. (1985a) derives age estimates statistically, makes use of age-progressive information obtained from many markers (both classic, including the pubic symphysis and cranial sutures, and newer introductions, such as the auricular surface and sternal rib ends), and provides assessments of observer error. Studies such as these ultimately brought about change in what was considered acceptable or conventional methodology through the reevaluation and, at times, rejection of presumed levels of accuracy and precision, reliability and repeatability, and applicability, and significance for the age estimation methods in use. To this point, Kemkes-Grottenthaler (2002) has declared that the rapid expansion of the field of forensic anthropology "undisputedly elevated classical individualization techniques to a new level of scientific inquiry" (p. 49). The establishment of the physical anthropology section of the American Academy of Forensic Sciences (AAFS) in 1972, the first publication of the *Journal of Forensic Sciences*, that same year, and the founding of the American Board of Forensic Anthropology (ABFA) in 1977 are testaments to a directed attention toward innovation, quality control, and consensus (Iscan 1988; Kerley 1978; Reichs 1995, 1998; Snow 1982; Weinker and Rhine 1989).

As Hoppa (2002) recalls, "the 1980s marked a pivotal point for paleodemography" (p. 15). In 1982, hefty criticism was first levied against age-at-death estimation in the context of past population studies and, in response, an intensive debate followed on the validity of the methods used for constructing age-at-death distributions. Bocquet-Appel and Masset (1982) argued that, since "the information conveyed by the age indicators is so poor," the estimation of vital rates for past skeletal populations is simply a product of "random fluctuations and errors of method" (p. 329). They advocated for the complete abandonment of age-related research and asserted that the estimation of adult skeletal age-at-death is a wholly unreliable endeavor because (1) target age-at-death distributions are heavily influenced by the distribution of the reference sample, (2) age-at-death estimates are biased, and (3) the correlation between skeletal age indicators and chronological age is unacceptably low (Bocquet-Appel 1986; Bocquet-Appel and Masset 1982, 1985). However, the accurate estimation of skeletal age-at-death is the key source of information for the reconstructions of mortality and fertility schedules, morbidity patterns, and the density, distribution, and age composition of skeletal populations. Approximations of age, as point estimates, or preferably, as ranges for forensic cases are an essential part of the biological profile and play an important role in assigning identity to unknown skeletal remains. For these reasons, a concentrated effort began for better explaining and improving the method and theory of age-at-death estimation (Hoppa and Vaupel 2002a; Konigsberg and Frankenberg 1992, 1994, 2002; Konigsberg et al. 1997). As Iscan (1989) remarks, "The 1980s has seen an unquestionable increase in interest in forensic anthropology as well as the introduction of the first aging methods from new sites in over 60 years" (p. 7). Using a large modern sample and multivariate statistics, Suchey et al. presented a revised method for age-at-death estimation from the pubic symphysis that simplified Todd's 10-phase method (1920) and produced the 6-phase system now commonly known as the Suchey–Brooks method (Brooks and Suchey 1990; Katz and Suchey 1986, 1989; Suchey and Katz 1998; Suchey et al. 1986). Lovejoy et al. (1985b) proposed an additional pelvic, phase-aging method using the auricular surface and incorporated cranial suture closure into their multiindicator MSA method (1985a). Cranial vault sutures were also reevaluated by both Meindl and Lovejoy (1985) and Masset (1989), whereas Mann et al. (1987, 1991) presented the utility of maxillary sutures for aging intact and fragmentary remains. Finally, Iscan et al. (1984, 1985) introduced a new method for estimating age from gross morphological changes observed at the sternal end of the fourth rib (Loth and Iscan 1989). The combined instances of conflict, recovery, and growth in opinions on *age* alongside these new efforts in revision, innovation, and refinement for *aging* have resulted in a continued movement toward codifying sound age estimation theory and its methodological application, the identification of best practices in the field, lab, and in publication, and the dissemination of the correct, consistent use, and interpretation of these ideas for both skeletal assemblages, and the individual skeleton in archaeological and forensic contexts.

AGING TODAY: CURRENT TRENDS IN ESTIMATION

Although skeletal biologists today are still interested in the identification of new indicators and new estimation formulae (Algee-Hewitt et al. 2008; Prince and Konigsberg 2008; Rissech et al. 2006), refining the approach to, or redefining the best combination of, known age-informative markers (Boldsen et al. 2002; Osborne et al. 2004; Shirley-Langley and Jantz 2010) and the mechanics of their practical application in the classroom, lab or field (Buikstra and

Ubelaker 1994; Moore-Jansen et al. 1994), the research focus of current forensic anthropologists is often guided by casework. We seek answers to new problems introduced by the contexts of the remains under study, we attempt to better define population- or group-specific characteristics that may aid in the identification of future unknowns, and we revisit those issues not yet, or adequately, resolved because of limitations imposed by reference samples, technology, and political or medico-legal circumstances. The estimation of age-at-death in forensic anthropology is following a course that is responding organically to the advances in technology (especially in computational statistics and imaging) and the demands of the ever-changing medico-legal and academic environments in which, we, as its practitioners, work. As a result, age-at-death estimation has reached subdisciplinary maturity. The questions posed are no longer framed by concerns for its basic utility, but reflect the need for refinement in two areas relevant to the role of the forensic anthropologist in the larger identification process: (1) diagnostics and (2) estimation and evidence.

First, how can we improve access to and increase availability of reference samples for method development and testing or specimen comparison, simplify access to the skeletal indicators required for analysis, and make use of new technologies for information retention and sharing? The contributions of *virtual anthropology* to age-at-death estimation are showing promise as viable solutions. Significant improvements in computerized modeling techniques, for example, are providing us with high-quality two- and three-dimensional reconstructions of bones, indicators, and the macroscopic features relevant to age estimation procedures (Zollikofer and Ponce de León 2005). Age information for each skeleton can be archived as a virtual reference collection in a convenient format that allows for economical and perpetual storage, repeated access to the *indicators* for reevaluation or new testing, and easy data sharing among researchers. Indeed, relationships have been found between individual chronological age and the stage of the indicators obtained from virtual reconstructions produced using multislice computed tomography (CT) and laser scans: Dedouit et al. (2008) applied the Iscan method to CT scans of right fourth ribs, Barrier et al. (2009) used the Lovejoy method on CT scans of the auricular surface, Ferrant et al. (2009) tested new CT scan-based criteria on multiple pelvic indicators, and Shirley et al. (2010) recently evaluated the utility of laser scans as surrogates for the actual boney indicators (pubic symphyses and auricular surfaces). Although there is an acknowledged loss of some detailed information and questionable difficulties in navigating equipment and software, additional testing of virtual samples is warranted, given the potential benefits to virtual autopsy, digital archiving, and data sharing of full specimens.

Second, after using known scoring systems to evaluate the macroscopic features that define a phase or stage, how do we extract the most accurate estimates given the complications of sex and population variability and limited reference samples? Moreover, what evidentiary contributions do our age-at-death estimates make to putative identifications when we consider the many new and changing requirements for expert testimony and the rules for the admissibility of evidence in courts, both at home and abroad? As the legal guidelines established by the judiciary body must be met for each method used, the value of an estimate will always be dependent on the expectations of the system, in which the testimony or admitted evidence of age-at-death is given. Although the rules of evidence vary by legal context (across systems, populations, and geography), the Daubert guidelines (following from the 1993 landmark case, Daubert v. Merrell Dow Pharmaceuticals, Inc.) support conservative admissibility criteria that should complement the level of expectations of most systems, including international humanitarian courts (Christensen 2004; Grivas and Komar 2008; Kimmerle et al. 2008a; Sapir 2002; Wiersema et al. 2009). Berg (2008), Kimmerle et al. (2008b), and Konigsberg et al. (2008), for example, address these issues in an international forensic context by testing the applicability of age-at-death estimation from the pubic symphysis using American, Asian, and Eastern European skeletal populations. These publications demonstrate how the choice of appropriate statistical methods and making informed decisions regarding reference samples and population parameters can provide well-founded estimates that meet current standards of evidence, identify areas of variability, and offer practical answers to the questions of age estimation in individual identification.

A combination of innovative statistical treatments, such as TA, Bayesian methods, and hazard analysis, allows for the estimation of the complete age-at-death distribution, and individual ages at death that are grounded in robust objective methods and supported by probabilities. In forensic applications, we can obtain convincing statistics that can be used as support of our contribution to the personal identification of the *unknown* remains (Steadman et al. 2006). Boldsen et al. (2002) have contributed significantly to the ease with which the anthropological community can apply the new TA method* to both individual cases of forensic significance and archaeological samples by providing a robust multiindicator, component-based scoring method for which Milner and Boldsen (2008) have

---

* The Transition Analysis method is discussed in more detail later in this chapter.

composed a detailed manual. The concomitant development of free-access, user-friendly software (e.g., the ADBOU Age Estimation Software by Milner and Boldsen and NPHASES program by Konigsberg) simplifies the execution of the statistical procedures required for estimating age using TA, including logistic regression (Boldsen et al. 2002) and the probit model (Konigsberg and Herrmann 2002). The open-source statistical software, R, can also be used to call to the *polr* (proportional odds logistic regression) function (included in the MASS package), with which we can conveniently fit a logistic or probit regression model to an ordered factor response. These programs are increasing both the accessibility of the method to the anthropological community and the use of this approach among forensic practitioners for routine casework. Wilson and Algee-Hewitt (2009) and Shirley et al. (2010) have demonstrated the accuracy of the estimates obtained and low degrees of inter- and intraobserver error; these researchers and practitioners attest to the ease of producing results using the TA coding manual and the ADBOU computer program. Using NPHASES, Kimmerle et al. (2008b), Berg (2008), and DiGangi et al. (2009) have also shown the applicability of this approach to human rights cases, questions of population-specific standards, and to testing new methods. The success of the current directions in age-at-death estimation suggests that future studies will continue to refine the research well underway using robust statistical methods (including Bayesian statistics, hazard models, TA procedures, and component coding systems), advanced digital techniques (virtual anthropology and biomedical imaging), and computational technologies (age estimation software), all of which promise for forensic casework (1) an increased accessibility to robust methods, (2) further ease in obtaining improved results, and (3) a greater degree of confidence in the conclusions that are drawn.

PROBLEMS WITH AGING: DIFFICULTIES IN ESTIMATING AGE-AT-DEATH

Chronological and biological age

In their study on the history of the human lifespan, Acsádi and Nemeskéri (1970) argued that, while both sex and age are the "two fundamental biological characteristics of man" (p. 100), the estimation of adult age-at-death from skeletal remains is more important to the reconstruction of population dynamics than the determination of sex. Despite the fact that accurate age estimation is a precondition for forensic identifications and paleodemographic reconstructions, they contended that "there are not many investigators who would treat age determination with a thoroughness that can withstand scientific criticism" (p. 100). We see this same sentiment later reiterated by Mays (1998), who claimed that "at present the lack of a wholly satisfactory technique for estimating age-at-death in adult skeletons from archaeological sites is one of the thorniest problems facing human osteoarchaeology" (p. 50). Therefore, age-at-death estimation is just as Howell (1976) has aptly characterized it as: "a subject which is simultaneously intensely interesting and devilishly difficult" (p. 25). Comments such as these are rooted in the very crux of the problem of determining skeletal age in any context such as: biological age and chronological age are not equivalent, *aging* is highly variable, and this variation is the significant source of method error.

Skeletally derived ages at death represent subjective assessments of the physiological state of the individual; these age estimates are inferred from an imperfect correlation between skeletal maturation stages (biological age) and calendar years since birth (chronological age). Variation in the timing of developmental and degenerative changes produces individual and population differences through time and across groups that confounds age estimation when *classic* phase-to-age criteria are applied to any case, without question (Kemkes-Grottenthaler 2002). Although Acsádi and Nemeskéri (1970) note that the regressive senescent processes (i.e., those changes that characterize what we call "aging" in the adult) begin at the transitional point between juvenile development and postmaturity heterostasis (when, e.g., the union of all epiphyses is complete), the qualification and quantification of all subsequent degenerative age-related morphological changes is not as straightforward. There is no question that senescence, or *aging*, is universal: all populations undergo age-related and age-progressive changes from birth to death regardless of sex, ancestry, temporal distribution, or geographical location. The human lifespan is, however, "an environmentally and genetically labile characteristic, and there are no species-specific lifespans … [U]nder different environmental conditions, different samples of the same species may have very different average and minimum lifespans" ([Harper and Crews 2000], pp. 495–496). The magnitude, rate, and proximal causes of senescent changes are not constants, and this plasticity permits differences across and within species (Finch 1994). There are, therefore, several notable sources of variation in morphology of importance to us as skeletal biologists: (1) vertical differences through time (e.g., evolutionary change between [sub]species and secular change within groups), (2) horizontal differences within and between populations (by sex and ancestry), and (3) within-subject variability in the timing of age-progressive changes observed throughout a single skeleton (asymmetry in bilateral indicators, disagreement between indicators from different regions).

While chronological age is independent of natural factors (i.e., the number of chronological age years increases systematically in accordance with the passing of calendar years in the individual's numerical *age* progression from birth to death), biological *aging* is an individual experience (Bryant and Pearson 1994). Factors that affect the timing of bone remodeling and degeneration and bring about the morphological changes upon which most methods for determining age are based can include genetic differences, diet, disease, endocrine status, climate, cultural practices, socio-economic status, and individual, or group behaviors. Consequently, an individual's living conditions greatly affect the condition of the individual's skeleton at death. Buckberry and Chamberlain (2002) extend this argument further by positing that two individuals living under the same or similar environmental conditions will age at varying rates. These emerging views on the interpersonal and within-subject plasticity of the senescent process improve our understanding of the biology of aging on the one hand (a concern expressed by Lovejoy et al. [1997]) but, on the other hand, call into question the requisite uniformitarian assumption held for any estimation of demographic parameters: that the patterns of age-progressive changes observed in one population/individual are not significantly different from what one would expect to observe in another population/individual regardless of time and space (Howell 1976). Arguments that bones and teeth do experience age-related morphological changes over the course of the human lifespan and, moreover, that each individual element of the skeleton undergoes degeneration at different rates and magnitudes that are dependent upon structure, function, location, and the unique stresses on the element under study, challenge the validity of the uniformitarian position that "observable biological markers of maturity [always] correspond to certain chronological ages" (Howell 1976, p. 27). Chronological (i.e., calendar) age and biological (i.e., physiological) age are never quite equal, and so age estimates are approximations based on observed similarity and concordance. This inherent variability in the biological process of human aging is often considered to be the "fundamental source of error for present osteological aging criteria" (Hoppa 2000, p. 185). Despite this problem of unresolved *error*, the physical anthropologist remains interested in obtaining data on age to better understand population structure, and the forensic anthropologist is specifically charged with providing a useful age estimate to the medico-legal community for human identification purposes. The challenge for age-at-death estimation is, therefore, to develop tools that tease out biological information from the skeleton that is most closely related to chronological age. Several phase-based aging methods are available, that, despite the overarching issues of variation and error, are regularly used to estimate chronological age because their approach captures the general progression, the senescent phases or stages of biological change through time.

Phase-aging

Even in the early years of studies on skeletal age-at-death, it was recognized that the postdevelopmental remodeling of osseous tissue, usually in the form of bony breakdown, was informative of age. Sashin (1930), for example, explained that the changes useful to age estimation from the auricular surface of the ilium are "progressive, and increase in extent and intensity with the age of the individual" (p. 909). It was determined early on that the age of the adult skeleton could be most profitably estimated by locating specific age-informative areas (i.e., sites that are invariant to environment or activity and undergo progressive, regular morphological change over time), by assessing these for the degree and pattern of macroscopic bony breakdown expressed over the lifetime, and by assigning this observed degeneration to a phase that is linked to chronological age. Techniques that make use of age-informative traits in this way, require practitioners to properly apply the method's predefined criteria. We are expected to (1) correctly identify the indicator/age-informative location on the bone; (2) correctly interpret the description of the bony changes of interest and the definition of the phases (i.e., what type of change characterizes each phase); (3) correctly follow the scoring procedures and employ them with consistency (i.e., with low inter- and intraobserver error); and (4) to correctly translate our categorical data on phase assignment into a chronological age estimate using published standards and, when appropriate, execute the statistical procedures necessary to calculate point estimates, age ranges, and measures of variation. Most conventional methods rely on this phase-to-age theory for adult age-at-death estimation. Popular examples include the following: estimating age from the pubic symphysis of the os pubis (e.g., Brooks 1955; Brooks and Suchey 1990; Gilbert and McKern 1973; McKern and Stewart 1957; Suchey and Katz 1986; Todd 1920, 1921a, 1921b), the auricular surface of the os coxae (Buckberry and Chamberlain 2002; Lovejoy et al. 1985b; Meindl and Lovejoy 1989; Osborne et al. 2004), the sternal ends of the ribs (Loth and Iscan 1989; Yoder et al. 2001), craniofacial suture closure (Brooks 1955; Key et al. 1994; Mann et al. 1987, 1991; Masset 1989; Meindl and Lovejoy 1985; Singer 1953; Todd and Lyon 1924), ossification of the medial clavicle (Black and

Scheuer 1996; McKern and Stewart 1957; Stevenson 1924; Todd and D'Errico 1928; Webb and Suchey 1985), and from patterns of dental wear (Brothwell 1989; Gustafson 1950; Miles 1963; Molnar 1971). Although many improvements have been made to these adult phase-aging methods over time (e.g., Suchey and Brooks' refinements to Todd's pubic aging standards, reducing the original scheme from 10 to 6 phases), difficulties unavoidably remain for any method that requires the observer to subjectively assess a skeletal element for *wear and tear* and then assign it to an ordinal phase category, for which age ranges have been previously determined from reference samples of *known* chronological ages (Schmitt et al. 2002). Problems with adult phase-aging methods are widely recognized among paleodemographers (Hoppa and Vaupel 2002a), but also have the following significant implication for forensic anthropologists:

1. Necessary indicators are often absent (due to taphonomic filters).
2. Morphological changes are not regular across and within all groups for the available indicators, and population-specific standards are not always available.
3. Methods are interpreted variably and subjectively applied, thereby producing inter- and intraobserver error.
4. Age intervals are broad (e.g., a 95% age range of 25–83 years for Suchey–Brooks females Phase V), and often the last phase is defined as an open-ended terminal interval such that the *old* are lumped together and assigned a collective age of, for example, 50+ years for Todd's pubic symphyseal Phase X and 60+ for Lovejoy's auricular surface Phase 8.
5. Phases often overlap (consider Suchey–Brooks female Phase IV, 26–70 years, vs. Phase V, 25–83 years).
6. Estimates are frequently biased, so ages for younger individuals are overestimated, and ages for older individuals are underestimated (this tendency is referred to as *attraction to the middle*).
7. Ages estimated for the target sample reflect the ages of the reference sample when improper priors are used (this phenomenon is called *age mimicry*, whereby the age estimate of the target sample mimics, or is shaped by the age distribution of the reference sample on which the method was developed).
8. Poor theoretical and statistical choices produce inaccurate age estimates, as well as misleading age-at-death distributions.

Choice of indicators

As Iscan (1989) reflects on the history of skeletal aging techniques, remarking that "from the 1920s on, most skeletal biologists have relied almost solely on the cranial sutures and pubic symphysis for age estimation in the adult" (p. 7), it can be argued that the parts of the skeleton selected as preferred age-informative indicators were not always chosen first for their accuracy in providing age estimates, but instead for practical and historical reasons. Crania are typically more resistant to taphonomic processes than are more fragile elements (e.g., sternal rib ends), exhibit obvious age changes (i.e., open vs. obliterated sutures), and have long been a subject of fascination among anthropologists for a variety of research questions (Boldsen et al. 2002; Kemkes-Grottenthaler 2002; Waldron 1987). This preferential treatment given to individual traits as a result of availability, logic, and tradition may result in choices that compromise biological reality. When the skeletal system is treated as a collection of isolated age indicators, we take up a view that denies the fact that developmental and degenerative patterns in the aging skeleton are influenced by many interrelated factors, and we tacitly accept a low or limited yield of age information from the isolated indicator. From assessing trends in publication, we can safely assume that it is common knowledge among skeletal biologists that age information can be obtained from many different parts of the skeleton, that these age-informative regions offer different types and levels of information, and that they contribute differently to making the final age estimate (i.e., some work better than others). For these reasons, it is often held as true that individual age indicators and methods can be combined, and that comprehensive approaches should produce superior results (Baccino et al. 1999; Boldsen et al. 2002; Saunders et al. 1992; Workshop of European Anthropologists 1980).

There are sound arguments for the use of multiple traits methods such as the MSA and TA methods because they can provide a more holistic picture of the aging process by including all possible morphological variation in the assessment of the individual's age (Acsádi and Nemeskéri 1970; Boldsen et al. 2002; Lovejoy et al. 1985a). The combination of several stand-alone age estimation methods in research[*] and in practice[†] has also

---

[*] See McKern and Stewart (1957) for testing many separate age-informative regions on the Korean War sample.
[†] See Houck et al. (1996) for the analysis of the Branch Davidian compound victims.

been encouraged as a means of correcting bias in any one method, accounting for population variability in one method, and evaluating if or how the senescent process varies throughout the single skeleton or skeletal assemblage (Aykroyd et al. 1999). These additive approaches have not, however, always yielded the best estimates. For some, research continues profitably using only a single, robust, skeletal, or dental age indicator (Buckberry and Chamberlain 2002; Prince and Konigsberg 2008; Prince et al. 2008). Nonetheless, method tests, assessments of bias, and tests of the assumption of conditional independence continue to produce conflicting results (Holman et al. 2002; Kemkes-Grottenthaler 1996). The use of multiple indicators does not effectively remove the issue of reference sample bias "by cancelling each other out" (Milner et al. 2008), combined traits do not contribute information on age-at-death in equal amounts, and conditional independence is probably not a fair assumption for all traits.* Given the complex architecture of the skeleton, age indicators are correlated with each other so that the information provided by one is not independent of another, and so age estimates are biased, and intervals are too narrow (Boldsen et al. 2002; Milner et al. 2008). However, the good results recently obtained using, for example, the TA method, suggest that the inconsistencies observed in the quality of multitrait estimates may be attributable to the specifics of individual research design, methodological choices, and statistical assumptions, and not to the underlying *theory* that supports a combined approach (Boldsen et al. 2002; Buikstra et al. 2006; Usher et al. 2000; Wilson and Algee-Hewitt 2009).

Choice of methods

The modal, phase-based design common among most age estimation methods is often criticized for failing to consider biological reality. Modal phase methods dictate that changes in skeletal morphology are assigned to discrete, ordered stages that progress through chronological time from young to old. Methods that use phases require the practitioner to subjectively assign the indicator to one of several artificially constructed categories that may bear little relationship to their actual distributions in any living or extinct population. This reliance on ordinal age-phase strategies results in additional problems that inhibit the accurate estimation of age-at-death. Accuracy and reliability is dependent on both the overall experience of the observer in recognizing the subtleties and variation in age-related morphological changes and on the observer's specific familiarity with the age-phase technique at hand. Assigning skeletal elements to subjectively determined categories leaves an open study to the possibility of introducing inter- and intraobserver error (Lynnerup et al. 1998). The morphological assessment of age change requires the categorization of three-dimensional skeletal features by following often ambiguous written descriptions, or by comparisons to limited *exemplary* photographs or casts. For example, any observer interested in evaluating the sternal end of the fourth rib for the purpose of making classification into a predefined age cohort, by assigning the rib to a phase, must take into account a host of morphological changes using only reference photographs or casts representing *classic* ribs by stages accompanied by descriptions of the features to be observed. Consider the Loth and Iscan (1989) description of Phase 4 (ages 24–32):

> There is noticeable increase in the depth of the pit, which now has a wide V-or narrow U-shape with, at times, flared edges. The walls are thinner, but the rim remains rounded. Some scalloping is still present, along with the central arc; however the scallops are not as well defined and the edges look somewhat worn down. The quality of the bone is fairly good, but there is some decrease in density and firmness (p. 112).

As the descriptive language used here is not universal among the publications of age estimation methods using other indicators, and the authors do not provide pictures or casts that account for every possible variation, observer interpretations are necessarily best guesses—what Paine and Boldsen (2002) call "investigator-specific, weightings of categorical observations" (p. 170). A significant portion of inter- and intraobserver error can be explained by the fact that the observations derived from these traditional methods cannot be quantified. Consider again Loth and Iscan's description of a Phase 4 rib end. Observers evaluate a suite of morphological characteristics (including the contour and depth of the pit, thickness of the walls, the shape of the wall edges, the regularity of the rim, the presence of bony outgrowths, and the condition of the entire bone), which in their multiplicity increase the potential for error. In addition, these characteristics are scored following the ambiguous qualitative language used in the Phase 4 description (e.g., *more irregular*, *some decrease*, and *still good*), which, in the absence of immediate comparative materials, do not facilitate quantification. This inherent subjectivity in age-phase assignment complicates

---

* See Boldsen et al. (2002) or Lucy and Pollard (1995) for more information.

age-at-death estimations by affecting repeatability. Even a reasonably experienced practitioner may not consistently choose the same age-phase assignment for a case scored at different times, and any two equally experienced observers may not interpret the method's criteria identically and, in turn, produce comparable estimates. Findings of considerable intra- and interobserver error discredit the reliability of the method and the accuracy of the conclusions drawn from the analysis (Schmitt et al. 2002). This issue is not unique to Loth and Iscan's (1989) method, nor is it limited to aging techniques using the thoracic region, but it is a problem that burdens all phase-based procedures for all skeletal elements.

As studies of the pubic symphysis by Todd (1920, 1921a, 1921b, 1921c, 1923), Brooks (1955), Brooks and Suchey (1990), and Meindl et al. (1985 represent the seminal work that relies on this traditional *age-phase* theory, let us also review the classification criteria in a phase from the Suchey–Brooks method. The description of Phase III (ages 21–53 for females and 21–46 for males) follows:

> Symphyseal face shows lower extremity and ventral rampart in process of completion. There can be a continuation of fusing ossific nodules forming the upper extremity and along the ventral border. Symphyseal face is smooth or can continue to show distinct ridges. Dorsal plateau is complete. Absence of lipping of symphyseal dorsal margin; no bony ligamentous outgrowths.

A quick read of this definition reveals problems similar to those noted for the rib, in Phase 4 but, what is of particular importance here is the risk confusion between morphological states of buildup and breakdown when we consider the characteristics of subsequent phases (Kimmerle et al. 2008c). Suchey and Katz (1998) even remark themselves that

> Phase III is somewhat problematic; it peaks in the mid to late twenties, but there are outlying cases trailing into the sixties. These outlying cases are probably interpretative errors on the part of Suchey and Brooks who may have confused buildup of the ventral rampart with breakdown of the ventral rampart (p. 221).

Although this intramethod ambiguity substantiates serious concerns for observer error, and reliability, it also serves to highlight the problem of forcing traits that are continually undergoing progressive change over time into discrete categories. Saunders et al. (1992) assert that weaknesses in language interpretation and picture correlation for the Lovejoy et al. (1985b) auricular surface method contributed to the high observer error obtained (19.3%). They attribute this inaccuracy to difficulties in assessing the characteristics examined (especially grain and density), and to the limited photographs, which do not sufficiently reflect the potential range of variation, and provide only a modal surface appearance for each phase. They also report that 50% of the means differed between two identical analyses conducted by the same observer for the lateral anterior system of Meindl and Lovejoy's (1985) ectocranial suture method. Kemkes-Grottenthaler (1996) posits that problems in the reliability of this method, and possibly for others not yet investigated, may be a direct result of the failure to take into account the sex-specific patterns of suture closure, and the differential rates of obliteration (what Todd and Lyon had already noted in 1924 as *lapsed union*). Here, then, the importance of both recognizing and taking into account the observed variation within and between individuals is paramount.

This problem of application is also compounded by the weakness often seen in the final age estimate. Phase-based aging methods typically generate wide-age intervals that have little practical use, especially when the intention is to arrive at age ranges with sufficient informational value for comparison with missing persons list, and for practical use in other medico-legal or research contexts. Consider, for instance, the "most frequently used" (Meindl et al. 1985, p. 29) and "most reliable" (Buikstra and Ubelaker 1994, p. 21) method for the pubic symphysis. Suchey–Brooks Phase V for females includes all ages from 25 to 83 years, for which the mean is 54 years (Brooks and Suchey 1990). This wide-range tells us very little about the real age distribution of the population, and similarly provides little help in assigning an age to an unknown individual for the purpose of forensic identification. The age ranges are so broad that only the very young (<25 years) and very old (>83 years) are mutually exclusive. Furthermore, since the relationship between the phase and chronological age is nonlinear, phases tend to overlap and vary in length, such that an age range associated with a particular phase is not exclusively tied to that phase, but may be attached to multiple phases. Suchey–Brooks Phases IV (26–70 years) and V (25–83 years) both describe the symphyseal morphology expected for females between the ages of 26 and 70. As a result, logical confusion may arise over assigning ambiguous skeletal elements to the *best* of the overlapping categories. Any such uncertainty increases the potential for interobserver error due to misclassifications, and valuable information on the individual is lost when overlapping age phases are employed (Love and Müller 2002). The issue of overlap is equally a problem for methods that use significantly smaller age ranges per phase and, consequently, more age categories (i.e., Todd's 10-phase compared to Suchey–Brooks collapsed 6-phase system). As Buckberry and Chamberlain (2002) observed, Lovejoy et al. (1985b) auricular surface method "oversimplifies the changes seen" (p. 232) by failing to recognize that the features of interest (e.g., porosity, surface texture, and marginal changes) develop and degenerate independently. The five year intervals in this system are *optimistically narrow* as they fail to capture the range of the variation in these features, and considerable overlap exists between different phases. This problem was restated recently by Osborne et al. (2004) who used the auricular surface for age estimation in forensic contexts. They show that only 33% of their target sample was accurately aged when using the five year intervals. They attribute this error in estimation to the fact that the mean ages of several of the intervals are not significantly different. Further, the use of these intervals with the constant width of five years presupposes that all individual age estimates have the same degree of error. This assumption is clearly problematic in light of the fact that every skeleton has its own degree of error, which is dependent on the individual's suite of indicators and state of preservation (Boldsen et al. 2002, p. 75). A decrease in the number of phases, a reduction in the size of age ranges, and/or the standardization of these intervals do not appear to adequately lessen observer confusion, and diminish the error associated with the larger theoretical problem of *quantification*, that is, assigning the biological age indicator to its actual chronological age stage (Paine and Boldsen 2002, p. 170).

The task of estimating age for older individuals is also shown to be problematic when using modal, phase methods. Individuals who have lived to old age have skeletons that have experienced the greatest amount and diversity of age-related remodeling and degeneration.[*] As these skeletons show the most bony breakdown, irregularity, and complex deterioration, their indicators are less easily assigned to age phases that are typically defined by a limited set of age-informative characteristics. This problem of ambiguity is often resolved by making the terminal age-phase an open-ended interval, to which all skeletons over a method-specific age are assigned.[†] This fix is not only unsatisfying to those anthropologists interested in working with skeletons of older individuals, but it is also wasteful of potentially important age information for personal identification. The inability to capture the individuals who are representative of the most extreme right-tail of the age-at-death distribution, results in a skewed mortality schedule for paleodemographic research, and an overly broad estimation of age for forensic casework. If the method makes use of an age indicator that is not sufficiently sensitive to the senescent changes that are particular to the target sample under analysis, then it is not helpful to studies that are concerned with evaluating the full variation of age-at-death within the population, or in determining the age of an obviously older individual.

While formulating criteria for the best choice of indicator and method is wise in theory, it is also important to consider the practical problem of preservation or recovery bias, which may confound attempts at upholding guidelines for best practice. Any skeletal sample is a small percentage of the population at large, and may be a poor representation of the once-living population. Sample composition is shaped by both human and natural factors that influence the condition, in which the remains are retained and recovered (Haglund and Sorg 1997, 2001; Larsen 1997). Standards for successful age estimation methods rely on indicators that are simultaneously the most informative of age and also subject to poor preservation: consider the more fragile, yet reliable pelvic indicators versus the more resistant, yet less effective, cranial sutures (Buikstra and Ubelaker 1994). The sample may be inherently biased by the singular or combined actions of poor preservation due to taphonomic processes, differential burial, and incomplete recovery (Waldron 1994; Walker et al. 1988; Wood et al. 1992). These taphonomic factors can compound method-based sources of error by placing limitations on the selection of the most robust age indicator, the use of multiindicator methods, the estimation of the target age-at-death distribution, and, in turn, the quality of the estimates produced, and the value of the information obtained from these results.

Error, bias, and statistics

Closely tied to these concerns of method and indicator choice, are the reports of bias in age estimates for the *target* sample resulting from the use of inappropriate *reference* samples or *priors*. This issue is what Meindl and Russell

---

[*] See Wood et al. (1992) for an explanation of the osteological paradox.
[†] See, for example, Todd's (1920, 1921a, 1921b, 1921c) method for the pubic symphysis, which lumps all individuals >50 into the terminal category, Phase X, or Lovejoy et al.'s (1985b) method for the auricular surface, which offers only a slight 10-year improvement by assigning all individuals over age 60 into the final Phase 8.

(1998) define as "Bias = Σ (estimated age−real age)/number of individuals, that is, sign is considered" (p. 383). We can better reach, and appreciate the source of this aging bias, and detangle the knotty statistical relationship between the estimated ages for the target sample of *unknowns*, and the documented ages of the reference sample of *knowns* by considering the following:

1. *The calibration problem*: The decision of when it is appropriate to use inverse or classical calibration for the estimation of age-at-death for unknown individuals or skeletal populations using regression techniques (Konigsberg et al. 1997, 1998)
2. *Attraction to the middle*: The tendency of uncompensated errors committed in the outer age categories to accumulate in the middle-aged categories (Masset 1989)
3. *Age mimicry*: The tendency for the estimated age-at-death distribution of the target sample to mimic the age-at-death distribution of the reference sample (Bocquet-Appel and Masset 1982)

When we conduct an analysis using inverse calibration, we regress the dependent variable (typically denoted as x) on the independent variable (represented by y). For age estimation contexts, we accordingly regress the fixed variable of age on the age indicator, and estimate age of an individual by ordinary least squares (OLS). For classical calibration, the reverse is true (OLS regression of y on x, followed by solving for x), so that instead we regress the indicator on age, and then solve for age. As inverse calibration involves regression toward the mean, any age-at-death estimate generated will naturally be skewed or *biased* in the direction of the mean age. This is what Masset (1989, 1982) has called the *attraction to the middle*, and what can be more simply defined in practice as the chronic tendency to underage older individuals, and to overage younger individuals so that, in many cases, the estimated ages more closely reflect the mean age than the chronological ages of the individuals for whom the estimates were made. As inverse calibration is a Bayesian method, we are framing our estimates of age for our unknown sample, in terms of probabilities that are informed, by what we know about ages at death from a documented reference collection; our analysis is said to be conditional upon prior information. Regression of age on the indicator requires us to assume correctly that the *target sample* (i.e., our unknown sample or case, comprised of the individual[s] for whom estimate[s] of age at death are desired) comes from the *same* (similar, comparable) age-at-death distribution as the *reference sample* (a documented collection of skeletons of known ages, used for method testing, and often referred to as a *prior*). Inverse calibration should only be employed in casework when a suitable reference sample or prior can be obtained for target skeleton(s); if an inappropriate prior is used, aging biases will result.

Although *age mimicry* was first discussed among paleodemographers working with skeletal populations from archaeological settings (Bocquet-Appel and Masset 1982), it also has significant implications for case analysis in forensic contexts. Age mimicry occurs when the age distribution obtained for the target, unknown skeletal sample is said to reflect the age distribution of the known reference sample (the skeletal series used to develop the estimation method). Although the distribution for the reference sample will not actually be duplicated, estimates will be nevertheless biased in the direction of the reference sample (Konigsberg and Frankenberg 1992). This *mimicry* is especially problematic, if a reference sample with a badly skewed age distribution has been used in the development of the method, as is the case for McKern and Stewart's (1957) method using the pubic symphysis. We know that this method used a reference population of American soldiers killed in the Korean War and, so, it has a rather unique, restricted age, sex, and race distribution: mostly young white American males, few of whom survived beyond the age 30. On account of this skewed distribution with an *oversampling* of young ages, any observed morphological changes in the pubic symphysis will appear characteristic of younger aged white males. The traits or components identified will in turn be mistakenly associated with aging in early adulthood, even if in the population-at-large the observed age characteristic is associated with individuals of middle or older age, age features will appear common in young men simply because more young men comprise reference sample. When this method is applied to other cases, like random *unknowns* selected from the distribution of current U.S. forensic cases, the ages estimated for these target skeletons will not be representative of their *true* chronological ages, but will be biased toward the skewed age-at-death distribution of the young soldiers making up the Korean War reference sample. The result is mimicry and age-at-death reconstructions that bear less, little, or no resemblance to the actual mortality profiles of the original populations, regardless of how well the practitioner applies the method. The trouble with mimicry is related to the method itself; the age distributions produced are merely artifacts of errors in method, as Bocquet-Appel and Masset (1982) have argued.

Maples' (1989) statement that "age determination is ultimately an art, not a precise science" (p. 323) is what Aykroyd et al. (1999) call a "cautionary tale for biological anthropologists when determining age in an adult skeleton" (p. 87). Although the importance of the *art* of *knowing* the skeleton, its system and elements, its variations and adaptations, and its patterns of senescence based on experience, should not be undervalued, Hoppa and Vaupel (2002b) remind us that "the concept of age estimation has, despite a variety of possible techniques, followed the same series of short steps" (p. 5): (1) assess skeletal morphology, (2) link skeletal morphology to chronological age through a reference collection, and (3) estimate age. Step 1 may well belong to the realm of art and experience, and so is subject to issues of error involving the observer(s), and the degrees of error associated with the skeletal elements themselves. Steps 2 and 3 bring to the fore, the last and perhaps most difficult problem to consider how to ground age estimation techniques in the most robust theoretical foundation, and in the best statistical practices. The considerable number of publications focused on testing, tweaking, and introducing new and improved osteological methods suggests that skeletal biologists recognize the inconsistency in estimation performance, and know that there is no single method, or best combination of estimation techniques that can produce ages at death for casework or paleodemographic reconstructions of age structures with 100% accuracy. The assignments of indicators to phases are best guesses, and the calculated ages from phase information and reference samples are best estimates. Results may be improved at two levels: (1) at the level of the observer, who can reduce bias or error by choosing and using methods thoughtfully, by acquiring more *aging* experience, and by replacing point estimates with age ranges that are sufficiently broad to capture true age, but narrow enough to aid in identification and (2) at the level of analysis by avoiding various statistical pitfalls (including compounding error using additive methods, using inappropriate reference samples, and ignoring issues of conditional independence).

Still, the nagging issue of the imperfect correlation between biological and chronological age promises a reality, in which the variability or error in age estimates can never be fully fixed. Consequently, the question of determining age-at-death should be framed in the context of probabilities (e.g., Gowland and Chamberlain 1999; Lucy et al. 1996; Schmitt and Broqua 2000). Despite the slight variation in indictors and methods, the practical application of a typical phase-based age estimation method to a reference population for method development yields a set of osteological data that demonstrate how skeletal morphology relates to known chronological age. The data tell what skeletons of which ages should fall into what phases, and forms the basis for morphological aging criteria (i.e., the rules and expectations for an age-phase method). This same osteological data can then be used to estimate the probability of observing a suite of skeletal characteristics conditional upon a known age. This is noted mathematically as $Pr(c|a)$, or the probability of c given a, where c represents the skeletal characteristic (as it is related to the morphological age-indicator phase, stage, or category), and a represents chronological age-at-death. Konigsberg and Frankenberg (1992, 1994, and 2002) have clearly explained, however, that the estimate of real interest is the age of an individual or group of individuals within a skeletal sample. To obtain the age-at-death estimates expected for forensic reports, and for modeling the mortality schedule of a population, we need instead to estimate $Pr(a|c)$: the probability that the unknown individual died at a particular age, a given evidence that his or her skeleton displays particular age-informative characteristics c. Although both probabilities, $Pr(c|a)$ and $Pr(a|c)$, appear at first glance to deal with the same things (ages, a, and aging characteristics, c), it is key to further calculations that we understand that they are not equal. $Pr(c|a)$ can be obtained from a thoughtfully chosen reference sample, whereas Bayes theorem must be used to calculate $Pr(a|c)$ from $Pr(c|a)$. While many of us would argue that Bayesian statistics are familiar to most anthropologists and may even "consider forensic anthropologists as being implicit Bayesians because they often do bring prior information to their cases" (Konigsberg et al. 2009, p. 84), solving for $Pr(a|c)$ is a complicated matter, as the calculations require information about the age-at-death distribution of the target population (finding $f[a]$), and a method to link the observed frequencies of age-related traits in the target population to the underlying probability distribution of these characteristics (defining the probability of these characteristics, $Pr[c]$). With these computational rules defined, we can see that $f(a)$, the probability distribution of ages at death of the skeletal under analysis, must be estimated before $Pr(a|c)$ can be calculated.* In Bayesian discussions, $f(a)$ is considered the prior distribution of ages at death, because it must be obtained before, or *prior to*, the estimation of $Pr(a|c)$. The consequence of this Bayesian approach is that, as Boldsen et al. (2002) explain, "Any individual age estimate is, in effect, a secondary by-product of the aggregate-analysis" (p. 77). Konigsberg and Frankenberg (1994) have identified this issue of order in estimation as *the* paradoxical problem in the implementation of the statistical methods necessary to obtain the desired age(s)-at-death estimates. Restating the guidelines for age-at-death estimation in a complex statistical framework begs a

---

* See the Rostock Manifesto in Hoppa and Vaupel (2002b).

theoretical question for practical application: if $f(a)$, the target age-at-death distribution, is required to generate the individual target ages, and these individual estimates can only be calculated for the target sample after obtaining $f(a)$, how can we produce the age-at-death distribution for the target population without the individual age estimates?

These statistical problems, however thorny, are not impassible and do not signal an end to profitable age estimation research, but instead call for change in focus—with attention being paid to the consequences of a poor theoretical and methodological framework. Solutions can be found in laying down improved statistical foundations and developing disciplinary protocol from these foundations that guides practitioners away from making poor choices in given conditions particular to the composition and context of forensic casework and individual research programs.

NEW SOLUTIONS TO SOME AGING PROBLEMS: IMPROVING AGE ESTIMATION

In 1982, a farewell was first bid to quantitative pursuits in age estimation in the face of what seemed to be insurmountable statistical hurdles: attraction to the middle-age mimicry, biased estimates, errors in method design and observer application, and poor theoretical foundations for age estimation methodology (Bocquet-Appel and Masset 1982). Since then skeletal biology has profited from a steady vested interest in age-at-death research that seeks to simultaneously isolate and remedy these aging problems in theory, method, and practice. The past three decades have been punctuated by significant high points in improved osteological and mathematical approaches that have been turning points in the shift from what was arguably an uncoordinated use of *standard* but unsatisfactorily defined age estimation techniques (Jackes 2000), toward a concerted effort in defining the best practices in the field (Ritz-Timme et al. 2000; Rösing et al. 2007). As senescent changes form the basis for adult age-at-death estimates, and the biology of aging is highly variable across populations, within [sub]groups and subjects, and through time and space, skeletally derived ages are always infected with some degree of error that can be improved but not removed. So for discussions of age estimation, based on the degenerative changes with maturity, we can collate the many problems (i.e., the eight points previously identified) associated with conventional adult aging methods, into the following "four basic analytical difficulties in adult age estimation, all of which become more critical in older skeletons" (Boldsen et al. 2002, p. 75).

1. How can we best tease out the most and optimal age-related information from the skeleton—by choosing one robust indicator or by combining many indicators distributed throughout the skeleton?
2. How can we most advantageously score areas informative of age to account for the complex architecture of the skeletal, and capture the most age-informative morphological variation in the indicators—using phase or component methods?
3. How can we obtain unbiased estimates of ages, given the potential obfuscatory effect of the reference sample on obtaining estimates that best reflect the true chronological ages of the target individuals, as well as the acknowledged problems of *attraction to the middle* and *age mimicry*?
4. How can we account for the uncertainty implicit within the estimation of adult ages at death?

The first two questions can be addressed by a discussion of the choice of indicator and method, whereas the last two can be resolved by the appropriate use of statistics.

Indicators and methods

The suitability of our approach to the estimation of age-at-death has both significant social and legal ramifications. Although in the past 30 years much has been written on the ways, aging methods and indicators affect our understanding of the lifespan and life history of past people, through the reconstruction of mortality schedules for ancient (i.e., archaeological, cemetery) populations in paleodemography (Howell 1982; Lovejoy et al. 1977; Molleson and Cox 1993; Ruff 1981), the more recent contributions of age-at-death estimates toward resolving issues of identification in medico-legal contexts are influenced by many of these same concerns, as well as by a set of disciplinary problems unique to forensic anthropology. No skeletal biologist would deny that the choice of both method and indicator influences how well we can produce age-at-death distributions for modern assemblages of many skeletal remains, and how much high-quality information we can bring to efforts in personal identification for unknown individuals in the aid of the medico-legal community. In response to these demands of forensic practice, we must strive for age estimation methods that fulfill a set of standardized criteria that uphold good science (Ritz-Timme et al. 2000). The *Daubert* ruling (*Daubert* v. *Merrell Dow Pharmaceuticals* 92–102, 509 U.S. 579, 1993) and its four requirements of admissibility serve as a general guide for standards in testing, publication, peer review, known error rate, and acceptance. According to Daubert, methods should be the following:

1. Rigorously tested, and data on accuracy and precision should be made available.
2. Known among the scientific community through peer review and publication.
3. Proven sufficiently successful to warrant acceptance as a best practice given by case conditions of use.

In order to better meet these standards, we first need to address the unique problems identified for the traditional methods for age-at-death estimation and respond specifically to the choice of indicator, and to the conventional use of phase-based approaches.

CHOOSING AN APPROPRIATE INDICATOR

Not all skeletal indicators of age perform equally well given the biological attributes of the samples (including sex, ancestry, pathology, life history, and modern versus ancient skeletons), particulars of the recovery conditions (forensic or archaeological, taphonomy, and degree of preservation, fragmentation, or commingling), and the tenets of aging methods, to which they are subjected (macro or microscopic, gross or metric, phase, or component based). The first step in choosing an appropriate indicator is to evaluate its applicability to two broad age cohorts—subadults and adults—and, sometimes, to two additional subdivisions—young adults and very old adults. The next step is to select from the available indicators those that have proven successful for use on skeletons that match that broad age class.

Methods applicable to trace juvenile, remains the developmental events associated with the formation and eruption of dentition, and the growth of the skeleton in utero, from birth through youth, and up to the point of maturity. On the other hand, adult aging methods capture the postdevelopmental degenerative changes that characterize individuals progressing, or *senescing*, from adulthood, to old age, to death. With the exception of the few sites of *late* epiphyseal union that begin their growth-related changes in youth, but do not complete their union prior to the second decade of life (e.g., some segments of sacrum and the sternal end of the clavicle), or the highly variable eruption of the third molars (e.g., anytime between teenage and early adult years, if at all), there is limited overlap between indicators that are helpful for estimating age in skeletons of *the growing* and *the grown-up*. This lack of overlap is attributable to the fundamental differences in the aging processes at work in subadults versus adults: development versus degeneration, growth versus senescence, and buildup versus breakdown, respectively (Hillson 1998; McKern and Stewart 1957; Scheuer and Black 2000; Shirley-Langley and Jantz 2010; Webb and Suchey 1985). Because there is not the same regularity in age-related changes in adulthood as in the synchronized development of youth, *and* because this irregularity in bony morphology increases with age, the selection of an appropriate indicator is most difficult for skeletons of postmaturity.

As many highly variable senescent changes occur throughout the adult skeleton, researchers have amassed many indicators to choose from, and continue to seek out new markers, or refine scoring techniques for traditional indicators that allow for better quantification of the subtleties of aging. Each of these markers has their own weaknesses, such that, choosing a good or *the best* indicator is as difficult as it is essential. For example, recall that few macroscopic, phase-based methods can estimate age accurately for individuals over 45–50 years because of the irregularity in degenerative change, apparent cessation in quantifiable activity in the indicator, or lack of the necessary sensitivity in the scoring systems to capture the subtleties in age-related change that may characterize skeletons of older years. Also remember that chronological and biological age cannot be linked without some degree of uncertainty, biological aging is highly variable, evidence suggests that patterns of skeletal aging are not consistent among or within groups, and that different standards are required for males and females and for individuals of different populations and life history.* Furthermore, many age estimation methods have been developed using reference populations (e.g., autopsy, donated, and museum collections) that have been sampled for a host of reasons, including their representation of the trait(s) of interest, but forensic casework and population-based research is to some degree always dependent on materials that have undergone taphonomic alteration, and are subject to preservation error and directional biases related to the effects of factors such as deposition, decomposition, and destruction (Komar and Grivas 2008; Meindl et al. 1990; Usher 2002; Wilson et al. 2007, 2008). From these few examples, we can see how important it is that the skeletal biologist makes use of a robust indicator for age estimation (Aykroyd et al. 1999). An appropriate age indicator should have the following:

---

* For more information, see Schmitt et al. (2002), Konigsberg et al. (2008), Berg (2008), and Hoppa (2000).

1. Be highly correlated with chronological age. This relationship should be validated by statistical means.[*]
2. Be invariant, so that the rate of change is consistent across all samples (Müller et al. 2002). It is neither easily subject to environmental alterations, nor significantly affected by metabolic and nutritional changes.
3. Exhibit age monotonicity (Milner et al. 2008). The indicator exhibits a clearly identifiable sequence of morphological change with age, and these changes occur within intervals, and not throughout the entire lifespan.
4. Be widely applicable across populations and between the sexes. If variation is found to exist, per-group corrective criteria should be developed and thoughtfully applied given what is reasonably known about the individual, for example, sex, ancestry, *race*, or ethnic affiliation.[†]
5. Be reasonably resistant to the effects of natural filters. Taphonomic processes may cause preservation bias and oversampling of certain age cohorts or social groups (Konigsberg and Frankenberg 1994; Waldron 1994; Walker et al. 1988; Wood et al. 1992).

The choice of the indicator and the method may be seen as steps in the age estimation process that are inextricably joined: if the state of the remains dictates what indicator(s) can be most reliably evaluated, then the approach to age estimation is tied to the methods developed for the available indicator(s). We are often faced with the conundrum of working with what we have. Klepinger (2006) has offered her view on *best* practice: "I am inclined to disregard cranial suture closure when other markers can be used. There are, however, occasions when the cranial sutures are all there is, and then the best option is to employ one of Nawrocki's (1998) regression formulae for various combinations of endocranial, ectocranial, and palatine suture closure" (p. 58). The obvious question is if a poor estimate is better than no estimate; the answer is most profitably framed in terms of observer experience and specifics of the case—so the decision to estimate is largely philosophical, situational, and variable. If many good indicators are available, then we are fortunate to have the option of being selective. In these instances, it is best to select robust indicators and methods that are free of the phase-aging problems and to combine these in such a way that captures the greatest amount of variation in the skeleton and, in turn, provides the strongest and most biologically informed estimates.

CHOOSING AN APPROPRIATE METHOD

To increase accuracy in age-phase techniques and broaden their applicability to diverse groups and situations, most methods have been rigorously tested, and improvements have been made accordingly. It is important that, as practitioners, we are not only educated in *the standards*, but that we are also active participants in the theoretical and technical conversations on the new directions in age estimation research. We should keep abreast of the advances in the method and theory of age estimation so that we do not necessarily choose the most common or most conventionally applied method when other approaches may be more successful or appropriate, given the constraints of the unknown remains recovered. Let us again take the pubic symphysis as an example of a standard indicator that is subject to both conventional and new methods. Although many forensic anthropology textbooks and lab manuals still provide instructions for Todd's system (Bass 2005; White and Folkens 2005), and it remains a part of the scoring procedure for casework (Buikstra and Ubelaker 1994; Forensic Databank data forms[‡]), over five decades of testing have shown that the revised, six-phase, Suchey–Brooks method produces more accurate results (after Suchey and Katz [1986] or Brooks and Suchey [1990]). For forensic casework in particular, Klepinger et al. (1992) have reported supportive findings from a test of the McKern–Stewart, Gilbert–McKern, and Suchey–Brooks methods on a large, diverse, contemporary autopsy sample: "The results suggest that the Suchey–Brooks methods are superior in forensic applications and that racially specific refinements for males should be used" (p. 763). Recent tests that include the pubic symphysis as part of the TA method have shown potential for greater improvement in results: Wilson and Algee-Hewitt (2009) scored known forensic cases from the WM Bass collection and obtained highly repeatable estimates that corresponded closely to known ages, and

---

[*] See Kemkes-Grottenthaler (2002, p. 56) for a summary table of coefficients of correlations between common age indicators and ages at death.
[†] See Ubelaker (2008), for a collection of studies that addresses the applicability of methods developed in the United States to individuals in the Balkan area and considers females separately. See Usher (2002) for available reference collections.
[‡] These forms are available at http://web.utk.edu/~fac/databank.shtml.

yielded small interobserver error, regardless of sex or ancestry; Buikstra et al. (2006) also found that the TA scoring procedures are "replicable and are effective in resolving historical issues" (p. 59). These data would suggest that there is merit in turning to traditional indicators as many have held fast for good reason, but that stepping outside of the traditional applications, and looking toward more methodologically rigorous and holistic techniques may yield improved results. The success of TA (as defined by Boldsen et al. [2002]) is in part attributable to the fundamental design of its scoring scheme, which does not make use of the typical age-phase approach but instead is component-based (Milner and Boldsen 2008). Before we can move forward and fully appreciate the contributions of new methods, we need first to revisit the weaknesses in using any phase-aging approach and consider specifically the problem of *fixing* the *phase* itself.

The phase-based approach assumes that the complex morphological changes observed in a single indicator, and representative of a single phase, are synchronized in their occurrence: they are progressive, unidirectional, and metamorphosing in unison. This is, however, rarely the case for our complex biological reality. Indeed, the range of morphology varies considerably at any given age because different parts of a single bone, region, or indicator change at different rates and in different ways. As phase-based methods require that all of the bony characteristics (i.e., the complex senescent changes) used to define a particular phase are fully observable in order to obtain an unambiguous classification of the indicator, there is (1) a loss of clarity in the indicator-to-phase assignment when the observed patterns shows variation away from the fixed definition of the phase, (2) a disregard for any age-informative changes that may occur as the bony features transition through stages of change *between* the phases, and (3) a failure to deal with the complications in scoring resulting from cases that span several phases by retaining some features of young age, or showing early progression in features indicative of older age (a problem noted for Suchey–Brooks phase III by Suchey and Katz [1998] and again by Kimmerle et al. [2008c]). To mitigate the problems of *phase* aging, we may choose to redefine our methods so that they use broad morphological categories. This can be done by reducing the number phases, relaxing the criteria to include more variation, and widening the estimated age intervals. Unfortunately, widening the intervals and accepting more variation simply introduces different obstacles to obtaining the best possible age estimates. In creating phases that cover large areas of the lifespan, like Suchey–Brooks phases IV (26–70 years) and V (25–83 years), we lose information on age, encounter the problem of overlapping phases, and reduce the method's utility in instances when, for example, estimated age ranges are to be used for trimming down missing persons' lists to only possible matches. A more effective approach may be to make use of methods that do not require the classification of the skeletal indicator into a phase. Methods using dental metric features, such as tooth root translucency and periodontal recession, do not rely on adult phase-aging techniques, and have the added benefits of narrower age ranges, and of capturing age-informative information for older individuals (Prince and Konigsberg 2008). Teeth are also durable, more resistant to taphonomic degradation than other bony indicators, and sometimes the only remains recovered at forensic case sites. Unfortunately, recent tests of interobserver error using dental metrics have shown significant differences in root height, periodontal recession, and translucency of the root taken by similarly trained observers (Kimmerle et al. 2008c). Consequently, the utility of the well-known and widely used Lamendin technique has been questioned given findings on interobserver error (Klepinger 2006; Prince and Ubelaker 2002). In addition, dental methods, like single indicator phase-methods, are limited regionally; when used alone, potential age information from the rest of the skeleton is disregarded. We are wise, therefore, to look forward to approaches that use a phase-free system, allow for flexibility in the areas analyzed, capture considerable variation when scoring, and allow for easy combination.

One such approach is the previously mentioned method of TA. TA is a robust alternative to phase-aging methods that permits the inclusion of multiple indicators when estimating age. Although TA is a new method (Boldsen et al. 2002), the basis for its scoring system can be traced back to McKern and Stewart's (1957) and Gilbert and McKern's (1973) component approaches for the pubic symphysis, which are in keeping with the common practice of scoring cranial sutures in segments, and obtaining a composite score (see Meindl and Lovejoy 1985). Unlike phase-aging methods (e.g., the Todd or Suchey–Brooks pubic symphysis methods, Lovejoy's method for the auricular surface, and Iscan's sternal rib end method) that rely on the classification of a single indicator in its entirety into one of several predefined categories, component scoring treats the skeletal marker as a complex of features. These features, the *components*, are scored individually, and each component is coded based on a series of stages. The estimation procedure underlying TA is a parametric method that provides information on the timing of the *transition* from one stage to the next, that is, the probability that an individual has attained a certain phase, given age and the mean age of transition. The general statistical procedure can be applied to any indicator, assuming that (1) phases do not overlap, (2) phases cannot be skipped, and (3) progression through phases is unidirectional (i.e., indicator states move from one phase to the next and never backward).

Milner and Boldsen (2008) have chosen three known age-informative markers (cranial sutures, pubic symphysis, and auricular surface) to include as part of the coding criteria for the TA method and have devised a component-based scoring system for each of these markers. For the pubic symphysis, for example, they identified five components such as: symphyseal relief, symphyseal texture, superior apex, ventral symphyseal margin, and dorsal symphyseal margin. Each of these components is coded separately given predefined characteristics, so that the first component of the pubic symphysis, symphyseal texture, is evaluated on the following four unidirectional stages: (1) smooth/fine grained, (2) coarse grained, (3) microporosity, and (4) macroporosity. Counting all the TA scoring criteria for left, right, and midline indicators, the skeleton is evaluated on 171 characteristics, spread out over 19 components. As such, a multicomponent method better captures the complex biological changes and large degree of variation in the human skeleton, and, in turn, provides a more holistic understanding of skeletal aging patterns. As the TA method is designed to incorporate all or some of the components and additional indicators may also be added, it provides a flexible scoring framework that may prove particularly useful in situations with fragmentary remains. Validation work by Boldsen et al. (2002) and independent analyses by Wilson and Algee-Hewitt (2009) have shown that the quality of the age estimates can be improved by simply increasing the number of indicators included in the analysis. Moreover, they found that the effect of knowing or changing sex and ancestry on the age estimates was reduced, as more indicators were included in the analysis. Boldsen et al. (2002) have also demonstrated that TA makes it possible for the user to code for *old age* traits, and to obtain age ranges for individuals comprising the right extreme tail of the age-at-death distribution (i.e., those cases that would otherwise have been lumped in an open-ended terminal phase such as "50+"). Being able to better address the skeletons of older cohorts is a critical step forward in our methodological progress, and in our ability, respond to changing demands in casework, as the elderly are the fastest growing subgroup of the United States population, and similar increase is noted for the very old (i.e., age 85+) in many other countries (Harper and Crews 2000). Adopting the TA component scoring, phase-free, multiindicator method allows us for the first time to combat at once several of the key issues with adult age-at-death estimation. We can deal more effectively with the entire mortality distribution, diverse samples, and remains of varying degrees of preservation in both forensic and archaeological contexts, in an easily applied and methodologically sound way.

Statistics

Quantitative methods in anthropology have always been hotly debated for their contributions to understanding skeletal characteristics and their relationship to worldwide population, subgroup, and individual levels of variation in modern, archaeological, and evolutionary contexts. Yet, never before have sound statistical approaches to understanding patterns of human variation and methods for detecting the unique biological attributes that identify each of us as individuals been more important or more profitable, particularly in the forensic sciences. The modern technological revolution not only enables computationally intensive quantitative approaches but, in doing so, also demands more precise tools, accurate methods, reliable applications, and greater accountability in scholarly output. The importance of science to the biological profile and for the estimation of age-at-death is apparent when we consider the increasing contributions of the community of forensic anthropologists and skeletal biologists to research in areas of medico-legal concern, the criteria of the three rules of law in the United States (the Federal Rule of Evidence 702, the *Frye* test, and the *Daubert* ruling), and the increasing role of forensic practice and testimony in singular cases, mass disasters, and international humanitarian work (Blau and Ubelaker 2009b; Klepinger 2006; Komar and Buikstra 2008; Steadman 2009). Ubelaker (2008) has aptly commented on our motivation for case-oriented research in response to forensic anthropological work in the former Yugoslavia: "stimulus consists of penetrating questions directed toward the forensic anthropologist expert witness in a legal proceeding. Such questioning of applied methodology in discussion of case testimony can represent a powerful incentive to address key issues that seem vulnerable" (p. 606). For these reasons, anthropologists have been working to refine probability-based techniques that are applicable to standard skeletal methods of personal identification and provide verifiable estimates and evidence (Christensen 2004, 2005; Konigsberg et al. 2008, 2009; Rodgers and Allard 2004; Steadman et al. 2006). The most gainful solutions to the issues affecting the estimation procedure and the quality of the estimates are found in improved statistical methods that address issues associated with regression analysis, reference populations, and the estimation of ages at death using a robust Bayesian approach.

PRIOR DISTRIBUTION PROBLEMS

Although regression in the form of inverse calibration is a common approach used in anthropology research, it is also among the most problematic and prone to misuse given that an appropriate prior distribution is required for unbiased age estimation. A good *reference sample* or *prior* should have an age-at-death distribution similar to the distribution of ages expected for the *target sample* (i.e., the unknown remains for whom the forensic anthropologist must develop an age estimate). Usher (2002) provides a detailed discussion of available skeletal collections, with known-age and sex that may be appropriate for use as informative priors given the different conditions and contexts of the unknown skeletons, and the goals of our individual research on age-at-death estimation.* Advances in collection development in the area of modern forensic cases through the growth of donated skeletal collections representing the diversity of the modern American population lessens this concern by providing comparative skeletal materials for use in U.S. contexts (e.g., the WM Bass donated and forensic skeletal collections housed at the University of Tennessee and the Maxwell museum's documented skeletal collection housed at the University of New Mexico). Age-at-death information from the University of Tennessee's Forensic Anthropology Data Bank (FDB) may also serve as an appropriate forensic prior as it stores a large sample of demographic and skeletal records for recent (since 1986) documented cases submitted by practicing forensic anthropologists from all across the United States (Ousley and Jantz 1998; http://web.utk.edu/~fac/databank.shtml). Although we may accept these reference samples as reasonable representations of contemporary American casework, their applicability to future casework, or to analyses conducted outside of the United States is unknown. As a cautionary example, Wilson and Algee-Hewitt (2009) tested the suitability of using U.S. national homicide data as an informative, prior for the cases documented in the FDB and found that, although homicides do contribute to the average forensic anthropologist's case load and both samples are similar in time and space, the age-at-death distributions of the two groups were not sufficiently similar for unbiased use. This suggests that making what may seem to be logical assumptions on sample similarity in the absence of appropriate testing and choosing *best fit* reference collections can have compromising effects on our age at death results. We have to agree with Usher (2002) that marrying populations with priors is "one of the most overlooked, but basic, sources of error in skeletal age estimation" (p. 29).

CALIBRATED APPROACHES

A more forward-thinking approach to this problem of reference samples is to devise statistical ways to circumvent the required external age-at-death distribution. Classical calibration can be used in place of inverse calibration when no suitable prior is available for the target sample, when age estimation is applied to an untested population, for which differences in senescent patterns may be a concern or are unknown, or when determining the structure of the age-at-death distribution of the target sample is not possible. Classical calibration assumes an uninformative prior for age-at-death and produces maximum likelihood estimates (MLE) of age by regressing the dependent variable (x, the age indicator) on the independent variable (y, the age), and then solving for age (Konigsberg et al. 1998). The prediction interval for classical calibration is larger than that from inverse calibration, so the estimated age ranges are wider, but the real age value is more likely to be captured within the expanded confidence interval and, importantly, the results are unbiased. In the context of applying TA, this is what Boldsen et al. (2002) considered a uniform prior distribution. This solution is not, however, without problems as this *uniform*, *flat*, and *uninformative* prior is not reflective of real mortality distributions: we must make the assumption that each skeleton in the target sample has an equal probability of being all ages (Hoppa and Vaupel 2002b). The uniform prior has been variably criticized for putting heavy weight on very old, and, so, less likely ages at death (Di Bacco et al. 1999), it was shown to overestimate old ages when compared to an informative prior (Boldsen et al. 2002), and it has been considered wasteful for disregarding all of the information on the age structure of the target sample (Konigsberg and Frankenberg 1992). In instances when a prior is in question, the most satisfying solution is to incorporate models of mortality (also known as *hazard models*) into our approach to the estimation of the probability distribution of the ages at death in the target population.† In

---

* See also the web-based "Reference Collection Database" at http://www2.potsdam.edu/usherbm/reference.
† See Hermann and Konigsberg (2002), Holman et al. (2002), and Love and Müller (2002) for more information about hazard modeling.

practical applications, when applying TA, age at death can be calculated from the age-at-death distribution estimated using hazard analysis and the ages of transition; the estimate for an individual skeleton can then be obtained by generating the highest posterior density for each stage. A series of publications by Konigsberg and colleagues (2008) (see DiGangi et al. 2009; Kimmerle et al. 2008b; Prince and Konigsberg 2008; Prince et al. 2008) have demonstrated the promise of this unbiased approach by obtaining reliable individual age-at-death estimates for international forensic casework, using a variety of scored indicators, including the pubic symphysis, first rib, and dental roots and wear.

GETTING BETTER WITH AGE: BEST PRACTICES IN AGE ESTIMATION

Returning to Maples' (1989) assertion that "age determination is ultimately an art" (p. 323), it is remarkable that the most productive approaches to our acknowledged troubles with aging are those that recognize the importance of precise science to concerns of skeletal biology, and actively engage its theoretical and biostatistical principles. A significant turning point in this paradigm shift from *art* to *science* came in 1999 and 2000, when the Laboratory of Survival and Longevity at the Max Planck Institute for Demographic Research in Rostock, Germany, held a workshop on mathematical modeling in paleodemography, whose goal was to reach a consensus on how best to tackle the unresolved issues in age-at-death estimation. A theoretical framework for the estimation of age distributions from skeletal populations emerged out of these meetings and its synthesized form became what is now called the *Rostock Manifesto*. The publication of this guide for age estimation in paleodemography marked not only the first collective effort to offer clear, codified, and statistically rooted solutions to many of the problems of estimating age-at-death from skeletal samples, but was also the first step in defining a sound theoretical framework for answering age-at-death questions, relevant to forensic anthropologists working in diverse, modern contexts (Hoppa and Vaupel 2002a, 2002b). The influence of the *Manifesto*'s guidelines is palpable in the response of the anthropological community to medico-legal questions of identification methodology. The organization of the 2004 American Academy of Forensic Sciences symposium titled "Estimation and Identification in American and International Populations" (Dallas, Texas) and the subsequent papers published in *Journal of Forensic Sciences* in 2008 clearly demonstrate the applicability of the *Manifesto*'s theory to contemporary concerns of significant humanitarian consequence. In response to the rapidly changing demands in forensic anthropological practice and output, we are witness to examples of *good practice* and recommendations for *best practice* in the application of laboratory methods for assessing age of skeletal remains, cadavers, and living humans (Houck et al. 1996; Ritz-Timme et al. 2000; Rösing et al. 2007; Schmitt et al. 2006; Steadman 2009). As Ritz-Timme et al. (2000) have explained, there is an immediate need "to find common standardization, calibration, and evaluation or to develop means of quality assurance for methods of age estimation … [e]fforts in these directions are necessary to guarantee quality standards and adequate answers to the important legal and social issue of age estimation in forensic medicine" (p. 129). For the United States, these concerns for principles of forensic practice are in the process of being addressed through the efforts of the newly formed *Scientific Working Group for Forensic Anthropology*, or SWGANTH (www.swganth.org; see also Adams and Lothridge 2000). The SWGANTH is a board of forensic professionals that "aims to develop consensus guidelines, that is, *best practices*, for the discipline of forensic anthropology, and to disseminate those guidelines to the broader forensic community … to identify and codify existing standards, or, where clear standards do not exist, to formulate and establish them … to examine targeted issues for the purpose of identifying what is best practice today and what paths should be followed in the future" (www.swganth.org).

As with any scientific study, our (age-at-death) results are only a good as our (skeletal) data. It is, therefore, essential that we are knowledgeable in the options available for the collection of age information from skeleton and are wise to the practices that may yield the most accurate and helpful data, and, in turn, the most reliable estimates that will ultimately contribute *age* to the case of individual identification. There are no adequate substitution for the original method reports, when learning how to properly identify indicators and areas of interest, how to apply the system of evaluation, and how to obtain the ages and associated ranges following the method's protocol. Nonetheless, when we also consider that our assessment may be communicated in the courtroom, it is essential that we understand the types and degrees of error associated with each technique, and can provide documented statistics on method accuracy and reliability. We may instead choose in the final *estimation* stage to employ alternative computational techniques (i.e., the TA method) that we know to be more statistically robust based on validation tests.

Although forensic anthropology is undergoing a paradigm shift toward rigorous scientific studies, we should embrace this [r]evolution and "direct the field's development into the most professional, efficient, and profitable pathway" (SWGANTH 2010). We must appreciate that methods once held as *gold standards* may be improved (given the introduction of new technologies, advances in statistics, or greater understanding of the proximate causes of aging, or the patterns of skeletal variation) or altered in response to the changing demands of forensic casework (accounting for secular change, population differences, and the expectations of the medico-legal community). We must also understand that these modifications may require the relearning of evaluation procedures, in order to obtain more accurate, reliable, and applicable results. Observer experience is a crucial factor in accurate age-at-death estimation; this is also true for any *applied* skeletal method that relies on gross visual assessment (i.e., sex and ancestry estimation, assessment of trauma, and pathologies). As forensic *practitioners*, we must *practice* in the osteology lab and make use of reference skeletons with known demographic data to measure our own proficiency across the full spectrum of available age estimation methods and assemble a reliable toolkit for case evaluation.

Regardless of age cohort (i.e., 0–3, 20–35, 50+ years) or general classification (young to old, juvenile, or adult), the theory for age-at-death estimation is founded on our knowledge that the skeleton undergoes changes throughout the lifetime, from our first days into our last years. These changes for the young skeleton are uniquely related to growth. The processes of bone formation and development occur along a unidirectional trajectory until homeostasis, or the cessation of growth, is achieved. This moment of stability (i.e., the theoretical point-in-time without skeletal change) marks the pivotal transition between the progress in the juvenile and the regress in the adult skeleton. Growth, therefore, is immediately followed by the lengthy period of remodeling and degeneration that characterizes the majority of our lifetime and continues throughout our postadolescent years from maturity until death. As the two broad age stages of our lifespan, subadulthood and adulthood, undergo oppositional bony activities in the form of skeletal buildup for the former and breakdown for the latter, the techniques that we use to assess individual age must consider the difference in these processes. Consequently, estimation methods for age-at-death are tailored to capture the types of morphological change that are distinctive of each stage.

Age estimation for the juvenile

As metamorphosis in the juvenile skeleton is inextricably tied to growth, biological age is inferred from the degree of maturation observed for bone (i.e., appearance and fusion of epiphyses, morphology of ossification centers, and skeletal dimensions), teeth (i.e., formation, eruption, and wear), or both tissue systems combined. The resulting physiological age estimate is translated into chronological age following published standards obtained mostly from modern clinical growth studies of living children.

As the growing skeleton undergoes regular changes, indicator events for age estimation occur sequentially, and are relatively consistent in timing and duration. Because of this regular progression and the short time period, in which human growth occurs relative to the complete lifespan, more accurate age-at-death estimates may be produced for immature skeletal remains than for adult lifespan. That is not to say, however, that age-at-death estimation methods for immature remains are without their own set of problems. The limited size and composition of reference samples, the frequent application of radiographic standards directly to skeletal and dental remains, the increased variability in the rate of aging with the increase in the age of the cohort, the inaccuracy in sex and ancestry estimation in prepubescent skeletons, the known sex and geographic variation in the rate of maturity, and recent secular changes in skeletal maturation in some populations, all represent sources of error in estimation and potential weaknesses in our current methods.

The most common and easily implemented approaches to age-at-death estimation for the juvenile skeleton involve the assessment of the epiphyses (e.g., appearance and/or timing of fusion), the development of dentition, and the use of skeletal dimensions.* Ossification centers, epiphyses, teeth, and osteometrics are convenient macroscopic indicators of age that can be easily accessed either by gross inspection of dry remains or observed radiographically. For example, age estimate ranges can be obtained by comparing the pattern of epiphyseal fusion throughout the skeleton.†

In the post-cranial skeleton, fusion occurs at the growth plate between a primary and secondary ossification center, called an epiphysis. This type of fusion follows a set of forward steps that facilitates interval-based estimation: the excretion of the growth hormone ceases, the growth plate ossifies, and the epiphysis (secondary ossification center)

---

* Thorough explanations of these and other methods, as well as the comparative age standards are available in Ubelaker (1987, 1989a, 1989b), Smith (1991), Buikstra and Ubelaker (1994), Hillson (1998), Saunders (2000), Scheuer and Black (2000, 2004), Lewis and Flavel (2006), Warren (1999) speak specifically to forensic contexts.
† For detailed charts, see Scheuer and Black (2000) or Buikstra and Ubelaker (1994).

is united with the metaphysic (primary ossification center), and longitudinal growth stops. We can obtain the age information necessary for constructing an estimated age range by (1) observing the suite of the potential points of union in the entire remains, (2) locating the most recent or *last* union, and (3) identifying the *next* union expected to occur. An age interval can be formed by taking the age standards associated with both the *last* and *next* sites, and assigning the lower age limit to the former and the upper-age limit to the latter. It is important to be aware that evaluations may vary based on the method of inspection (i.e., dry bones versus X-rays). When using radiographs, for example, Cope (1946) observed persistent fusion lines on the distal femur and proximal tibia in an individual over age 71; whereas Hillson (1998) noted that the developing dentition will appear later on film than in direct inspection of dissected tissues, as enamel and dentine are first deposited as a poorly mineralized matrix. White and Folkens (2005) also caution us that *regular* events do not mean *constant* events, and, thus, it is important to make use of the most applicable standards in order to reduce the confounding effects of inconsistencies related to secular change, sex, and population differences.

Let us consider, for example, the case of sex and rates of fusion. Female skeletons are known to mature faster than males, so that epiphyseal fusion typically occurs at a younger chronological age in girls. As the expected ages per bone region or by epiphysis are known to vary between the sexes, any estimate that relies on pooled sex standards will be less informative of true age. In other words, combining males and females increases the amount of variation, which in turn necessitates broader intervals. If we take age data for fusion of the humerus, as reported by Scheuer and Black (2000), we see that the distal epiphysis fuses for females at 11–15 and 12–17 years for males. A simply combined sex interval would estimate age at 11–17 years, although, this range should capture the true age, regardless of sex, it encompasses a considerable portion of the juvenile lifespan and, in its breadth, may be unhelpful in making a personal identification. The opposite problem is of equal concern. Although these sex-specific standards are available from documented clinical studies, as yet, we are not able to estimate sex (or ancestry) from the subadult skeleton with any statistical certainty, at least for the prepubescent years. Even when sexually dimorphic features are recognized (i.e., those features that differ significantly between males and females), the degree of overlap between males and females is often too great to recommend their use for consistent sex identification (e.g., Holcomb and Konigsberg 1995, for the sciatic notch). An incorrect assignment of sex would erroneously truncate the age interval estimated from the sex-specific epiphyseal information, thereby making the interval too low for a *true* female or too high for a *true* male. Reporting such a narrow interval may result in the exclusion of the correct individual from the list of possible matches.

Recent studies have also demonstrated that different epiphyses provide different levels of accuracy and precision for different individuals. This information attests to considerable variation in developmental timing across ancestry groups and secular trends in maturation (Crowder and Austin 2005; Schaefer and Black 2005; Shirley-Langley and Jantz 2010). Any of our estimates (wide or narrow) may be compromised further by these population-specific patterns and maturation trends (Klepinger 2001) without the appropriate choice of reference standards, and improvements in our ability to estimate sex and ancestry for juvenile remains (Buck and Strand Vidarsdóttir 2004; Cardoso 2008; Cardoso and Saunders 2008). As skeletal development is ultimately an individual process, factors affecting the single skeleton may also result in the following unexpected patterns: for some individuals complete fusion may lag, extend into adulthood, or never be fully achieved. For this reason, the extent of closure is evaluated using a progressive scoring system: unfused or nonunion, fusing or uniting, fully fused or complete union. Despite this variation in union, the assessment of epiphyseal fusion is preferable to epiphyseal appearance for practical purposes. For example, the common clinical method of assessing hand and foot maturation based on the appearance of secondary ossification centers are not advisable for recovered skeletal remains (e.g., Greulich and Pyle 1959; Tanner Whitehouse System, TW-2 1975, or TW-3 2001). As separate partially formed epiphyses are small and irregular, they may be damaged easily or overlooked in the field, and are often difficult to reassociate with a particular bone in the lab.

Methods estimating age from dentition follow a similar process to the evaluation of epiphyseal union. Two popular dental methods proposed by Demirjian (1978, 1990) and Moorrees et al. (1963a, 1963b) require (1) the identification of the physiological state or the tooth development stage by direct visual assessment or from radiographs and (2) the association of the observed state or a calculated *maturity score* (Smith 1991), with known age standards obtained from reference samples. Most skeletal biologists recognize, however, that dental development more closely reflects chronological age than skeletal maturation. This is likely due to the fact that markers of skeletal maturity are more easily influenced by external factors such as activity, social and physiological stress, and nutrition, whereas tooth formation is under greater genetic control (Cardoso and Padez 2008; Saunders 2000). Although sex and population variability in teeth is known (Chaillet et al. 2005; Demirjian and Levesque 1980; Tomkins 1996), dental development is typically accepted to be quite regular regardless of subgroup (i.e., females are, on average, only 1–6 months more advanced in development than males) and published dental charts have been shown to be reliable, easily interpreted, and simply applied.* Ubelaker's (1987, 1989a, 1989b) revision of Schour and Massler's (1941) original dental development diagram, although originally designed for use on native Americans materials, is now a recognized worldwide standard (Buikstra and Ubelaker 1994; Hillson 1998). Age estimates from tooth development are preferred to those derived from the assessment of alveolar emergence or gingival eruption as they consider the whole development of the dentition and, so, avoid potential estimation errors related to variation in the timing and sequence of singular tooth eruption, or sex and population differences (Hillson 1998).

Subadult age may also be calculated using osteometrics, including individual bone (diaphyseal) lengths and total skeletal size (Fazekas and Kosa 1978; Scheuer and Black 2000). This method proves less reliable as bone size increases, and there is considerable metric variation among populations (Saunders 2000; Ubelaker 1989a, 1989b). Its utility is therefore limited to skeletons of young ages (<10 years) and is best and most widely used for fetal, infant, and early postnatal skeletons and in cases, for which population-specific reference data are not available (Byers 2005).

Age estimation for the adult

The identification of *adult* in biological or osteological terms is often tricky because there exists a brief window, during which the young adult may be subject simultaneously to the progressive and regressive changes that characterize skeletal immaturity and maturity, respectively. Evaluation of the *adult* skeleton may include, therefore, the examination of the degree of union at sites whose fusion is known to extend into postadolescent years and the assessment of late-stage dental development. Recent work on the utility of these markers of young adulthood (e.g., clavicle, vertebrae, ribs, sacrum, anterior iliac crest, and third molar) in forensic casework has yielded mixed results (Albert 1998; Kunos et al. 1999; Mincer et al. 1993; Rios et al. 2008; Shirley-Langley and Jantz 2010; Webb and Suchey 1985). Shirley-Langley and Jantz's (2010) reanalysis of the age-related change in the medial epiphysis (sternal end) of the clavicle shows promising results for quantifying this change using a robust statistical approach and simplifying the scoring system considerably. They also provide convincing evidence for secular change in the maturation of the clavicle, as well as sexual dimorphism in the onset of fusion (with females maturing earlier). Their results demonstrate the importance of using appropriate, modern skeletal standards for the reliable use of the sites of late-stage development for age estimation in forensic cases. Patterns of age-related fusion in the sacrum reported by Rios et al. (2008) also demonstrate the same sex difference in fusion. This accumulating evidence suggest that the accelerated maturation observed in young females in ontogeny holds true for the developmental changes occurring in the early adult years, and so should encourage careful study of sex differences in other maturation indicators, as well as inspire further research in sex estimation from the growing skeleton. Rios et al. (2008) do not obtain, however, the same level of precision in age estimates as those offered by other adult maturation methods, and, thus, they advise that sacral fusion is best used for the initial assignment of skeletons to broad age groups. These results may be attributed to the later ages at which final fusion was observed in this and other studies (into the third decade by Rios et al. [2008]; but even as late as 54 years by Belcastro et al. [2008]), as we have observed that skeletal variation increases with the increase in the age at which the event occurs.

Like the late-fusing bone markers, the development of third molar (M_3) has also been used to provide estimates of age that span the transitional young adult period, from approximately 15 years into the second decade. However, M_3 has been long recognized to be the most variable tooth in size and shape, in timing of formation and eruption, and even in its presence or absence. For these reasons, the M_3 has been tested for use in modern forensic odontology casework. Results show that M_3 has little (if any) sexual dimorphism in its timing, that the degree of associated variation is in keeping with the amount expected, given the positive direct relationship between variation in dental development and increasing age, and that distinct root development stages exist for white Americans and can be identified using diverse samples and independent examiners (Garn et al. 1962; Hillson 1998; Mincer et al. 1993). Despite these findings, the considerable variation in the M_3 does not recommend its use as a sole age indicator. More accurate dental estimation methods are available that provide coverage for this same period of early adulthood, offer flexibility in tooth choice, and have been shown to be successful in forensic cases with target ages spanning the adult lifetime. These include age estimation from dental wear patterns (Miles 1962; 1963, 1978; Prince et al. 2008; Smith 1984a,

---

* For a list of population-specific dental studies see Lewis and Flavel (2006, p. 246).

1984b), and dental metric features for single-rooted teeth, including root translucency and periodontal recession (Lamendin et al. 1992; Prince and Konigsberg 2008; Prince and Ubelaker 2002).

Although Shirley-Langley and Jantz (2010) note that "young adults constitute a large portion of forensic casework" (p. 571), most conventional adult age-at-death estimation methods are less concerned with patterns of late development. Instead, they focus on tracing the progressive, degenerative changes in skeleton that characterize senescence throughout the entire postmaturation period of adulthood, which constitutes the largest segment of our total lifespan and the most variable in age-related morphology. Standard osteological methods for obtaining adult age estimates involve the macroscopic, visual assessment of bony features using predefined criteria. When the goal is identification, as in the forensic case, the typical protocol may include the following steps: (1) score all available elements using either phase and/or component methods to obtain individual or combined indicator results, (2) supplement the indicator-based assessment with an overall evaluation of *age* using generalized degenerative changes distributed throughout the skeleton (which may include scoring for degrees of joint wear, macro- and microporosity, and/or osteophytic development), and (3) come to a consensus age based on multiple points of evidence, preferably using robust Bayesian methods and obtaining probability-based estimates of the individual's age-at-death (see the recommendations by Buikstra and Ubelaker [1994], Baccino and Schmitt [2006], Komar and Buikstra [2008], Ritz-Timme et al. [2000], or Rösing et al. [2007], or Komar and Buikstra [2008]). We may choose to reverse the order of steps 1 and 2 and first evaluate the overall appearance of the skeleton to obtain a broad age classification (young adult, middle age, and old age), which can then be used to select the most appropriate indicator and method (Aiello and Molleson 1993). More specialized techniques are also available and sometimes necessary, but often they are more expensive, invasive, and destructive (e.g., histological or molecular methods). An expert in these advanced laboratory methods should be consulted when the condition of the remains (e.g., highly fragmented, burned) or the circumstances of the forensic case warrants additional methods or secondary confirmation (Marks et al. 2009; Robling and Stout 2008; Stout 2009).

For method development, testing, and validation, and when the focus of the analysis is not identification at the level of the individual, we should restructure our protocol to incorporate blind scoring, multiple observers, and repeated measures. We should also pay careful attention to our choice of reference collection (modern, forensic, archaeological, and multiethnic), sampling methods (randomly chosen, or a subset), data collection decisions (lefts, rights, or both sides for bilateral indicators), approaches to missing data (elimination by case or by variable), and treatment of pathological cases (inclusion as normal variation or exclusion as an outlier). Thorough evaluation and careful documentation is also essential to any skeletal analysis as reassessment may not be possible, be it on account of reburial under the Native American Graves Protection and Repatriation Act (NAGPRA) or the release of remains to the next-of-kin, following a positive identification or the closing of the case. In all circumstances, we should initially score the indicators or features individually without reference to other age-informative areas or to the partner indicator (in the case of bilateral traits) and collect as much raw data as available during the preliminary, and often only, skeletal assessment. *Post hoc* manipulation of the information collected (e.g., data cleaning, sample partitioning, and variable selection) and the choice of statistical methods allow us to refocus the final analysis toward questions of particular interest and to obtain unbiased estimates.

In order to best capture only the age-related information contained in the adult skeleton, we rely on age markers that have been identified, tested, and validated for (1) their progressive, unidirectional, and regular changes; (2) their coverage of the entire adult lifespan; and (3) their invariance to the effects of individual environment and life-ways (such as activity, weight, muscularity, and stress). Some assumptions regarding these indicators may at times be relaxed when certain other conditions are met, or when corrections have been made, as long as this is acknowledged properly. Conventional, phase-aging methods typically make use of one or some combination of the following skeletal indicators: pubic symphyses, auricular surface, sterna rib ends, and cranial sutures. New areas of interest (e.g., for the acetabulum, see Rissech et al. 2006, 2007; Rouge-Maillart et al. 2007) and combinations of many low-information *subjective* traits are also being continually proposed and tested (Algee-Hewitt et al. 2008).

As Meindl et al. (1985) have argued that "the anatomical feature most frequently used for determination of age-at-death in both demographic and forensic osteology is the pubic symphyseal face … it is frequently the only source of information used for age estimation, even when used in conjunction with other criteria it is universally considered to be the most reliable" (p. 29), it is not surprising then that estimation methods that make use of the pubic symphysis are often the anthropologist's first choice. Buikstra and Ubelaker (1994) explain that "morphological changes in the pubic symphyseal face are considered to be among the most reliable criteria for estimating age-at-death in human remains" (p. 21). This confidence in the pubic symphysis is likely based in part on (1) the many studies that demonstrate the relatively consistent and recognizable morphological developments in the pubic symphysis as it progresses through a series of irreversible age-related changes (e.g., Brooks and Suchey 1990; Suchey and Katz 1998; Todd 1920); (2) continued interest in making refinements to its use (Boldsen et al. 2002; Brooks 1955; Garmus 1990; Hanihara and Suzuki 1978; Konigsberg et al. 2008; McKern and Stewart 1957; Meindl et al. 1985; Snow 1983; Zhang 1982); and (3) the considerable amount of data now accumulated for assessing its use in both well-documented and emerging methods, for example, the Todd, McKern, and Stewart, Suchey–Brooks, complex MSA, and, only recently, the TA method (Kemkes-Grottenthaler 1996; Klepinger et al. 1992; Suchey et al. 1986; Wilson and Algee-Hewitt 2009).

Few skeletal biologists concerned with age estimation would disagree that the contribution of greatest effect is the scoring system defined by Suchey, Katz and Brooks in the mid-1980s. Validation tests of the Todd (1920), McKern and Stewart (1957), and Gilbert and McKern (1973) methods lead Katz and Suchey (1986) to revise Todd's 10-phase system using a large ($n = 1225$), modern, and diverse sample of documented remains from the Los Angeles medical examiner's office. They simplified Todd's scoring method by collapsing phases I through III and phases IV, V, VII, and VIII into two categories, made changes to select features of interest in each stage, and expanded the estimated age range for each phase. These modifications produced the six-phase system now commonly known as the Suchey–Brooks method (Brooks and Suchey 1990; Suchey and Katz 1998; Suchey et al. 1986). The widespread use of this method is facilitated by the availability of casts that provide examples of the range of variations observable in each phase. These three-dimensional models offer the practitioner tangible comparative material that aids in reducing subjectivity in phase assignment (Klepinger et al. 1992). It is important to note, however, that this method often fails to capture the variability that characterizes certain groups (Klepinger 2006)—a conclusion recently supported by Schmitt's (2004) test of the Suchey–Brooks method on an identified Asian sample. Here, both bias and inaccuracy increased with age, and true age was underestimated. Nevertheless, the Suchey–Brooks method remains "the symphyseal technique of preference for U.S. anthropologists today" (Komar and Buikstra 2008, p. 142), and has emerged "throughout the world as the standard for estimating age from the pubic symphysis" (Kimmerle et al. 2008b, p. 558).

Lovejoy et al. (1985b) proposed an alternative method for the estimation of age from the pelvis using the gross morphology of the auricular surface. As the auricular surface method was developed using archaeological materials and anatomical specimens from the Hamann–Todd collection, it has received attention in both bioarchaeological and forensic research. Although it is often considered the more difficult of the two pelvic methods in interpretation and correct application, it remains popular for its use of a region of the skeleton that is more resistant to taphonomic damage. In their original study, Lovejoy et al. (1985b) described eight phases of age-informative changes defined by characteristics related to the progressive breakdown of the surface, including billowing, granularity, density, and porosity. The method's scheme assumes that changes first occur at five year intervals between ages 20 and 49 years and then broaden to 10 year intervals from ages 50 to 59 years before terminating with an open-ended category for individuals older than age 60 (i.e., 60+). Although this later terminal phase allows for slightly better estimates of old age than the 50+ phase of the Suchey–Brooks method, the eight stages and their narrow age ranges also fail to account sufficiently for variability in morphology. Osborne et al. (2004) recently proposed a revision of this original configuration, in a similar spirit to Brooks, Katz, and Suchey treatment of Todd's 10-phase system, by reducing the number of phases from 8 to 6 and widening the prediction intervals for each phase. Osborne and colleague's modified method shows great promise for forensic applications as it was validated, in part, using the modern and well-documented WM Bass skeletal collection, returned acceptable levels of interobserver error, and found that sex and ancestry had little effect on the age estimates.

Iscan et al. (1984, 1985) have extended this same phase-based approach to the thoracic region by proposing age estimation criteria for use with the medial (sternal) surface of the fourth rib. The original method involved the classification of the right fourth rib into one of eight modal phases, with sex-specific standards defined individually for white and black Americans. Loth and Iscan (1989) have subsequently reported that ribs 3 and 5 follow the same age progression through phases 1–8 and may be treated using the same *fourth rib* criteria; more recently, however, Yoder et al. (2001) have proposed the use of a composite rib score instead of relying on a single rib. The increased flexibility in rib choice offers improved potential for use in settings in which the *ideal* rib end may not be preserved. However, additional large sample studies are necessary for continued improvement in the rib-scoring criteria and in the accuracy of the estimates produced.

Without question, the most problematic of all adult aging methods in the estimation of reliable and reasonably focused ages at death is the degree of closure of the cranial and maxillary sutures. Although this topic has generated consistent publications throughout the past century since Todd and Lyon's (1924) initial studies, none of these tests and revisions have offered convincing methods or solutions that correct for the significant problems of accuracy, precision, and the confounding effects of secular change (Mann et al. 1987, 1991; Masset 1989; Nawrocki 1998; Meindl and Lovejoy 1985; Milner and Boldsen 2008; Singer 1953). Buikstra and Ubelaker (1994) propose that individuals should be assigned to broad age classes based on an overall assessment (young adult, 20–34 years; middle adult, 35–49 years; old adult, 50+ years) and that weight should be given to age estimates from postcranial features over cranial suture scores. Repeated attempts have been made to improve methodology by obtaining composite scores for vault and laterals anterior sites (Lovejoy et al. 1985a; Meindl and Lovejoy 1985), by scoring only the sutures from regions with good prediction rates (Milner and Boldsen 2008), or by choosing sutures that represent closure patterns for many areas of the skull, including endocranial, ectocranial, and palatine sutures (for the standard on combined scoring sites, see Buikstra and Ubelaker 1994), and by subjecting these to proper statistical analyses (e.g., for regression equations, see Nawrocki 1998; for TA, see Boldsen et al. 2002). Nevertheless, they all have made but weak improvements and have yielded very mixed results. The best decision regarding the evaluation of cranial sutures may ultimately rest in the context of the case: the preservation of other age indicators, the need for external validation of other methods, the contribution of the cranial suture scores to the final age estimate, and the evidentiary standards required for the age estimate should all be weighed carefully.

ADDING UP THE YEARS: CHAPTER SUMMARY OF AGE-AT-DEATH ESTIMATION

The estimation of age at death in forensic anthropology is an essential step in (1) the formulation of a biological profile and (2) the contribution of evidence toward making the identification for unknown skeletal remains of medico-legal significance. The theory for the estimation of age-at-death is predicated on our knowledge that our skeleton undergoes changes throughout our lifetime that reflect *maturation* in the young (bone growth and development) and *senescence* in the old (bone breakdown and degeneration). Current age estimation methods are tailored to capture the variations in skeletal morphology that are unique to each of these two broad age classes. Choice of indicator and method is guided by both the context of the materials under study (including the maturity of the skeleton, its state of preservation, the available indicators, and information about sex and population affinity), as well as the ultimate goal of the analysis (evidence for legal testimony regarding the individual or a series of remains, method development, or method testing). There is, therefore, no single and consistent *best choice* for the age-related indicator or method; nevertheless, we can identify criteria for *best practice*. We should choose techniques for their accuracy and precision; we must be cognizant of the acknowledged problems with age estimation methods and well-informed about the implementation of possible solutions; we should be experienced in osteological practices, knowledgeable of our own weaknesses, and proactive in maintaining high professional standards of proficiency; we should meet current medico-legal expectations for our estimation technique and promote the use of rigorous science in the production of our results; we must embrace interdisciplinary collaboration and seek aid from experts when the conditions of our casework call for assistive studies or confirmatory analyses. Finally, we are wise to remember that age-at-death estimation, like any other gross morphological assessment of the skeleton, produces estimates, and estimates can always be improved through future research and enhanced methodologies.

Review questions

1. How is age-at-death important to the forensic casework? Is the goal of the estimation process different for the skeletal biologist as forensic anthropologist than as paleodemographer?
2. Why should we say that we *estimate* age-at-death instead of *determine* it?
3. Provide a brief explanation of the major problems with age-at-death estimation? Discuss the ways in which these difficulties have been or can be overcome. What smart choices can be made?
4. What difficulties still lie ahead and what future improvements can be made to the method and theory of age estimation?
5. What defines a *good* reference sample? Why do we need a *prior*? What can we do in its absence?
6. How does our approach to *aging* differ between subadult and adult skeletons?
7. What is unique about estimating age in young adults? What is the problem with very old adult skeletons?

8. What defines a *good* indicator and a *good* method?
9. Discuss the difference between phase and component scoring procedures. Is one approach better than the other? If so, under what conditions? Why?
10. What is TA? How can it contribute to improved age-at-death estimation?
11. What do we mean by *group variation*? Does sex or population affect the way we estimate age-at-death or the results produced?
12. How do biological and chronological age differ? Which of these ages do we estimate? Which is most useful in medico-legal contexts?
13. Do our genes or our environment affect our biological age? If so, which or both? In what way?
14. Why are standards important for age-at-death estimation? Have legal expectations affected the way we obtain and report ages? If so, which rulings? How?
15. Provide a brief outline of the history of age estimation. How have philosophic and temporal trends in anthropology and biology affected this scholarship?
16. What is the *Rostock Manifesto*? Explain its guidelines.
17. What are $Pr(a|c)$, $Pr(c|a)$, $f(a)$? What do we need practically and statistically to obtain an age estimate?
18. When estimating age for the individual skeleton, should we provide law enforcement with a point estimate or an age range? How are these different in terms of the information that they provide? What effect can our choice have on making an identification? In the courtroom?
19. What new technologies are contributing the ways, in which we collect age-informative data, how we use this data to estimate age, and our confidence in the results produced?
20. Is estimating age-at-death an *art* or a *science* or both? Why?

References

Acsádi G and Nemeskéri J. 1970. *History of Human Life Span and Mortality*. Budapest, Hungary: Akadémiai Kiadó.

Adams DE and Lothridge KL. 2000. Scientific Working Groups. Forensic science communications 2: www.fbi.gov/hq/lab/fsc/backissu/july2000/swgroups.htm.

Aiello L and Molleson T. 1993. Are microscopic ageing techniques more accurate than macroscopic ageing techniques. *J Archaeol Sci* 20:689–704.

Albert MA. 1998. The use of vertebral ring epiphyseal union for age estimation in two cases of unknown identity. *Forensic Sci Int* 97:11–20.

Algee-Hewitt BFB, Weisensee KE, and Milner GR. 2008. Age is subjective: A non-traditional method of age estimation for the adult skeleton. Paper presented at the *Annual Meeting of the American Academy of Physical Anthropologists*, Columbus, OH.

American Board of Forensic Anthropology, 2010. ABFA home page: www.theabfa.org.

Armelagos GJ. 2003. Bioarchaeology as anthropology. In SD Gillespie and D Nichols (Eds.), *Archaeology Is Anthropology*, pp. 27–41, Archaeological Papers of the American Anthropological Association Series, No. 13. Arlington, VA: American Anthropological Association.

Armelagos GJ. 2008. Biocultural anthropology at its origins: Transformation of the new physical anthropology in the 1950s. In AJ Kelso (Ed.), *The Tao of Anthropology*, pp. 269–288. Gainesville, FL: University of Florida Press.

Armelagos GJ, Carlson DS, and Van Gerven DP. 1982. The theoretical foundations and development of skeletal biology. In F Spencer (Ed.), *A History of American Physical Anthropology*, pp. 305–328. New York: Academic Press.

Armelagos GJ and Van Gerven DP. 2003. A century of skeletal biology and paleopathology: Contrast, contradictions, and conflict. *Am Anthropol* 105:51–62.

Aykroyd RG, Lucy D, Pollard AM, and Roberts CA. 1999. Nasty, brutish, but not necessarily short: A reconsideration of the statistical methods used to calculate age at death from adult human skeletal and dental age indicators. *Am Antiq* 64:55–70.

Baccino ED and Schmitt A. 2006. Determination of adult age at death in forensic context. In A Schmitt, E Cunha, and J Pinheiro (Eds.), *Forensic Anthropology and Medicine: Complementary Sciences from Recovery to Cause of Death*, pp. 259–280. Totowa, NJ: Humana Press.

Baccino E, Ubelaker DH, Hayek LA, and Zerilli A. 1999. Evaluation of seven methods of estimating age at death from mature human skeletal remains. *J Forensic Sci* 44:931–936.

Barrier PF, Dedouit F, Braga J et al. 2009. Age at death estimation using multislice computed tomography reconstructions of the posterior pelvis. *J Forensic Sci* 54:773–778.

Bass WM. 2005. *Human Osteology,* 5th edn. Columbia, MO: Missouri Archaeological Society.

Belcastro MG, Rastelli E, and Mariotti V. 2008. Variation of the degree of sacral vertebral body fusion in adulthood in two European modern skeletal collections. *Am J Phys Anthropol* 135:149–160.

Berg GE. 2008. Pubic bone age estimation in adult women. *J Forensic Sci* 53:569–577.

Black S and Scheuer L. 1996. Age changes in the clavicle from the early neonatal period to skeletal maturity. *Int J Osteoarchaeol* 6:425–434.

Blau S and Ubelaker DH (Eds.). 2009a. Forensic anthropology and archaeology: Introduction to a broader view. In *Handbook of Forensic Anthropology and Archaeology*, pp. 21–25. Walnut Creek, CA: Left Coast Press.

Blau S and Ubelaker DH. 2009b. *Handbook of Forensic Anthropology and Archaeology.* World Archaeology Congress Research Handbooks in Archaeology. Walnut Creek, CA: Left Coast Press.

Bocquet-Appel J-P. 1986. Once upon a time: Paleodemography. *Mitteilungen der Berliner Gesellschaft für Anthropologie, Ethnologie und Urgeschichte* 7:127–133.

Bocquet-Appel J-P and Masset C. 1982. Farewell to paleodemography. *J Hum Evol* 11:321–333.

Bocquet-Appel J-P and Masset C. 1985. Paleodemography: Resurrection or ghost? *J Hum Evol* 14:107–111.

Boldsen JL, Milner GR, Konigsberg LW, and Wood JW. 2002. Transition analyses: A new method for estimating age from skeletons. In RD Hoppa and JW Vaupel (Eds.), *Paleodemography: Age Distributions from Skeletal Samples*, pp. 73–106. Cambridge, NY: Cambridge University Press.

Brooks ST. 1955. Skeletal age at death: The reliability of cranial and pubic age indicators. *Am J Phys Anthropol* 13:567–597.

Brooks S and Suchey JM. 1990. Skeletal age determination based on the os pubis: A comparison of the Acsádi-Nemeskéri and Suchey-Brooks methods. *Hum Evol* 5:227–238.

Brothwell D. 1989. The relationship of tooth wear to aging. In MY Iscan (Ed.), *Age Markers in the Human Skeleton*, pp. 303–316. Springfield, MA: Charles C Thomas.

Bryant LJ and Pearson JD. 1994. Modeling the variability in longitudinal patterns of aging. In DE Crews and RM Garruto (Eds.), *Biological Anthropology and Aging. Perspectives on Human Variation over the Lifespan*, pp. 373–393. Oxford, NY: Oxford University Press.

Buck TJ and Strand Vidarsdóttir U. 2004. A proposed method for the identification of race in sub-adult skeletons: A geometric morphometric analysis of mandibular morphology. *J Forensic Sci* 49:1159–1164.

Buckberry JL and Chamberlain AT. 2002. Age estimation from the auricular surface of the ilium: A revised method. *Am J Phys Anthropol* 119:231–239.

Buikstra JE. 1977. Biocultural dimensions of archaeological study: A regional perspective. In RL Blakely (Ed.), *Biocultural Adaptation in Prehistoric America*, pp. 67–84. Athens, Greece: University of Georgia Press.

Buikstra JE, Milner GR, and Boldsen JL. 2006. Janaab' Pakal: The age-at-death controversy re-revisited. In V Tiesler and A Cucina (Eds.), *Janaab' Pakal of Palenque: Reconstructing the Life and Death of a Maya Ruler*, pp. 48–59. Tucson, AZ: University of Arizona Press.

Buikstra JE and Ubelaker DH. 1994. *Standards for the Data Collection from Human Skeletal Remains*. Research Series 44. Fayetteville, NC: Arkansas Archeological Survey.

Byers S. 2005. *Introduction to Forensic Anthropology: A Textbook*. Boston, MA: Allyn & Bacon Publishers.

Cardoso HFV. 2008. Sample-specific (universal) approaches for determining the sex of immature human skeletal remains using permanent tooth dimensions. *J Archaeol Sci* 35:158–168.

Cardoso HFV and Padez C. 2008. Changes in height, weight, BMI and in the prevalence of obesity among 9 to 11 year-old affluent Portuguese schoolboys, between 1960 and 2000. *Ann Hum Biol* 35:624–638.

Cardoso HFV and Saunders SR. 2008. Two arch criteria of the ilium for sex determination of immature skeletal remains: A test of their accuracy and an assessment of intra- and inter-observer error. *Forensic Sci Int* 178:24–29.

Chaillet N, Nystrom M, and Demirjian A. 2005. Comparison of dental maturity in children of different ethnic origins: International maturity curves for clinicians. *J Forensic Sci* 50:1164–1172.

Christensen AM. 2004. The impact of Daubert: Implications for testimony and research in forensic anthropology (and the use of frontal sinuses in personal identification). *J Forensic Sci* 49:427–430.

Christensen AM. 2005. Testing the reliability of frontal sinuses in positive identification. *J Forensic Sci* 50:18–22.

Clark JDG. 1972. *Star Carr: A Case Study in Bioarchaeology*. Reading, MA: Addison Wesley.

Cope Z. 1946. Fusion lines of bones. *J Anat* 25:280–281.

Crowder C and Austin D. 2005. Age ranges of epiphyseal fusion in the distal tibia and fibula of contemporary males and females. *J Forensic Sci* 50:1001–1007.

Daubert v. Merrell Dow Pharmaceuticals, 113 S.Ct. 2786; 1993 U.S. LEXIS 4408.

Dedouit F, Bindel S, Gainza D et al. 2008. Application of the Iscan method to two- and three-dimensional imaging of the sternal end of the right fourth rib. *J Forensic Sci* 53:288–295.

Demirjian A. 1978. Dentition. In F Falkner and J Tanner (Eds.), *Human Growth*, Vol. 2, pp. 413–444, New York: Plenum Press.

Demirjian A. 1990. Dentition. In F Falkner and JM Tanner (Eds.), *Human Growth: A Comprehensive Treatise*, Vol. 2, pp. 269–297. New York: Plenum.

Demirjian A and Levesque GY. 1980. Sexual differences in dental development and prediction of emergence. *J Dent Res* 59:1110–1122.

Di Bacco M, Ardito V, and Pacciani E. 1999. Age-at-death diagnosis and age-at-death distribution estimate: Two different problems with many aspects in common. *Int J Anthropol* 14:161–169.

DiGangi EA, Bethard JD, Kimmerle EH, and Konigsberg LW. 2009. A new method for estimating age-at-death from the first rib. *Am J Phys Anthropol* 138:164–176.

Enlow DH. 1996. *Essentials of Facial Growth*. Philadelphia, PA: W.B. Saunders Company.

Fazekas I and Kosa K. 1978. *Forensic Fetal Osteology*, 1st edn. Budapest, Hungary: Akademiai Kiado Publishers.

Ferrant O, Rougé-Maillart C, Guittet L et al. 2009. Age at death estimation of adult males using coxal bone and CT scan: A preliminary study. *Forensic Sci Int* 186:14–21.

Finch CE. 1994. *Longevity, Senescence, and the Genome*. Chicago, IL: University of Chicago Press.

Garmus AK. 1990. The determination of the age of an individual by the morphological signs of the *pubic* symphysis. *Sudebno-Meditsinskaia Ekspertiza* 33:22–24.

Garn SM, Lewis AB, and Vicinus JH. 1962. Third molar agenesis and reduction in the number of other teeth. *J Dent Res* 41:717.

Gilbert BM and McKern TW. 1973. A method for aging the female Os pubis. *Am J Phys Anthropol* 38:31–38.

Gowland R and Chamberlain A. 1999. The use of prior probabilities in aging perinatal skeletal remains: Implications for the evidence of infanticide in Roman Britain. *Am J Phys Anthropol* 28:138–139.

Greulich WW and Pyle SI. 1959. *Radiographic Atlas of Skeletal Development of the Hand and Wrist.* Stanford, CA: Stanford University Press.

Grivas CR and Komar DA. 2008. Kumho, Daubert, and the nature of scientific inquiry: Implications for forensic anthropology. *J Forensic Sci* 53:771–776.

Gustafson G. 1950. Age determination on teeth. *J Am Dent Assoc* 41:45–54.

Haglund WD and Sorg MH. 1997. *Forensic Taphonomy: The Postmortem Fate of Human Remains.* Boca Raton, FL: CRC Press.

Haglund WD and Sorg MH. 2001. *Advances in Forensic Taphonomy: Method, Theory, and Archaeological Perspectives.* Boca Raton, FL: CRC Press.

Hanihara K and Suzuki T. 1978. Estimation of age from the pubic symphysis by means of multiple regression analysis. *Am J Phys Anthropol* 48:233–240.

Harper GJ and Crews DE. 2000. Aging, senescence, and human variation. In S Stinson, B Borgan, R Huss-Ashmore, and D O'Rourke (Eds.), *Human Biology: An Evolutionary and Biocultural Perspective,* pp. 465–506. New York: Wiley-Liss.

Herrmann NP and Konigsberg LW. 2002. A re-examination of the age-at-death distribution of Indian Knoll. In RD Hoppa and J Vaupel (Eds.), *Paleodemography: Age Distributions from Skeletal Samples,* pp. 243–257. Cambridge, NY: Cambridge University Press.

Hillson S. 1998. *Dental Anthropology.* Cambridge, NY: Cambridge University Press.

Holcomb SMC and Konigsberg LW. 1995. A statistical study of sexual dimorphism in the human fetal sciatic notch. *Am J Phys Anthropol* 97:113–125.

Holman DJ, Wood JW, and O'Connor KA. 2002. Estimating age-at-death distributions from skeletal samples: Multivariate latent-trait approach. In RD Hoppa and JW Vaupel (Eds.), *Paleodemography: Age Distributions from Skeletal Samples,* pp. 193–221. Cambridge, NY: Cambridge University Press.

Hoppa RD. 2000. Population variation in osteological aging criteria: An example from the pubic symphysis. *Am J Phys Anthropol* 111:185–191.

Hoppa RD. 2002. Paleodemography: Looking back and thinking ahead. In RD Hoppa and JW Vaupel (Eds.), *Paleodemography: Age Distributions from Skeletal Samples,* pp. 9–28. Cambridge, NY: Cambridge University Press.

Hoppa RD and Vaupel JW. 2002a. *Paleodemography: Age Distributions from Skeletal Samples.* Cambridge, NY: Cambridge University Press.

Hoppa RD and Vaupel JW. 2002b. The Rostock Manifesto for paleodemography: The way from stage to age. In RD Hoppa and JW Vaupel (Eds.), *Paleodemography: Age Distributions from Skeletal Samples,* pp. 1–8. Cambridge, NY: Cambridge University Press.

Houck MM, Ubelaker DH, Owsley DW et al. 1996.The role of forensic anthropology in the recovery and analysis of Branch Davidian compound victims: Assessing the accuracy of age estimations. *J Forensic Sci* 41:796–801.

Howell N. 1976. Toward a uniformitarian theory of human paleodemography. In RH Ward and KM Weiss (Eds.), *The Demographic Evolution of Human Populations,* pp. 25–40. New York: Academic Press.

Howell N. 1982. Village composition implied by a paleodemographic life table: The Libben site. *Am J Phys Anthropol* 59:263–269.

Iscan MY. 1988. Rise of forensic anthropology. *Yearb Phys Anthropol* 31:203–230.

Iscan MY (Ed.). 1989. Assessment of age at death in the human skeleton. In *Age Markers in the Human Skeleton,* pp. 5–18. Springfield, MA: Charles C Thomas.

Iscan MY and Loth SR. 1989. Osteological manifestations of age in the adult. In MY Iscan and KAR Kennedy (Eds.), *Reconstruction of Life from the Skeleton,* pp. 23–40. New York: Alan R. Liss.

Iscan MY, Loth SR, and Wright RK. 1984. Age estimation from the rib by phase analysis: White males. *J Forensic Sci* 29:1094–1104.

Iscan MY, Loth SR, and Wright RK. 1985. Age estimation from the rib by phase analysis: White females. *J Forensic Sci* 30:853–863.

Jackes M. 2000. Building the bases for paleodemographic analysis: Adult age determination. In AM Katzenberg and SR Saunders (Eds.), *Biological Anthropology of the Human Skeleton*, pp. 417–466. New York: Wiley-Liss.

Katz D and Suchey JM. 1986. Age determination of the male os pubis. *Am J Phys Anthropol* 69:427–435.

Katz D and Suchey JM. 1989. Race differences in pubic symphyseal aging patterns in the male. *Am J Phys Anthropol* 80:167–172.

Kemkes-Grottenthaler A. 1996. Critical evaluation of osteomorphognostic methods to estimate adult age at death: A test of the "complex method." *Homo* 46:280–292.

Kemkes-Grottenthaler A. 2002. Aging through the ages: Historical perspectives on age indicator methods. In RD Hoppa and JW Vaupel (Eds.), *Paleodemography: Age Distributions from Skeletal Samples*, pp. 48–72. Cambridge, NY: Cambridge University Press.

Kerley ER. 1978. Recent developments in forensic anthropology. *Yearb Phys Anthropol* 20:160–173.

Key CA, Aiello LC, and Molleson T. 1994. Cranial suture closure and its implications for age estimation. *International Journal of Osteoarchaeology* 4:193–207.

Kimmerle EH, Jantz RL, Konigsberg LW, and Baraybar JP. 2008a. Skeletal estimation and identification in American and East European populations. *J Forensic Sci* 53:524–532.

Kimmerle EH, Konigsberg LW, Jantz RL, and Baraybar JP. 2008b. Analysis of Age-at-death estimation through the use of pubic symphyseal data. *J Forensic Sci* 53:558–568.

Kimmerle EH, Prince DA, and Berg GE. 2008c. Inter-observer variation in methodologies involving the pubic symphysis, sternal ribs, and teeth. *J Forensic Sci* 53:594–600.

Klepinger LL. 2001. Stature, maturation, variation and secular trends in forensic anthropology. *J Forensic Sci* 46:788–790.

Klepinger LL. 2006. *Fundamentals of Forensic Anthropology*. Hoboken, NJ: Wiley-Liss.

Klepinger LL, Katz D, Micozzi MS, and Carroll L. 1992. Evaluation of cast methods for estimating age from the os pubis. *J Forensic Sci* 37:763–770.

Komar D and Buikstra J. 2008. *Forensic Anthropology Contemporary Theory and Practice*. Oxford, NY: Oxford University Press.

Komar D and Grivas C. 2008. Manufactured populations: What do contemporary reference skeletal collections represent? A comparative study using the Maxwell Museum Documented Collection. *Am J Phys Anthropol* 137:224–233.

Konigsberg LW, Algee-Hewitt BFB, Steadman DW. 2009. Estimation and evidence in forensic anthropology. *Sex Race* 139:77–90.

Konigsberg LW and Frankenberg SR. 1992. Estimation of age structure in anthropological demography. *Am J Phys Anthropol* 89:235–256.

Konigsberg LW and Frankenberg SR. 1994. Paleodemography: "Not quite dead." *Evol Anthropol* 3:92–105.

Konigsberg LW and Frankenberg SR. 2002. Deconstructing death in paleodemography. *Am J Phys Anthropol* 117:297–309.

Konigsberg LW, Frankenberg SR, and Walker RB. 1997. Regress what on what? Paleodemographic age estimation as a calibration problem. In RR Paine (Ed.), *Integrating Archaeological Demography: Multidisciplinary Approaches to Prehistoric Population*, pp. 64–88. Carbondale, IL: Southern Illinois University Press.

Konigsberg LW, Hens SM, Jantz LM, and Jungers WL. 1998. Stature estimation and calibration: Bayesian and maximum likelihood perspectives in physical anthropology. *Yearb Phys Anthropol* 41:65–92.

Konigsberg LW and Herrmann NP. 2002. Markov chain Monte Carlo estimation of hazard model parameters in paleodemography. In RD Hoppa and JW Vaupel (Eds.), *Paleodemography: Age Distributions from Skeletal Samples*, pp. 222–242. Cambridge, NY: Cambridge University Press.

Konigsberg LW, Herrmann NP, Wescott DJ, and Kimmerle EH. 2008. Estimation and evidence in forensic anthropology: Age-at-death. *J Forensic Sci* 53:541–557.

Kunos CA, Simpson SW, Russell KF, and Hershkovitz I. 1999. First rib metamorphosis: Its possible utility for human age-at-death estimation. *Am J Phys Anthropol* 110:303–323.

Lamendin H, Baccino E, Humbert JF, Tavernier JC, Nossintchouk RM, and Zerilli A. 1992. A simple technique for age estimation in adult corpses: The two criteria dental method. *J Forensic Sci* 37:1373–1379.

Larsen CS. 1997. *Bioarchaeology: Interpreting Behavior from the Human Skeleton*. Cambridge Studies in Biological Anthropology 21. Cambridge, NY: Cambridge University Press.

Lewis M and Flavel A. 2006. Age assessment of child skeletal remains in forensic contexts. In A Schmitt, E Cunha, and J Pinheiro (Eds.), *Forensic Anthropology and Medicine: Complementary Sciences from Recovery to Cause of Death*, pp. 243–257. Totowa, NJ: Humana Press.

Loth SR and Iscan MY. 1989. Morphological assessment of age in the adult: The thoracic region. In MY Iscan (Ed.), *Age Markers in the Human Skeleton*, pp. 105–136. Springfield, MA: Charles C Thomas.

Love B and Müller H-G. 2002. A solution to the problem of obtaining a mortality schedule for paleodemographic data. In RD Hoppa and J Vaupel (Eds.), *Paleodemography: Age Distributions from Skeletal Samples*, pp. 73–106. Cambridge, NY: Cambridge University Press.

Lovejoy CO, Meindl RS, Mensforth RP, and Barton TJ. 1985a. Multifactorial determination of skeletal age at death: A method and blind test of its accuracy. *Am J Phys Anthropol* 68:1–14.

Lovejoy CO, Meindl RS, Pryzbeck TR, and Mensforth RP. 1985b. Chronological metamorphosis of the auricular surface of the ilium: A new method for the determination of adult skeletal age at death. *Am J Phys Anthropol* 68:15–28.

Lovejoy CO, Meindl RS, Pryzbeck TR, Mensforth RP, Barton TJ, Heiple KG, and Kotting D. 1977. Paleodemography of the Libben site, Ottawa County, Ohio. *Science* 198:291–293.

Lovejoy CO, Meindl RS, Tague RG, and Latimer B. 1997. The comparative senescent biology of the human pelvis and its implications for the use of age-at-death indicators in the human skeleton. In R Paine (Ed.), *Integrating Anthropological Demography: Multidisciplinary Approaches to Prehistoric Population*, pp. 43–63. Carbondale, IL: Center for Archaeological Investigations.

Lovejoy C, Owen C, Mensforth RP, and Armelagos GJ. 1982. Five decades of skeletal biology as reflected in the American journal of physical anthropology. In *F Spencer (Ed.)*, *A History of American Physical Anthropology 1930–1980*, pp. 329–336. New York: Academic Press.

Lucy D, Aykroyd RG, Pollard AM, and Solheim T. 1996. A Bayesian approach to adult human age estimation from dental observations by Johanson's age changes. *J Forensic Sci* 41:189–194.

Lucy D and Pollard AM. 1995. Further comments on the estimation of error associated with the Gustafson dental age estimation method. *J Forensic Sci* 40:222–227.

Lynnerup N, Thomsen JL, and Frohlich B. 1998. Intra- and inter-observer variation in histological criteria used in age at death determination based on femoral cortical bone. *Forensic Sci Int* 91:219–230.

Mann RW, Jantz RL, Bass WM, and Willey PS. 1991. Maxillary suture obliteration: A visual method for estimating skeletal age. *J Forensic Sci* 36:781–791.

Mann RW, Symes SA, and Bass WM. 1987. Maxillary suture obliteration: Aging the human skeleton based on intact or fragmentary maxilla. *J Forensic Sci* 32:148–157.

Maples WR. 1989. The practical application of age estimation techniques. In MY Iscan (Ed.), *Age Markers in the Human Skeleton*, pp. 319–324. Springfield, MA: Charles C Thomas.

Marks MK, Marden K, and Mileusnic-Polchan D. 2009. Forensic osteology of child abuse. In DW Steadman (Ed.), *Hard Evidence: Case Studies in Physical Anthropology,* 2nd edn, pp. 205–220. Upper Saddle River, NJ: Prentice Hall.

Masset C. 1989. Age estimation on the basis of cranial sutures. In MY Iscan (Ed.), *Age Markers in the Human Skeleton*, pp. 71–103. Springfield, MA: Charles C Thomas.

Mays S. 1998. *The Archaeology of Human Bones*. New York: Routledge.

McKern TW. 1957. Estimation of skeletal age from combined maturational activity. *Am J Phys Anthropol* 15:399–408.

McKern T and Stewart T. 1957. Skeletal age changes in young American males. Technical Report EP-45. Natick, MA: Headquarters, Quartermaster Research and Development Center, Environment Protection Research Division.

Meindl RS and Lovejoy CO. 1985. Ectocranial suture closure. A revised method for the determination of skeletal age at death based on the lateral-anterior sutures. *Am J Phys Anthropol* 68:57–66.

Meindl RS and Lovejoy CO. 1989. Age changes in the pelvis: Implications for paleodemography. In MY Iscan (Ed.), *Age Markers in the Human Skeleton*, pp. 137–168. Springfield, MA: Charles C Thomas.

Meindl RS, Lovejoy CO, Mensforth RP, and Walker RA. 1985. Revised method of age determination using the os pubis, with a review and tests of other current methods of pubic symphyseal aging. *Am J Phys Anthropol* 68:29–45.

Meindl RS and Russell KF. 1998. Recent advances in method and theory in paleodemography. *Ann Rev Anthropol* 27:375–399.

Meindl RS, Russell KF, and Lovejoy CO. 1990. Reliability of age at death in the Hamann Todd Collection: Validity of subselection procedures used in blind tests of the summary age technique. *Am J Phys Anthropol* 83:349–357.

Miles AEW. 1962. Assessment of the ages of a population of Anglo-Saxons from their dentitions. *Proc R Soc Med* 55:881–886.

Miles AEW. 1963. Dentition and the estimation of age. *J Dent Res* 42:255–263.

Miles AEW. 1978. Teeth as an indicator of age in man. In PM Butler and KA Joysey (Eds.), *Development, Function and the Evolution of Teeth*, pp. 455–462. London, U.K.: Academic Press.

Milner GR and Boldsen JL. 2008. *Transition Analysis Age Estimation Skeletal Scoring Procedure*, 7.v.07.

Milner GR and Buikstra JE. 2006. Skeletal biology: Northeast. In DH Ubelaker (Ed.), *Handbook of North American Indians: Environment, Origins, and Population*, Vol. 3, pp. 630–639. Washington, DC: Smithsonian Institution Press.

Milner GR, Wood JW, and Boldsen JL. 2008. Advances in paleodemography. In MA Katzenberg and SR Saunders (Eds.), *Biological Anthropology of the Human Skeleton*, 2nd edn, pp. 561–600. New York: Wiley-Liss.

Mincer HH, Harris EF, and Berryman HE. 1993. A.B.F.O. study of third molar development and its use as an estimator of chronological age. *J Forensic Sci* 38:379–390.

Molleson T and Cox M. 1993. The spitalfields project, Vol. 2. In *The Middling Sort*. York, U.K.: Council for British Archaeology.

Molnar S. 1971. Human tooth wear, tooth function and cultural variability. *Am J Phys Anthropol* 34:175–190.

Moore-Jansen PM, Ousley SD, and Jantz RL. 1994. Data Collection Procedures for Forensic Skeletal Material. Report of Investigations 48. Knoxville, TN: Department of Anthropology, University of Tennessee.

Moorrees CFA, Fanning EA, and Hunt EE. 1963a. Formation and resorption of three deciduous teeth in children. *Am J Phys Anthropol* 21:205–213.

Moorrees CFA, Fanning EA, and Hunt EE. 1963b. Age variation of formation stages for ten permanent teeth. *J Dent Res* 42:1490–1502.

Moss ML. 1968. A theoretical analysis of the functional matrix. *Acta Biotheoretica* 18:1–4.

Moss ML and Young RW. 1960. A functional approach to craniology. *Am J Phys Anthropol* 18:281–292.

Müller HG, Love B, and Hoppa RD. 2002. A semiparametric method for estimating demographic profiles from age indicator data. *Am J Phys Anthropol* 117:1–14.

NamUS (National Missing and Unidentified Persons System). 2010. U.S. Department of Justice. www.namus.gov.

Nawrocki SP. 1998. Regression formulae for estimating age at death from cranial suture closure. In KJ Reichs (Ed.), *Forensic Osteology: Advances in the Identification of Human Remains*, 2nd edn, pp. 276–292. Springfield, MA: Charles C Thomas.

Osborne DL, Simmons TL, and Nawrocki S. 2004. Reconsidering the auricular surface as an indicator of age at death. *J Forensic Sci* 49:1–7.

Ousley SD and Jantz RL. 1998. The forensic data bank: Documenting skeletal trends in the United States. In KJ Reichs (Ed.), *Forensic Osteology*, 2nd edn, pp. 441–458. Springfield, MA: Charles C Thomas.

Paine RR and Boldsen JL. 2002. Linking age-at-death distributions and ancient population dynamics: A case study. In RD Hoppa and JW Vaupel (Eds.), *Paleodemography: Age Distributions from Skeletal Samples*, pp. 169–180. Cambridge, NY: Cambridge University Press.

Prince DA, Kimmerle EH, and Kongisberg LW. 2008. A Bayesian approach to estimate skeletal age-at-death utilizing dental wear. *J Forensic Sci* 53:588–593.

Prince DA and Konigsberg LW. 2008. New formulae for estimating age-at-death in the Balkans utilizing Lamendin's dental technique and Bayesian analysis. *J Forensic Sci* 53:578–587.

Prince DA and Ubelaker DH. 2002. Application of Lamendin's adult dental aging technique to a diverse skeletal sample. *J Forensic Sci* 47:107–116.

Reichs KJ. 1995. A professional profile of diplomates of the American board of forensic anthropology: 1984–1992. *J Forensic Sci* 40:176–182.

Reichs KJ. 1998. Forensic anthropology: A decade of progress. In KJ Reichs (Ed.), *Forensic Osteology: Advances in the Identification of Human Remains*, pp. 13–38. Springfield, MA: Charles C. Thomas.

Rissech C, Estabrook GF, Cunha E, and Malgosa A. 2006. Using the acetabulum to estimate age at death of adult males. *J Forensic Sci* 51:213–229.

Rissech C, Estabrook GF, Cunha E, and Malgosa A. 2007. Estimation of age at death for adult males using the acetabulum, applied to four Western European populations. *J Forensic Sci* 52:774–779.

Ritz-Timme S, Cattaneo C, and Collins MJ. 2000. Age estimation: The state of the art in relation to the specific demands of forensic practice. *Int J Leg Med* 113:129–136.

Robling AG and Stout SD. 2008. Methods of determining age at death using bone microstructure. In MA Katzenberg and SR Saunders (Eds.), *Biological Anthropology of the Human Skeleton,* 2nd edn, pp. 149–171. New York: Wiley-Liss.

Rodgers T and Allard T. 2004. Expert testimony and positive identification of human remains through cranial suture patterns. *J Forensic Sci* 49:203–207.

Rios L, Weisensee KE, and Rissech C. 2008. Sacral fusion as an aid in age estimation. *Forensic Sci Int* 180:111–117.

Rösing FW, Graw M, Marré B et al. 2007. Recommendations for the forensic diagnosis of sex and age from skeletons. *Homo* 58:75–89.

Rouge-Maillart C, Jousset N, Vielle B, Gaudin A, and Telmon N. 2007. Contribution of the study of acetabulum for the estimation of adult subjects. *Forensic Sci Int* 171:103–110.

Ruff CB. 1981. A reassessment of demographic estimates for Pecos Pueblo. *Am J Phys Anthropol* 54:147–151.

Sapir GI. 2002. Legal aspects of forensic science. In R Saferstein (Ed.), *Forensic Science Handbook*, Vol. I, 2nd edn, pp. 2–24. Upper Saddle River, NJ: Prentice Hall.

Sashin D. 1930. A critical analysis of the anatomy and the pathological changes of the sacro-iliac joints. *J Bone Joint Surg* 28:891–910.

Saunders SR. 2000. Subadult skeletons and growth related studies. In MA Katzenberg and SR Saunders (Eds.), *Biological Anthropology of the Human Skeleton*, pp. 135–161. New York: Wiley.

Saunders SR, Fitzgerald C, Rogers T, Dudar JC, and McKillop H. 1992. A test of several methods of skeletal age estimation using a documented archaeological sample. *Can Soc Forensic Sci J* 25:97–118.

Schaefer MC and Black SM. 2005. Comparison of ages of epiphyseal union in North American and Bosnian skeletal material. *J Forensic Sci* 50:777–784.

Scheuer L and Black S. 2000. *Developmental Juvenile Osteology*. London, U.K.: Academic Press.

Scheuer L and Black S. 2004. *The Juvenile Skeleton*. London, U.K.: Elsevier.

Schmitt A. 2004. Age-at-death assessment using the os pubis and the auricular surface of the ilium: A test on an identified Asian sample. *Int J Osteoarchaeol* 14:1–6.

Schmitt A and Broqua C. 2000. Application pratique de la méthode d'estimation de l'âge au décès de *Schmitt* et *Broqua*. *Bull Mem Soc Anthropol* 16:115–120.

Schmitt A, Cunha E, and Pinheiro J (Eds.). 2006. *Forensic Anthropology and Medicine: Complementary Sciences from Recovery to Cause of Death*. Totowa, NJ: Humana.

Schmitt A, Murail P, Cunha E, and Rougé D. 2002. Variability of the pattern of aging on the human skeleton: Evidence from bone indicators and implications on age at death estimation. *J Forensic Sci* 47:1–7.

Schour I and Massler M. 1941. The development of the human dentition. *J Am Dent Assoc* 28:1153–1160.

Scientific Working Group for Forensic Anthropology (SWGANTH). 2010. Homepage: www.swganth.org.

Shirley NR, Algee-Hewitt BFB, and Wilson RJ. 2010. The reality of aging virtually: A test of transition analysis on pelvic laser scans. Paper presented at the *Annual Meeting of the American Academy of Physical Anthropologists*, Albuquerque, NM.

Shirley-Langley N and Jantz RL. 2010. A Bayesian approach to age estimation in modern Americans from the clavicle. *J Forensic Sci* 55:571–583.

Singer R. 1953. Estimation of age from cranial suture closure. *J Forensic Med* 1:52–59.

Smith BH. 1984a. Patterns of molar wear in hunter-gatherers and agriculturalists. *Am J Phys Anthropol* 63:39–56.

Smith BH. 1984b. Rate of molar wear: Implications for developmental timing and demography in human evolution. *Am J Phys Anthropol* 63:220.

Smith BH. 1991. Standards of human tooth formation and dental age assessment. In M Kelly and C Larsen (Eds.), *Advances in Dental Anthropology*, pp. 143–168. New York: Wiley-Liss.

Snow CC. 1982. Forensic anthropology. *Ann Rev Anthropol* 11:97–131.

Snow CC. 1983. Equations for estimating age at death from the pubic symphysis: A modification of the McKern-Stewart method. *J Forensic Sci* 28:864–870.

Steadman DW (Ed.). 2009. *Hard Evidence: Case Studies in Forensic Anthropology*, 2nd edn. Upper Saddle River, NJ: Prentice Hall.

Steadman DW, Adams B, and Konigsberg L. 2006. The statistical basis for positive identifications in forensic anthropology. *Am J Phys Anthropol* 131:15–26.

Stevenson P. 1924. Age order of epiphyseal union in man. *Am J Phys Anthropol* 7:53–93.

Stewart TD. 1979. *Essentials of Forensic Anthropology: Especially as Developed in the United States*. Springfield, MA: Charles C Thomas.

Stout SD. 2009. Small bones of contention. In DW Steadman (Ed.), *Hard Evidence: Case Studies in Physical Anthropology*, 2nd edn, pp. 239–247. Upper Saddle River, NJ: Prentice Hall.

Suchey JM and Katz D. 1986. Skeletal age standards derived from an extensive multiracial sample of modern Americans. Paper presented at the *Fifty-Fifth Annual Meeting of the American Association of Physical Anthropologists*, Albuquerque, NM.

Suchey JM and Katz D. 1998. Applications of pubic age determination in a forensic setting. In KJ Reichs (Ed.), *Forensic Osteology: Advances in the Identification of Human Remains*, 2nd edn, pp. 204–236. Springfield, MA: Charles C Thomas.

Suchey JM, Wiseley DV, and Katz D. 1986. Evaluation of the Todd and McKern-Stewart methods for aging the male os pubis. In KJ Reichs (Ed.), *Forensic Osteology: Advances in the Identification of Human Remains*, pp. 33–67. Springfield, MA: Charles C Thomas.

Tanner J, Healy RM, Goldstein H, and Cameron N. 2001. *Assessment of Skeletal Maturity and Prediction of Adult Height (TW3 Method)*. London, U.K.: W. B. Saunders.

Tanner J, Whitehouse R, Cameron N, Marshall W, Healy M, and Goldstein H. 1975. *Assessment of Skeletal Maturity and Prediction of Adult Height (TW2 Method)*, 2nd edn. London, U.K.: Academic Press.

Todd TW. 1920. Age changes in the pubic bones. I. The male white pubis. *Am J Phys Anthropol* 3:285–334.

Todd TW. 1921a. Age changes in the pubic bones. II. The pubis of the male Negro-white hybrid. II. The pubis of the white female IV. The pubis of the female Negro-white hybrid. *Am J Phys Anthropol* 4:4–70.

Todd TW. 1921b. Age changes in the pubic bones. V. Mammalian pubic bone metamorphosis. *Am J Phys Anthropol* 4:333–406.

Todd TW. 1921c. Age changes in the pubic bones. V. The interpretation of variations in the symphyseal area. *Am J Phys Anthropol* 4:407–424.

Todd TW. 1923. Age changes in the pubic bones. VII. The anthropoid strain in human pubic symphysis of the third decade. *J Anat* 57:274–294.

Todd TW. 1930. Age changes in the pubic bones. VII. Roentgenographic differentiation. *Am J Phys Anthropol* 14:255–271.

Todd T and D'Errico J. 1928. The clavicular epiphyses. *Am J Anat* 41:25–50.

Todd TW and Lyon DW. 1924. Endocranial suture closure: Part I. Adult males of white stock. *Am J Phys Anthropol* 7:325–384.

Tomkins RL. 1996. Human population variability in relative dental development. *Am J Phys Anthropol* 99:79–102.

Ubelaker DH. 1987. Estimating age at death from immature human skeletons: An overview. *J Forensic Sci* 23:1254–1263.

Ubelaker DH. 1989a. *Human Skeletal Remains: Excavation, Analysis, Interpretation,* 2nd edn. Washington, DC: Taraxacum.

Ubelaker DH. 1989b. The estimation of age at death from immature human bone. In MY Iscan (Ed.), *Age Markers in the Human Skeleton*, pp. 55–69. Springfield, MA: Charles C Thomas.

Ubelaker DH. 2008. Issues in the global applications in methodology of forensic anthropology. *J Forensic Sci* 53:606–607.

Ubelaker DH. 2009. Historical development of forensic anthropology: Perspective from the United States. In D Ubelaker and S Blau (Eds.), *Handbook of Forensic Anthropology and Archaeology*, pp. 76–86. Walnut Creek, CA: Left Coast Press.

Usher B. 2002. Reference samples: The first step in linking biology and age in the human skeleton. In R Hoppa and J Vaupel (Eds.), *Paleodemography: Age Distributions from Skeletal Samples*, pp. 29–47. London, U.K.: Cambridge University Press.

Usher BM, Boldsen JL, and Holman D. 2000. Age estimation at Tirup cemetery: An application of the transition analysis method. *Am J Phys Anthropol* 30:307.

Waldron T. 1987. The relative survival of the human skeleton: Implications for paleodemography. In A Boddington, AN Garland, and RC Janeway (Eds.), *Death, Decay, and Reconstruction: Approaches to Archaeology and Forensic Science*, pp. 55–64. Manchester, U.K.: Manchester University Press.

Waldron T. 1994. *Counting the Dead: The Epidemiology of Skeletal Populations*. New York: John Wiley & Sons.

Walker PL, Johnson J, and Lambert P. 1988. Age and sex biases in the preservation of human skeletal remains. *Am J Phys Anthropol* 76:183–188.

Warren MW. 1999. Radiographic determination of developmental age in fetuses and stillborns. *J Forensic Sci* 44:708–712.

Washburn SL. 1951. The new physical anthropology. *Trans New York Acad Sci Ser II* 13:298–304.

Washburn SL. 1953a. The new physical anthropology. *Yearb Phys Anthropol* 7:124–130.

Washburn SL. 1953b. The strategy of physical anthropology. In AL Kroeber (Ed.), *Anthropology Today*, pp. 714–727. Chicago, IL: University of Chicago Press.

Web of Science. 2010. *ISI Web of Knowledge*. Thomson Reuters: apps.isiknowledge.com

Webb P and Suchey J. 1985. Epiphyseal union of the anterior iliac crest and medial clavicle in a modern multiracial sample of American males and females. *Am J Phys Anthropol* 68:457–466.

Weinker C and Rhine S. 1989. A professional profile of the physical anthropology section membership, American Academy of Forensic Sciences. *J Forensic Sci* 34:12.

White T and Folkens P. 2005. *The Human Bone Manual*. New York: Elsevier.

Wiersema J, Love JC, and Naul LG. 2009. The influence of the Daubert guidelines of anthropological methods of scientific identification in the medical examiner setting. In DW Steadman (Ed.), *Hard Evidence: Case Studies in Forensic Anthropology,* 2nd edn, pp. 80–90. Upper Saddle River, NJ: Prentice Hall.

Wilson RJ and Algee-Hewitt BFB. 2009. [Inter]facing age: A test of the ADBOU age estimation software in a forensic context. Paper presented at the *Annual Meeting of the American Academy of Physical Anthropologists*, Chicago, IL.

Wilson RJ, Algee-Hewitt BFB, and Jantz LM. 2007. Demographic trends within the Forensic Anthropology Center's body donation program. Paper presented at the *Annual Meeting of the American Academy of Physical Anthropologists*, Philadelphia, PA.

Wilson RJ, Algee-Hewitt BFB, and Jantz LM. 2008. Age-at-death distributions in the W.M. Bass Donated Skeletal Collection, Part II: Modeling mortality in a Body Donation Program. Paper presented at the *Annual Meeting of the American Academy of Physical Anthropologists*, Columbus, OH.

Wood JW, Milner GR, Harpending HC, and Weiss KM. 1992. The osteological paradox: Problems of inferring prehistoric health from skeletal samples. *Curr Anthropol* 33:343–370.

Workshop of European Anthropologists. 1980. Recommendations for age and sex diagnoses of skeletons. *J Hum Evol* 9:517–549.

Yoder C, Ubelaker DH, and Powell JF. 2001. Examination of variation in sternal rib end morphology relevant to age assessment. *J Forensic Sci* 46:223–227.

Zhang Z. 1982. A preliminary study of estimation of age by morphological changes in the symphysis pubis. *Acta Anthropol Sinica* 1:132–136.

Zollikofer CPE and Ponce de León MS. 2005. *Virtual Reconstruction: A Primer in Computer-Assisted Paleontology and Biomedicine*. New York: Wiley.

Glossary

CHAPTER 1

American Academy of Forensic Sciences (AAFS): the professional organization with which most forensic anthropologists are affiliated in the United States. The AAFS is comprised of 11 sections and publishes the *Journal of Forensic Sciences*.

American Association of Physical Anthropologists (AAPA): the leading professional organization for physical anthropologists consisting of paleoanthropologists, primatologists, and forensic anthropologists. The AAPA publishes the *American Journal of Physical Anthropology* and the *Yearbook of Physical Anthropology*.

American Board of Forensic Anthropology (ABFA): a nonprofit organization that provides a program of certification in forensic anthropology. Diplomates must demonstrate an ongoing record of practice and research in the field of forensic anthropology and engage in continuing education.

Anthropology: the discipline that studies all aspects of what it means to be human (culture, language, history and origins, and biology).

Biological profile: the four primary components of a person's physical identity (phenotype) that forensic anthropologists ascertain from the skeleton: age, sex, ancestry, and stature. The biological profile helps law enforcement in the search for missing persons.

Four-field approach: the study of the four subfields of anthropology in order to gain a more holistic understanding of humans and our ancestors (cultural anthropology, biological anthropology, linguistic anthropology, and archaeology).

International Association of Forensic Sciences (IAFS): the only worldwide association of academics and practicing professionals from various forensic science disciplines. The IAFS holds meetings every three years.

National Commission on Forensic Science: a commission created by the Department of Justice (DOJ) and the National Institute of Standards and Technology (NIST) upon the recommendation of the NAS report (2009). The commission seeks to promote scientific validity, reduce fragmentation, and improve federal coordination of forensic science.

Native American Graves Protection and Repatriation Act (NAGPRA): An Act enacted in 1990 that requires federal agencies and institutions that receive federal funding to return Native American remains and cultural items to lineal descendants and culturally affiliated Indian tribes. NAGPRA also establishes processes for the excavation or discovery of Native American cultural items and makes it a crime to traffic in Native American human remains without the right of possession.

Organization of Scientific Area Committees (OSAC): the overarching committee that consists of five scientific area committees (SACs). The OSAC coordinates development of standards and guidelines to improve quality and consistency of work in the forensic science community.

Scientific Working Group for Forensic Anthropology (SWGANTH): a scientific working group consisting of a number of committees that recommend and disseminate guidelines for best practice, quality assurance, and quality control in forensic anthropology.

CHAPTER 2

Accessioned: to record the addition of; in forensic matters or cases, the addition of new evidentiary material.
Antemortem: occurring before death.
Articular: of or relating to joints.
Bilateral asymmetry: different or unequal size or shapes on the right versus left side of the body.
Biocultural: incorporation of a biology and culture to understand how the two influence each other and interpretations within forensic anthropology.

Desiccated: dried out, dehydrated, or mummified.
Faunal: pertaining to animals.
Interment: the method of burial of a corpse.
Medico-legal: something involving both medical and legal aspects.
Minimum number of individuals: a calculation of the minimum number of individuals represented in a skeletal assemblage.
Osseous: consisting of bone.
Viable: capable of working successfully, in this case, capable of sustaining life.

CHAPTER 3

Aerial survey: an examination of a specific area above the ground; sometimes conducted with manned vehicles (such as airplanes and helicopters) or through digital imagery with unmanned devices (such as drones and satellites).
Archaeological record: an account of natural and cultural events and transformation processes at a single location as documented only by the physical environment.
Archaeological tool kit: the suite of knowledge and resources used by a forensic archaeologist to effectively process a site. Can also refer to the actual assemblage of equipment used by a forensic archaeologist.
Articulated: the positioning of human skeletal elements relative to each other as they would normally exist in a living human.
Azimuth: a method of mapping whereby points are plotted using their distance and clockwise bearing relative to a single reference point, such as the site datum. Azimuth degrees range from 0° to 360°.
Baseline: a specific transect placed at a scene that is used as a reference line for various data collection activities such as measurements, pedestrian surveys, excavation, and so on.
Burial: specifically, a human body that has been deliberately placed subsurface and covered, usually with sediment. In a general sense, any assemblage of human remains found under or near the ground surface.
Cache: a localized deposit or assemblage of something of interest (such as a cache of weapons or a cache of tools).
Cartesian grid system: a method of mapping whereby points are plotted using northings' and eastings' measurements taken from a North and East baseline. The measurements are taken at a perpendicular distance from each baseline (thus, requiring that each baseline is aligned straight, true, and perpendicular to each other).
Chain of custody: a critical step in evidence management whereby the disposition of an item is tracked from initial collection at a crime scene to its accession into a crime laboratory (as well as any subsequent transportation to different locations).
Contamination: any kind of process, activity, or physical contact that can potentially damage, destroy, and/or compromise the forensic significance and context of an item by commingling it with data from elsewhere.
Context: the relationship and association of an item or piece of data relative to another item or piece of data at a recovery scene. An item's location in time and space relative to another item's location in time and space. While context is the most important information that can be derived from a recovery scene, it is also the most *easily lost* kind of information, especially when best practices are not adhered to.
Controlled deconstruction: systematic processing of a site conducted with adequate documentation using transparent and repeatable processes with particular attention to maximizing contextual data collection while minimizing contamination and alteration of the scene.
Diachronic: change over time.
Disarticulated: the positioning of human skeletal elements relative to each other in a way that they could not possibly exist in a living human.
Disturbance processes: all taphonomic activities that can influence human remains after its primary deposition including cultural, mechanical, natural, biological, and chemical activities.
Evidence management: the controlled process and administration of forensically significant items that include identifying, documenting, collecting, protecting, transporting, and curating.
Exhumation: specifically, the act of removing a deliberately buried skeleton from its perceived resting place, usually a cemetery. In a general sense, the removal of a body from any burial location. Also called *disinterment*.
Extended: a skeletal position whereby most of the limbs are stretched out to their maximum length. Modern traditional American burials are extended.

Feature: a specific assemblage of data that appear localized to itself and separate from the surrounding matrix. Examples can include a termite mound, a grass stain, a patch of differential soil disturbance, and a cache of tools. A human burial and associated grave can be considered as a feature.

Flexed: a skeletal position whereby most of the limbs are tightly bent at the joints.

Forensic archaeology: the application of theories and methods used in traditional research-based archaeology to process modern scenes of a medico-legal nature (usually with skeletal remains).

Forensic value: the significance of a site, an item, or an assemblage of skeletal remains with regard to its medico-legal context.

Global positioning system: a process whereby every location on earth is recognized by unique coordinates, as determined by the triangulation of that location using multiple orbital and/or suborbital satellites. The coordinates are obtained with a GPS-receiving device, using one of many available coordinate formats in reference to one of many available reference data.

Ground penetrating radar: a remote sensing method of distinguishing different densities below the ground surface through the measurement of administered radar pulses.

***in situ*:** in its original place. A burial *in situ* means the skeletal remains have not been moved from the exact location where they were found.

Latitude–Longitude: a type of coordinate format system used to recognize a specific location on earth in reference to its place on the latitudinal and longitudinal axes of the globe.

Law of Association: a concept whereby the distance of items within a deposit "and" the integrity of the strata they are found in determine the reliability of any observed spatial and chronological relationship between the two objects.

Law of superposition: a fundamental principal of stratigraphy whereby the older, more undisturbed strata are generally found on the bottom and the younger, more disturbed strata are generally found on the top. Combined with the Law of Association, an object found in a stratum close to the surface might logically be interpreted as having been deposited more recently than an object found in a stratum far deeper in the ground.

Magnetometer: a remote sensing device capable of recognizing metallic signatures below the ground surface using magnetic forces. A magnetometer focuses on the earth's magnetic field while a metal detector focuses on the magnetic field around a localized object.

Military grid reference system (MGRS): a type of coordinate format system used to recognize a specific location on earth in reference to its place on the MGRS of the globe.

Pedestal: a method of excavation by which a specific item (e.g., a bone or material item) is left in place embedded in its matrix while the surrounding sediment is removed and leveled. This leaves the item slightly raised as the now-level area around it is lower.

Pedestrian survey: a systematic and controlled search of an area wherein a team of people walks across the scene in transects, and visually searches for data and evidence of forensic value.

Preliminary assessment: the initial step in scene processing where the forensic anthropologist makes broad determinations of which methods should be used.

Primary deposit: an undisturbed disposition of forensic evidence whereby the *in situ* location of an item is the same place where that item was originally placed in the ground. For example, a body that is buried in the ground and subsequently discovered by law enforcement in that same location is considered to be a primary deposit.

Prone (face down): a skeletal position where the majority of the body is facing down toward the ground.

Provenience: the exact location of an item (or a piece of data) in space; its three-dimensional location at a recovery scene.

Reference datum: a physical location, usually marked by a fixed point or solid item, whereby all other items mapped are based on that location. Data taken from the reference datum can include distance measurements, azimuths, and depth.

Remote sensing: a suite of methods used to collect data (both on the ground surface and below it) from a distance (i.e. without actually touching it) by means of various devices that *scan* the area in question. Remote sensing methods include metal detection, ground penetrating radar, resistivity, and conductivity.

Resistivity and conductivity: a remote sensing method of distinguishing different densities below the ground surface through the measurement of administered electricity.

Scene documentation: the act of preserving methodological practices of scene recovery as well as preserving the contextual data that are recovered and collected at a crime scene, using written notes, photography, and digital data collection.

Scientific integrity: it is the adherence to the best practices of the scientific method with transparency and accountability for all actions; requires sound and accepted methods and high ethical standards regardless of the nature of the work.

Screening: a process where sediment is placed through a mesh or sieve of a certain size to separate particles larger than that size. In burial recovery, the sediment that makes up the burial fill should be screened through at least a ¼-inch mesh (preferably smaller) to find small bone elements and material items that might be evidentiary.

Secondary deposit: a disturbed disposition of forensic evidence whereby the *in situ* location of an item is *not* the same place where that item was originally placed in the ground. For example, a body that is buried in the ground, years later to be washed away to a different location by flood waters, and subsequently discovered by law enforcement in that different location is considered to be a secondary deposit.

Sediment: a layer of earthly material found on the ground or below the ground placed there by natural or cultural practices. Burials are commonly found within various types of sediment, including soil, sand, and concrete.

Semi-flexed: a skeletal position whereby many of the limbs are loosely bent at the joints.

Shovel test pit: the product of a subsurface testing method whereby a small area (usually the diameter of a shovel) is excavated to variable depths (usually the depth of the shovel) to examine its contents and collect information below the ground surface.

Site formation: the process by which a physical location (usually, but not always, outdoor) is transformed in significance by a cultural event (e.g., when a crime takes place). In reference to forensic anthropology, it is when an assemblage of human skeletal remains is first deposited at a location.

Site transformation: all the natural and cultural processes that happen at a location after the initial site formation. These processes can be slow and long term (such as a trickle of water flowing into a grave) or a one-time, sudden event (such as a coyote digging through a grave). Site transformation processes can significantly alter the contextual data of the primary deposition.

Skeletal position: a description of the physical layout of a human skeleton (or skeletal elements). Can include extended, semi-flexed, flexed, supine, and prone.

Soil: a layer of sediment (such as silt, sand, or clay) usually in the higher strata of the ground that contains heavy organic qualities.

Strata: differing layers of sediment as deposited either by natural or cultural forces (stratum is singular). The strata are distinguished by type of sediment, color, density, moisture, inclusions, texture, and particle size.

Stratigraphy: the analysis and interpretation of differing strata at a particular location relative to each other. Stratigraphy considers when each stratum was deposited relative to each other, as well as how each are related to the formation of the crime scene.

Subsurface sites: scenes where the majority of the data and evidence are below the surface of the ground.

Subsurface testing: a process of collecting data from a specific area below the ground surface.

Supine (face up): a skeletal position where the majority of the body is facing up toward the sky.

Surface sites: scenes where the majority of the data and evidence are on or very near to the surface of the ground.

Systematic methods: transparent and coordinated processes of best practice that are conducted with step-by-step procedures and can be repeated.

Taphonomy: specifically, the "study of the laws of burial". In a general sense, a description and understanding of the processes that can affect human remains after original deposition.

Theodolite: may also be called a "transit." A highly accurate device used to measure the bearing of an object or location from a specified reference point. A transit and theodolite differ in mechanics, but are sometimes used synonymously. While the origins of the device are hundreds of years old, digital and laser-assisted versions are now the norms.

Triangulation: a method of mapping whereby points are plotted using the triangular vertices that are formed relative to each other. For example, mapping an unknown point can be done by measuring the distance of that point from where two other known points are located, as long as the distance between the two known points is also measured.

Uniformitarianism: a fundamental principal of nature whereby properties that affect the earth (such as gravity, erosion, flooding) act in the present day in the same way as they did in the past.

Universal Transverse Mercator (UTM): a type of coordinate format system used to recognize a specific location on earth in reference to its place on the UTM system of the globe.

Unsystematic methods: unrepeatable processes that are conducted largely at random with no semblance of control, structure, or holistic view.

CHAPTER 4

Carnassial teeth: teeth that have evolved in carnivores and some other species to shear past each other for slicing soft tissue and bone-like scissors.

Commensal: wild species that live among human habitations; examples include mice and rats.

Crenellated margins: the jagged edges left behind on carnivore-gnawed bone as they splinter away fragments.

Cylinders: hollow tubes of bone created as scavengers remove the proximal and distal epiphyses of a long bone.

Dispersal: the spread of remains and related evidence through multiple taphonomic processes, including scavenger action and water transport.

Edge polish: forms on freshly exposed cortical bone margins that become worn down by repeated tooth wear from additional carnivore gnawing.

Ethology: the study of living animal behavior; this research is one aspect of taphonomy.

Furrows: elongated tooth marks that penetrate through the cortical bone into the cancellous bone or marrow cavity; they are at least three times long as they are wide.

Gastric corrosion: found on small bones and fragments consumed by carnivores, whether retrieved from vomit or feces. It is characterized by the thinning of edges and a sculpted appearance to bone surfaces and may include windowing (see below).

Gripping marks: unpatterned tooth pits and scores often left on long bone shafts as a carnivore shifts the element within its mouth to reposition it for gnawing on the epiphyses or for transport.

Osteophagia: consumption of bone by herbivores (including cattle, deer, etc.) likely in response to dietary deficiencies. The grinding of the bone between their broad cheek teeth can leave a characteristic "Y-fork" shape to long bone ends.

Parallel grooves: tooth marks left by rodents as they gnaw with their paired incisors on dry bone.

Pedestaling: taphonomic formation caused by a rodent gnawing into an epiphysis or other area of thin cortical bone, leaving behind untouched areas of bone that therefore have a pedestaled appearance.

Pits: small tooth marks (usually oval or irregular) that do not penetrate the cortical bone; they are less than three times as long as they are wide.

Provisioning: transport of food by adults to their young. As a taphonomic process, it is one reason why remains' dispersal can be so extensive.

Punctures: tooth marks similar in form to pits but with deeper perforations that do penetrate the cortical bone; they are less than three times as long as they are wide.

Scores: elongated tooth marks that do not penetrate through the cortical bone into the cancellous bone or marrow cavity; they are at least three times long as they are wide.

Taphonomy: the study of entire series of changes that organisms go through as portions of their remains return to the inorganic sphere.

Windowing: the creation of small holes (or expansion of existing foramina) in bones by a variety of taphonomic processes, including gastric corrosion.

CHAPTER 5

Connective tissue: one of the four types of tissues in the human bodies. Connective tissues support or connect other tissues. Connective tissues are comprised of specialized cells embedded in an extracellular matrix.

Eversion: a movement of the foot that rotates the plantar surface of the foot laterally, away from the midline of the body.

Hyaline cartilage: transparent cartilage found on many joint surfaces, in the nose, ears, trachea, and larynx. Hyaline is the most abundant of the three types of cartilage in the human body.

Monocytes: the largest of the three major types of white blood cells.

Mononucleic: a cell that has a single nucleus.

Osteogenic: concerned with bone production, growth, or repair.
Osteoprogenitor cells: cells that arise from mesenchymal stem cells in the bone marrow. Osteoprogenitor cells give rise to osteoblasts.
Skeletogenesis: the embryological process of skeleton formation.
Type I collagen: the most abundant form of collagen in body. Type I collagen is found in bone, where it is the structural protein in the extracellular space of this connective tissue.
Vestibulocochlear complex: a series of specialized structures in the inner ear that regulate balance/equilibrium and hearing. The vestibulocochlear complex is innervated by the vestibulocochlear, or auditory, nerve (cranial nerve VIII).
Zooarchaeology: the study of faunal remains, or the remains left by animals after they die.

CHAPTER 6

Buccal: next to or toward the cheek.
Cementoenamel junction (CEJ): the junction between the enamel and cementum on a tooth; also known as the cervical line. The CEJ separates the enamel of the crown from the cementum of the tooth root.
Cementum: connective tissue forming the external layer of the tooth root.
Cervix: a narrow or constricted portion of a tooth in the region of the junction of the crown and root; also known as the "neck" of the tooth.
Crown: the portion of the tooth that is covered with enamel and normally is visible in the oral cavity.
Dentin: hard, yellowish tissue underlying the enamel and cementum; dentin constitutes the majority of the tooth crown and root.
Diphyodonts: an animal with two successive sets of teeth (deciduous and permanent).
Distal: away from the midline.
Enamel: white, external surface of the crown of the tooth; hardest substance in the human body.
Furcation: location on multirooted teeth where the root trunk divides into separate roots.
Gingiva: soft tissue that covers the alveolar process of the maxilla and mandible; surrounds the cervical portion of the tooth.
Incisal: the biting surface of the incisors (equates to the occlusal surface of the posterior teeth).
Interdental septum: the alveolar bone between two teeth.
Interradicular septum: the alveolar bone between two or more roots of a single tooth.
Labial: next to or toward the lip; also referred to as the facial surface.
Lingual: next to or toward the tongue.
Mesial: towards the midline.
Occlusal: the biting surface of a posterior tooth.
Odonotogenesis: the formation or development of teeth.
Pulp: soft tissue located in the center of the tooth crown and root.
Root: the portion of the tooth embedded in the alveolar process (bone) and covered with cementum.
Root trunk: the portion of a multirooted tooth between the cervical line and the bifurcation or trifurcation of the separate roots.
Splanchnocranium: the facial skeleton.

CHAPTER 7

Adipocere: a waxy substance formed by the anaerobic hydrolysis of fat during the decomposition process; adipocere is sometimes called "grave wax."
Bench notes: notes that accompany a forensic case report. They are not included as part of the official report document but are subject to subpoena and examination in court.
Biohazardous materials: biological substances that pose a threat to living organisms or to the environment.
Curate: the process of organizing materials (skeletal remains) for later use or observation.
Radiographic images: medical images such as radiographs (X-rays) and computed tomography (CT) scans that can be used to visualize the inside of a body, identify radiopaque foreign objects, and facilitate positive identification.

CHAPTER 8

Discriminant function: a mathematical function that maximizes the differences between two or more groups in multivariate space.

Gender: the social role or characterization of biological sex to which an individual ascribes or wishes to be perceived.

Gestalt: an impression of whole form that is derived from a summation of its parts but cannot be teased apart. The visual assessment of sex or ancestry from a skeletal Gestalt is the product of many years of experience and is difficult to quantify.

Morphology: the shape of a skeletal feature/trait or skeletal element.

Scientific Working Group for forensic ANTHropology (SWGANTH): a governmental working group assigned to determine best practice for the field of forensic anthropology. The working group was funded primarily through the joint POW/MIA Accounting Command and the Federal Bureau of Investigation. This working group has been superseded by the working group for anthropology through NIST (the National Institute of Standards and Technology).

Scoring system: a system by which morphological states of a skeletal feature are segregated using concise definitions. Scores of sexually dimorphic traits are devised as graded expressions (e.g., score = 1, 2, 3, …) starting from a gracile expression and progressing to a robust expression of the feature.

Sectioning point: a mathematical point that maximizes the differences between two groups in one dimension for a given feature or skeletal measurement.

Sexual dimorphism: differences in size and shape between males and females of a given species. Forensic anthropologists are concerned with sexual dimorphism of skeletal features.

CHAPTER 9

Assortative mating: a form of mating (sexual selection) in which individuals select mates who are phenotypically (and therefore genotypically) more similar as opposed to a random mating pattern in which no selection is involved.

Craniometric data: skeletal data consisting of measurements of the cranium.

Morphoscopic traits: skeletal features with ordinal grades of expression used for estimating ancestry or sex.

Multivariate statistical methods: statistical methods that analyze many variables simultaneously to determine how the variables are related to one another and test their combined predictive power to separate and classify groups.

Osteometric data: skeletal data consisting of bone measurements.

Quantitative traits: measurable phenotypes (traits) that are contributed to polygenic effects, or the effects of multiple genes.

Reference data: a dataset consisting of known values that is used to determine unknown parameters.

Secular changes: changes in human biology that occur over a relatively short period of time primarily due to environmental variables (e.g., increased stature in populations throughout the world).

CHAPTER 10

Articulation: the joint between two bones.

Dental attrition: the wear and tear caused by tooth-to-tooth contact. This occurs naturally with age and results in the loss of enamel on the incisive/occlusal surfaces of the teeth.

Eburnation: a polishing of the articular surface of a bone that occurs from the bone-on-bone rubbing associated with osteoarthritis; the subchondral bone becomes a smooth ivory-like mass that sometimes contains ridges or striae.

Enthesophyte: bony projections that form at the attachment of a ligament or tendon.

Hyaline cartilage: one of the three types of cartilage in the body; hyaline cartilage contains a significant portion of collagen and is usually found lining the bone surfaces at joint articulations. The nose, ears, larynx, and trachea also contain hyaline cartilage.

Laryngeal cartilages: the nine cartilages that comprise the larynx (or voice box), including the unpaired thyroid and cricoid cartilages. The thyroid cartilage has a midline prominence that is recognizable in the neck of some individuals as the Adam's apple. The thyroid and cricoid cartilages may ossify in older individuals.

Osteoarthritis: also called "degenerative joint disease," osteoarthritis is the most chronic condition of the joints. The protective hyaline cartilage on the ends of the joints wears down, causing painful bone-on-bone rubbing at the articulation. Osteophytes develop around the periphery of the joint, resulting in further pain and stiffening.

R-squared: also known as the coefficient of determination; in a linear regression, a statistical measure of how close the data are fitted to the regression line. R-squared expresses the percentage of the response variable that is explained by the linear model.

Subchondral bone: the layer of bone located immediately beneath (or deep to) the articular cartilage.

CHAPTER 11

Allometric changes: changes in body shape are not proportional to changes in body size, such that two organisms of the same size may not be the same shape.

Anatomical method: a method of estimating stature that requires assembling the skeleton in anatomical position, applying a correction factor for soft tissue, and measuring height. Elements used for the anatomical method are: cranium, vertebral column (minus the atlas), sacrum, femur, tibia, talus, and calcaneus.

Confidence interval: the interval that expresses the probability that the true parameter (i.e. stature), given the value(s) of the independent variable(s) (i.e. bone measurements), lies within the interval predicted by the regression line. A 95% confidence interval means that there is a 95% probability that the true linear regression line of the population lies within the confidence interval of the regression line calculated from the sample data (measurements).

Forensic stature (FSTAT): self-reported stature such as that on a driver's license or passport.

Informative prior: prior probability distributions are used in Bayesian statistical inference. An informative prior conveys specific information about the variable(s) in question and influences predictions/results based on this information.

Measured stature (MSTAT): stature that is measured directly (i.e. from the body and not from photos) and taken systematically following guidelines.

Osteometric board: an instrument used to measure long bones. One end of the board has an immovable upright plate that sits at a right angle to the base of the board. The measuring scale is on the base of the board. The opposite end of the board is a movable endplate. The bone is placed on the board against the immovable upright endplate, and the movable plate is adjusted to measure the length of the bone.

Prediction interval: an interval in which future observations are predicted to fall based on the regression equation, given a specific probability.

R^2 value: also known as the *coefficient of determination*. In regression analysis, this value indicates how well the regression line approximates or fits the data. The value ranges between 0 and 1 (or between 0% and 100%); the higher the R^2, the better the regression model fits the data.

Regression theory: a mathematical method first proposed by Karl Pearson in 1899. Regression analysis uses the correlation between long bone length and living stature to derive a linear equation that predicts stature from long bone length.

Secular change: short-term evolutionary change.

Supernumerary vertebrae: a common congenital anomaly of the spine in which extra vertebrae are present.

CHAPTER 12

Angulation (Dental): deviation of the tooth from normal alignment.//
Ankylosis: joint fusion, usually by osseous bridging.//
Antemortem: occurring before the death event.//
Arthropathy: any disease or condition affecting a joint.//
Compression: squeezing, compacting, or crushing force.//
Cutback: area of osteoclastic activity on the subadult diaphysis near the metaphysis where the periosteum is being lifted during normal growth.//
Dental agenesis: the failure to form a tooth, especially noted in the third molars (wisdom teeth).//
Diastema: a space between two teeth, especially noted in the maxillary front incisors.

Differential diagnosis: the process of identifying all possible conditions that could cause the observed skeletal defect or lesion.
Dislocation: displacement of elements of a joint from their normal position. This may be temporary or permanent.
Eburnation: a polished-looking area on a synovial joint surface indicating increased density in response to persistent erosion.
Erosion: effect of OA in which the cartilage is damaged enough to allow attrition of the underlying bone surface
Etiology: the origin or cause of a disease determined by observed features.
Flexion: bending force.
Fracture: partial or complete breaking of a bone from stress or impact.
Gemination (Dental): the initial growth of a tooth bud.
Heterotopic dentition: a tooth that develops and/or erupts in the wrong position or wrong orientation.
Idiopathic: an adjective describing a condition for which the etiology is not understood with certainty.
Infection: the invasion of pathogenic agent.
Luxation: complete dislocation of a joint.
Myositis ossificans: development of bone within the connective tissue, usually of major muscles, at the site of trauma.
Necrosis: cell death due to interrupted blood supply.
Pathognomonic: specific to a certain disease of condition.
Perimortem: occurring at or around the time of death.
Porosity: perforations in the bone surface. When observed in subchondral bone, porosity results from erosion related to OA; when observed in cortical bone, porosity is caused by hypervascularization related to inflammation.
Postmortem: occurring after death.
Pseudoarthrosis: false joint.
Ribbon matching: attributing an observed lesion to a particular pathological condition on the basis of superficial visual comparison (usually to photographs of severe cases depicted in atlases).
Rotation (Dental): dislocation of a tooth in a clockwise or counterclockwise position.
Shearing: two forces working in opposite directions.
Subluxation: partial dislocation of a joint.
Supernumerary dentition: development of more teeth than expected in the normal dental arcade.
Surface osteophytes: development of exostoses on the joint surface indicative of OA.
Tension: a stretching or pulling force.
Teratogenic: interference with normal embryonic development.
Torsion: twisting.
Winging (Dental): rotation of associating teeth in opposite direction, especially noted in the maxillary and mandibular first incisors.
Woven bone: the interlaced osseous structures formed in the fibrous matrix.

CHAPTER 13

Anisotropic: the property of being directionally dependent.
Axial loading: application of force along the long axis of a bone.
Bone biomechanics: the study of the mechanical laws relating to the material, structural, and functional properties of bone.
Bony callus: a temporary formation during the process of bone healing consisting of a mass of fibroblasts and chondroblasts.
Contrecoup fracture: a fracture occurring at a location approximately opposite to the point of impact.
Ectocranial: the outer cortical surface of the skull; the outer skull table.
Endocranial: the inner cortical surface of the skull; the inner skull table.
Greenstick fractures: incomplete fractures that indicate a high collagen content in bone; especially common in children.
Hinge fracture: a fracture across the middle cranial fossa that separates the skull base into anterior and posterior halves.
Kerf: the walls and floor of a cut mark.
Scientific Working Group for Forensic Anthropology (SWGANTH): a group of committees that provides consensus best-practice guidelines and establishes minimum standards for the forensic anthropology discipline.

Formed in 2008 as a co-sponsorship of the Federal Bureau of Investigation (FBI) and the Department of Defense Central Identification Laboratory (DOD CIL), the SWGANTH committees are being incorporated into the Organization of Scientific Area Committees (OSAC).

Tensile loading: loading that induces maximum stress by stretching or pulling.

Viscoelastic: a property of materials that exhibit both viscous and elastic characteristics when undergoing deformation; the viscoelastic properties of bone are largely velocity dependent.

Woven bone: bone with haphazard organization of collagen fibers that are mechanically weak; also referred to as immature bone; most common during development.

CHAPTER 14

Generalized distance (see Mahalabonis distance): a unitless measure of the standard Euclidean distance between a point x and a data distribution that is based on the mean, variance, and covariance matrix of the variables.

Geometric mean: a value that indicates the central tendency or typical value of a set of numbers; the nth root of a product of n numbers.

Isometric: having equal dimensions or measurements; bones that are isometric increase in size at the same rate.

Least squares regression: a regression analysis that minimizes the sum of the squared deviations between the data and the model in order to estimate the unknown parameters.

Linear discriminant function analysis: a statistical analysis that uses the linear combination of one or more continuous independent variables to predict a categorical independent variable (i.e. assign it to a class).

Linear regression: a method for modeling the linear relationship between a dependent variable (x) and one or more independent variables (y) by finding the best fitting straight line through the points.

Mahalanobis distance (see generalized distance): a unitless measure of the standard Euclidean distance between a point x and a data distribution that is based on the mean, variance, and covariance matrix of the variables.

Secular change: biological changes that occur over a relatively short period of time (generations) that are thought to be the result of environmental factors (e.g., increasing stature with improvements in nutrition and healthcare).

CHAPTER 15

Abiotic: nonbiological in nature.

Actualistic studies: field experimental studies directed at understanding site formation processes, as borrowed from archaeology. Research that strives to provide an analogy for forensic investigations conducted in the field rather than a laboratory environment and without the relative strict control of variables.

Adipocere: a waxy, soap-like substance formed by the breakdown of fatty acids under moist, alkaline, and anaerobic conditions.

Algor mortis: the cooling of the body following death.

Autolysis: self-digestion, the breakdown of the cells as a result of their own enzymes.

Biotic: biological in nature.

Bloating: the distention of soft tissue resulting from the gaseous by-products of bacterial activity during decomposition.

Contact pallor: a pale area of the body resulting from compression against or contact with an object that has pushed all of the blood out from within the area and prevents pooling of blood within the area, associated with livor mortis.

Desiccation (also see mummification): the process by which tissue preserves through drying.

Diagenesis: the breakdown or conversion of the constituent parts of a material, as borrowed from geology. The physiochemical changes to bone after soft tissue decomposition.

Differential decomposition: atypical decomposition associated with the premature, irregular, or disproportional decomposition between body regions.

Livor mortis: the hypostatis or pooling of blood following death, resulting in purplish-red discoloration of the skin (lividity).

Longitudinal studies: studies that observe research subjects over a period of time to study how specific variables change.

Marbling: patterned discoloration of the skin visible during decomposition resulting from micro-organisms released within the circulatory vessels.
Microbe: a micro-organism (e.g., bacterium) that contributes to the decomposition process; associated with microbial.
Mummification (also see desiccation): the process by which tissue preserves through drying.
Necrophagous: feeding on decomposing remains.
Postmortem interval (PMI): also referred to as the time since death, the amount of time between death and discovery of deceased carrion.
Putrefaction: decomposition of organic material caused by microbial activity often producing foul-smelling byproducts.
Rigor mortis: the temporary stiffening of the human body following death that occurs soon after death, resulting from the lack of ATP to undo the protein binding associated with muscle contraction.
Saponification: anaerobic chemical reaction (hydrolysis or hydrogenation) of fats in the presence of an alkaline during decomposition that produces adipocere.
Skeletonization: complete soft tissue decomposition of a carrion in which only the hard tissues (e.g., teeth and bone) remain.

CHAPTER 16

n/a

CHAPTER 17

Genocide: any of the following acts committed with an intent to destroy, in whole or in part, a national, ethnical, racial or religious group, such as the following:
  (a) Killing members of the group;
  (b) Causing serious bodily or mental harm to members of the group;
  (c) Deliberately inflicting on the group conditions of life calculated to bring its physical destruction in whole or in part;
  (d) Imposing measures to prevent births within the group;
  (e) Forcibly transferring children of the group to another group.
  (Article 2 of the Convention of the Prevention and Punishment of the Crime of Genocide)
Humanitarian forensic action: support in the search for recovery, analysis, identification, and management of the remains of large numbers of victims resulting from armed conflict, disasters, migration, and other situations, within a humanitarian scope.
Human rights: rights inherent to all human beings, whatever nationality, place of residence, sex, national or ethnic origin, color, religion, language, or any other status. Humans are all equally entitled to human rights without discrimination. These rights are all interrelated, interdependent, and indivisible. Universal human rights are often expressed and guaranteed by law, in the forms of treaties, customary international law, general principles, and other sources of international law. International human rights law lays down obligations of governments to act in certain ways or to refrain from certain acts, in order to promote and protect human rights and fundamental freedoms of individuals or groups (United Nations definition).
Mass disaster: threatening event, or probability of occurrence of a potentially damaging phenomenon within a given time period and area. These include natural hazards (earthquakes, landslides, tsunamis, volcanic activity, avalanches, floods, wildfires, cyclones, disease epidemics, insect/animal plagues, etc.) and technological or "man-made" hazards (conflicts, famine, displaced populations, industrial accidents, transport accidents) (IFRC definition).
Mass grave: a burial with at least three individuals, typically placed in burial in a careless or disrespectful manner (e.g., no marking of the grave, commingling of remains, without respecting cultural norms).
Missing person: a person whose whereabouts are unknown to his/her relatives and/or who, on the basis of reliable information, has been reported missing in accordance with the national legislation in connection with an international or noninternational armed conflict, a situation of internal violence or disturbances, natural catastrophes or any other situation that may require the intervention of a competent State authority (ICRC definition).

Truth and reconciliation commission: commission created in the scope of transitional justice consequently to a dictatorship, civil war, or internal unrest, aiming at a national reconciliation. Such commissions usually investigate human rights violations, and help traumatized societies, and sometimes are tasked to address to issue of missing persons (specific mandates vary among contexts).

CHAPTER 18

Archaeofaunal: dealing with assemblages of animal remains recovered from archaeological sites.

Calcined bone: burned bone that is white in color and very brittle. Calcined bone has lost all water and collagen, and has been reduced to its mineral constituents.

Cremains: the cremated remains of an individual. Cremains can be created by accidental fires or funerary processes (e.g., cremation).

Dermis: the layer of skin deep to the epidermis, which is the most superficial layer. The dermis contains blood vessels, nerve endings, sweat glands, and hair follicles.

Epidermis: the outer, most superficial layer of skin composed of epithelial cells. The epidermis is nonvascular and nonsensitive (contains no blood vessels or nerves).

Macroscopic methods: methods of assessing the biological profile by visual examination of gross morphological features of the skeleton.

Pugilistic posturing: the position assumed by a burned body in which the more powerful flexor muscles contract due to the heat exposure and cause the joints to flex. This posture makes the body assume a boxer-like position, thus the pugilistic posture is sometimes called boxer's pose.

Spectrophotometer: a device that measures the intensity of light that is transmitted or emitted by particular substances.

Index

Note: Page numbers followed by 'f' and 't' refer to figures and tables, respectively.

2nd-degree of burning, 349
4th-degree of burning, 349
5th-degree of burning, 350
10-phase Todd method, 185
14-stage coding system, 177

A

AAFS. *See* American Academy of Forensic Sciences (AAFS)
AAPA. *See* American Association of Physical Anthropologists (AAPA)
ABFA. *See* American Board of Forensic Anthropology (ABFA)
Abiotic organisms, 278
Accessioned materials, 31
Accumulated degree days (ADD), 290, 297
Accumulated degree hours (ADH), 290
Acellular extrinsic fiber cementum, 114
Acetabular destruction, 223f
Acromion, 101
Active decomposition stage (bloat), postmortem changes, 283–284
Actualistic research, 297
Actualistic studies, 295
Adenosine triphosphate (ATP), 279
Adipocere, 127, 276, 277f
Adult age-at-death, 178–190
    acetabulum, 189
    age estimation in young adults, 180–181, 180f, 181f
    age indicators, 189–190
    ages and ancestries, 185
    archaeological collections, blacks and whites, 184
    auricular surface, 183–185, 183f, 184f
    cranial and palate sutures, 181–183, 182f
    degenerative changes, 178–179
    description, traits, 176, 189
    gross morphology, fourth sternal rib end, 187–188, 188f
    medial clavicular epiphysis and sacral segment (S1–S2). *See* Epiphyses
    morphological change, 179
    phase system structure, 187
    pubic symphysis, gross morphology, 183f, 185–187, 186f
    sacroiliac joint, 176, 184
    skeletal age indicators, 188
Advanced decomposition stage, postmortem changes, 284
Adventive species, 289
Aerial surveys, 41
Afibrillar cementum, 114
Age estimation, 175–190
    adult age-at-death estimation, 178–190
        age estimation in young adults, 180–181, 180f, 181f
        age indicators, 189–190
        cranial and palate sutures, 181–183, 182f
        gross morphology, 183–188, 183f, 184f, 186f, 188f
    description, 176
    subadult age-at-death, 176–178
        dental development, calcification, and eruption, 177
        epiphyses, appearance and union, 178, 179f
        fetal age estimation, 177–178, 178f
Age indicators, 189–190
Agenesis, teeth features, 216
Algor mortis, 281, 287
Alkaline soils, 279
Allometric changes, 199
Alphanumeric scoring system, 177
Amelogenesis imperfecta, 122
American Academy of Forensic Sciences (AAFS), 4, 11
    description, 4
    forensic science disciplines, 5
    physical anthropology section members, 13f
    sections percentage, status membership, 6f
American Academy of Forensic Sciences Humanitarian, 340
American Association of Physical Anthropologists (AAPA), 5, 18
American Board of Forensic Anthropology (ABFA), 4, 8, 11
American Journal of Physical Anthropology (AJPA), 9
Amphiarthroses, 212
Anatomical method, 326
    stature estimation, 196–197
        requirements, 197
Anatomical specimens, 29
Ancestry, 353
Ancestry estimation, 164–169
    in biological profile, importance, 164–165
        missing individuals, searching, 164
        population-specific data, 164–165
    forensic anthropology, ancestry in, 165–166
    practice, 166–169
        ancestry estimation in forensic anthropology, 168–169
        craniometric data, 166–167
        estimation of ancestry, 167–168, 168f
        forensic anthropologists, 166
Angulation, teeth features, 216
Animal scavenging, taphonomic process
    bone destruction, 57–58
    carnivore damage, skeletal remains
        biting force, 65
        bone movement, 60–61
        distal human femur, 63
        Odocoileus virginianus, 65f, 67
        taxonomic grouping, 67
        tooth impression shape, 67
        tooth punctures, 66
        vertebrate skeletons modification, 60

Animal scavenging, taphonomic process (*Continued*)
    forensic anthropology, 58
    rodent damage, skeletal remains
        Erethizon dorsatum, 69, 69f
    types, osseous alteration, 58
Anisotropic, 235
Ankylosing spondylitis, 212, 213f
Ankylosis, 214
Anomalies, 321
    congenital, 215–220
    dental, 218f
Antemortem, 199, 206, 212
    body modifications, 28
    defect, 233f
    radiographs, 138
    trauma, 232, 232f
    vs. postmortem, 318
Anthropology
    defined, 4
    research facilities, 12–14
    subdisciplines, 5f
Anthropology Research Facility (ARF), 296, 279
    description, 12
    geographic locations and climate, decomposition process, 14
    information, donors, 12
    open-minded UT administration, 12
    TSD estimation, 12
Argentine Forensic Anthropology Team, 341
Arthritic joint destruction, 210f
Arthritic lipping, 134
Arthritis. *See* Osteoarthritis
Arthropathies, 210–212
Arthropathy, 210
Articular cartilage, 86
Articular deterioration, 210
Articular surfaces, 25
Articulation, 181
Artistic reproduction, 325
Assortative mating practice, 167
Auditory ossicles, 93
Auricular surface, gross morphology, 183–185, 183f, 184f
Autolysis, 274–275
    tissue decomposition order, 277t
Automobile fires, 349
Avulsion fracture, 245f
Axial skeletal elements, 95–100
    hyoid, 95
    ribs, 99–100
    sacrum, 99
    sternum, 99, 100f
    vertebrae
        cervical classification, 96–98
        components, 96
        description, 95
        differentiating, 98–99
        transverse process, 96
        typical cervical, thoracic and lumbar vertebra, 95, 98f
Axial skeletal epiphyses, 178
Azimuths, 44

B

Bass Collection, 15
Bayesian approach, 189
Benign tumors, 221f
Bioarchaeology and Forensic Anthropology Association (BARFAA), 7
Biocultural approach, 23
Biological profile, 4, 136–138
    importance of ancestry estimation, 164–165
        missing individuals, searching, 164
        population-specific data, 164–165
Biotic organisms, 278
Bite-mark analysis, 112
Blind tests, 152
Bloating, 276, 276f
Blunt-force trauma, 353–354
    axial and appendicular skeleton, 244–246
    cranium and facial skeleton, 242–243
Body orientations and anatomical terminology, 88
Bone
    biomechanics, 234–235
    as connective tissue, 82–84
    fever, 49
    fragment displacement, 240f
    growth, 86–88
    infections, 224–225, 225f
    inflammation types, 224t
    repair and remodeling, 215f
    stability of, 318
    staining, 285
    structure, 84–86
    types and joints, 86
Bone-healing process, 213–214
    impediment factors, 214
Bony callus, 232
Boot-strapping analyses, 155
Breast bone, 99
Buccal cusp, 118
Buckle fracture, 234f, 244, 246
Burning, degrees of, 349–350, 349f
Burn line fractures, 356
Butterfly fracture, 245f

C

Cache, 42
Cadaver decomposition island (CDI), 292
Cadaver length, 195–196
Calcaneus tarsal, 105–106
Calcined (white) bone, 356–357
Calliphoridae, 289
Calotte, 89
Calvaria, 89
Cancellous bone, 84
Carnivore damage to skeletal remains, 60–67
    characteristics, 65
Carpals, 102–103
Cartilaginous joints, 86, 87f
Caudal–cranial fusion, 181
Cellular intrinsic fiber cementum, 114
Cellular mixed fiber cementum, 114
Cementoenamel junction (CEJ), 113

Cementum, 113–114
Central Identification Laboratory, DoD (DoD CIL), 16
Central Identification Laboratory (CILHI) in Hawaii, 322
Centroid, 155
Cephalic index, 165
Cervical vertebrae, 96–98
Charred (blackened), 356–357
Cheek bones, 94
Chopping trauma, 246
Circumferential lamellar bone, 85
Clefting, 122
Close-up facial view, 322
Coccyx, 99
Coding systems, 359
Collar bone, 100
Combined DNA Index System (CODIS), 165
Commission on Missing Persons (CMP), 337
Compact bone, 84–85
Comparative radiography, 315–319, 316f, 317f, 319f
    antemortem *vs.* postmortem, 318
    applications, 316
    frontal sinuses, 316f
    MSCT, 316
    spinal and pelvic features, 317f
    therapeutic procedures, 318
Complex for Forensic Anthropology Research (CFAR), 298
Composite scoring system, 181, 187
Computed tomography (CT), 15, 183, 249, 316
Congenital skeletal anomalies, 217f
Contamination of evidences, 50
Controlled deconstruction, 39
Coronal
    dentin, 113
    plane, 88
    suture, 90
Cortical bone, 84–85, 85f
Costal and thyroid cartilage ossifications, 190
Costovertebral joints, 212
Cranial and palate sutures, 181–183
Cranial bone, 90
    delamination of, 355
Craniofacial morphology, 166–167
Craniofacial superimposition, 322–323, 322f
Craniometric data, 166–167
Craniosynostosis, 215
Cranium and mandible, 88–95
    definition, 88
    description, 88
    parameters, biological profile, 89
    skull bones. *See* Skull, bones
    terminology, 89–91
        endocranial and ectocranial cortical bone layers, 90
        fontanelles, infant, 90, 90f
        landmarks, 90, 90f
        splanchocranium and calvaria, 89
        vault sutures, 90
Cremains, 350
Cremation, 350
Crime scene methodology
    archaeological record, 36
    controlled deconstruction, 38–39
    evidence management
        access control, control areas, 51
        accurate and labeling things, 50
        consequences, mishandling, 50
        contamination, 50
        definition, chain of custody, 50–51
        description, 50
        fingerprints analysis, 50
        forensic anthropologists, 39
    excavation and recovery
        bone fever avoidance, 49
        burial and type of deposition, 48
        burial stages, 49
        data collection and potential damage, 45
        description, 48
        disturbance processes, 48–49
        law of association, 48
        law of superposition, 48
        position documentation, 46
        preservation condition, 49
        sample section drawing, soil stratigraphy, 48, 48f
        screening process, 49
        skeletal positioning, 47
        stratigraphy and uniformitarianism, 48
        traditional archaeological technology, 49
    forensic anthropologists, 36
    law enforcement personnel, 36
    mapping. *See* Mapping, crime scene methodology
    natural environment materials, 36–37
    in past
        CSI effect, 38
        Easter Egg Hunt, 38
        photographs and documentation methods, 37
        scientific evidence, 37
        scientific integrity, 37–38
        systematic and unsystematic approach, 38
    processing
        diagrams, walking path, 42f
        investigators line, 42f
        preliminary assessment, 41
        teamwork, 41–42
        types, sites, 39
        visual pedestrian survey, 41–42
    scene documentation. *See* Scene, documentation, crime scene methodology
    shapes and sizes, 36
    site formation
        mechanical excavator and rappelling equipment, 43
        remote-sensing devices, 43
        skeletal cases, 43
        subsurface site, 43
        surface, 43
Criminal legal process, 336
Cross-validation analyses, 155
Crow-Glassman scale (CGS), 352
Crown, 113, 117–118, 120, 122
Crumbly variety, adipocere, 275
Cultural indicators, 28
Cuneiform tarsal, 107

D

Darroch and Mosimann's shape variables, 264
Daubert decision, 325
Daubert v. Merrell-Dow Pharmaceuticals, Inc., 325
Decay rates, 299t
Deciduous (baby/primary) dentition, 115
Deciduous quadrants, 116
Deciduous teeth, 120, 122
Decomposition, 274–280, 352
    effects on soil, 292t
    Rates, 300t
    research, 297–301
    variables
        arthropod activity, 280
        depositional factors, 279–280
        humidity, 279
        rates, factors, 278t
        temperature, 278
        trauma/bonding agents, 280
        water, 279
    variables affecting, 278–280, 278t, 279f
        abiotic, 278
        biotic, 278
        burial depth, 280
        cultural, 278
        environmental, 278
        humidity, 279
        physical barriers, 280
        proximity, 280
        rainfall, 279
        scavenger activity, 280
        soil pH, 279
        temperature, 278–279
        water availability, 279
Decompositional Odor Analysis (DOA), 14
Defense POW/MIA Accounting Agency (DPAA), 130
Dehydration, 352
Delicate bones, 128
Dental anomalies, 218f
    environmental, 122
    genetic, 122
    and pathologies, 122
Dental attrition, 189
Dental calcification, 176–177
Dental code, 116
Dental development, calcification, and eruption, 177
Dental-facial complex, 112–113
Dental maturity, 176
Dental morphology, 112, 116–122, 119f, 120f, 121f
    tooth anatomy and, 112–114, 113f, 114f
Dental notation system, FDI, 117f, 118f
Dental numbering systems, 115–116, 115f, 116f, 117f, 118f
    description, 115
    FDI, 116, 117f, 118f
    UNS, 115–116
Dentin, 113
Dentition
    adult age-at-death. *See* Adult age-at-death
    ancestry determination
        Carabelli's cusp, mesial-lingual surface, 371, 371f
        classification, nonmetric dental traits, 371
        crenulated molars, 371, 371f
        shovel-shaped incisors, 371, 371f
    assistance, forensic odontologists, 112
    biological profile estimation, 111
    dental anomalies and pathologies, 122
    dental morphology
        canines, 117–118
        classification, incisors, 117
        crown symmetry, 117
        cusps classification, 118, 120
        definition, lingual fossa, 117
        diphyodonts, 116
        grinding function, 118
        homodonts, 116
        human dental formula, 116, 119f, 120
        mandibular right second premolar, 118, 120
        maxillary and mandibular molars, 120
        mesial surface, marginal groove and bifurcated root, 118
        premolars, 118
    dental numbering. *See* Dental numbering systems
    environmental dental anomalies, 122
    forensic settings, 112
    genetic dental anomalies, 122
    hardest substances, human body, 112
    immense information, 112
    physical anthropologists, 112, 365
    post-mortem longevity, 365
    radiographs, 112
    sex estimation, 372–373
    skeletal biologists, 112
    subadult age-at-death estimation, 365–367
    tooth anatomy and dental morphology
        cementum, 114
        cross-sectional, 113f
        crown and root, 113
        dental-facial complex, 112–113
        dental nomenclature, 114f
        dentin, 113–114
        enamel, 113
        gingiva (gums), 114
        groups, dental-facial complex, 113
        interdental and interradicular septum, 113
        pulp organ, 114
    variations, 365
Department of defense central identification laboratory (DoD CIL), 16
Dermatoglyphics (fingerprint analysis), 9
Dermis, 349
Desiccated remains, 28
Desiccation, 278
Determination, PMI
    chemistry, 292
    forensic botany, 290–296
    forensic entomology, 289–290
    morphoscopic techniques, 293–296
Developmental Juvenile Osteology, 177
Diagenesis, 279
Diagnostic imaging technology, 207
Diaphysis, 86
Diarthrodial joint, 86
Diastema, teeth features, 216

Dichotomy, 339
Diffuse idiopathic skeletal hyperostosis (DISH), 211, 211f
Diphyodonts, 116
Disaster Victim Identification (DVI), 339
Discoloration stage, postmortem changes, 281–283
Discriminant function analysis (DFA), 154–155, 166, 168
Disease processes, 220–225
    bone infections, 224–225, 225f
    infectious diseases, 223
    intrinsic disease processes, 221, 222f
Dislocation, trauma, 214–215, 216f
Distobuccal cusp, 120
Distolingual cusp, 120
Disturbance processes, 49
DNA analysis, 127, 145, 165, 358
DNA and forensic anthropology, 323–324
    biological profile, 323–324
    identification purposes, 323
Documentation, skeletal analysis, 146
Documentation of crime scene. *See* Scene, documentation, crime scene methodology
Documented donated skeletal collections, 14–15
Domestic homicides, 335
Dry bone, 128, 177, 354

E
Eburnation, 189, 210
Ectocranial surface, 236
Electron microscope with energy dispersive spectroscopy (SEM-EDS), 25
Endosteum, 84
Enthesophytes, 183
Entomotoxicology, 290
Environmental dental anomalies, 122
Epidermis, 349
Epiphyseal plate, 87, 178, 207
Epiphyses, 86–87
    appearance and union, 178, 179f
        age-at-death estimation methods, 178
        human osteology, 177
        immature skeleton, 178
        union timing, skeletal epiphyses, 179f
    medial clavicular and sacral segment (S1–S2)
        clavicular maturation, 180
        late-fusing maturation site, 180
        scoring systems, 180
        sternal end, 180
Erosion, 210
Ethmoid bone, 93

F
Facial approximation, 325–326
Facial imagery methods, 326
Family Reference Samples (FRS), 165
Faunal remains, 25, 25f
FBI Files (TV program), 7
FDI (Fédération Dentaire Internationale) System, 115–116
    deciduous dentition, 118f
    permanent dentition, 117f
Federal Bureau of Investigation (FBI), 16, 130, 145, 165
    biomedical research companies, 15
    DoDCIL, 16
    forensic anthropology, 4
    human skeletal material identification, 10
    non-fiction programs, 7
Federal Rules of Evidence, 324–325
Femur, 104
    growth and development, 88f
    head, 156
Femur/stature ratio, 198
Fetal age estimation, 177–178, 178f
Fetuses viability, 29–31
Fibrous callus formation, 214
Fibrous joints, 86, 87f
Fibula, 105
Finite element analysis (FEA), 249
Fire, 348
    and commingling, 347–360
        anthropological research, 350–352, 351f
        commingling, 357–359
        degrees of burning, 349–350, 349f
        fire, 348
        at the fire scene, 352–353
        fractures, 354–357, 354f, 355f, 356f, 357f
        in laboratory, 353–354, 353f
        research and novel methods, 359–360, 360f
    dynamics, 348–349
    triangle/tetrahedron, 348
Fire-induced fracture, 354
Fire-sustaining fuel, 348
First-degree of burning, 349
Five-tiered system, 349
Flat bones, 86
Flexion/bending, 213
Floating ribs, 99
Fluorosis, 122
Fondebrider, Luis, 341
Fontanelles, 90, 90f
Fordisc, 197
Fordisc 3 (FD3), 256, 258–267
    advanced topics, 263–265
        Howells' data, 263–264
        options, 264–265
    analytical options, 264, 264f
    basic, 258
    computational burden, 257
    cranial and postcranial measurements, 258
    criticisms, 265–266
    entry screen for unknown case, 258f
    forensic anthropology data bank, 257
    forensic case, 257
    geometric mean, 264
    Howells' data, 263–264
    limitations, 266–267
    linear regression and discriminant functions, 256
    LOO cross-validate option, 265
    manners of death distribution, 257f
    measurement entry blanks, 258f
    measurement errors, 264
    opportunities, 267

Fordisc 3 (FD3) (*Continued*)
    options screen, 264f
    output for
        cranial analysis, 259t
        postcranial analysis, 262t
    overview, 258
    postcranial screen, 261f
    sex and ancestry from
        cranial measurements, 258–260
        postcranial measurements, 260–261
    statistical methods, 256
    stature, 261–263
    stature. *See* Stature
    stature screen, 263f
    stepwise option, 265
    typicality probs option, 265
FORDISC 3.0, 136
FORDISC 3.1, 166
FORDISC program, 130, 155–156
Forensic anthropologist, 4, 31, 176, 315, 323
    in the international law context, role, 341–342
    medical examiner's setting between 1980 and 2010, 13f
Forensic anthropology, 318. *See also* Age estimation
    AAFS sections percentage, status membership, 6, 6f
    ancestry in, 165–166
        problem, 168–169
    ARF. *See* Anthropology Research Facility (ARF)
    best practice, 16–17
    component, 4
    consolidation period (1939–1971)
        FBI's law, 10
        human osteology, 11
        paper assignment, 10–11
    defined, 4
    developmental events, 8
    DNA and, 323–324
    documented donated skeletal collections, 14–15
    Dr. Bass's views, 12
    educational and employment opportunities
        description, 8
        employment agencies, 8
        genetics and biomechanics, 8
        NAS report, 8
    ethics, practice and research, 15–16
    formative period (early 1800s–1938)
        AJPA and AAPA, 9
        anatomy laboratory, 8
        Dorsey's assertions, 9
        epiphyseal union and stature estimation, 9
        Hamann–Todd collection, 10
        human skeleton information, 9
        origin, 8
        personal identification, 9
        photographs and anthropometric measurements, 10
        physical contributions, 9
        sex indicators, 9
        theoretical foundations, 9–10
    Hollywood myths
        characteristic skeletal features, 7
        educational requirements, 7
        public jurors and members, 8
        television dramas, 7
    human skeletal variation, 7
    implication, 325
    in international arena, 17–18
    and investigations of international law, developments, 340
    mass disasters. *See* Mass, disasters
    modern period (1972–present)
        AAFS, 11
        ABFA, 11
        coroner's and medical examiner's offices, 12
        medical examiner (ME) setting, 12
        physical/forensic anthropologists, 11
    odontologists, 5
    outdoor decomposition research facilities, 18
    overview, 112
    public eye, 7
    subfields, 4, 5
    SWGANTH. *See* Scientific Working Group for Forensic Anthropology (SWGANTH)
    taphonomy, 7
    technique, 317
Forensic Anthropology Center (FAC), 14, 130, 133
Forensic Anthropology Data Bank (FDB), 170, 257
    anatomical and donated collections, 257
    description, 257
    homicide and suicide, 257, 257f
    sequence, event, 257
Forensic Anthropology Research Facility (FARF), 14, 15
Forensic Anthropology Society of Europe (FASE), 17
Forensic archaeology, 36
Forensic botany, 290–296
    bone diagenesis, 292
    chemistry, 292–293
    morphoscopic techniques, 293–296
Forensic Data Bank (FDB), 130
Forensic discrimination (Fordisc), 12
Forensic entomology, 289–290
Forensic equations, 197
Forensic investigation of international crimes, frameworks, 336–340
    forensic anthropology as a service to prosecution, 338–339
    humanitarian forensic action, 337–338, 338f
    mass fatalities and emergency action, 339–340
    preliminary notes on human rights and mass graves, 336–337
Forensic odontology, 112, 318
Forensic pathologists, 5
Forensic pathology, 5, 286–289
Forensic Science Standards Board, 16
Forensic stature (FSTAT), 199
Forensic taphonomy, 4
    carnivore damage to skeletal remains, 60–67
    definition, 58
    gnawing sources, 70
    overview, 57–60
    rodent damage to skeletal remains, 67–70
Four-field approach, 7
Four-level system, 352
Fourth sternal rib end, gross morphology, 187–188, 188f
Fracture, 354–357, 354f, 355f, 356f, 357f
    avulsion, 245f
    buckle, 234f, 244, 246
    butterfly, 245f

clavicle, 244
contrecoup, 242
gunshot wound, 238f
gutter, 239, 239f
hinge, 242
LeFort I, II, and III, 243, 244f
ontrecoup, 243f
perimortem, 234f
radiating, 237
trauma, 213–214
Fragmentation index (FI), 360
Frankfurt horizontal, 88
Fresh stage, postmortem changes, 281, 281f
Frontal bone, 91, 92f
Frontal sinuses, 316f
Frye Standard, 324
Frye v. United States, 324

G

Gaseous by-products, 276
Gastric corrosion, 67
Gemination, teeth features, 216
General acceptance test, 324
General Electric Co v. Joiner, 325
Geneva Conventions, 336
Gestalt, 157
Gingivae (gums), 114
Gloving effect, 275f
Gomphosis, 113
Grave wax. *See* Adipocere
Greenstick fractures, 233
Gripping marks, 66
Gross morphology
    auricular surface, 183–185, 183f, 184f
        posterior ilium, 183
        retroauricular changes, 183, 184f
        transverse organization, 183
    fourth sternal rib end, 187–188, 188f
        ossified cartilage, 187
        rib end casts, 187–188
        sternal end, 187
    pubic symphysis, 183f, 185–187, 186f
        age-at-death estimation methods, 185
        description, 185
        formation and deterioration, rim, 185
        pubic face texture, 185
        rim erosion, 185
        Suchey–Brooks pubic symphyseal changes, 185, 186f, 187
Ground-penetrating radar, 43
Gunshot wound fractures, 238f
    in rib, 240f
Gutter fracture, 239, 239f

H

Hair mat, 275
Hamann–Todd Collection, 10
Hanging cadaver, outdoor context, 294–295, 295f
Harsh chemotherapy, 176
Hartnett method, 187
Haversian bone, 85
Haversian canals, 85
Haversian system, 84f, 85
Heat-affected bone, 134
Hematoma, 214
Heritability, 166–167
Heterotopic dentition, teeth features, 216
High-velocity trauma, 236–241, 236f
Hinge fracture, 242
Hispanic reference standards, 164
Homicide, 257
Homodonts, 116
Howells' data, 263–264
Howship's lacunae, 83, 83f, 85
Human bone micromorphology, 26
Human Identification Laboratory, 322
Humanitarian forensic action, 337–338, 338f
Humanitarian-oriented investigations, 340
Human micromorphology, 26
Human osteology, 82, 129
    body orientations and anatomical terminology
        "above/toward the head" and "below/away from the head," 88
        medial and lateral, 88
        position and planes, 88
    connective tissue
        Haversian system, 83–84, 84f
        Howship's lacunae, 83, 83f
        lining cells, 84
        mechanical stress, 84
        organic and inorganic components, 82–83
        osteoblasts, 83
        osteoclasts, 83
        osteocytes, 83–84
    cranium and mandible. *See* Cranium and mandible
    description, 82
    fibrous and cartilaginous joints, 86
    growth, 86, 88f
    joints, 86
    long and short bones, 86
    postcranial skeleton
        axial skeletal elements, 95–100
        lower limb, 104–108
        pelvic girdle, 103–104
        shoulder girdle, 100–101
        upper limb, 101–103
    structure
        canals and bays, 85
        density/porosity, dynamics, 85
        forms, adults, 84
        locations, trabecular and cortical, 84–85
        periosteum and endosteum, 84
        primary and secondary, 85–86
    synovial joints and classifications, 86
Human remains, 25, 26f
Human rights/humanitarian, 336
Human Rights Resource Center, 340
Human rights violations, 336
Human skeletal sex estimation
    description, 144
    DNA, 145
    documentation
        approaches and elements, 146
        SWGANTH, 146

Human skeletal sex estimation (*Continued*)
    methods disagreement, 157
    metric approach
        analysis and evaluation, 152
        centroids, 155
        classification accuracy, 155
        complex multidimensional space, 155
        computer-generated functions, 154
        cranial measurement, 155
        data distributions, 153, 153f
        data spread and accuracy trade-offs, 153
        FORDISC 3.0, 155, 156
        leave-one-out/boot-strapping analysis, 155
        quantitative measures, 153
        statistical methods, 153
        Terry collection, 154
        univariate analysis method, 153
    population-specific standards, 156–157
    sex determination
        metric approaches, 152–156, 153f, 154f
        morphological approaches. *See also* Morphological approach, sex determination
    sexual dimorphism, 144–145
    sex *versus* gender, 144
The Human Skeleton in Forensic Medicine, 11
Human *versus* nonhuman bones, 25, 26f
Humerus, 101
Hyaline cartilage plate, 178
Hydroxyapatite, 25

I
Idiopathic arthropathy, 211
Idiopathic disorders, 222f
Impaction, teeth features, 216
Indeterminate state, 157
Indian Ocean tsunami, 339
Indian Residential Schools, 341
Infection, 214
Infectious diseases, 223
    acetabular destruction, 223f
    to affect skeleton, 223t
    vertebral collapse, 224f
Inferior nasal conchae, 94
Informative prior, 199
Inter- and intraobserver error tests, 147
Interdental septum, 113
Intermediate cementum, 114
International Association of Dental Research, 116
International Association of Forensic Sciences (IAFS), 5, 17, 18
International Commission on Missing Persons, 338
International Committee of the Red Cross (ICRC), 337
International Conference of Governmental and Non-Governmental Experts, 337
International Criminal Court (ICC), 339
International Criminal Tribunal for Rwanda, 338
International Criminal Tribunal for the former Yugoslavia (ICTY), 338
International Humanitarian Law (IHL), 337
Intertubular dentin, 113–114
Intramedullary canal, 86
Intrinsic disease processes, 221, 222f
    categories, 221t

Intrinsic variables, bone, 235
Inversion, 352
    teeth features, 216
Irregular bones, 86
Ischiopubic ramus, 147
Ischium, 104
Isometric size, 264

J
Jackknife method, 260
Johnson, Jean, 135, 138
Joiner decision, 325
Joints, 86

K
Kerley method, 11
Keyhole defect, 236, 237f

L
Lambdoidal suture, 90
Large-scale commingling, 357
Laryngeal cartilages, 190
Latin American Forensic Anthropology Association, 340
Law of association, 48
Law of superposition, 48
Least squares regression, 261–262
Leave-one-out (LOO)
    analyses, 155
    method, 260, 265
LeFort I, II, and III fractures, 243, 244f
Lie detector, 324
Limb bones, 88
Lincoln Index (LI), 359
Linear discriminant functions, 256
Linear regression equation, 197
Lingual fossa, 117
Linguistic anthropology, 4, 5f
Lipping, 210
Lividity, 287–288, 289f
Livor mortis, 281, 287–288
Load-deformation curve. *See* Stress-strain curve
Loading scenarios, force direction, 234f
Lovejoy approach, 136
Lower limb, 104–108
Low-velocity trauma, 241–249
    blunt-force
        axial and appendicular skeleton, 244–246
        cranium and facial skeleton, 242–243
    sharp-force, 246–249
Lumbar vertebrae, 99
Luxation, 214

M
Machine learning technique, 168
Magnetic resonance imaging (MRI), 249, 316
Magnetometers, 43
Mahalanobis distance (D^2), 256, 260
Malnutrition, 221
Malocclusion, 122
Mandibular canal, 318
Mandibular dentition, 116
Mandibular molars, 120

Manners of death distribution in FDB, 257f
Manubrium, 99
Mapping, crime scene methodology
   definition, provenience, 45
   large-scale and medium-scale, 44
   provenience and context, 45
   sample sketch
      burial, 46f
      tools, 44
      topographic variation, 45f
Marbling, 282, 282f
Mass
   disasters, 336
   fatalities and emergency action, 339–340
   grave, 336–337
Massachusetts Medical Society, 9
Mastication, 112
Mastoid process, 152
Mathematical method, stature estimation, 196–197
Maxilla, 92f, 94–95
Maxillary
   canines, 117
   dentition, 116
Maxwell Donated Collection, 15
Maxwell Museum Documented Skeletal Collection, 15
Measured stature (MSTAT), 199
Medial condyle, 104
Medial epiphyseal ossification, 180
Medico-legal death investigation, 314
Medico-legal significance, 23
Medullary cavity, 86
Mesenchymal cells, 83
Mesiolingual cusp, 118
Metacarpals, 103
Metal brushes, 127
Metallic poisons, 316
Metatarsals, 107
Microdontia, 122
   teeth features, 216
Minimum number of individuals (MNI), 27, 358–360
Minimum number of skeletal elements (MNE), 358, 360
Missing element estimation, 197
Missing person, 164, 336–337
Mitochondrial DNA (mtDNA), 321, 323
Mixed dentition period, 120
Morphological approach, sex determination
   description, 146
   pelvis
      accuracy, 146
      description, 146
      male and female comparison, 146
      preauricular sulcus, 148, 149f
      pubic and subpubic regions, 146–147
      sciatic notch, 148
      "traits of Phenice," 147
   postcranial bones, 152
   scoring systems, 146
   skull
      accuracy rates, 149, 152
      crania, male and female, 148
      discriminant functions, 152
      glabellar region, 150, 151f
      mandibles, 151f, 152
      mastoid scores, 150
      nuchal crest, male and female, 150f
      scoring systems, 149
      statistical approach, 152
      supraorbital margin, 150, 151f
Most likely number of individuals (MLNI), 359
Mountain, Desert, and Coastal Forensic Anthropologists (MD&C), 7
Mountain, Swamp, and Beach Forensic Anthropologists (MS&B), 7
Muffle furnace, 350
Multislice computed tomography (MSCT), 316
Multivariate statistical methods, 167
Mummification, 278
Munsell charts, 351
Myositis ossificans, 214

N

Nasal bone, 92f, 94
National Academy of Sciences (NAS), 8, 18
National Commission on Forensic Science, 16
Native American Graves Protection and Repatriation Act (NAGPRA), 15, 24
Navicular tarsal, 106–107
Necrophagous species, 289
Necrosis, 210
Necrotic femoral head, 218f
Neoplasms. *See* Tumors
Neurocranium, 89
Neutral soils, 279
Noncommunicable diseases, 221, 222f
Nongreasy bone, 354
Nonhuman mammals, 26
Nonimaged records comparison, 319–322, 320f
Nonsegmentation, 219
Normal skeletal variants, 219f
Northeast Forensic Anthropology Association (NEFAA), 7
Nuclear DNA analysis, 323

O

Occupational Safety and Health Administration's (OSHA), 126
Odontoblastic processes, 113
Odontoblasts, 114
Office of Management and Budget (OMB), 169
Olecranon foramina, 219
Olecranon fossa, 101, 152
Omnivorous species, 289
Ontrecoup fracture, 243f
Open fires, 349
Options, advanced topics FD3, 264–265
   LOO cross-validate, 265
   measurement errors, check for, 264–265
   stepwise, 265
   typicality probs, 265
Orbital breadth (OBB), 260
Orfila, Matthieu Joseph Bonaventure, 195
Organization of Scientific Area Committees (OSAC), 16
Ossification, 87, 178
Osteoarthritis (OA), 176, 210
   classification, 211

Osteoblasts, 83, 206
Osteoclasts, 83, 206
Osteocytes, 83–84
Osteogenesis, 87
Osteological identification, 315
Osteometric board, 196, 196f
Osteometric data, 166
Osteometric sorting, 358
Osteonal bone, 26
Osteoprogenitor cells, 83

P

Palatine
    bone, 94
    process, 136
Parietal
    bones, 92, 92f
    foramina, 219
Pars defect, 134
Passalacqua's method, 188
Patella, 104–105
Patellar notch, 219
Pathological conditions, 321
    as individuating traits, 205–225
        classification, 210–225
        problems to diagnose, 207–209
Patina, 354–355, 355f
Pedestaled remains, 49
Pelvic girdle, 103–104
Pelvis, 146–148, 147f, 148f, 149f
Perimortem, 212
    fractures, 234f
    trauma, 138, 233
Periodontal ligament, 114
Periosteal remodeling, 206–207
Personal identification, methods, 314–326
    additional considerations, 324–326
        evidence, admissibility of, 324, 325
        facial approximation, 325–326
    current techniques and practices, 315–323
        comparative radiography, 315–319, 316f, 317f, 319f
        craniofacial superimposition, 322–323, 322f
        nonimaged records comparison, 319–322, 320f
    DNA and forensic anthropology, 323–324
    genetic traits, 315
    humanitarian, 314
    skeletal information, 315
Personal protective equipment (PPE), 126
Phalanges, 103, 107–108
Phalanx, 25
Phenice method, 146–147
Photographic superimposition/video superimposition, 322
Physical/biological anthropology, 4
Pioneering virtual research, skeletal trauma, 249
Piophilidae, 289–290
Platymeric index, 166
Plexiform bone, 85–86
Polynesian skull, 169
Population-specific data/method, 164–165
Population-specific standards, 156–157
Porosity, 210–211
Positive-by-lay persons, 314

Postcranial bones, 152
Postcranial skeleton, 95–108, 96f
    axial skeletal elements, 95–100
        hyoid, 95
        ribs, 99–100
        sacrum, 99
        sternum, 99, 100f
        vertebrae, 95–99
    lower limb, 104–108
        femur, 104
        fibula, 105
        metatarsals, 107
        patella, 104–105
        phalanges, 107–108
        tarsals, 105–107
        tibia, 105
    pelvic girdle, 103–104
    shoulder girdle, 100–101
        clavicle, 100
        scapula, 100–101
    upper limb, 100–103
        carpals, 102–103
        humerus, 101
        metacarpals, 103
        phalanges, 103
        radius, 101
        ulna, 101
Postmortem, 112, 212–213
    change, 280–285
        active decomposition stage (bloat), 283–284, 298f
        advanced decomposition stage, 284
        anatomical changes, 280–281
        autolysis, 274–275, 283f
        discoloration stage, 281–283
        early appearance, marbling, 282f
        fresh stage, 281, 281f
        initial discoloration, body, 282f
        initial skeletonization, 284
        multi-year exposure, elements, 286f
        skeletonization stage, 285
        skeletonized remains, 286f
        stages, 280–281
    damage, 233f
    radiograph, 138, 177
    trauma, 233
Postmortem interval (PMI), 274, 321
    Algor mortis, 287
    approaches, 290
    chemistry
        biochemical methods, 296
        CDI, 292
        mandible, root mass, 291f
        observation-based methods, 291
        time intervals, 291, 292t
    colonel shy, 287, 288f
    determination, 285–297
    entomotoxicology, 290
    forensic botany, 290–296
        bone diagenesis, 292
        chemistry, 292–293
        morphoscopic techniques, 293–296
    forensic entomology, 289–290

  developmental stage, 290
  ecological-based research, 290
  flies, 289
 forensic entomology
 forensic pathology, 286–289
 livor mortis/lividity, 287–288
 methods, 285
 morphoscopic techniques, 293–296
  case-specific variables, 294–295
  experimental, hanging cadaver, 294–295
  human-based experiments, 296
  TBS, 293
  thumb calculation, 297
 muscle contraction, 288
 rigor mortis, 288
Post-processing care and handling, 128–130, 128f, 129f
Presumptive identification methods, 337
Pruning, 283
Pseudarthrosis, 214
Putrefaction, 274–276
 tissue decomposition order, 277t

R
R^2 values, 198
Radiating fracture, 237
Radiocarbon dating, 28
Radiograph, 24, 112
Radiographic identification, 315–316
Radiographic images, 126
Radiopaque, 24
Ratio method, stature estimation, 197
Regional certification systems, 340
Regional/national forensic teams, 338
Regression approach, stature estimation, 196–197
Regression theory, 196
Remodeling process, 86
Remote sensing, 43
Research and novel methods, 359–360, 360f
Residential fires, 349
Resistivity and conductivity devices, 43
Ribbon matching, 207
Ribs, 99–100
Rigor mortis, 281, 288
Rodent damage to skeletal remains, 67–70
Root, 113
 trunk, 122
Rotation, teeth features, 216
Rotator cuff syndrome, 214–215
Royal Anthropological Institute, 340
R-squared values, 181
Rule 702, 324–325
The rule of ten, 278

S
Sacroiliac joints, 183, 212
Sacroiliac morphology, 184
Sacrum, 99, 181
Sagittal plane, 88
Sagittal suture, 90
Saponification, 276
Sarcophagidae, 289
Saw mark kerf features, 248f

Scallop pattern, 187
Scalpel/knife, 126
Scapula, 100–101
Scapular notch, 219
Scene
 documentation, crime scene methodology
  basics, 39
  description, 39
  exercise, 40
  goal, 39
  information and evidence loss, 40
  photography, 40–41
  time differential, 39
 and field methods, 133–134
Scientific Area Committees (SACs), 17t, 19
Scientific Working Group for Forensic Anthropology (SWGANTH), 59, 144–146, 157, 196–197, 241, 315
 "best practice model", 16, 18
 description, 16
 documents, 146
 DoD CIL and FBI, 16
 guidelines, 157
 joint venture, 144–145
 practice suggestions, 16
Scoring systems, 65, 146, 149
Screening, 49
Sectioning point method, 154
Secular change, 166, 198, 267
Self digestion, 274–275
Semi-flexed position, 46
Sex and ancestry
 cranial measurements
  case identification number, 259–260
  case UT91-42, 258–260
  classification, unknown skull, 260
  cranial analysis, 259t
  standard deviations, 260
  typ F and R, 260
 postcranial measurements
  case UT91-42, 260–261, 262t
  cranial screen, 260, 261f
Sex determination
 methods disagree, 157
 metric approaches, 152–156, 153f, 154f
 morphological approaches, 146–152
  pelvis, 146–148, 147f, 148f, 149f
  postcranial bones, 152
  skull, 148–152, 150f, 151f
 skeletal, 144
 subadult, 145
Sexual dimorphism, 164
 anthropologists, 144–145
 bell curves overlap, 145
 biological states, 144
 definition, human skeleton, 145
 gorillas, 144
 hormones control, 144
 population's variation, 145
 SWGANTH, 144
Sex *versus* gender, 144
Shape option, 264
Sharp-force trauma, 246–249

Shelf wear, 30f
Short bones, 86
Shot dispersal pattern, 241f
Shoulder
    blade, 100
    girdle, 100–101
        clavicle, 100
        scapula, 100–101
Situational fractures, 357, 357f
Skeletal biologists, 112
Skeletal examination and documentation, 125–138
    appendix 7.A, 132–134, 133f
        scene and field methods, 133–134
    appendix 7.B, 135–138, 137f
        biological profile, 136–138
    case report, 130–131, 131f
    post-processing care and handling, 128–130, 128f, 129f
    skeletal preparation, 126–128
        dry bone, 128
        soft tissue removal, 126–128, 127f
Skeletal healing complications, 216f
Skeletal maturity, 176
Skeletal morphology, 176
Skeletal pathology, 134
Skeletal preparation, 126–128
    dry bone, 128
    soft tissue removal, 126–128, 127f
Skeletal responses to insult, 206–207, 207f
Skeletal sex determination, 144
Skeletal system, 144
Skeletal trauma analysis, 231–249
    bone biomechanics, 234–235
    high-velocity trauma, 236–241, 236f
    low-velocity trauma, 241–249
        axial and appendicular skeleton, blunt-force trauma, 244–246
        cranium and facial skeleton, blunt-force trauma, 242–243
        sharp-force trauma, 246–249
    overview, 232
    pioneering virtual research, 249
    traumatic events, timing, 232–234
Skeletal variants, 219, 219f
Skeletonization stage, postmortem changes, 285
Skin slippage, 275
Skull, 88–89, 148–152, 150f, 151f
    bones, 91–95
        auditory ossicles, 93
        ethmoid, 93
        frontal, 91, 92f
        inferior nasal conchae, 94
        lacrimals, 93
        mandible, 92f, 95
        maxilla, 92f, 94–95
        nasals, 92f, 94
        occipital, 92f, 93
        palatines, 94
        parietals, 92, 92f
        sphenoid, 92f, 93
        temporals, 92, 92f
        vomer, 92f, 94
        zygomatics, 92f, 94
    sutures, 92f
Small-scale commingling, 357
Society of Forensic Anthropologists (SOFA), 8
Sociocultural anthropology, 4
Soft paste-like variety, adipocere, 275
Soft tissues, 314
Soft- to medium-bristle brush, 127–128
Southeast Texas Applied Forensic Science Facility (STAFS), 14
Spectrophotometer, 351
Sphenoid, 92f, 93
Spheno-occipital synchondrosis, 182
Spinal and pelvic features, 317f
Splanchnocranium, 89, 112–113
Spondylolisthesis, 134
Spongy bone, 84
Squamosal suture, 90
Stand-alone method, 188
Standardized scoring system, 148
Statistical approach, 152
Stature estimation, 195–199
    antemortem records, 198–199
    bivariate plot, 263
    cadaver length, 196
    least squares regression, 261–262
    limb bones, 198
    long bone length and proportions, 196
    mathematical *vs.* anatomical, 196
    medical anthropologist, neuroanatomy, 196
    methods, 196–198
        anatomical, 247
        femur/stature ratio, 198
        Fordisc computer program, 197
        forensic anthropology, 197
        medical anthropologist, neuroanatomy, 196
        multiple regression formulae, 197
        R^2 value, 198
        skeletal measurement, 197
        SWGANTH, 196–197
    middle-aged and older adults, 198
    monograph, 196
    osteometric board, long bones, 196, 196f
    overview, 195–196
    paleoanthropologists, 198
    postcranial screen, 261f
    screen, 261f, 263f
    skeletal measurements, 195–196
    subadult body proportions and allometric changes, 199
    types, 261–262
    upper panel, 261–262
Stellate lesions, 209f
Step fractures, 355
Sternal body, 99
Sternal foramen, 99, 219
Sternum, 99, 100f
Stevenson, Paul, 9
Stocking effect, 275f
Stopping power, 236
Straightforward process, 314
Stratigraphy, 48
Stratum, 48
Stress coat experiments, 242

Stress-strain curve, 235f
Subadult, 176
    aging technique, 176
    sex determination, 145
Subadult age-at-death estimation, 176–178
    dental development, calcification, and eruption, 177
    epiphyses, appearance and union, 178, 179f
    fetal age estimation, 177–178, 178f
    skeletal/dental development, 176
Subchondral bone, 176
Subluxation, 214
Subpubic concavity, 146–147
Subsurface sites, 43
Suchey, Judy, 11
Suchey–Brooks approach, 185, 186f, 187
Suchey–Brooks method, 134, 136
Sue, Jean Joseph, 195
Suicide, 257
Superficial fracture pattern, 355
Supernumerary
    dentition, teeth features, 216
    teeth, 115–116
    vertebrae, 197
Suppression efforts, 348
Surface osteophytes, 210
Sutures, 90
SWGANTH (Scientific Working Group for Forensic Anthropology), 196–197
Synchondroses, 86
Synovial joint, 86, 87f
Syphilis, 223
Systolic blood pressure deception testing, 324

T

Talocalcaneal height, 197
Talus tarsal, 106
Taphonomic changes, 214f
Taphonomic differences, 27
Taphonomic process, 164, 274
Taphonomy, 57, 134
Tardieu spots, 288
Tarsals, 105–107
Teeth
    archaeological, 25, 28
    features of, 216
Temporal bones, 92, 92f
Tennessee Bureau of Investigation (TBI), 132–133
Tensile loading, 234
Tentative identification, 314, 326
Theodolite systems, 44
Therapeutic procedures, 318
Third-degree of burning, 349
Thoracic gunshot wound, case example of, 238t–239t
Thoracic vertebrae, 98–99
Three-dimensional approximations, 326
Thumbnail fractures, 354
Tibia, 105
Time since death (TSD)
    decomposition research
        active decomposition stages, 283–284
        adipocere, 276
        anaerobic fermentation, 276
        autolysis, tissue decomposition, 276, 277t
        bloating, 276
        carrion models, 298
        decay rates, dry climates, 298, 299t
        environmental conditions, 278
        gloving effect, 275, 275f
        mummification, 278
        opportunistic scavenging, 300
        processes, 274–275
        putrefaction, 274
        rates, cold climates, 300t
        thawing, 299
        tissue decomposition, putrefaction, 276, 277t
        VFA, 276–277
        warm/moist climates, decay rates, 298, 299t
Tissue
    decomposition order, 276, 277t
    depth method, 325–326
Tooth anatomy, 113–114, 113f, 114f
    and dental morphology, 112–114
Topinard's ratios, 196
Torsion/twisting, 213
Total body score (TBS), 293
    head and neck, 293t
    limbs, 294t
    trunk, 293t
Trabecular bone, 84, 85f
Trabecular lattice pattern, 318
Traits of Phenice, 147
Transformations, 264
Trauma, 134, 136, 137f, 212–215
    antemortem, 232
    blunt-force, 242–246, 242f
    chopping, 246
    high-velocity, 236–241
    low-velocity, 241–249
    perimortem, 233
    postmortem, 233
    sharp-force, 246–249
    types, 213–215
        dislocation, 214–215
        fracture, 213–214
Trauma/taphonomic, 189
Trepanematosis, 223
Triangulation, 44
Trotter and Gleser's formulae, 197–198
Trotter mstats, 261
True ribs, 99
Truth and Reconciliation Commissions (TRC), 339, 341
Tumors, 220
Two-dimensional approximations, 326
Type I collagen, 82–83

U

Ulna, 101
Uniformitarianism, 48
Univariate analysis method, 153
Universal Numbering System (UNS), 115
    deciduous dentition, 115, 116f
    permanent dentition, 115, 115f
Upper limb, 100–103

V

Van't Hoff's rule, 278
Vascular etching, 219
Vertebrae, 95–99
Vertebral arch, 96
Vertebral body, 96
Vertebral collapse, 224f
Vertebral column, 95, 99
Vertebral osteophytes, 190
Viable fetus, 29–31
Viscoelastic material, 235
Visual
    pair matching, 358
    pedestrian survey, 41–42
    recognition, 314
    sex assessment, 150
Volatile fatty acid (VFA), 276–277
Volkmann's canals, 85
Vomer, 92f, 94

W

Walker method, 152
Wear and tear of joint, 211
Wilks' lambda, 265
William Shy, Colonel: case study, 287, 288f
Windowing, 67
Winging, teeth features, 216
Winter vs summer decomposition, 278–279, 279f
World Health Organization, 116
Wormian bones, 219
Woven, 207
    bone, 232
Wrinkling, 283

X

Xiphoid process, 99
X-ray fluorescence (XRF) analyzer, 25
X-rays, 316

Z

Zonal method, 359
Zooarchaeologists, 358
Zooarchaeology, 82
Zygomatic bone, 92f, 94
Zygomatic process, 148